Combating Plastic Pollution in Terrestrial Environment

S. Suriyanarayanan · Shivaraju H. P. ·
Kandiah Pakeerathan
Editors

Combating Plastic Pollution in Terrestrial Environment

Challenges and Strategies for a Sustainable Future

Springer

Editors
S. Suriyanarayanan
JSS Science and Technology University
Mysuru, Karnataka, India

Kandiah Pakeerathan
Department of Agricultural Biology
Faculty of Agriculture
University of Jaffna
Kilinochchi, Sri Lanka

Shivaraju H. P.
Department of Environmental Sciences
JSS Academy of Higher Education and Research
Mysuru, Karnataka, India

ISBN 978-981-96-2342-6 ISBN 978-981-96-2343-3 (eBook)
https://doi.org/10.1007/978-981-96-2343-3

This Springer imprint is published by the registered company Springer Nature Singapore Pte Ltd.
The registered company address is: 152 Beach Road, #21-01/04 Gateway East, Singapore 189721, Singapore

Foreword

In recent decades, the world has witnessed an alarming increase in plastic pollution, particularly in terrestrial environments. The consequences of plastic pollution are far-reaching, affecting ecosystems, human health, and the global economy. This growing challenge requires urgent attention, innovative strategies, and a concerted global effort to mitigate its devastating effects.

The publication *Combating Plastic Pollution in Terrestrial Environment—Challenges and Strategies for a Sustainable Future* is a timely and significant contribution to this pressing issue. It combines the expertise and insights of scientists, researchers, and environmentalists from various regions, offering a comprehensive overview of the sources, impacts, and potential solutions to plastic pollution. The diverse range of topics covered in this book, from the hazards of microplastics in agriculture to sustainable waste management approaches, highlights the complexity of the problem and the multifaceted strategies needed to address it.

This publication's relevance extends beyond environmental concerns to directly supporting the achievement of the United Nations Sustainable Development Goals (SDGs), particularly SDG 12 (Responsible Consumption and Production), SDG 14 (Life Below Water), and SDG 15 (Life on Land). By addressing plastic waste management and promoting sustainable practices, this book contributes to minimizing plastic pollution's adverse effects on terrestrial and aquatic ecosystems. Furthermore, the innovative technologies and community-based initiatives presented here offer actionable pathways for policymakers, industries, and communities to embrace more sustainable practices and reduce plastic's harmful footprint on our planet.

Each chapter in this book underscores the current state of plastic pollution and points toward innovative technologies, community-driven initiatives, and policy measures that can drive sustainable change. This publication's interdisciplinary approach is commendable, as it bridges the gap between science, technology, and community engagement, making it a valuable resource for researchers, policymakers, and environmental advocates alike.

The contributions from nine countries including India provide a diverse perspective on how these countries are tackling the challenges of plastic pollution through various innovative technologies, management strategies, energy and waste valorisation, biological solutions and chemical and physical characterisation of different types of plastics.

I am confident that this book will inspire a deeper understanding of the global plastic pollution crisis and encourage the scientific community and policymakers in the Non-Aligned Movement and other developing countries to take decisive steps toward a sustainable future. This publication paves the way for a cleaner, healthier, and more resilient world by addressing the challenges and exploring viable strategies for combating plastic pollution.

It is an honor to contribute this foreword to such an important and timely work. I commend the editors, S. Suriyanarayanan, Shivaraju H. P., and Kandiah Pakeerathan, as well as the contributors from across the globe, for their dedication and commitment to this critical cause. I am confident that this publication will serve as a cornerstone in the global effort to combat plastic pollution and inspire meaningful action toward safeguarding our planet for future generations.

September 2024

Prof. Dr. B. Suresh
Pro-Chancellor,
JSS Academy of Higher Education & Research
Mysuru 570015 - India

Preface

Plastic waste has become a critical global concern that is receiving the attention of scientists, decision-makers, and the general public. Over the past six decades, the world has produced approximately 8300 million tonnes of virgin plastics from which over 5800 million tonnes of wastes have been generated. All these wastes were managed through incineration, landfilling, release to the environment, and recycling. These waste materials have severely impacted the environment and global health. The accumulation of plastics in the environment has negatively impacted human health, agriculture, and ecosystems, which requires an immediate action.

Many studies have been conducted focusing on the plastic pollution in marine environment, whereas impacts on the terrestrial systems such as freshwater habitats, soil and groundwater have received less attention. The amount of plastics generated, consumed and released is more in the terrestrial environment rather than the marine ecosystem. Despite the increasing amount of research, there are still a number of important issues that need to be addressed. These include investigations into the sources of plastic pollution, the transport of these materials into various environments, the hazards it poses to ecosystems and human health, and possible ways to reduce emissions of plastic waste in the terrestrial settings. In terrestrial plastic pollution, the smaller plastics/microplastics are produced by a variety of sources, such as landfills, farming, and industry, and cause an extensive pollution of terrestrial habitats.

Plastics pollution significantly alter the physicochemical properties of soil, affecting bulk density, pore space, capillarity, soil moisture levels, evaporation, and evapotranspiration. These changes can disrupt the hydrological cycles, impair plant growth, and harm soil organisms, ultimately disturbing the food chains including the ecological services. Research studies demonstrated that plastics in the environment can also adsorb contaminants such as heavy metals, poisonous organic compounds and other emerging contaminants like antibiotics. These compounds pose a threat to the ecosystems especially the living things that inhabit them.

Inadequate waste management systems and the increasing use of plastics have led to significant plastic contamination in the terrestrial environment, particularly in developing nations.

The global plastic waste crisis is being exacerbated by ineffective policy measures aimed at limiting single-use plastics (SUPs). Poor enforcement of regulations and a lack of social consensus on reducing the plastic usage have led to negative consequences, with the impact often spilling over into neighbouring countries.

In this context, we understand how critical it is to address these changes carefully and comprehensively. As a result, we are delighted to present an edited book titled *Combating Plastic Pollution in Terrestrial Environment: Challenges and Strategies for a Sustainable Future* that emphasizes on different issues related to plastics in developing countries and their approaches in tackling this important problem. The book containing 25 chapters from 9 countries highlights various aspects of terrestrial plastic pollution and its management strategies, and the environmental and health hazards across the different regions. The key studies include the microbiological degradation of plastic materials, production of wood-plastics composites, risks of plastic wastes and the community-driven initiatives to tackle the plastic pollution, microplastics pollution in agricultural settings and freshwater ecosystem, use of technology for effective sorting methods, and use of waste plastics for sustainable practices provide a deep understanding of approaches to mitigate the plastic pollution in the terrestrial environment. To effectively reduce the plastic use, new and innovative business models are required. It is also recommended that a global plastic treaty, extended producer responsibility (EPR) schemes, import duties on plastic materials, reducing the trade barriers on substitutes, phasing out of fossil fuel subsidies and the exploration and adoption of bio-plastic as an alternative to traditional plastic and improved waste management infrastructure are to be implemented to address this global issue. We believe that the insightful research and best practices presented in this book would be a useful resource material for scientists and researchers interested in these emerging issues. We hope that this book successfully reaches its target audience.

Editors would like to express their gratitude to Dr. Amitava Bandopadhyay, Director General, NAM S&T Centre and their Staff Members for all of the technical and administrative assistance provided in the publication of this book. We are grateful to Prof. Dr. B. Suresh, Director of Technical Education at JSS Mahavidyapeetha, and Pro-Chancellor of the JSS Academy of Higher Education and Research (Deemed to be University), Mysuru for penning the "Foreword" of this book. We also thank Dr. Loyola D'Silva, Executive Editor at Springer Nature, Singapore and Springer Nature Staff Members for helping to make this project a success.

Mysuru, India S. Suriyanarayanan, Ph.D.
Mysuru, India Shivaraju H. P., Ph.D.
Jaffna, Sri Lanka Kandiah Pakeerathan, Ph.D.

Introduction

Plastic pollution in terrestrial environment has rapidly escalated into a global crisis, capturing the attention of scientists, government officials, policymakers, etc., and affected the lives of the common people. The World Economic Forum has consistently identified environmental degradation, including plastic pollution, as one of the top challenges facing humanity. With over 380 million tons of plastic produced annually, the adverse effects of this waste are profoundly felt, particularly in developing nations where waste management systems are mostly inadequate.

The challenge of addressing plastic pollution in terrestrial environment is multifaceted. Regulatory frameworks often lag behind the rapid pace of plastic production and consumption. Many existing policies are outdated, failing to cope with the evolving landscape of plastic use and waste management. Strengthening these frameworks and enhancing enforcement mechanisms are essential for curbing plastic waste generation and promoting sustainable alternatives.

Moreover, technological innovations in waste management and recycling present promising pathways to mitigate plastic pollution. Advanced materials and processes can enhance recycling efficiency and reduce the environmental footprint of plastic production and subsequent disposal. Community engagement and education also play pivotal roles in fostering behavioral changes that support sustainable consumption practices. By raising awareness about the impacts of plastic pollution, we can encourage individuals and communities to adopt more responsible habits, such as reducing single-use plastics and participating in local recycling initiatives.

Plastics, designed for durability and convenience, have unfortunately become ubiquitous in our ecosystems. From urban centers to remote rural areas, plastic waste is not only an eyesore but also a significant threat to biodiversity, wildlife, and human health. The fragmentation of larger plastic items into microplastics poses a severe risk, as these particles can infiltrate soil and water systems, impacting both terrestrial and aquatic life.

The extraction and disposal of plastics have accelerated with increasing population growth, urbanization, and changing consumption patterns, exacerbated by inadequate recycling infrastructures. The consequences of this pollution are alarming, leading to significant ecological damage and raising public health concerns about the presence of microplastics in food and drinking water.

To address these urgent issues, the *Centre for Science and Technology of the Non-Aligned and Other Developing Countries (NAM S&T Centre), New Delhi,* is pleased to present this comprehensive publication aimed at disseminating relevant knowledge and innovative strategies for combating plastic pollution.

This book features contributions from scientists and researchers across 9 countries, viz., Egypt, India, Iraq, Malaysia, Mauritius, Myanmar, Pakistan, Sri Lanka, and Taiwan - highlighting the pressing challenges posed by plastic waste and showcasing effective management techniques.

Through its 25 chapters, the book provides an extensive overview of the sources and impacts of plastic pollution, exploring microplastics in freshwater systems, the hazards of micro-plastic contamination in agriculture, and the innovative reuse of plastic waste. The chapters delve into various strategies for combating plastic pollution, including microbial degradation, community initiatives, and advanced technological solutions, thereby offering a holistic view of the issue.

This publication not only serves as a resource for understanding the scope of plastic pollution but also aims to inspire action and collaboration among researchers, policymakers, environmental professionals, and students. The methods discussed in this book contribute significantly to achieving global Sustainable Development Goals (SDGs), particularly SDG-12 on "Responsible Consumption and Production".

I extend my heartfelt gratitude to all the contributors for their dedication and expertise, which have made this compilation possible. I specially thank the editorial team—Dr. S. Suriyanarayanan (India), Dr. Shivaraju H. P. (India), and Dr. Kandiah Pakeerathan (Sri Lanka) for their relentless efforts in reviewing and refining the chapters, ensuring that the information presented is both accurate and impactful.

I am immensely grateful to Prof. B. Suresh, Pro-Chancellor and Director (TED), JSS Academy of Higher Education (JSS AHER), Mysuru for kindly agreeing to our request to write the "Foreword" of this book in spite of his very busy schedule. We are thankful to Prof. Suresh and Dr. Surinder Singh, Former Vice Chancellor, JSS AHER for their encouragement and support towards S&T collaboration between JSS AHER and the NAM S&T Centre. We are also thankful to Prof. Vishal Gupta, Dean (Academics), JSS AHER for his help and support in many ways in our partnership with JSS AHER.

I am thankful to Dr. Loyola D'Silva, Executive Editor, Springer Nature, Singapore for his kind support and guidance towards bringing out this publication and Ms. Shalini Monica C. Selvam, Production Editor, Springer Nature, Trichy, India for managing all the technical and administrative responsibilities for the publication process.

I take this moment to express my sincere gratitude to Mr. Madhusudan Bandyopadhyay, Senior Adviser, NAM S&T Centre for his kind advice and guidance in various stages of planning and execution of this bookproject. Further, I am also thankful to Ms. Abhirami Ramdas, Programme Officer, NAM S&T Centre for her significant contributions in coordinating this book project. I also acknowledge the help and support received from the entire team of the NAM S&T Centre, especially

Mr. Pankaj Buttan, Data Processing Manager; Mr. Rahul Kumra, Assistant Administrative Officer; Mr. Sunil Kumar, Accounts Manager; Mr. Jayakumaran, Public Relation Manager; and Ms. Nidhi Utreja, Programme Officer for their support in bringing out this publication.

Amitava Bandopadhyay, Ph.D.
Director General
NAM S&T Centre
New Delhi, India

Contents

Part I
Sources and Overview of Plastic Pollution

Chapter 1
Source and Transport of Microplastics in Freshwater Systems in Asia—A Systematic Review

Mohamad Hafiz Halim, Laila Zulaifah Zulkipli, Masni Mohd Ali, Mohd Talib Latif, and Khairiatul Mardiana Jansar

Abstract Millions of tonnes of plastic are created annually, and in the aquatic environment, microplastics are a new pollutant that has steadily gained awareness in the Asian region. Although much more research on microplastics in freshwater has been published in recent years, there are still questions about the potential sources and routes of microplastics into freshwater ecosystems. Thirty-three publications on the sources and transport of microplastics in freshwater ecosystems were evaluated using a systematic quantitative literature review approach. We assessed the findings relating to the shape and colour of microplastics reported in water and sediments. This review found that microplastics are frequently detected in various freshwater bodies in six regions, including rivers, estuaries, lakes, and wetlands. The results of this review indicated that research on microplastics in Asia has gained attention since 2016, with a gradual increase in the number of studies from 2019 onwards. The results showed that more than half of the studies focused on China. It was discovered that microplastics infiltrate freshwater ecosystems either from direct or indirect sources. The typical shapes of microplastics identified are fibre, fragment, film, foam, pellet, and sheet. The most common colours of microplastics found in freshwater are transparent, blue, red, white, and black. Fishing tools, textiles and clothing, beauty products, vehicles, and product packaging are examples of direct sources, while some indirect sources are aquacultural, agricultural, domestic sewage, and tourism. Microplastics can be transported to water bodies by surface runoff, fragmentation of larger-size plastics, and erosion of the surrounding land. Moreover, this study revealed that only a few studies discussed the transportation of microplastics in sediments compared to water. This implies the need for further research on the transportation of microplastics in sediments to holistically address this new water pollutant.

M. H. Halim · L. Z. Zulkipli · M. M. Ali · M. T. Latif · K. M. Jansar (✉)
Department of Earth Sciences and Environment, Faculty of Science and Technology, Universiti Kebangsaan Malaysia, Bangi, Selangor, Malaysia
e-mail: mardiana@ukm.edu.my

S. Suriyanarayanan et al. (eds.), *Combating Plastic Pollution in Terrestrial Environment*,
https://doi.org/10.1007/978-981-96-2343-3_1

Graphical Abstract

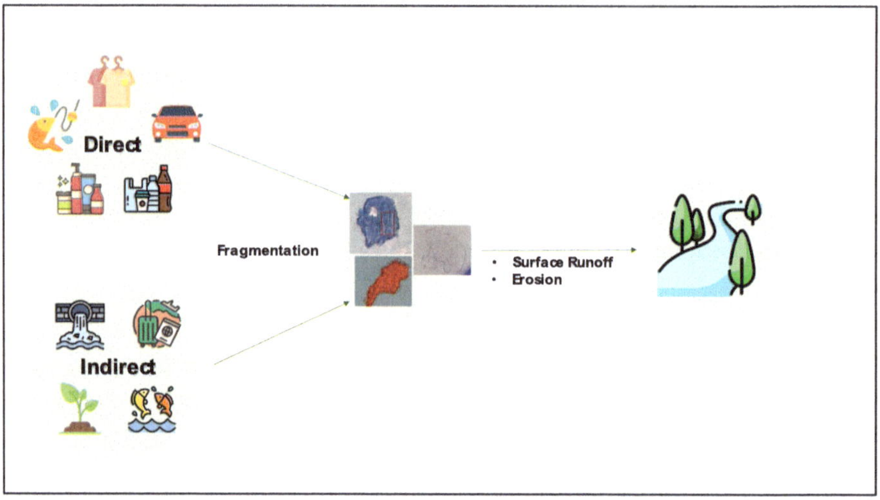

Keywords Microplastics · Fate · Behaviour · Sources · Freshwater · Morphology

1.1 Introduction

Millions of tons of plastic are produced globally each year, reaching about 367 million tonnes in 2020 (Plastic Europe, 2019). Since their successful synthesis in the early twentieth century, plastics have become essential in daily life, agriculture, and industry (Li et al., 2021). Over time, large plastic pieces degrade into microplastics, particles smaller than 5 mm (Shi et al., 2021). As a result, microplastic pollution has become a critical global environmental issue.

Microplastics are either produced from big plastic particles by various mechanical and photooxidative processes, known as secondary microplastics; or are purposefully created at such small sizes for specific applications, known as primary microplastics (Hidalgo-Ruz et al., 2012). According to recent studies, microplastics, which are plastic particles with a diameter of less than 5 mm, have recently been identified as potential contaminants in aquatic and terrestrial ecosystems (Wong et al., 2020). These contaminants are so abundant in nature that many aquatic organisms unintentionally consume them, including zooplankton (Botterell et al., 2019; Cole et al., 2013), shrimp (Wang et al., 2019), fish (Halstead et al., 2018; Merga et al., 2020; Shi et al., 2021; Tata et al., 2020; Zhang et al., 2020a), seabirds, turtles, and mammals (Alimba & Faggio, 2019; Di Beneditto & Ramos, 2014). It is now acknowledged that marine species may suffer adverse effects if they consume microplastics, including mechanical harm, olfactory toxicity, immunotoxicity, neurotoxicity, slow growth, decreased fecundity, and even increased incidence of death (Cannon et al., 2016;

Nadal et al., 2016; Shi et al., 2021; Tang et al., 2020). Microplastics are easily transmitted through and bioaccumulated in the food chain, and they may eventually be transported into the human body, causing a risk to human health (Rochman et al., 2015). Recent studies reveal that microplastics are prevalent in drinking water and certain food products (Koelmans et al., 2019; Kwon et al., 2020). Gruber et al. (2023) reported that an average adult could ingest about 5 g of plastic weekly, equivalent to the weight of a credit card, through common foods and beverages. Several studies have found microplastics in the human body, including the male testis (Zhao et al., 2023) and placenta (Ragusa et al., 2021). Microplastic exposure may lead to health problems like intestinal injury, liver infection, metabolic disorders, and inflammation, potentially contributing to chronic diseases (Li et al., 2023). However, many scientific questions about their toxicity remain unanswered and must be further explored.

The freshwater ecosystems on Earth, which include lakes, ponds, rivers, springs, bogs, and wetlands, are aquatic ecosystems. The distribution of freshwater on Earth is uneven. All inland bodies of water, such as rivers, streams, wetlands, subsurface waters, and estuaries, are considered freshwater habitats. Because they do not include saltwater environments, "inland water ecosystems" are sometimes called "freshwater ecosystems." According to preliminary studies, microplastics are widely distributed in freshwater habitats worldwide, including beaches, water bodies, and the sediments of lakes, rivers, and estuaries (Lebreton et al., 2017; Schmidt et al., 2017; Wagner et al., 2014; Wang et al., 2017; Zbyszewski & Corcoran, 2011; Zhang et al., 2015, 2016; Zhao et al., 2015). Furthermore, microplastic pollution in many freshwater bodies is as severe as or worse than in marine environments (Wu et al., 2018). So, it is essential to pay attention to microplastic pollution in inland freshwater systems (Eerkes-Medrano et al., 2015; Li et al., 2018; Lin et al., 2018).

Microplastics come in a variety of shapes and densities and originate from a range of sources (Crawford & Quinn, 2017; Hidalgo-Ruz et al., 2012). These characteristics influence their distributions and environmental effects. The small size of microplastics is a significant factor in their distribution in that they cannot easily be filtered and can enter the central irrigation system and river areas. The effluent containing microplastics causes water pollution and becomes a threat to aquatic life, thus harming local residents who use rivers in their daily activities (Sillanpää & Sainio, 2017). In contrast to marine habitats, our understanding of microplastics in freshwater ecosystems is still relatively poor (Lambert & Wagner, 2018; Yao et al., 2020).

This review describes the present level of knowledge about the fate and behaviour of microplastics in freshwater systems and highlights knowledge gaps. The review will specifically focus on (1) the potential sources of microplastics in freshwater systems, (2) how microplastics are transported in freshwater ecosystems, and (3) the characteristics of microplastics. This information will help to make appropriate recommendations for future studies regarding microplastic pollution in freshwater.

1.2 Material and Methods

1.2.1 Systematic Quantitative Literature Review

We opted for a systematic quantitative literature review methodology as it can identify trends and fill in knowledge gaps (Pickering & Byrne, 2013). A Preferred Reporting Items for Systematic Reviews and Meta-Analyses (PRISMA) 2020 statement was used to establish this present review (Page et al., 2021). A PRISMA flow diagram is shown in Fig. 1.1.

1.2.2 Literature Search

Relevant articles were obtained from Web of Science, ScienceDirect, Scopus, and PubMed. The search strings used in each database were microplastic AND freshwater AND abundance OR distribution. All publications were retrieved to the search date, November 11, 2021, and were included with no publication date limitation. The literature search returned 2155 records from ScienceDirect, 100 from Web of Science, 88 from Scopus, and 100 from PubMed. These records were carefully screened to ensure the data met the desired criteria. We solely focused on publications investigating microplastic abundance or distribution in freshwater systems, specifically in Asia, producing a shortlist of 33 publications that were then reviewed.

1.3 Results

1.3.1 Geographic and Temporal Analysis of Scientific Publications

The temporal distribution of the research articles for this study is presented in Fig. 1.2. The oldest article on microplastics in freshwater environments was published in 2016; the most recent was published in 2021 (7 articles). Since 2019, the annual number of publications has risen dramatically. This rising trend is continuing, demonstrating unequivocally that microplastics are new emerging environmental pollutants. Twelve articles found in 2020 alone investigated microplastics in freshwater environments. The surge in microplastic research could be attributed to the increased awareness and concerns among researchers, driven by growing recognition of the environmental impact of microplastics on ecosystems and human health. Additionally, interdisciplinary collaboration, funding initiatives, and increased recognition by governments and international organizations have further propelled research efforts to understand and mitigate microplastic pollution. For instance, Erasmus+ funded the Ecomarine

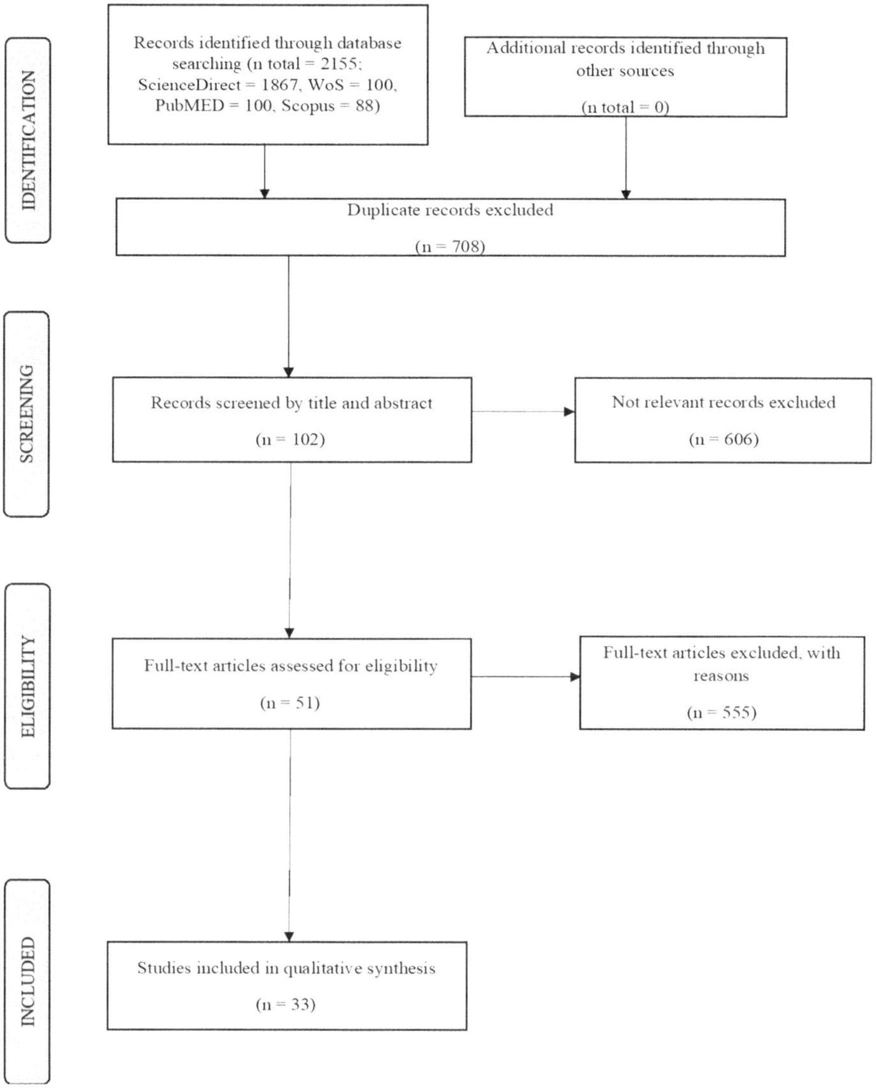

Fig. 1.1 Preferred Reporting Items for Systematic Reviews and Meta-Analysis (PRISMA) diagram of systematic quantitative literature review selection process. n = number of publications

project on marine debris monitoring in Asia (Symplexis, 2024), exemplifying the support for such initiatives.

Figure 1.3 depicts the spatial distribution of research on microplastics in freshwater at the country level. The reviewed studies typically chose a water system and collected samples from several points on the same system. Microplastic studies were conducted in many Asian countries since microplastics have been identified as a

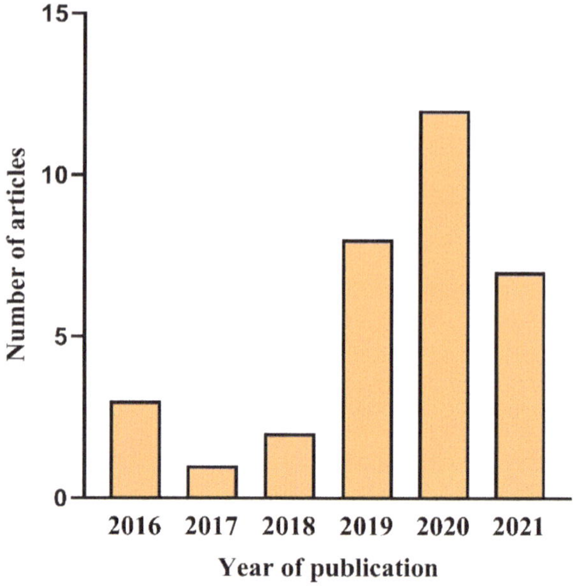

Fig. 1.2 Annual distribution of microplastics research in freshwater environments throughout Asia

developing pollutant that impacts the aquatic environment. The study's 33 publications were spatially distributed among six countries. It was discovered that 24 reviewed works focus on China, followed by India and Iran with 4 and 2 articles, respectively. China and the rest of Asia's countries differ significantly in their research findings on microplastics in freshwater. Some articles had overlapping research locations, such as the Dongting Lake in China, which implies that the lake attracts more scientific attention from Chinese researchers (Hu et al., 2020; Wang et al., 2018; Yin et al., 2020).

Several important rivers and lakes have been the subject of extensive scientific investigation, such as the Yangtze River 4 articles (Li et al., 2019; Su et al., 2018;

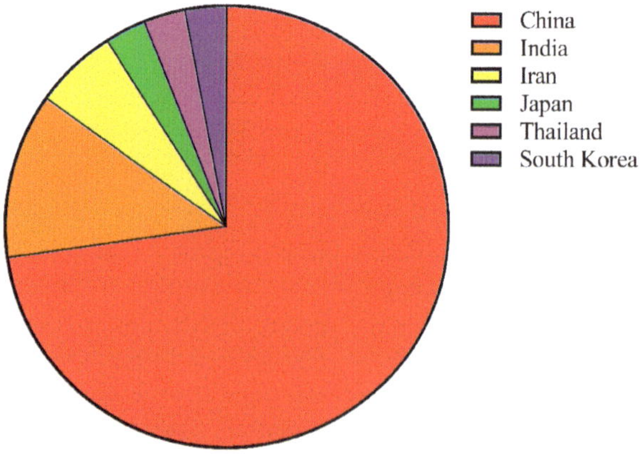

Fig. 1.3 Spatial distribution of microplastics research in Asia

Wang et al., 2017, 2018), Pearl River, three articles (Lin et al., 2018; Wu et al., 2020; Yan et al., 2019), Fuhe River, Haihe River, Maozhou River, Qin River, Tuojiang River, and Zhangjiang River, one article respectively (Liu et al., 2021; Pan et al., 2020; Wu et al., 2020; Zhang et al., 2020b; Zhou et al., 2020; Zhou et al., 2021). For India, only one research article focused on a lake in a high-altitude area, Renuka Lake (Ajay et al., 2021). Other articles focused on lakes and groundwater from coastal Southern India in areas such as Vembanad Lake (Sruthy & Ramasamy, 2017) and Veeranam Lake (Bharath et al., 2021), each with a single case study. For South Korea, one article focused on urban rivers, such as the Han River (Park et al., 2020). For Japan, an article was discovered that concentrates on four rivers: the Awano River, Ayagi River, Asa River, and Majime River (Kabir et al., 2021; Kataoka et al., 2019). For Iran, two articles focused on wetland areas, such as Anzali Wetland (Rasta et al., 2020) and Hashilan Wetland (Abbasi, 2021), with one case each. For Thailand, the main river body explored was the Chao Praya River (Ta & Babel, 2020).

1.3.2 Target Discipline of Journals of Reviewed Articles

The 33 publications on freshwater microplastics were published in ten different journals, which published various research articles using a transdisciplinary approach (Fig. 1.4). Figure 1.5 displays the major disciplines for the 33 articles that were published. More than two-thirds of the articles were discovered to have appeared in publications with a strong connection to environmental research. This demonstrates a great desire for current information on the state of microplastics in the environment on a regional or local level. Research on microplastics is receiving much attention. This review is currently much more critical due to this need or demand.

1.3.3 Microplastics in Freshwater Ecosystems

Of the 33 publications that were reviewed, 17 articles indicated the occurrence of microplastics in both water and sediment samples. In contrast, 11 and 5 articles of the studies examined microplastics in water and sediment samples, respectively. According to Fig. 1.2, microplastic contamination in freshwater environments was first documented in Asia in 2016 in Vembanad Lake, India (Sruthy & Ramasamy, 2017).

We retrieved articles on microplastics from various types of freshwater environments for this review study. Riverine ecosystems generally got the most attention, with 15 articles of studies reporting microplastics in rivers, followed by lakes with nine articles. Figure 1.6 shows the microplastic studies from different freshwater systems. By determining the concentrations and examining the morphology, the studies evaluating microplastics gave detailed pictures of microplastic contamination in the research areas. Microplastic research focused more on rivers than other

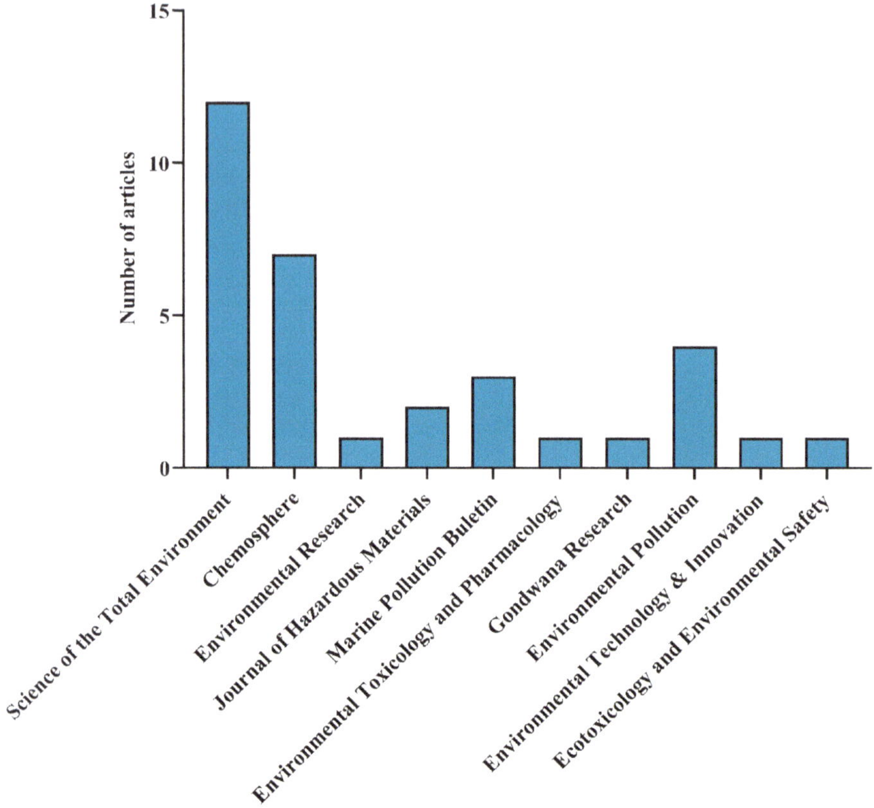

Fig. 1.4 Journals for reviewed articles

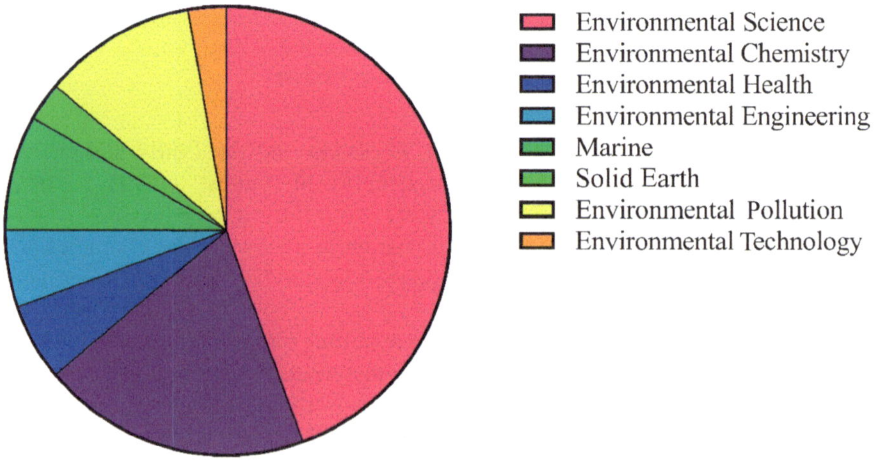

Fig. 1.5 Primary disciplines of journals for selected articles

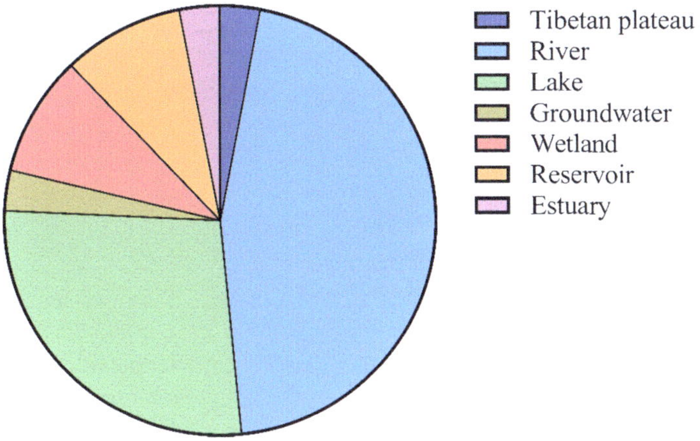

Fig. 1.6 Focus on freshwater systems in studies

water bodies due to their role as primary pathways for plastic pollutants, from terrestrial ecosystems to diverse aquatic environments like reservoirs, estuaries, and coastal waters. Rivers are a major source of microplastic contamination, accounting for 98% of the total amount of microplastics found in the marine environment. According to earlier research, the amount of microplastic pollution increases with proximity to a river (Suteja & Purwiyanto, 2022). Since there are a lot of people living next to rivers, it is imperative that microplastics be thoroughly and continuously assessed throughout river basins in order to track their exposure to aquatic life and, ultimately, human health (Bratovcic et al., 2022). Figure 1.3 shows that China is the country in Asia where microplastics in freshwater systems are most thoroughly investigated and where microplastics in reservoirs receive more attention than in other countries (Di & Wang, 2018; Di et al., 2019; Tan et al., 2019).

Table 1.1 summarizes the various units used in the examined studies to report microplastic amounts in water and sediment. There were 17 papers on water that indicated the amount of microplastic particles per volume of water (particles/L or m^3), whereas three research studies gave the number of particles per surface area (particles/m^2 and km^2). The reporting units have been more reliable for sediment samples. For instance, the units of particles per mass (particles/kg or 100 g) were used in 13 of the 14 investigations that reported the occurrence of microplastics in sediment. The remaining one study used a different unit (particle/L). The different usage of reporting units may be due to the different techniques employed during sampling. For example, studies that used bulk sampling reported microplastic concentrations in particles/m^3(Feng et al., 2020; Zhou et al., 2020; Zhu et al., 2022), while microplastics were reported in particles/km^2 for mantra trawl (Bharath et al., 2021) and particles/m^2 for plankton net techniques (Hu et al., 2020),

Table 1.1 Units used to report average microplastic concentrations by number of articles

Reporting units	Water	Sediment
Particles/L	5	
Particles/km^2	1	
Particles/m^2	2	
Particles/m^3	11	
Particles/100g		1
Particles/kg		12
Particles/L		1

1.3.4 Morphology of Microplastics

The articles that assessed microplastics conducted morphological analyses to identify shape, colour, and size. It is essential to know about their physical and chemical properties to identify the sources of microplastics and evaluate any possible environmental and ecotoxic hazards. Figure 1.7 shows the shapes of microplastics found in the studies' water and sediment samples. The form of microplastics is considered to be a crucial indicator when working out where they come from and how they break down (Wu et al., 2019). Among the 33 studies reviewed, 27 articles and 21 articles of the reviewed articles found fibre shapes in water and sediment, respectively. In contrast, 23 articles and 17 articles of the studies reported fragments in both water and sediment samples, respectively. As a result, both water and sediments had the highest concentrations of fibres and fragments. More than half of the water sample studies reported foam-shaped microplastics. In contrast, more than 30% of sediment sample studies reported foam shapes, followed by film (water: 17 articles, sediments: 13 articles), pellets (water: 12 articles, sediments: 9 articles), and sheets (water: 2 articles, sediment: 2 articles). Domestic sewage is then recognized as one of the primary sources of microplastic fibre in water and sediment (Yan et al., 2021). Laundry and everyday cleaning chemicals frequently release fibre into aquatic environments (Wen et al., 2018). Additionally, fragments from fishing nets have also been considered a source of fibres (Di & Wang, 2018). In general, fibres are described as long, thin, uniformly-diameter microplastic particles (Hidalgo-Ruz et al., 2012). While some researches refer to it as a line or filament instead, all 33 of the examined articles utilized the term "fibre." As such, the comparison of papers in this review is facilitated. According to Yan et al. (2021), the majority of fragments come from the degradation of common plastic products like packaging, containers, cosmetics, and cleaning supplies. Fragments were reported to have a higher abundance in water than sediment ($n = 28$). For instance, according to Li et al. (2019), fragments accounted for 6.79% of microplastics in surface water and 5.23% in sediment samples. To support previous evidence, Ta and Babel (2020) also reported that microplastics recovered from water samples were more than 50% fragments and more than 30% from sediment samples in the Chao Praya River. Spherical particles are referred to as granules, beads, and pellets. Particularly difficult is comparing these

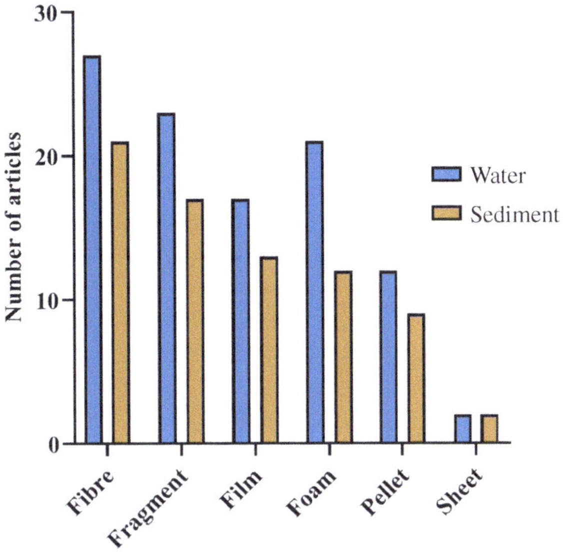

Fig. 1.7 Microplastic shapes in studies

concepts between different studies. Therefore, all spherical particles are categorized as pellets in this study.

The documented microplastic forms may be impacted by the sample and processing techniques. For example, microfibres may pass through trawl or net techniques when bigger mesh sizes are utilized (Dris et al., 2015). According to a recent study, the filtration process's filter design can have an impact on how efficiently different-shaped microplastics are retained (Cai et al., 2020). For example, a nylon filter with the same particle size maintained the fibres nearly entirely, but a polycarbonate filter only kept 62% of the fibres. On a polycarbonate filter, however, more fragments (80%) remained in situ compared to a cotton fibre filter (54%).

In this study, 26 out of the 33 studies provided details on the colours of the microplastics, and 6 articles of them differentiated six colours or more, as shown in Fig. 1.8. The studies report colour as a percentage of all microplastics found. White, black, red, green, blue and yellow were typical colours identified in microplastics in the studies. Prominent colours varied from one study to the next, depending on the location of the study. In a few studies ($n = 2$), coloured microplastics were referred to as "coloured" or "multicolour" without indicating the specific colour. The colour of microplastics is important to identify the sources of microplastics (Hartmann et al., 2019). For instance, coloured microplastics denote their potential sources as packaging, bottles, cosmetics, and clothes (Stolte et al., 2015). The majority of studies indicate colour as a percentage of all microplastics found. Figure 1.9 demonstrates that of the total microplastics detected in water samples, blue plastic microplastics accounted for 19 articles, followed by white (18 articles) and red (16 articles) microplastics. White microplastics comprised 12 articles of all microplastics in sediment samples, whereas the number of articles reported of blue microplastics dropped

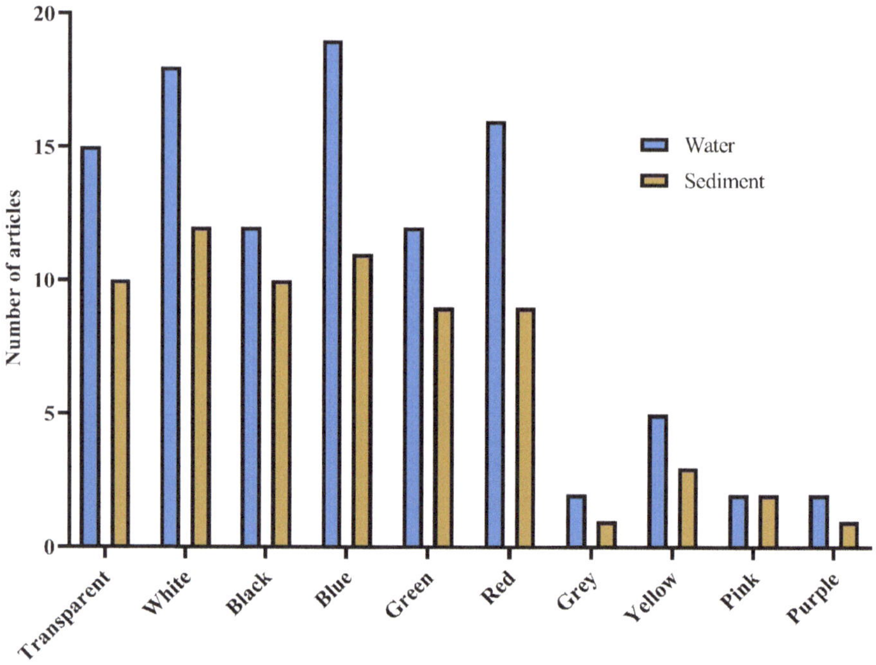

Fig. 1.8 A number of Microplastic colours were identified

to 11. The hydrogen peroxide solution may have caused discolouration during the laboratory work (Nuelle et al., 2014; Yuyao, 2021). Fragments were also observed to be changing from green to light blue, and fibres were observed to be fading from blue to translucent. Plastic is broken down into microplastics by UV radiation, which also alters the colour. Therefore, figuring out whether the colours found through study are the original colours of the plastics could be difficult. The quality of the microscopy, the background and lighting of the microscope, alterations in the surroundings over time, and individual characteristics can also affect an individual's perception of colour (Lu et al., 2021). To have a deeper understanding of microplastic discolouration, more research is necessary.

1.3.5 Source and Transportation of Microplastics in Freshwater Systems

Sources of microplastics in the examined publications were investigated for this review study, and the results are summarized in Fig. 1.10a and b. Microplastics can come from sources that are direct or indirect. Among the 33 studies, 28 provide information about microplastic sources for water sample studies. Less than 50% of

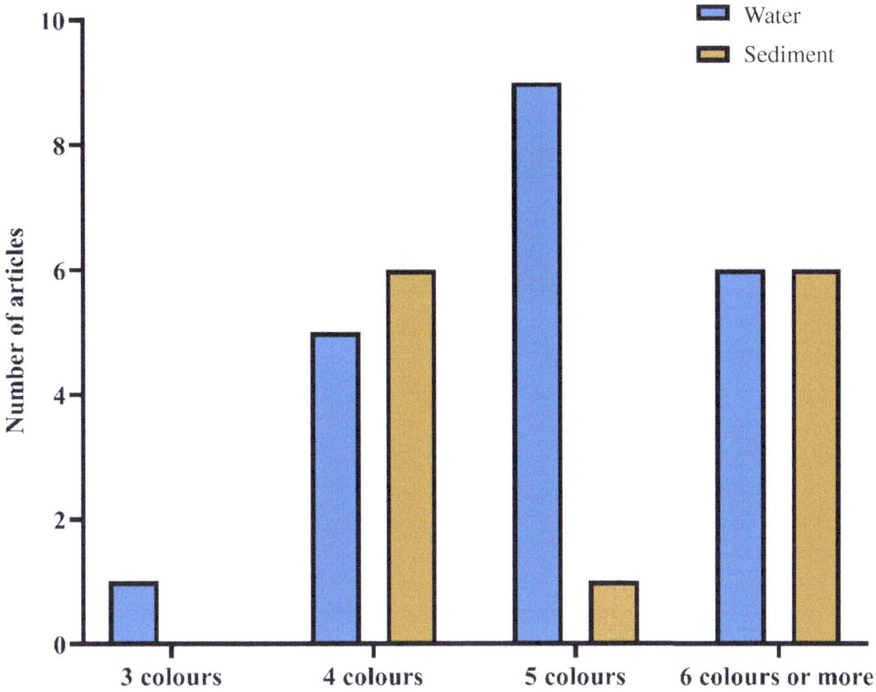

Fig. 1.9 Classification of microplastics based on colours

microplastic studies reported the sources of microplastics for sediment samples (16 out of 33).

Figure 1.10a describes different direct sources of microplastics identified in the review research work. Four out of 33 articles were identified where authors mentioned product packaging and vehicles as the sources of microplastics in water samples in their studies. Meanwhile, three analysed articles pointed to cosmetics as the primary source of microplastics in freshwater water samples. According to Napper et al. (2015), face scrubs consist of between 1000 and 19,000 microplastic particles/mL, which vary widely in size, shape, and colour. Microbeads and microplastics are commonly found in scrubs and other personal care products. Leachate from landfills and wastewater treatment facilities (Kataoka et al., 2019) are the two main ways that microbeads from personal care items are introduced into the aquatic environment (Golwala et al., 2021; Silva et al., 2021). According to Ustabasi and Baysal (2019), 0.4–1.0% polyethylene was present in 20% of 20 different toothpaste brands in Istanbul. The finer sizes of microbeads will end up in aquatic environments. The fewest potential sources found in the studies are textiles or clothing and fishing tools, with one study each. However, fishing tools and textiles or clothing were not found in sediment studies. It should be noted that only six studies provided information on sediment direct sources. In addition, 1 article of the reviewed articles

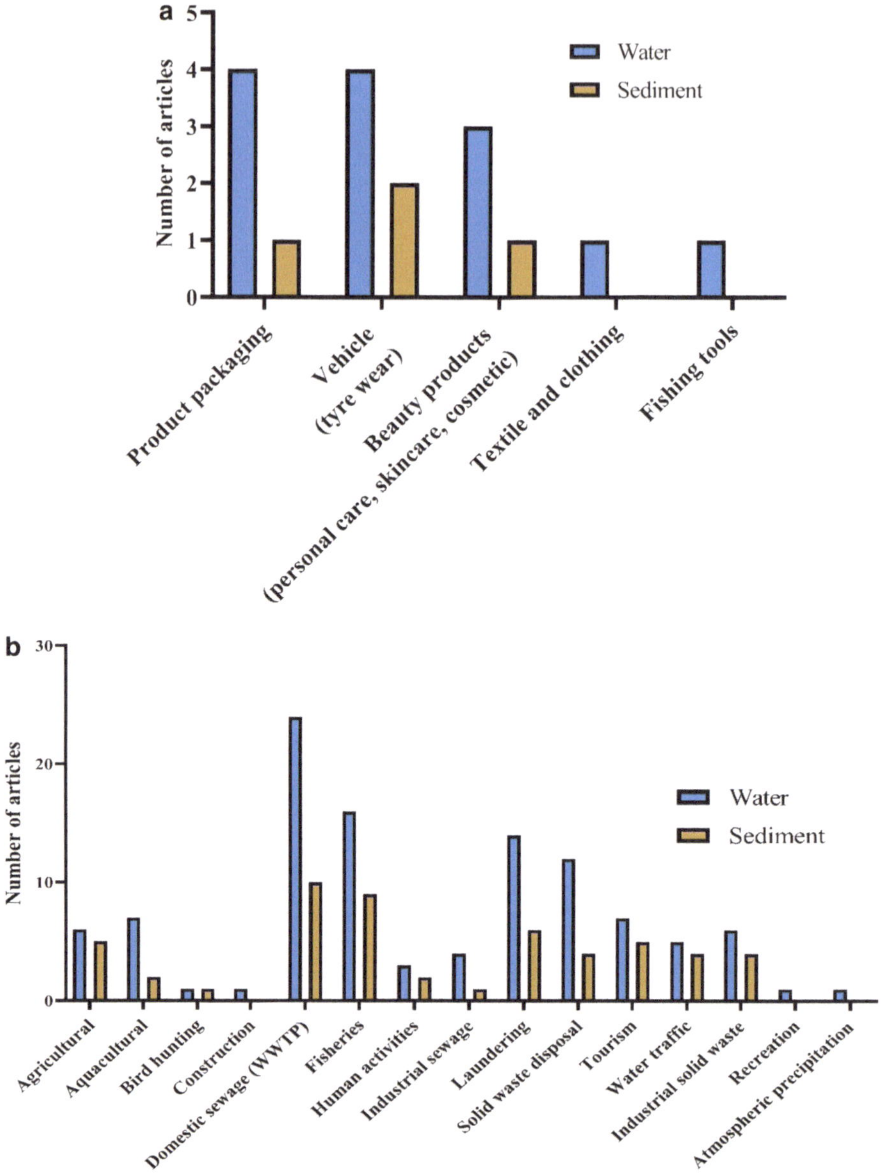

Fig. 1.10 **a** Direct sources of microplastics. **b** Indirect sources of microplastics

identified product packaging and beauty products as direct sources of microplastics in sediments.

Further, 15 indirect sources were identified across 33 reviewed articles for water samples. Simultaneously, 13 were found for sediments. Construction, recreation, and atmospheric precipitation were not mentioned in sediment studies. The most abundant sources identified were domestic sewage for both water (n: 24) and sediments (n: 10), as seen in Fig. 1.10b.

Among 33 studies on microplastics in freshwater, the transportation of microplastics in water was reported in 19 articles, while only 11 reported transportation for sediments. Feng et al. (2020) reported that microplastic in sediments was likely to transfer into the surrounding rivers and lakes by runoff and agricultural irrigation. Out of 33 studies on water, eight articles mentioned surface runoff, and 1 article mentioned irrigation. In addition, microplastics in water are more likely to concentrate in sediments due to weak hydrodynamic forces than to migrate with water flow (Yin et al., 2020). Therefore, five articles found microplastic transportation in sediments, compared to 4 articles that found microplastics being transported in water, as seen in Fig. 1.11. A study conducted by Cai et al. (2020) showed that microplastics were identified in the atmospheric fallout from 3 sites of the air monitoring system at the Dongguan Environmental Monitoring Central Station. This study supports the idea that atmospheric fallout can be one of the ways in which microplastics are transported. However, in this review, only 1 article reported atmospheric fallout in water.

1.4 Conclusions

Freshwater microplastic research is becoming more prevalent, showing a significant interest in this new topic that is becoming more widely understood in Asia. The findings of this systematic review study demonstrate that the majority of research focuses on documenting the presence of microplastics in the freshwater system, either in the sediment or the water. The heterogeneity of studies on freshwater microplastics in Asia presents challenges to a comprehensive understanding due to various factors. The reported amounts and properties of microplastics in various samples and freshwater locations have shown considerable variability and discrepancies due to changes in sample collection, experiment design, data analysis, and findings. The reviewed studies exhibit diversity in part due to the utilization of diverse reporting units and techniques for morphological analysis. Variations in research priorities, funding availability, and institutional support across regions also influence the scope of studies. Addressing these challenges requires standardization of methodologies, harmonization of data reporting practices, and collaborative efforts among researchers and policymakers to establish consistent frameworks for assessing freshwater microplastic pollution in Asia to produce comparable results. This study also revealed that only a few studies extended their discussions to the transport of microplastics, especially in sediment.

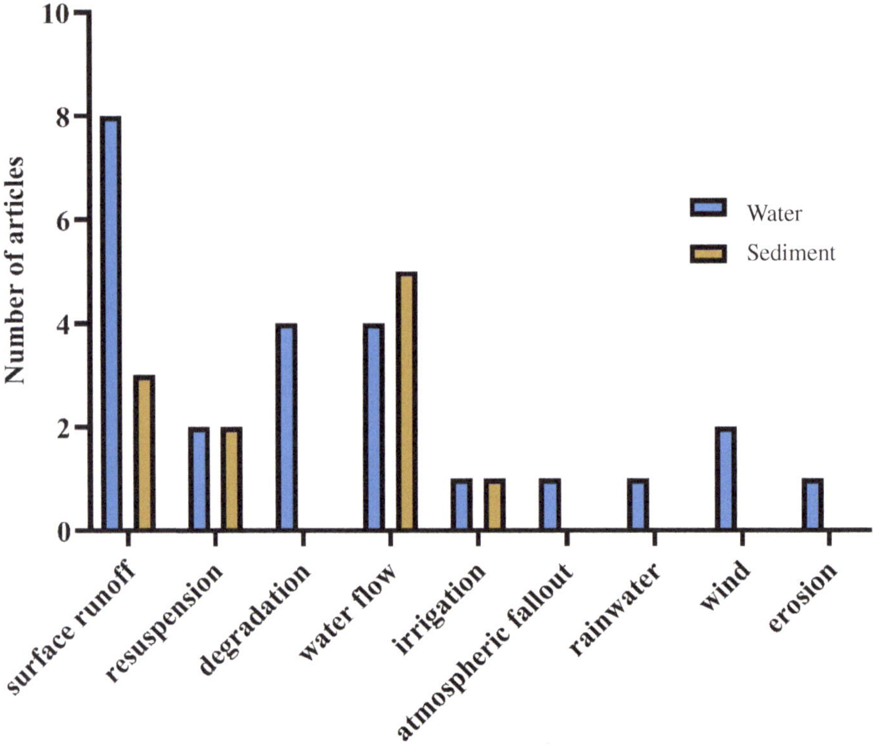

Fig. 1.11 Type of microplastics transportation based on different studies

Acknowledgements We thank the staff of Microplastic Laboratory, Department of Earth Sciences and Environment, Faculty of Science and Technology, Universiti Kebangsaan Malaysia, for their support and cooperation. The main author is a Ph.D. candidate supported by the Faculty of Science and Technology, University Kebangsaan Malaysia. This research was partially supported by a University Research Grant (TAP-K016136 & GUP-2021-056).

References

Abbasi, S. (2021). Prevalence and physicochemical characteristics of microplastics in the sediment and water of Hashilan Wetland, a national heritage in NW Iran. *Environmental Technology & Innovation, 23*, 101782. https://doi.org/10.1016/j.eti.2021.101782.

Ajay, K., Behera, D., Bhattacharya, S., Mishra, P. K., Ankit, Y., & Anoop, A. (2021). Distribution and characteristics of microplastics and phthalate esters from a freshwater lake system in Lesser Himalayas. *Chemosphere, 283*, 131132. https://doi.org/10.1016/j.chemosphere.2021.131132.

Alimba, C. G., & Faggio, C. (2019). Microplastics in the marine environment: Current trends in environmental pollution and mechanisms of toxicological profile. *Environmental Toxicology and Pharmacology, 68*, 61–74. https://doi.org/10.1016/j.etap.2019.03.001.

Bharath, K. M., Srinivasalu, S., Natesan, U., Ayyamperumal, R., Kalam, N., Anbalagan, S., Sujatha, K., & Alagarasan, C. (2021). Microplastics as an emerging threat to the freshwater ecosystems of Veeranam lake in south India: A multidimensional approach. *Chemosphere, 264*, 128502. https://doi.org/10.1016/j.chemosphere.2020.128502.

Botterell, Z. L. R., Beaumont, N., Dorrington, T., Steinke, M., Thompson, R. C., & Lindeque, P. K. (2019). Bioavailability and effects of microplastics on marine zooplankton: A review. *Environmental Pollution, 245*, 98–110. https://doi.org/10.1016/j.envpol.2018.10.065.

Bratovcic, A., Ajith, N., & Arumugam, S. (2022). *Microplastics Pollution in Rivers* (pp. 21–40). https://doi.org/10.1007/978-981-16-8440-1_2.

Cai, H., Chen, M., Chen, Q., Du, F., Liu, J., & Shi, H. (2020). Microplastic quantification affected by structure and pore size of filters. *Chemosphere, 257*, 127198. https://doi.org/10.1016/j.chemosphere.2020.127198.

Cannon, S. M. E., Lavers, J. L., & Figueiredo, B. (2016). Plastic ingestion by fish in the Southern Hemisphere: A baseline study and review of methods. *Marine Pollution Bulletin, 107*(1), 286–291. https://doi.org/10.1016/j.marpolbul.2016.03.057.

Cole, M., Lindeque, P., Fileman, E., Halsband, C., Goodhead, R., Moger, J., & Galloway, T. S. (2013). Microplastic ingestion by zooplankton. *Environmental Science & Technology, 47*(12), 6646–6655. https://doi.org/10.1021/es400663f.

Crawford, C. B., & Quinn, B. (2017). *Microplastic Pollutants*.

Di Beneditto, A. P. M., & Ramos, R. M. A. (2014). Marine debris ingestion by coastal dolphins: What drives differences between sympatric species? *Marine Pollution Bulletin, 83*(1), 298–301. https://doi.org/10.1016/j.marpolbul.2014.03.057.

Di, M., Liu, X., Wang, W., & Wang, J. (2019). Manuscript prepared for submission to environmental toxicology and pharmacology pollution in drinking water source areas: Microplastics in the Danjiangkou Reservoir, China. *Environmental Toxicology and Pharmacology, 65*, 82–89. https://doi.org/10.1016/j.etap.2018.12.009.

Di, M., & Wang, J. (2018). Microplastics in surface waters and sediments of the Three Gorges Reservoir, China. *Science of the Total Environment, 616–617*, 1620–1627. https://doi.org/10.1016/j.scitotenv.2017.10.150.

Dris, R., Gasperi, J., Rocher, V., Saad, M., Renault, N., & Tassin, B. (2015). Microplastic contamination in an urban area: A case study in Greater Paris. *Environmental Chemistry, 12*(5), 592–599. https://doi.org/10.1071/EN14167.

Eerkes-Medrano, D., Thompson, R. C., & Aldridge, D. C. (2015). Microplastics in freshwater systems: A review of the emerging threats, identification of knowledge gaps and prioritisation of research needs. *Water Research, 75*, 63–82. https://doi.org/10.1016/j.watres.2015.02.012.

Feng, S., Lu, H., Tian, P., Xue, Y., Lu, J., Tang, M., & Feng, W. (2020). Analysis of microplastics in a remote region of the Tibetan Plateau: Implications for natural environmental response to human activities. *Science of the Total Environment, 739*, 140087. https://doi.org/10.1016/j.scitotenv.2020.140087.

Golwala, H., Zhang, X., Iskander, S. M., & Smith, A. L. (2021). Solid waste: An overlooked source of microplastics to the environment. *Science of the Total Environment, 769*, 144581. https://doi.org/10.1016/j.scitotenv.2020.144581.

Gruber, E. S., Stadlbauer, V., Pichler, V., Resch-Fauster, K., Todorovic, A., Meisel, T. C., Trawoeger, S., Hollóczki, O., Turner, S. D., Wadsak, W., Vethaak, A. D., & Kenner, L. (2023). To waste or not to waste: Questioning potential health risks of micro- and nanoplastics with a focus on their ingestion and potential carcinogenicity. *Exposure and Health, 15*(1), 33–51. https://doi.org/10.1007/s12403-022-00470-8.

Halstead, J. E., Smith, J. A., Carter, E. A., Lay, P. A., & Johnston, E. L. (2018). Assessment tools for microplastics and natural fibres ingested by fish in an urbanised estuary. *Environmental Pollution, 234*, 552–561. https://doi.org/10.1016/j.envpol.2017.11.085.

Hartmann, N. B., Hüffer, T., Thompson, R. C., Hassellöv, M., Verschoor, A., Daugaard, A. E., Rist, S., Karlsson, T., Brennholt, N., Cole, M., Herrling, M. P., Hess, M. C., Ivleva, N. P., Lusher, A. L., & Wagner, M. (2019). Are we speaking the same language? Recommendations for a definition and categorization framework for plastic debris. *Environmental Science & Technology, 53*(3), 1039–1047. https://doi.org/10.1021/acs.est.8b05297.

Hidalgo-Ruz, V., Gutow, L., Thompson, R. C., & Thiel, M. (2012). Microplastics in the marine environment: A review of the methods used for identification and quantification. *Environmental Science and Technology, 46*(6), 3060–3075. https://doi.org/10.1021/es2031505.

Hu, D., Zhang, Y., & Shen, M. (2020). Investigation on microplastic pollution of Dongting Lake and its affiliated rivers. *Marine Pollution Bulletin, 160*, 111555. https://doi.org/10.1016/j.marpolbul.2020.111555.

Kabir, A. H. M. E., Sekine, M., Imai, T., Yamamoto, K., Kanno, A., & Higuchi, T. (2021). Assessing small-scale freshwater microplastics pollution, land-use, source-to-sink conduits, and pollution risks: Perspectives from Japanese rivers polluted with microplastics. *Science of the Total Environment, 768*, 144655. https://doi.org/10.1016/j.scitotenv.2020.144655.

Kataoka, T., Nihei, Y., Kudou, K., & Hinata, H. (2019). Assessment of the sources and inflow processes of microplastics in the river environments of Japan. *Environmental Pollution, 244*, 958–965. https://doi.org/10.1016/j.envpol.2018.10.111.

Koelmans, A. A., Mohamed Nor, N. H., Hermsen, E., Kooi, M., Mintenig, S. M., & De France, J. (2019). Microplastics in freshwaters and drinking water: Critical review and assessment of data quality. *Water Research, 155*, 410–422. https://doi.org/10.1016/j.watres.2019.02.054.

Kwon, J. H., Kim, J. W., Pham, T. D., Tarafdar, A., Hong, S., Chun, S. H., Lee, S. H., Kang, D. Y., Kim, J. Y., Kim, S. B., & Jung, J. (2020). Microplastics in food: A review on analytical methods and challenges. *International Journal of Environmental Research and Public Health, 17*(18), 6710. https://doi.org/10.3390/ijerph17186710.

Lambert, S., & Wagner, M. (2018). Microplastics are contaminants of emerging concern in freshwater environments: An overview. In M. Wagner & S. Lambert (Eds.), *Freshwater Microplastics: Emerging Environmental Contaminants?* (pp. 1–23). Springer International Publishing. https://doi.org/10.1007/978-3-319-61615-5_1.

Lebreton, L. C. M., van der Zwet, J., Damsteeg, J.-W., Slat, B., Andrady, A., & Reisser, J. (2017). River plastic emissions to the world's oceans. *Nature Communications, 8*(1), 15611. https://doi.org/10.1038/ncomms15611.

Li, J., Liu, H., & Paul Chen, J. (2018). Microplastics in freshwater systems: A review on occurrence, environmental effects, and methods for microplastics detection. *Water Research, 137*, 362–374. https://doi.org/10.1016/j.watres.2017.12.056.

Li, L., Geng, S., Wu, C., Song, K., Sun, F., Visvanathan, C., Xie, F., & Wang, Q. (2019). Microplastics contamination in different trophic state lakes along the middle and lower reaches of Yangtze River Basin. *Environmental Pollution, 254*, 112951. https://doi.org/10.1016/j.envpol.2019.07.119.

Li, X., Liang, R., Li, Y., Zhang, Y., Wang, Y., & Li, K. (2021). Microplastics in inland freshwater environments with different regional functions: A case study on the Chengdu Plain. *Science of the Total Environment, 789*, 147938. https://doi.org/10.1016/j.scitotenv.2021.147938.

Li, Y., Tao, L., Wang, Q., Wang, F., Li, G., & Song, M. (2023). Potential health impact of microplastics: A review of environmental distribution, human exposure, and toxic effects. *Environment & Health, 1*(4), 249–257. https://doi.org/10.1021/envhealth.3c00052.

Lin, L., Zuo, L. Z., Peng, J. P., Cai, L. Q., Fok, L., Yan, Y., Li, H. X., & Xu, X. R. (2018). Occurrence and distribution of microplastics in an urban river: A case study in the Pearl River along Guangzhou City, China. *Science of the Total Environment, 644*, 375–381. https://doi.org/10.1016/j.scitotenv.2018.06.327.

Liu, Y., You, J., Li, Y., Zhang, J., He, Y., Breider, F., Tao, S., & Liu, W. (2021). Insights into the horizontal and vertical profiles of microplastics in a river emptying into the sea affected by intensive anthropogenic activities in Northern China. *Science of the Total Environment, 779*, 146589. https://doi.org/10.1016/j.scitotenv.2021.146589.

Lu, H.-C., Ziajahromi, S., Neale, P. A., & Leusch, F. D. L. (2021). A systematic review of freshwater microplastics in water and sediments: Recommendations for harmonisation to enhance future study comparisons. *Science of the Total Environment, 781*, 146693. https://doi.org/10.1016/j.scitotenv.2021.146693.

Merga, L. B., Redondo-Hasselerharm, P. E., Van den Brink, P. J., & Koelmans, A. A. (2020). Distribution of microplastic and small macroplastic particles across four fish species and sediment in an African lake. *Science of the Total Environment, 741*, 140527. https://doi.org/10.1016/j.scitotenv.2020.140527.

Nadal, M. A., Alomar, C., & Deudero, S. (2016). High levels of microplastic ingestion by the semipelagic fish bogue Boops boops (L.) around the Balearic Islands. *Environmental Pollution, 214*, 517–523. https://doi.org/10.1016/j.envpol.2016.04.054.

Napper, I. E., Bakir, A., Rowland, S. J., & Thompson, R. C. (2015). Characterisation, quantity and sorptive properties of microplastics extracted from cosmetics. *Marine Pollution Bulletin, 99*(1–2), 178–185. https://doi.org/10.1016/j.marpolbul.2015.07.029.

Nuelle, M. T., Dekiff, J. H., Remy, D., & Fries, E. (2014). A new analytical approach for monitoring microplastics in marine sediments. *Environmental Pollution, 184*, 161–169. https://doi.org/10.1016/j.envpol.2013.07.027.

Page, M., McKenzie, J., Bossuyt, P., Boutron, I., Hoffmann, T., Mulrow, C., Shamseer, L., Tetzlaff, J., Akl, E., Brennan, S., Chou, R., Glanville, J., Grimshaw, J., Hróbjartsson, A., Lalu, M., Li, T., Loder, E., Mayo-Wilson, E., McDonald, S., & Moher, D. (2021). The PRISMA 2020 statement: An updated guideline for reporting systematic reviews. *BMJ, 372*, n71. https://doi.org/10.1136/bmj.n71.

Pan, Z., Sun, Y., Liu, Q., Lin, C., Sun, X., He, Q., Zhou, K., & Lin, H. (2020). Riverine microplastic pollution matters: A case study in the Zhangjiang River of Southeastern China. *Marine Pollution Bulletin, 159*, 111516. https://doi.org/10.1016/j.marpolbul.2020.111516.

Park, T.-J., Lee, S.-H., Lee, M.-S., Lee, J.-K., Lee, S.-H., & Zoh, K.-D. (2020). Occurrence of microplastics in the Han River and riverine fish in South Korea. *Science of the Total Environment, 708*, 134535. https://doi.org/10.1016/j.scitotenv.2019.134535.

Pickering, C., & Byrne, J. (2013). The benefits of publishing systematic quantitative literature reviews for PhD candidates and other early-career researchers. *Higher Education Research & Development, 33*, 534–548. https://doi.org/10.1080/07294360.2013.841651.

Plastic Europe. (2019). *Plastics—The facts 2019*. Retrieved November 11, 2019, from https://plasticseurope.org/application/files/9715/7129/9584/FINAL_web_version_Plastics_the_facts2019_14102019.pdf.

Ragusa, A., Svelato, A., Santacroce, C., Catalano, P., Notarstefano, V., Carnevali, O., Papa, F., Rongioletti, M. C. A., Baiocco, F., Draghi, S., D'Amore, E., Rinaldo, D., Matta, M., & Giorgini, E. (2021). Plasticenta: First evidence of microplastics in human placenta. *Environment International, 146*, 106274. https://doi.org/10.1016/j.envint.2020.106274.

Rasta, M., Sattari, M., Taleshi, M. S., & Namin, J. I. (2020). Identification and distribution of microplastics in the sediments and surface waters of Anzali Wetland in the Southwest Caspian Sea. *Northern Iran. Marine Pollution Bulletin, 160*, 111541. https://doi.org/10.1016/j.marpolbul.2020.111541.

Rochman, C. M., Tahir, A., Williams, S. L., Baxa, D. V., Lam, R., Miller, J. T., Teh, F. C., Werorilangi, S., & Teh, S. J. (2015). Anthropogenic debris in seafood: Plastic debris and fibers from textiles in fish and bivalves sold for human consumption. *Scientific Reports, 5*, 14340. https://doi.org/10.1038/srep14340.

Schmidt, C., Krauth, T., & Wagner, S. (2017). Export of plastic debris by rivers into the sea. *Environmental Science & Technology, 51*(21), 12246–12253. https://doi.org/10.1021/acs.est.7b02368.

Shi, W., Sun, S., Han, Y., Tang, Y., Zhou, W., Du, X., & Liu, G. (2021). Microplastics impair olfactory-mediated behaviors of goldfish Carassius auratus. *Journal of Hazardous Materials, 409*, 125016. https://doi.org/10.1016/j.jhazmat.2020.125016.

Sillanpää, M., & Sainio, P. (2017). Release of polyester and cotton fibers from textiles in machine washings. *Environmental Science and Pollution Research, 24*(23), 19313–19321. https://doi.org/10.1007/s11356-017-9621-1.

Silva, A. L. P., Prata, J. C., Duarte, A. C., Soares, A. M. V. M., Barceló, D., & Rocha-Santos, T. (2021). Microplastics in landfill leachates: The need for reconnaissance studies and remediation

technologies. *Case Studies in Chemical and Environmental Engineering, 3*, 100072. https://doi.org/10.1016/j.cscee.2020.100072.
Sruthy, S., & Ramasamy, E. V. (2017). Microplastic pollution in Vembanad Lake, Kerala, India: The first report of microplastics in lake and estuarine sediments in India. *Environmental Pollution, 222*, 315–322. https://doi.org/10.1016/j.envpol.2016.12.038.
Stolte, A., Forster, S., Gerdts, G., & Schubert, H. (2015). Microplastic concentrations in beach sediments along the German Baltic coast. *Marine Pollution Bulletin, 99*(1), 216–229. https://doi.org/10.1016/j.marpolbul.2015.07.022.
Su, L., Cai, H., Kolandhasamy, P., Wu, C., Rochman, C. M., & Shi, H. (2018). Using the Asian clam as an indicator of microplastic pollution in freshwater ecosystems. *Environmental Pollution, 234*, 347–355. https://doi.org/10.1016/j.envpol.2017.11.075.
Suteja, Y., & Purwiyanto, A. I. S. (2022). The role of rivers in microplastics spread and pollution. In M. Sillanpää, A. Khadir, & S. S. Muthu (Eds.), *Microplastics Pollution in Aquatic Media: Occurrence, Detection, and Removal* (pp. 65–88). Springer Singapore. https://doi.org/10.1007/978-981-16-8440-1_4.
Symplexis. (2024). *Building a comprehensive mechanism for preserving marine ecosystems and life from the negative consequences of climate change and the disposal of plastic debris*. https://symplexis.eu/ecomarine/.
Ta, A. T., & Babel, S. (2020). Microplastic contamination on the lower Chao Phraya: Abundance, characteristic and interaction with heavy metals. *Chemosphere, 257*, 127234. https://doi.org/10.1016/j.chemosphere.2020.127234.
Tan, X., Yu, X., Cai, L., Wang, J., & Peng, J. (2019). Microplastics and associated PAHs in surface water from the Feilaixia Reservoir in the Beijiang River, China. *Chemosphere, 221*, 834–840. https://doi.org/10.1016/j.chemosphere.2019.01.022.
Tang, Y., Zhou, W., Sun, S., Du, X., Han, Y., Shi, W., & Liu, G. (2020). Immunotoxicity and neurotoxicity of bisphenol A and microplastics alone or in combination to a bivalve species. *Tegillarca Granosa. Environmental Pollution, 265*, 115115. https://doi.org/10.1016/j.envpol.2020.115115.
Tata, T., Belabed, B. E., Bououdina, M., & Bellucci, S. (2020). Occurrence and characterization of surface sediment microplastics and litter from North African coasts of Mediterranean Sea: Preliminary research and first evidence. *Science of the Total Environment, 713*, 136664. https://doi.org/10.1016/j.scitotenv.2020.136664.
Ustabasi, G. S., & Baysal, A. (2019). Occurrence and risk assessment of microplastics from various toothpastes. *Environmental Monitoring and Assessment, 191*(7), 438. https://doi.org/10.1007/s10661-019-7574-1.
Wagner, M., Scherer, C., Alvarez-Muñoz, D., Brennholt, N., Bourrain, X., Buchinger, S., Fries, E., Grosbois, C., Klasmeier, J., Marti, T., Rodriguez-Mozaz, S., Urbatzka, R., Vethaak, A. D., Winther-Nielsen, M., & Reifferscheid, G. (2014). Microplastics in freshwater ecosystems: What we know and what we need to know. *Environmental Sciences Europe, 26*(1), 12. https://doi.org/10.1186/s12302-014-0012-7.
Wang, J., Wang, M., Ru, S., & Liu, X. (2019). High levels of microplastic pollution in the sediments and benthic organisms of the South Yellow Sea, China. *Science of the Total Environment, 651*, 1661–1669. https://doi.org/10.1016/j.scitotenv.2018.10.007.
Wang, W., Ndungu, A. W., Li, Z., & Wang, J. (2017). Microplastics pollution in inland freshwaters of China: A case study in urban surface waters of Wuhan, China. *Science of the Total Environment, 575*, 1369–1374. https://doi.org/10.1016/j.scitotenv.2016.09.213.
Wang, W., Yuan, W., Chen, Y., & Wang, J. (2018). Microplastics in surface waters of Dongting Lake and Hong Lake, China. *Science of the Total Environment, 633*, 539–545. https://doi.org/10.1016/j.scitotenv.2018.03.211.
Wen, X., Du, C., Xu, P., Zeng, G., Huang, D., Yin, L., Yin, Q., Hu, L., Wan, J., Zhang, J., Tan, S., & Deng, R. (2018). Microplastic pollution in surface sediments of urban water areas in Changsha, China: Abundance, composition, surface textures. *Marine Pollution Bulletin, 136*, 414–423. https://doi.org/10.1016/j.marpolbul.2018.09.043.

Wong, G., Löwemark, L., & Kunz, A. (2020). Microplastic pollution of the Tamsui River and its tributaries in northern Taiwan: Spatial heterogeneity and correlation with precipitation. *Environmental Pollution, 260*, 113935. https://doi.org/10.1016/j.envpol.2020.113935.
Wu, C., Zhang, K., & Xiong, X. (2018). *Microplastic pollution in inland waters focusing on Asia* (pp. 85–99). https://doi.org/10.1007/978-3-319-61615-5_5.
Wu, P., Huang, J., Zheng, Y., Yang, Y., Zhang, Y., He, F., Chen, H., Quan, G., Yan, J., Li, T., & Gao, B. (2019). Environmental occurrences, fate, and impacts of microplastics. *Ecotoxicology and Environmental Safety, 184*, 109612. https://doi.org/10.1016/j.ecoenv.2019.109612.
Wu, P., Tang, Y., Dang, M., Wang, S., Jin, H., Liu, Y., Jing, H., Zheng, C., Yi, S., & Cai, Z. (2020). Spatial-temporal distribution of microplastics in surface water and sediments of Maozhou River within Guangdong-Hong Kong-Macao Greater Bay Area. *Science of the Total Environment, 717*, 135187. https://doi.org/10.1016/j.scitotenv.2019.135187.
Yan, M., Nie, H., Xu, K., He, Y., Hu, Y., Huang, Y., & Wang, J. (2019). Microplastic abundance, distribution and composition in the Pearl River along Guangzhou city and Pearl River estuary, China. *Chemosphere, 217*, 879–886. https://doi.org/10.1016/j.chemosphere.2018.11.093.
Yan, Z., Chen, Y., Bao, X., Zhang, X., Ling, X., Lu, G., Liu, J., & Nie, Y. (2021). Microplastic pollution in an urbanized river affected by water diversion: Combining with active biomonitoring. *Journal of Hazardous Materials, 417*, 126058. https://doi.org/10.1016/j.jhazmat.2021.126058.
Yao, L., Hui, L., Yang, Z., Chen, X., & Xiao, A. (2020). Freshwater microplastics pollution: Detecting and visualizing emerging trends based on Citespace II. *Chemosphere, 245*, 125627. https://doi.org/10.1016/j.chemosphere.2019.125627.
Yin, L., Wen, X., Du, C., Jiang, J., Wu, L., Zhang, Y., Hu, Z., Hu, S., Feng, Z., Zhou, Z., Long, Y., & Gu, Q. (2020). Comparison of the abundance of microplastics between rural and urban areas: A case study from East Dongting Lake. *Chemosphere, 244*, 125486. https://doi.org/10.1016/j.chemosphere.2019.125486.
Yuyao, X. (2021). Microplastic pollution in Chinese urban rivers: The influence of urban factors. *Resources, Conservation, and Recycling, 173*, 105686. https://doi.org/10.1016/j.resconrec.2021.105686.
Zbyszewski, M., & Corcoran, P. L. (2011). Distribution and degradation of fresh water plastic particles along the beaches of Lake Huron, Canada. *Water, Air, & Soil Pollution, 220*(1), 365–372. https://doi.org/10.1007/s11270-011-0760-6.
Zhang, C., Wang, S., Sun, D., Pan, Z., Zhou, A., Xie, S., Wang, J., & Zou, J. (2020a). Microplastic pollution in surface water from east coastal areas of Guangdong, South China and preliminary study on microplastics biomonitoring using two marine fish. *Chemosphere, 256*, 127202. https://doi.org/10.1016/j.chemosphere.2020.127202.
Zhang, K., Gong, W., Lv, J., Xiong, X., & Wu, C. (2015). Accumulation of floating microplastics behind the Three Gorges Dam. *Environmental Pollution, 204*, 117–123. https://doi.org/10.1016/j.envpol.2015.04.023.
Zhang, K., Su, J., Xiong, X., Wu, X., Wu, C., & Liu, J. (2016). Microplastic pollution of lakeshore sediments from remote lakes in Tibet plateau, China. *Environmental Pollution, 219*, 450–455. https://doi.org/10.1016/j.envpol.2016.05.048.
Zhang, L., Liu, J., Xie, Y., Zhong, S., Yang, B., Lu, D., & Zhong, Q. (2020b). Distribution of microplastics in surface water and sediments of Qin river in Beibu Gulf, China. *Science of the Total Environment, 708*, 135176. https://doi.org/10.1016/j.scitotenv.2019.135176.
Zhao, Q., Zhu, L., Weng, J., Jin, Z., Cao, Y., Jiang, H., & Zhang, Z. (2023). Detection and characterization of microplastics in the human testis and semen. *Science of the Total Environment, 877*, 162713. https://doi.org/10.1016/j.scitotenv.2023.162713.
Zhao, S., Zhu, L., & Li, D. (2015). Microplastic in three urban estuaries, China. *Environmental Pollution, 206*, 597–604. https://doi.org/10.1016/j.envpol.2015.08.027.
Zhou, G., Wang, Q., Zhang, J., Li, Q., Wang, Y., Wang, M., & Huang, X. (2020). Distribution and characteristics of microplastics in urban waters of seven cities in the Tuojiang River basin, China. *Environmental Research, 189*, 109893. https://doi.org/10.1016/j.envres.2020.109893.

Zhou, Z., Zhang, P., Zhang, G., Wang, S., Cai, Y., & Wang, H. (2021). Vertical microplastic distribution in sediments of Fuhe River estuary to Baiyangdian Wetland in Northern China. *Chemosphere, 280*, 130800. https://doi.org/10.1016/j.chemosphere.2021.130800.

Zhu, Z., Wei, H., Huang, W., Wu, X., Guan, Y., & Zhang, Q. (2022). Occurrence of microplastic pollution in the Beibu Gulf, the Northern South China Sea [original research]. *Frontiers in Marine Science, 8*, 821008. https://doi.org/10.3389/fmars.2021.821008.

Chapter 2
Hazards of Microplastic Pollution in Agriculture

Kandiah Pakeerathan

Abstract Microplastics (MPs) are small pieces of plastic debris that are less than 5 mm in size. MP pollutants come from a variety of sources, including: physio-chemical degradation of general plastic debris, farm application of MPs contaminated sewage sledges, plastic sheeting that is used to protect crops from weeds and pests, application of plastic-coated fertilizers and pesticides, plastic-based packaging, such as bags and bottles, can end up in agricultural fields when they are blown away from landfills or other waste disposal sites. MPs can have a number of negative effects on agriculture. They can reduce crop yields by blocking the uptake of water and nutrients, contaminate food crops, which can pose a risk to human health, harm soil organisms which play an important role in nutrient cycling. Therefore, the hazards of microplastic pollution in agriculture are a growing concern. Combatting with MP pollution is really a challengeable task, unless, all the government, private and public sector groups work together with the well-planned strategies. Therefore, this chapter intends to present the hazardous effect of MPs in various biotic components of the agroecosystem; and the legislative policy frameworks and advanced depth scientific research needed to mitigate from the MP hazards in terrestrial environment.

Keywords Agriculture · Microplastics · MPs · Pollution · Soil enzyme activities · Terrestrial environment

2.1 Introduction

Plastic is a product from petroleum-based raw resources. Its production has increased by 4% per year since its invention in the 1950s. Current global plastic production is 390.7 million metric tons (Statista, 2024). Plastic is considered one of the both, most welcomed extremely versatile, and indispensable goods as well as highly

K. Pakeerathan (✉)
Department of Agricultural Biology, Faculty of Agriculture, University of Jaffna, Kilinochchi, Sri Lanka
e-mail: pakeerathank@univ.jfn.ac.lk

S. Suriyanarayanan et al. (eds.), *Combating Plastic Pollution in Terrestrial Environment*,
https://doi.org/10.1007/978-981-96-2343-3_2

hated human inventions because its waste cannot degrade in the environment (Wang et al., 2022a). Among the several classes of plastic such as macroplastics, mesoplastics, microplastics (MPs), and nano-plastics, MPs may be polyamide, polystyrene, polyvinyl chloride, polyethylene, and other modified forms whose diameter ranges from 100 μm to 5 mm (Ullah et al., 2021; Yu et al., 2022). The used MPs and formed MPs while degrading thrown plastic wastes into the terrestrial environment are accumulating continuously and create problems significantly to all life forms (Bläsing and Amelung 2018).

Particularly, agriculture soils are polluted with MPs when open dumping of MP beads from personal care products, utilize plastic debris contaminated sludge and compost products, plastic film mulching, partially or untreated sewage, fertilizer and pesticide packaging wastes, wastewater irrigation, other sources (Kumar et al., 2020; Maddela et al., 2023). Scientific studies reveal that prevalence and persistence of MPs in Agricultural soil ecosystem alter the soil physicochemical properties, nutrient cycling, soil enzyme activities, composition of soil microorganisms, flora and fauna (de Souza Machado et al., 2019; Sajjad et al., 2022; Uwamungu et al., 2022). Its sole toxicological effects as well as the combined effect with other toxic substances hinder the growth, reproduction, immunity level, existence and survival of the soil biota that lead to severe fertility loss of the potential agricultural soils (Igalavithana et al., 2022). When improper dumping of abandoned plastic wastes into the terrestrial environment causes severe problems such as the collapse of the natural biogeochemical cycle in terrestrial ecosystems, the food chain, blocks the water passages leading to stagnation of water in the wastewater ditches that serve as breeding for mosquitoes (Zhang et al., 2020).

The major challenges of plastic management are lack of public awareness on the MPs, poor understanding on the impact of MPs in ecosystems and human health; improper monitoring mechanism of recycling and disposal of plastic wastes; unavailability of cheaper alternative biodegradable plastics; and inconsistencies in existing policy and legislative measures of different countries (Hanun et al., 2022). Therefore, to combat from MP pollution at national/regional/global level, holistic approach within the context of UN SDGs is needed.

2.2 Sources of MPs in Agricultural Lands

The prime sources of MPs contaminate agricultural soils, as mentioned in Fig. 2.1, application of processed sewage sludge (biosolids) as soil amendments and compost processed from urban waste recycling process, the extensive use of plastic film as mulches, components of micro irrigation system, water pipes, plastic greenhouse covers, packaging materials, synthetic fertilizers and pesticides (Maddela et al., 2023; Pérez-Reverón et al., 2022b; Yu et al., 2022). Rain fall and wind are seasonal sources to bring large amount of MPs to the agriculture land.

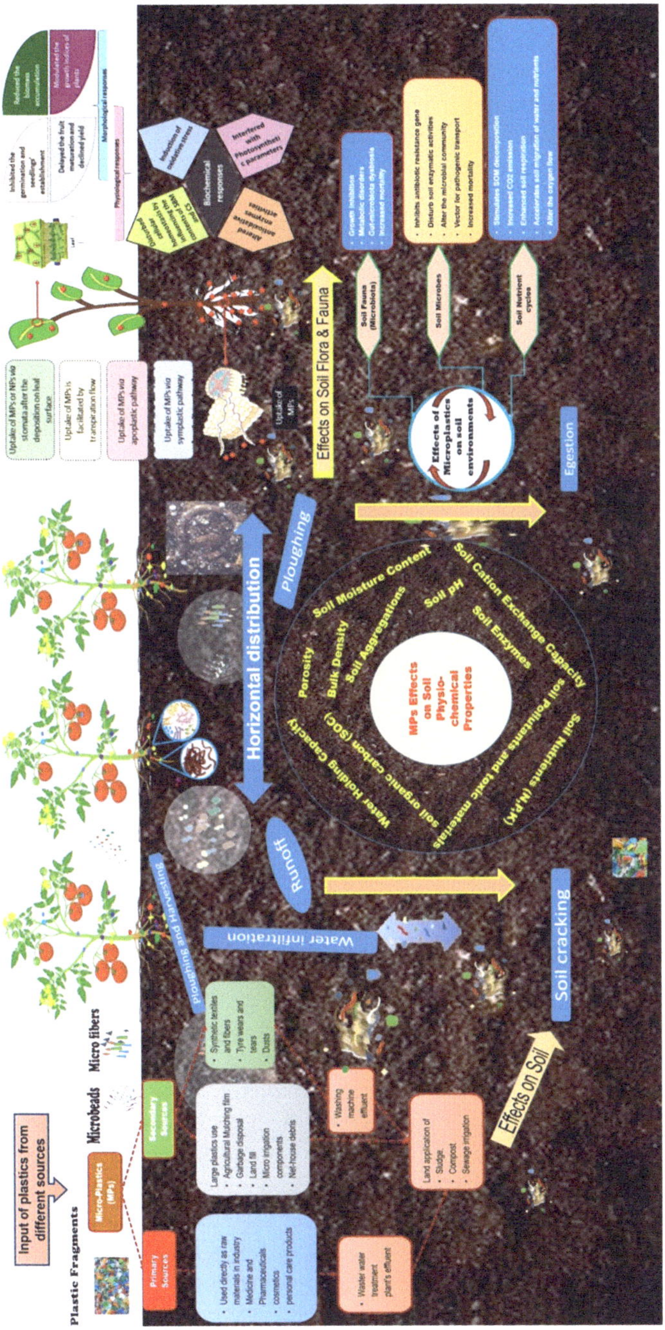

Fig. 2.1 Sources, fate and hazardous impact of microplastics in agricultural soils

2.2.1 Sewage and Sludge Compost as a Source of MPs

The studies show that up to 99% of MPs in wastewater are retained in the sludge (Nizzetto et al., 2016a). The use of sewage sludge as manure for farming or the storage in landfills are important pathways and source for microplastics to enter the agriculture soil (Mason et al., 2016; Kallenbach et al. 2022). In sewage and sledge synthetic fibers are a major source of microplastics (Corradini et al., 2019). Regardless of whether the sewage is treated or not nearly 1.5×10^4–4.5×10^6 particles, still daily leak into the immediate environment (Carr et al., 2016). The estimated quantity of the MPs present in the sludge (i.e., activated sludge, digested sludge, or MBR sludge) after treatment can vary with the concentrations ranging from 1000 to 183,000 particles/kg (dry weight) according to the finding (Wu et al., 2022). Among the countries investigated higher concentrations of MPs were detected in Spanish (165,000–183,000 particles/kg) (Edo et al., 2020), and Finland (27,300 particles/kg) (Lares et al., 2018), respectively. Partially treated or untreated sewage is used for irrigation of around 20 million hm^2 of farmland in 50 countries (Yu et al., 2022). Moreover, it is reported that approximately 35 billion to 2.2 trillion pieces of MPs input annually into the agriculture land through the application of compost in Germany (Weithmann et al., 2018). Single-use plastics, poorly maintained landfills and its indiscriminate disposal in farm lands and residential areas are primary and secondary sources of MPs in Bangladesh (Himu et al., 2022).

Continuous application of sludge to the agricultural land in European and North American countries leads to pollute the land with approximately 63,000–430,000 and 44,000–300,000 tons of MPs annually (Gionfra, 2018). Nizzetto et al. (2016a) reported that the quantity of MPs in sewage and non-sewage irrigated croplands are 5190 and 2030 pieces kg^{-1}, respectively. Moreover, European farmland's top fertile soil layer receives 4000 MPs granules and 670 MPs fibers per kg of sewage sludge application (Leed and Smithson 2019).

2.2.2 Wastewater Irrigation as a Source of MPs

Freshwater is a praised limited natural resource. As the agriculture sector is the huge user of all kind of fresh water sources, therefore, for the crop cultivation the use of treated wastewater could be a feasible sustainable option to safeguard the drinking water and minimize the environmental issues (Singh, 2021). Synthetic microfiber and tumble dryers generated during the production and processing of textiles are being released in farmland ecosystems is considered as a major input of MPs to agricultural soils, with unknown consequences for sustainability and food security (Nizzetto et al., 2016b). Moreover, synthetic micro-fibers from household washing machines and debris emerge from personal care product's usage, such as shoe polish, toothpaste, shampoo, soap, and facial scrubs are also the important urban sources of these MPs in waste water (Yu et al., 2022).

van den Berg et al. (2020) reported that MPs concentrations in sewage irrigated agriculture soil in East Spain are 5190 pieces per kg in comparison to non-sewage irrigated farmlands consisting of 2030 pieces kg^{-1}. But around the globe the various researchers reported that the concentration of the MPs in the water irrigated land could vary from ~79 to 238 items·kg^{-1} in Fuerteventura, Spain (Pérez-Reverón et al., 2022b); ~7100–42,960 items·kg^{-1} in southwestern China (Zhang and Liu 2018); ~600–10,400 items·kg^{-1} in the Metropolitan region of Chile (Corradini et al., 2019); ~96,000–690,000 items·kg^{-1} in suburban soil of Central China (Zhou et al., 2019); ~320–12,560 items·kg^{-1} in suburb Wuhan, Central China (Chen et al., 2020); and ~263–571 items·kg^{-1} in coastal plain of Hangzhou Bay, East China (Zhou et al., 2020). The amount of MPs accumulated in a particular farm land(s) depends on the number of years wastewater being used for irrigation, from which source the wastewater originated and the efficiency of the water treatment plants used for purification.

2.2.3 Plastic Film Mulches and Micro Irrigation Components as a Source of MPs

In modern agriculture, various sophisticated cost-effective methods are in widespread application and plastic mulching is one among them intensively being used to conserve the soil moisture, pest management, provide congenial microclimate and weed management to facilitate the crop growth. Since 1960s plastic mulches are being used and land area being covered by plastic mulch has increased to 22 million hectares in 2017 (Abbate et al., 2023). According to Zhang et al. (2019) around 40,000 km^2 agriculture land has been covered under plastic mulch in Europe. Yu et al. (2020); (Zhou et al., 2020) cited that around the globe, 128,000 km^2 of farm lands is being cultivated under plastic mulch, therefore, the accumulation of MPs in those soils would be more than twice that in non-mulching soils. Unfortunately, adverse climatic parameters such as high temperature, solar radiation and moisture, and intensive use of agrochemicals contain sulfur, halogen, iron, and chlorine hastening the deterioration of plastic films physically and chemically in the farm lands. Even though, ultraviolent stabilizers and hindered amine light stabilizers like Ni quenchers, UVASIL 816, and UVASIL 229 used to reduce the degradation of plastic mulches, but environmental pollutants such as hydrocarbon, nitrogen oxides, sulfur oxides, and particulate enhance the degradation (Sa'adu and Farsang 2023).

MPs form from the degradation of mulches and irrigation components like PVC pipes considered as an emerging pollutant that pose deleterious effects on soil fauna and flora when exposed (Santini et al., 2023). Santini et al. (2023) reported that during the degradation process of plastic mulches and micro-irrigation components release toxic heavy metals like metals Cu and Pb hinder the seed germination, root proliferation and plant biomass production by altering the plant physiological process. Plastic films contain 20–60% phthalates, and while degradation toxic phthalates are being released into the soil slowly and alter the soil microbial carbon and nitrogen, enzyme activities and microbial diversity (Wang et al., 2016).

In China, 32 years of plastic mulch-based agriculture contributed 33–56% to the total microplastics of 7183–10,586 particles/kg in 0–100 cm depth. Therefore, long-term use of plastic mulch films made up of low-density polyethylene can be a potential source of MPs in modern agriculture and accumulation of these plastic residues causes enduring pollution of soils (Li et al., 2022).

2.2.4 *Synthetic Fertilizers and Pesticides as Sources of MPs*

Synthetic fertilizers and pesticides are important inputs of commercial agriculture. To increase the efficiency of the soil-based pesticides, particularly, fumigants, and to prevent the fertilizer leak plastics are being used. Open dumping of fertilizer bags and pesticide containers are considered as a source for MPs. Moreover, MPs present and formed during the degradation process have an ability to adsorb heavy metals and persistent organic pollutants (POPs) such as polybrominated diphenyl ethers (PBDEs), organochlorine pesticides such as dichlorodiphenyltrichloroethane (DDT) and polycyclic aromatic hydrocarbons (PAHs) such as phenanthrene (PHE), and help them to persist in the soil for long term and cause deleterious effect (Hüffer et al., 2019; Pérez-Reverón et al., 2022a). The presence of MPs made of PE reduces the natural retention capacity of the soil, and therefore, benefits the transport of POPs, which could eventually provoke the contamination of groundwater (Hüffer et al., 2019). When exposed to weather forces such as ultraviolet (UV) radiation, temperature, soil acidity and alkalinity or oxidation processes, the pollutants adsorbed by the MPs are released into the soil and can hazardous to soil microbial activity due to their mutagenic, endocrine and carcinogenic characteristics (Hüffer et al., 2019; Pérez-Reverón et al., 2022a).

2.3 Hazards of MPs in Agriculture Soils

Soil health and fertility in the agriculture soil is decided by the physical, chemical properties and microbial composition of soil. Once MPs reached the soil, accumulation of MPs substantially alter the soil physio-bio-chemical properties such as structure, texture, bulk density, organic carbon content colour, fertility, pH, water holding capacity (WHC), cation exchange capacity (CEC), composition of soil microbiota and its enzymatic activities (Medyńska-Juraszek and Jadhav 2022; de Souza Machado et al., 2018b; Zhao et al., 2021). Due to the hydrophobicity nature, high adsorption capacity, diversified shape and polymer type, MPs may alter the water movement and availability and water holding capacity of soil (de Souza Machado et al., 2018b; Jing et al., 2023; Zhang et al., 2023; Zhao et al., 2021).

Physical and biological degradation of MPs release toxic substances, like bisphenol-A, PBDEs and phthalates (PAEs), cause deleterious effect on soil fauna and rhizosphere microbiota, and phyto-toxicity on plants (Meng et al., 2021). Further, blocking of soil micro-pores by MPs and direct intake of MPs by the plants are considered as riskier to the environment and human being.

2.3.1 Impact of MPson Soil Enzymatic Activity

Soil fertility of the agriculture fields depends on the biological composition and the enzymatic activity of the soil. Therefore, the influence the soil enzymatic activities by MPs has been systematically studied by several researchers in the modified experimental conditions. However, published literatures argue that MPs can inhibit or can activate the extracellular enzyme activities in the agriculture soils depending on the type of MPs, size of the MPs, shape of the MPs and exposure time of the MPs to the microbes, composition of the microbes present in the soils and physical and chemical properties of the soil (Fei et al., 2020; Shah et al., 2023; Yu et al., 2020). Wang et al. (2016) reported that the soil accumulated plastic film much debris released phthalate that declined soil microbial C and N, enzyme activities, AWCD value, and Shannon–Weaver and Simpson indices by about 28.9–76.2, 14.9–59.0, 4.9–22.7, 23.0–42.0 and 1.8–18.7%, respectively, when increasing the phthalate concentration to 0–2.02 mg kg^{-1} in the different trails. Further, fluorescein diacetate hydrolysis (FDAse) is an important enzyme and indicator of the microbial activity and short-term changes in soil quality. Yu et al. (2020) reported that MPs tremendously hinder fluorescein diacetate hydrolysis (FDAse) and dehydrogenase activity by 1.6–30.7 and 14.9–59.0%, respectively. In the nutrient cycles of C, N and P, β-glucosidase and cellobiohydrolase; dehydrogenase and leucine-aminopeptidase; and alkaline-phosphatase, respectively, are the important enzymes involved in transformation of those nutrients from non-available to available form. All enzyme actives are decreased significantly post-exposure to MPs (Awet et al., 2018).Fei et al. (2020); (de Souza Machado et al., 2018a) showed that increase in the concentration of MPs like Polypropylene and polyester, inhibit FDAse activities, urease (URE) and β-glucosidase (GLU).

2.3.2 Hazards of MPs on Microorganisms

Microplastics (MPs) can be ingested by microorganisms, and their effects on these organisms are still being studied. Some studies have shown that MPs can have negative effects on microorganisms. For example, they can disrupt the metabolism of microorganisms, leading to changes in their growth and activity. MPs can also increase the production of reactive oxygen species (ROS) in microorganisms, which can damage their cells (Guo et al., 2022).In addition, MPs can carry pollutants, such as heavy metals and organic chemicals. These pollutants can also have negative effects on microorganisms, making them more susceptible to disease and death. Composition of the microbial community is a sensitive indicator of soil health, therefore, any alteration in species diversity and number can create both temporary and long-term changes in soil health (Sajjad et al., 2022).

(Fei et al., 2020) proved that spiking of 1 and 5% (w/w) of polyethylene (PE) and polyvinyl chloride (PVC) based MPs declined the richness and diversity of the Sphingomonadaceae and Xanthobacteraceae bacterial communities and inhibited the biodegradation of xenobiotics in the soil.

However, other studies have shown that MPs can have positive effects on microorganisms. For example, they can provide a surface for microorganisms to attach to and grow on (Huang et al., 2019). MPs can also increase the diversity of microbial communities as they can be processed into energy by some bacteria (i.e., *Rhodococcusruber*, and *Actinomadura* sp.)(Parsaeimehr et al., 2023). Fei et al. (2020) reported that PE treated soils compared to the PVC treated soils the relative abundances of families Burkholderiaceae increased significantly after MPs addition, suggesting the bacteria associated with nitrogen fixation stimulated by the MPs input. Mover, the PICRUStanalysis demonstrated that membrane transporter was a sensitive prediction functional gene of microplastics exposure in the soil (Fei et al., 2020). The overall effects of MPs on microorganisms are complex and depend on a number of factors, including the type of MP, the concentration of MPs, and the type of microorganism. More research is needed to fully understand the effects of MPs on microorganisms and the implications for the environment (Guo et al., 2022; Parsaeimehr et al., 2023).

2.3.3 Hazards of MPs on Soil Fauna

Soil fauna play a vital role in a number of ecosystem processes, including: decomposition of organic matter; nutrient cycling; soil aeration; water infiltration, etc. (Briones, 2018). The MPs on soil fauna can have a number of negative consequences for the environment. When soil fauna ingests either by mistaking them for food or by ingesting them along with other food items, MPs can block the digestive system, leading to starvation or death, intestinal damages, reduced growth and development, damage to organs and tissues, death and other adverse effects, histopathological damage, oxidative stress, DNA damage, genotoxicity, reproductive toxicity, neurotoxicity, metabolic disorders, and gut microbiota dysbiosis can occur (Wang et al., 2022b). In turn, activity of soil animals can transport, fragment, and even degrade MPs, and further influence the fate of MPs in soil. In addition to direct toxicity, MPs can also indirectly harm soil fauna by

1. Disruption of the food chain: MPs can disrupt the food chain by making it difficult for soil fauna to find food. This is because MPs can coat food items, making them less attractive to soil fauna.

2. Alteration of the soil environment: MPs can alter the soil environment in a number of ways. For example, MPs can increase the acidity of the soil, which can make it difficult for soil fauna to survive.
3. Increased risk of infection: MPs can also increase the risk of infection in soil fauna. This is because MPs can act as a vector for pathogens, which can then infect soil fauna.
4. Chemical leaching: MPs can leach chemicals into the soil, which can further harm soil fauna. These chemicals can include endocrine disruptors, heavy metals, and other pollutants.
5. Physical damage: MPs that are sharp or rough can also cause physical damage to soil fauna. This can lead to injuries, infections, and even death.
6. Altered behavior: MPs can alter the behavior of soil fauna. This can include changes in feeding behavior, movement patterns, and social interactions.

Lin et al. (2020) reported that higher quantity of MPs reduced the abundance of oribatid mites $-15.3 \pm 5.7\%$, Dipteran larvae ($-30.5 \pm 9.3\%$), Lepidopteran larvae ($-41.5 \pm 12.2\%$), Hymenoptera (ants; $-62.5 \pm 7.5\%$) and nematodes ($-19.7 \pm 3.4\%$) in their experiments. Further, community structure of microarthropods was also significantly affected by the MPs.

Among 400 additives used to increase the polymer flexibility, BPA and polybrominated diphenyl ethers (PBDEs) in MPs negatively impact on poison, growth and survival rate of the fauna as well as the gut microflora, behaviors of soil animals, such as crawling, scraping, chewing, and excretion, can lead to vertical and horizontal migration of MPs in soil (Ding et al., 2022; Guo et al., 2022). In addition, particle size, shape, polymer types and concentration of MPs are also adversely impact on soil animals, for example, many studies pointed out that earthworms, springtails, nematodes, collembola, and snail. MPs cause intestinal damage that affects their survival, growth, and reproduction (Selonen et al., 2020; Wang et al., 2022b).

Therefore, the hazards of MPs to soil fauna are a serious concern. More research is needed to understand the full extent of the problem and to develop effective mitigation strategies.

2.3.4 Harmful Effect of MPs on Plants

MPs can be found in a variety of environments, including the soil, air, and water. The deleterious impact of MPs in plants has recently become a hot research area. Considerable research focused on MPs but very little has been done to impact on terrestrial plants (Roy et al., 2022). MPs uptake by roots and subsequent translocation can intervene plant physiology varying capacity. When MPs reach plant surfaces from any of the environment can have following harmful effects:

1. **Reduced germination and growth**: MPs can block the pores in seeds, preventing them from germinating (Roy et al., 2022). They can also damage the roots of plants, which can lead to reduced growth (Khalid et al., 2020). Bosker et al. (2019)

found that 8 h of exposure of seeds to 4800 nm size MPs at the concentration of 103–107 particles ml^{-1} reduced the seed germination by 78%. Significant differences in root growth were also observed after 24 h due to physical blockage of the pores in the seed capsule by MPs and then accumulation on root hairs. Polycyclic aromatic hydrocarbons from the personal care products disturb seed germination, and root growth of *Lepidium sativum*, *Sorghum saccharatum*, and *Sinapis alba* (Lipińska et al., 2015). Further, reported that nano- and microplastics caused adverse effects on plant biomass accumulation and length in lettuce, wheat, corn, and rice irrespective of the polymer type and size used (Zantis et al., 2023). This effect is due to the reduction of chlorophyll levels and upregulated ROS or free radicals and stress respondents like antioxidant enzymes).

2. **Reduced photosynthesis and production**: MPs can accumulate heavy metals, reflect sunlight, and absorb sunlight, which can reduce the amount of photosynthesis that plants can perform. A study by Hamid et al. (2010) on *Phaseolus vulgaris* reported that accumulation of Pb on crop decrease in the chlorophyll protein, carbohydrate metabolism, DNA and RNA content in the plant. This can lead to reduced growth and productivity. de Souza Machado et al. (2018b) found that increased concentration of different MP particles like Polypropylene (PP), Polyamide (PA), and HDPE reduced the root and fruit biomass of *Allium fistulosum* (spring onions). In another study, Hernández-Arenas et al. (2021) proved that MPscontaminatedsewage sludge soil delayed tomato fruit production but enhanced the growth.
3. **Increased stress**: MPs can cause stress in plants by preventing the absorption of the required quantity of water, and nutrients, which can make them more susceptible to disease and pests.
4. **Accumulation of toxins**: Domestic effluent application to the agriculture and long-term use of synthetic fertilizers and pesticides can add huge quantity of heavy metals to the soil. MPs present in the soil can accumulate such toxins around the rhizosphere, which can be harmful to plants. Alengebawy et al. (2021) reported that heavy metals like Cd Hg and Pb can cause physiological problems, such as DNA damage and destroying root and shoot systems.
5. **Accumulation of MPs in plants**: Microplastics are transferred to humans through the food chain by consuming food contaminated with microplastics. Oliveri Conti et al. (2020) reported that accumulation of MPs and nano-plastics in edible fruits and vegetables is an emerging diet risk to the general population, as people are consuming fruits and and vegetables as fresh. Accumulation of MPs and NPs are high in perennial plants due to long time exposure to the MPs contaminated environment. Further a research study from Turkey reported that the children are consuming Estimated Daily Intake (EDI) of 68.24 particles kg^{-1} day^{-1} through tomato consumption while Estimated Annual Intake (EAI) MP through tomato was nearly 398,520 particles individual—1 $year^{-1}$ (Aydın et al., 2023). Oliveri Conti et al. (2020) reported in their studies that apples, carrots, and lettuce are carrying MPs in their body may harm the human and other animate consumers.

6. **Physical damage**: MPs can physically damage plant cells, which can lead to reduced growth and productivity. Liu et al. (2023) found that application of polypropylene (PP) and rubber crumb (RC) caused the root cell damage on peanut, therefore, symbiotic association of rhizobium and peanut was collapsed which led to low nitrogen uptake and impaired the soil nitrogen cycling.
7. **Chemical toxicity**: MPs can leach chemicals and heavy metals which produce free radicals that can elevate the intracellular levels of reactive oxygen species (ROS), causing oxidative stress, which causes damage to the biological molecules (e.g., proteins, nucleic acids, lipids, and enzymes), which can be harmful to plants (Wu et al., 2016).
8. **Changes in soil properties**: MPs can change the physical and chemical properties of soil, which can make it more difficult for plants to grow. de Souza Machado et al. (2019)reported that polyester fibers and polyamide beads trigger plant traits and function by altering soil structure and water dynamics.
9. **Changes in the microbiome**: MPs can disrupt the microbiome of the soil and change in the microbial enzymes, i.e., phosphatase, urease and β-glucosidase effects the uptake of different substances by plant roots thereby, affecting nutrient cycling and organic matter decomposition which can make it more difficult for plants to get the nutrients they need (Awet et al., 2018).

The harmful effects of MPs on plants are a growing concern, and *In-Vitro* and *In-Vivo* findings proved that MPs can have a significant impact on plant health. However, wide range of in-depth scientific researches are needed to understand the full extent of the problem and it is important to take steps to minimize the exposure of plants to these harmful particles (Khalid et al., 2020).

2.4 Challenges of MPs Management in Agricultural Lands

The rising use of agricultural plastic is alarming the need of better monitoring and regulation of the quantities of MPs that leak into the environment as well as emphasizing the need of sustainable international treaties and national policy regulations to manage the MP pollution in Agro-Ecosystem (FAO, 2021; Fuschi et al., 2022; Kibria et al., 2023). The major challenges of plastic management in the developing countries are

(a) Budding stage of public awareness on the MPs, Effects of MPs in agricultural soils, ecosystems and human health;
(b) Improper monitoring mechanism of recycling and disposal of plastic wastes;
(c) Unavailability of cheaper alternative biodegradable plastics;
(d) Discrepancies or partiality in implementation of plastic banning policy and existing legislative measures (Pandey et al., 2023).

As per the Basal convention of 1989 and the ban amendment entered into force on 5 December 2019, national and international policies and legislations need to be developed and imposed to create barriers to prevent the entry of MPs and associated toxic materials into the agricultural lands. Anyhow, global and regional data on annual generation of MP waste, strategies adopted for reducing reusing and recycling of MPs waste, the lack of systematic evaluation of MPs leaked into the agroecosystem, the qualitative and quantitative evaluation of fate and impact of MPs under different soil-crop-management systems, exposure risk of MPs to soil biota are very minimal. Therefore, a call for action on multi-faceted regional level holistic approach within the context of UN SDGs, and strengthen the advanced research and data collection on regional and national level MP generation, impact of MPs on beneficial microbes in agricultural lands, degree of toxicity in quantity, type, long and short exposure, combined effect of MPs with soil persistent organic pollutants and heavy metals and environmental fate of soil MPs, comprehensive effects of different types of MP mixtures on the soil-microbe-plant system are timely needed to combat the problem of MP pollution and understand the true comprehensive effects on soil environment under long-term conditions from the developing countries (Yu et al., 2022).

Management of MPs in agricultural lands is a complex and challenging task. There are a number of challenges that can be encountered in managing MPs in agricultural lands include:

1. The difficulty of identifying and surveying MPs
2. The difficulty of controlling MPs without harming other organisms
3. The cost of management
4. The potential for conflict between farmers and public perceptions, attitudes, and behavioral preferences toward MPs and conservationists
5. Breaching of policy formulated for the resistance and or banning the use of single-use plastics and 3R methods proposed.

2.5 Combating Strategies of MPs in Agricultural Lands and Future Directions

Production of the huge amount of plastic worldwide and increasing trend of use in agricultural purposes pose great challenges to control. Production use control is the best acceptable method to combat MPs pollution (Lamichhane et al., 2023).

Creating public awareness everywhere specially from the nursery/kindergarten schools to uneducated to highly educated society's every level special focus to limit unnecessary use of single-use of plastic items like water bottles, plastic shopping bags, straws, and utensils and its impact on when frequently use and its disposal methods. Several volunteer groups, like London student group named "Tyre Collective" is working against tire wear emissions; another group called Waste Bazaar (revolutionize in Africa) and Tontonton (plastic community from Vietnam), are creating awareness ICT-based platforms to reducing MP sources. In Indonesia, the Siklus community works door-to-door canvassing to enable residents to fulfill daily necessities without using plastic packaging (Hanun et al., 2022).

The government should take special focus on separate collection of plastics from degradable garbage and implement advanced recycling systems to reduce the MPs leak in to the farmlands through sewage sledge and compost.

County specific legislative and administrative processes have been taken around the world to mitigate the MP pollution. Sustainable international treaties and national policy regulations to manage the MP pollution in Agro-Ecosystem would be an ideal to work all scientists in a common platform to fighting against this MP hazards. For example, since 2015, the United Nations Environment Programme (UNEP) is in operational to wipeout the use of MPs in cosmetics; the European Commission has also listed MP pollution as a key area of concern; China has already issued the "Plastic Restriction Order" and "Opinions on Further Strengthening the Control of Plastic Pollution" in 2008 and 2020, respectively (Yu et al., 2022). Many counties have formulated very strict plastic regulation policies, and breaching of these policies are frequent due to the lack of effectiveness and administrative lapses in responsible authority's supervision.

In addition to the above, there are a number of things that can be done to mitigate MPs pollution from agricultural soil. These include:

(a) Reduce the use of single-use plastics: Single-use plastics, such as plastic bags and straws, are a major source of MPs pollution. By reducing the use of these plastics, we can help to reduce the amount of MPs that end up in the environment.
(b) Choose products that are free of microbeads: Many cosmetics and personal care products contain microbeads. By choosing products that are free of microbeads, we can help to reduce the amount of MPs that are released into the environment.
(c) Recycle plastic whenever possible: Recycling plastic helps to keep it out of the environment. When plastic is recycled, it is turned into new products, which helps to reduce the demand for new plastic.
(d) Support businesses that are working to reduce MPs pollution: There are a number of businesses that are working to reduce MPs pollution. By supporting these businesses, we can help to make a difference.

By taking these suggestions positively and tactfully, MPs pollution can be mitigatable to protect the terrestrial environment.

2.6 Summary

Microplastics are tiny plastic particles, that can reach the agricultural soil or farm land in various pathways. Application of sewage sludge or biosolids, irrigation of waste water industries, plastic mulches, pesticides and fertilizers. The presence of MPs in agriculture soil can have several potential impacts such as altering soil physio-biochemical properties which can reduce soil fertility, affect microbial activity, impede nutrient, and water uptake; which hinder the seed germination, plant growth, and root development, and ecological consequences. Addressing MP pollution in agriculture soils requires a multi-faceted approach to protect both human health and the environment. By implementing appropriate measures, we can minimize the introduction and spread of MPs in agricultural systems, thereby safeguarding our food supply and ecosystem health. Promoting sustainable and organic farming practices can minimize the use of synthetic pesticides and fertilizers, reducing the potential for MP contamination. Further research and monitoring is needed to better understand the extent and impact of MP pollution in agriculture. Monitoring programs can help identify hotspots and guide targeted mitigation strategies.

References

Abbate, C., Scavo, A., Pesce, G. R., Fontanazza, S., Restuccia, A., & Mauromicale, G. (2023). Soil bioplastic mulches for agroecosystem sustainability: A comprehensive review. *Agriculture, 13*(1), 197. https://doi.org/10.3390/agriculture13010197.

Aydın, R. B., Yozukmaz, A., Şener, İ, Temiz, F., & Giannetto, D. (2023). Occurrence of microplastics in most consumed fruits and vegetables from turkey and public risk assessment for consumers. *Life, 13*(8), 1686. https://doi.org/10.3390/life13081686.

Alengebawy, A., Abdelkhalek, S. T., Qureshi, S. R., & Wang, M. Q. (2021). Heavy metals and pesticides toxicity in agricultural soil and plants: Ecological risks and human health implications. *Toxics, 9*(3), 42. https://doi.org/10.3390/toxics9030042.

Awet, T. T., Kohl, Y., Meier, F., Straskraba, S., Grün, A. L., Ruf, T., Jost, C., Drexel, R., Tunc, E., & Emmerling, C. (2018). Effects of polystyrene nanoparticles on the microbiota and functional diversity of enzymes in soil. *Environmental Sciences Europe, 30*(1), 11. https://doi.org/10.1186/s12302-018-0140-6.

Bläsing, M., & Amelung, W. (2018). Plastics in soil: Analytical methods and possible sources. *Science of the Total Environment, 612*, 422. https://doi.org/10.1016/j.scitotenv.2017.08.086.

Briones, M. J. I. (2018). The serendipitous value of soil fauna in ecosystem functioning: The unexplained explained. *Frontiers in Environmental Science, 6*, 149. https://doi.org/10.3389/fenvs.2018.00149.

Carr, S. A., Liu, J., & Tesoro, A. G. (2016). Transport and fate of microplastic particles in wastewater treatment plants. *Water Research, 91*, 174–182. https://doi.org/10.1016/j.watres.2016.01.002.

Chen, Y., Leng, Y., Liu, X., & Wang, J. (2020). Microplastic pollution in vegetable farmlands of suburb Wuhan, central China. *Environmental Pollution, 257*, 113449. https://doi.org/10.1016/j.envpol.2019.113449.

Corradini, F., Meza, P., Eguiluz, R., Casado, F., Huerta-Lwanga, E., & Geissen, V. (2019). Evidence of microplastic accumulation in agricultural soils from sewage sludge disposal. *Science of the Total Environment, 671*, 411–420. https://doi.org/10.1016/j.scitotenv.2019.03.368.

de Souza Machado, A. A., Lau, C. W., Kloas, W., Bergmann, J., Bachelier, J. B., Faltin, E., Becker, R., Görlich, A. S., & Rillig, M. C. (2019). Microplastics can change soil properties and affect plant performance. *Environmental Science & Technology, 53*(10), 6044–6052. https://doi.org/10.1021/acs.est.9b01339.

de Souza Machado, A. A., Lau, C. W., Till, J., Kloas, W., Lehmann, A., Becker, R., & Rillig, M. C. (2018a). Impacts of microplastics on the soil biophysical environment. *Environmental Science & Technology, 52*(17), 9656–9665. https://doi.org/10.1021/acs.est.8b02212.

de Souza Machado, A. A., Lau, C. W., Till, J., Kloas, W., Lehmann, A., Becker, R., & Rillig, M. C. (2018b). Impacts of microplastics on the soil biophysical environment. *Environmental Science and Technology, 52*(17), 9656–9665. https://doi.org/10.1021/acs.est.8b02212.

Ding, J., Liu, C., Chen, Q., Zhang, Z., Han, J., Liang, A., Zhu, D., Wang, H., Lv, M., & Chen, L. (2022). Extractable additives in microplastics: A hidden threat to soil fauna. *Environmental Pollution, 294*, 118647. https://doi.org/10.1016/j.envpol.2021.118647.

Edo, C., González-Pleiter, M., Leganés, F., Fernández-Piñas, F., & Rosal, R. (2020). Fate of microplastics in wastewater treatment plants and their environmental dispersion with effluent and sludge. *Environmental Pollution, 259*, 113837. https://doi.org/10.1016/j.envpol.2019.113837.

FAO. (2021). Assessment of agricultural plastics and their sustainability—A call for action. Rome

Fei, Y., Huang, S., Zhang, H., Tong, Y., Wen, D., Xia, X., Wang, H., Luo, Y., & Barceló, D. (2020). Response of soil enzyme activities and bacterial communities to the accumulation of microplastics in an acid cropped soil. *Science of the Total Environment, 707*, 135634. https://doi.org/10.1016/j.scitotenv.2019.135634.

Fuschi, C., Pu, H., MacDonell, M., Picel, K., Negri, M., & Chen, J. (2022). Microplastics in the great lakes: Environmental, health, and socioeconomic implications and future directions. *ACS Sustainable Chemistry & Engineering, 10*(43), 14074–14091. https://doi.org/10.1021/acssuschemeng.2c02896.

Gionfra, S. (2018). Plastic pollution in soil. Institute for European Environmental Policy.

Guo, X., Cui, Q., Ma, S., Xiao, C., & Yang, Z. (2022). Soil microplastics: We need to pay more attention. *Water Emerging Contaminants & Nanoplastics, 1*(2), 9. https://doi.org/10.20517/wecn.2022.06.

Hamid, N., Bukhari, N., & Jawaid, F. (2010). Physiological responses of Phaseolus vulgaris to different lead concentrations. *Pakistan Journal of Botany, 42*(1), 239–246.

Hanun, J. N., Hassan, F., & Sidik, F. (2022). The existence of microplastics as an emerging concern in daily routines and the implications of global mitigation efforts. Indonesia post-pandemic outlook: Environment and technology role for Indonesia development 135.

Hernández-Arenas, R., Beltrán-Sanahuja, A., Navarro-Quirant, P., & Sanz-Lazaro, C. (2021). The effect of sewage sludge containing microplastics on growth and fruit development of tomato plants. *Environmental Pollution, 268*, 115779. https://doi.org/10.1016/j.envpol.2020.115779.

Himu, M. M., Afrin, S., Akbor, M. A., Bakar Siddique, M. A., Uddin, M. K., & Rahman, M. M. (2022). Assessment of microplastics contamination on agricultural farmlands in central Bangladesh. *Case Studies in Chemical and Environmental Engineering, 5*, 100195. https://doi.org/10.1016/j.cscee.2022.100195.

Huang, Y., Zhao, Y., Wang, J., Zhang, M., Jia, W., & Qin, X. (2019). LDPE microplastic films alter microbial community composition and enzymatic activities in soil. *Environmental Pollution, 254*, 112983. https://doi.org/10.1016/j.envpol.2019.112983.

Hüffer, T., Metzelder, F., Sigmund, G., Slawek, S., Schmidt, T. C., & Hofmann, T. (2019). Polyethylene microplastics influence the transport of organic contaminants in soil. *Science of the Total Environment, 657*, 242–247. https://doi.org/10.1016/j.scitotenv.2018.12.047.

Igalavithana, A. D., Mahagamage, M. G. Y. L., Gajanayake, P., Abeynayaka, A., Gamaralalage, P. J. D., Ohgaki, M., Takenaka, M., Fukai, T., & Itsubo, N. (2022). Microplastics and potentially toxic elements: Potential human exposure pathways through agricultural lands and policy based countermeasures. *Microplastics, 1*(1), 102–120. https://doi.org/10.3390/microplastics1010007.

Jing, X., Su, L., Wang, Y., Yu, M., & Xing, X. (2023). How do microplastics affect physical properties of silt loam soil under wetting & drying cycles? *Agronomy, 13*(3), 844. https://doi.org/10.3390/agronomy13030844.

Kallenbach, E. M. F., Rødland, E. S., Buenaventura, N. T., & Hurley, R. (2022). Microplastics in terrestrial and freshwater environments. In: M. S. Bank (Ed.), *Microplastic in the Environment: Pattern and Process*. Springer International Publishing, Cham (pp. 87–130). https://doi.org/10.1007/978-3-030-78627-4_4.

Khalid, N., Aqeel, M., & Noman, A. (2020). Microplastics could be a threat to plants in terrestrial systems directly or indirectly. *Environmental Pollution, 267*, 115653. https://doi.org/10.1016/j.envpol.2020.115653.

Kibria, M. G., Masuk, N. I., Safayet, R., Nguyen, H. Q., & Mourshed, M. (2023). Plastic waste: Challenges and opportunities to mitigate pollution and effective management. *International Journal of Environmental Research, 17*(1), 20. https://doi.org/10.1007/s41742-023-00507-z.

Kumar, M., Xiong, X., He, M., Tsang, D. C. W., Gupta, J., Khan, E., Harrad, S., Hou, D., Ok, Y. S., & Bolan, N. S. (2020). Microplastics as pollutants in agricultural soils. *Environmental Pollution, 265*, 114980. https://doi.org/10.1016/j.envpol.2020.114980.
Lamichhane, G., Acharya, A., Marahatha, R., Modi, B., Paudel, R., Adhikari, A., Raut, B. K., Aryal, S., & Parajuli, N. (2023). Microplastics in environment: Global concern, challenges, and controlling measures. *International Journal of Environmental Science and Technology: IJEST, 20*(4), 4673–4694. https://doi.org/10.1007/s13762-022-04261-1.
Lares, M., Ncibi, M. C., Sillanpää, M., & Sillanpää, M. (2018). Occurrence, identification and removal of microplastic particles and fibers in conventional activated sludge process and advanced MBR technology. *Water Research, 133*, 236–246. https://doi.org/10.1016/j.watres.2018.01.049.
Leed, R., & Smithson, M. (2019). Ecological effects of soil microplastic pollution. *Science Insights, 30*, 70–84. https://doi.org/10.15354/si.19.re102.
Li, S., Ding, F., Flury, M., Wang, Z., Xu, L., Li, S., Jones, D. L., & Wang, J. (2022). Macro- and microplastic accumulation in soil after 32 years of plastic film mulching. *Environmental Pollution, 300*, 118945. https://doi.org/10.1016/j.envpol.2022.118945.
Lin, D., Yang, G., Dou, P., Qian, S., Zhao, L., Yang, Y., & Fanin, N. (2020). Microplastics negatively affect soil fauna but stimulate microbial activity: Insights from a field-based microplastic addition experiment. *Proceedings of the Royal Society B: Biological Sciences, 287*(1934), 20201268. https://doi.org/10.1098/rspb.2020.1268.
Lipińska, A., Wyszkowska, J., & Kucharski, J. (2015). Diversity of organotrophic bacteria, activity of dehydrogenases and urease as well as seed germination and root growth Lepidium sativum, Sorghum saccharatum and Sinapis alba under the influence of polycyclic aromatic hydrocarbons. *Environmental Science and Pollution Research International, 22*(23), 18519–18530. https://doi.org/10.1007/s11356-015-5329-2.
Liu, Y., Xu, F., Ding, L., Zhang, G., Bai, B., Han, Y., Xiao, L., Song, Y., Li, Y., Wan, S., & Li, G. (2023). Microplastics reduce nitrogen uptake in peanut plants by damaging root cells and impairing soil nitrogen cycling. *Journal of Hazardous Materials, 443*, 130384. https://doi.org/10.1016/j.jhazmat.2022.130384.
Maddela, N. R., Ramakrishnan, B., Kadiyala, T., Venkateswarlu, K., & Megharaj, M. (2023). Do microplastics and nanoplastics pose risks to biota in agricultural ecosystems? *Soil Systems, 7*(1), 19. https://doi.org/10.3390/soilsystems7010019.
Mason, S. A., Garneau, D., Sutton, R., Chu, Y., Ehmann, K., Barnes, J., Fink, P., Papazissimos, D., & Rogers, D. L. (2016). Microplastic pollution is widely detected in US municipal wastewater treatment plant effluent. *Environmental Pollution, 218*, 1045. https://doi.org/10.1016/j.envpol.2016.08.056.
Medyńska-Juraszek, A., & Jadhav, B. (2022). Influence of different microplastic forms on pH and mobility of Cu(2+) and Pb(2+) in soil. *Molecules (Basel, Switzerland), 27*(5), 1744. https://doi.org/10.3390/molecules27051744.
Meng, J., Xu, B., Liu, F., Li, W., Sy, N., Zhou, X., & Yan, B. (2021). Effects of chemical and natural ageing on the release of potentially toxic metal additives in commercial PVC microplastics. *Chemosphere, 283*, 131274. https://doi.org/10.1016/j.chemosphere.2021.131274.
Nizzetto, L., Bussi, G., Futter, M. N., Butterfield, D., & Whitehead, P. G. (2016a). A theoretical assessment of microplastic transport in river catchments and their retention by soils and river sediments. *Environmental Science: Processes & Impacts, 18*(8), 1050–1059. https://doi.org/10.1039/C6EM00206D.
Nizzetto, L., Futter, M., & Langaas, S. (2016b). Are agricultural soils dumps for microplastics of urban origin? *Environmental Science and Technology, 50*, 10777. https://doi.org/10.1021/acs.est.6b04140.
Oliveri Conti, G., Ferrante, M., Banni, M., Favara, C., Nicolosi, I., Cristaldi, A., Fiore, M., & Zuccarello, P. (2020). Micro- and nano-plastics in edible fruit and vegetables. The first diet risks assessment for the general population. *Environmental Research, 187*, 109677. https://doi.org/10.1016/j.envres.2020.109677.

Pandey, P.A.-O., Dhiman, M., Kansal, A., & Subudhi, S. P. (2023). Plastic waste management for sustainable environment: techniques and approaches. *Waste Disposal & Sustainable Energy, 5*, 205–222. https://doi.org/10.1007/s42768-023-00134-6.

Parsaeimehr, A., Miller, C. M., & Ozbay, G. (2023). Microplastics and their interactions with microbiota. *Heliyon, 9*(4), e15104. https://doi.org/10.1016/j.heliyon.2023.e15104.

Pérez-Reverón, R., Álvarez-Méndez, S. J., Kropp, R. M., Perdomo-González, A., Hernández-Borges, J., & Díaz-Peña, F. J. (2022a). Microplastics in agricultural systems: Analytical methodologies and effects on soil quality and crop yield. *Agriculture, 12*(8), 1162. https://doi.org/10.3390/agriculture12081162.

Pérez-Reverón, R., González-Sálamo, J., Hernández-Sánchez, C., González-Pleiter, M., Hernández-Borges, J., & Díaz-Peña, F. J. (2022b). Recycled wastewater as a potential source of microplastics in irrigated soils from an arid-insular territory (Fuerteventura, Spain). *Science of the Total Environment, 817*, 152830. https://doi.org/10.1016/j.scitotenv.2021.152830.

Roy, T., Dey, T. K., & Jamal, M. (2022). Microplastic/nanoplastic toxicity in plants: An imminent concern. *Environmental Monitoring and Assessment, 195*(1), 27. https://doi.org/10.1007/s10661-022-10654-z.

Sa'adu, I., & Farsang, A. (2023). Plastic contamination in agricultural soils: a review. *Environmental Sciences Europe, 35*(1), 13. https://doi.org/10.1186/s12302-023-00720-9.

Sajjad, M., Huang, Q., Khan, S., Khan, M. A., Liu, Y., Wang, J., Lian, F., Wang, Q., & Guo, G. (2022). Microplastics in the soil environment: A critical review. *Environmental Technology & Innovation, 27*, 102408. https://doi.org/10.1016/j.eti.2022.102408.

Santini, G., Memoli, V., Vitale, E., Di Natale, G., Trifuoggi, M., Maisto, G., & Santorufo, L. (2023). Metal release from microplastics to soil: Effects on soil enzymatic activities and spinach production. *International Journal of Environmental Research and Public Health, 20*(4), 3106. https://doi.org/10.3390/ijerph20043106.

Selonen, S., Dolar, A., Jemec Kokalj, A., Skalar, T., Parramon Dolcet, L., Hurley, R., & van Gestel, C. A. M. (2020). Exploring the impacts of plastics in soil—The effects of polyester textile fibers on soil invertebrates. *Science of the Total Environment, 700*, 134451. https://doi.org/10.1016/j.scitotenv.2019.134451.

Shah, T., Ali, A., Haider, G., Asad, M., & Munsif, F. (2023). Microplastics alter soil enzyme activities and microbial community structure without negatively affecting plant growth in an agroecosystem. *Chemosphere, 322*, 138188. https://doi.org/10.1016/j.chemosphere.2023.138188.

Singh, A. (2021). A review of wastewater irrigation: Environmental implications. *Resources, Conservation and Recycling, 168*, 105454. https://doi.org/10.1016/j.resconrec.2021.105454.

Statista. (2024). Retrieved October 20, 2023, from https://www.statista.com/statistics/282732/global-production-of-plastics-since-1950/.

Ullah, R., Tsui, M.T.-K., Chen, H., Chow, A., Williams, C., & Ligaba-Osena, A. (2021). Microplastics interaction with terrestrial plants and their impacts on agriculture. *Journal of Environmental Quality, 50*(5), 1024–1041. https://doi.org/10.1002/jeq2.20264.

Uwamungu, J. Y., Wang, Y., Shi, G., Pan, S., Wang, Z., Wang, L., & Yang, S. (2022). Microplastic contamination in soil agro-ecosystems: A review. *Environmental Advances, 9*, 100273. https://doi.org/10.1016/j.envadv.2022.100273.

van den Berg, P., Huerta-Lwanga, E., Corradini, F., & Geissen, V. (2020). Sewage sludge application as a vehicle for microplastics in eastern Spanish agricultural soils. *Environmental Pollution, 261*, 114198. https://doi.org/10.1016/j.envpol.2020.114198.

Wang, C., Tang, J., Yu, H., Wang, Y., Li, H., Xu, S., Li, G., & Zhou, Q. (2022a). Microplastic pollution in the soil environment: Characteristics, influencing factors, and risks. *Sustainability, 14*(20), 13405. https://doi.org/10.3390/su142013405.

Wang, J., Lv, S., Zhang, M., Chen, G., Zhu, T., Zhang, S., Teng, Y., Christie, P., & Luo, Y. (2016). Effects of plastic film residues on occurrence of phthalates and microbial activity in soils. *Chemosphere, 151*, 171–177. https://doi.org/10.1016/j.chemosphere.2016.02.076.

Wang, Q., Adams, C. A., Wang, F., Sun, Y., & Zhang, S. (2022b). Interactions between microplastics and soil fauna: A critical review. *Critical Reviews in Environmental Science and Technology, 52*(18), 3211–3243. https://doi.org/10.1080/10643389.2021.1915035.

Weithmann, N., Möller, J. N., Löder, M. G. J., Piehl, S., Laforsch, C., & Freitag, R. (2018). Organic fertilizer as a vehicle for the entry of microplastic into the environment. *Science Advances, 4*(4), eaap8060. https://doi.org/10.1126/sciadv.aap8060.

Wu, X., Cobbina, S. J., Mao, G., Xu, H., Zhang, Z., & Yang, L. (2016). A review of toxicity and mechanisms of individual and mixtures of heavy metals in the environment. *Environmental Science and Pollution Research, 23*, 8244–8259. https://doi.org/10.1007/s11356-016-6333-x.

Wu, X., Zhao, X., Chen, R., Liu, P., Liang, W., Wang, J., Teng, M., Wang, X., & Gao, S. (2022). Wastewater treatment plants act as essential sources of microplastic formation in aquatic environments: A critical review. *Water Research, 221*, 118825. https://doi.org/10.1016/j.watres.2022.118825.

Yu, H., Fan, P., Hou, J., Dang, Q., Cui, D., Xi, B., & Tan, W. (2020). Inhibitory effect of microplastics on soil extracellular enzymatic activities by changing soil properties and direct adsorption: An investigation at the aggregate-fraction level. *Environmental Pollution, 267*, 115544. https://doi.org/10.1016/j.envpol.2020.115544.

Yu, H., Zhang, Y., Tan, W., & Zhang, Z. (2022). Microplastics as an emerging environmental pollutant in agricultural soils: Effects on ecosystems and human health. *Frontiers in Environmental Science, 10*, 855292. https://doi.org/10.3389/fenvs.2022.855292.

Zantis, L. J., Borchi, C., Vijver, M. G., Peijnenburg, W., Di Lonardo, S., & Bosker, T. (2023). Nano- and microplastics commonly cause adverse impacts on plants at environmentally relevant levels: A systematic review. *Science of the Total Environment, 867*, 161211. https://doi.org/10.1016/j.scitotenv.2022.161211.

Zhang, B., Yang, X., Chen, L., Chao, J., Teng, J., & Wang, Q. (2020). Microplastics in soils: A review of possible sources, analytical methods and ecological impacts. *Journal of Chemical Technology & Biotechnology, 95*(8), 2052–2068. https://doi.org/10.1002/jctb.6334.

Zhang, G. S., & Liu, Y. F. (2018). The distribution of microplastics in soil aggregate fractions in southwestern China. *Science of the Total Environment, 642*, 12–20. https://doi.org/10.1016/j.scitotenv.2018.06.004.

Zhang, X., You, S., Tian, Y., & Li, J. (2019). Comparison of plastic film, biodegradable paper and bio-based film mulching for summer tomato production: Soil properties, plant growth, fruit yield and fruit quality. *Scientia Horticulturae, 249*, 38–48. https://doi.org/10.1016/j.scienta.2019.01.037.

Zhang, Z., Wu, X., Liu, H., Huang, X., Chen, Q., Guo, X., & Zhang, J. (2023). A systematic review of microplastics in the environment: Sampling, separation, characterization and coexistence mechanisms with pollutants. *Science of the Total Environment, 859*, 160151. https://doi.org/10.1016/j.scitotenv.2022.160151.

Zhao, T., Lozano, Y. M., & Rillig, M. C. (2021). Microplastics increase soil pH and decrease microbial activities as a function of microplastic shape, polymer type, and exposure time. *Frontiers in Environmental Science, 9*, 675803. https://doi.org/10.3389/fenvs.2021.675803.

Zhou, B., Wang, J., Zhang, H., Shi, H., Fei, Y., Huang, S., Tong, Y., Wen, D., Luo, Y., & Barceló, D. (2020). Microplastics in agricultural soils on the coastal plain of Hangzhou Bay, east China: Multiple sources other than plastic mulching film. *Journal of Hazardous Materials, 388*, 121814. https://doi.org/10.1016/j.jhazmat.2019.121814.

Zhou, Y., Liu, X., & Wang, J. (2019). Characterization of microplastics and the association of heavy metals with microplastics in suburban soil of central China. *Science of the Total Environment, 694*, 133798. https://doi.org/10.1016/j.scitotenv.2019.133798.

Chapter 3
Plastic Waste Management in India: A Design Thinking Approach

C. Gajendran, D. Geetha, P. P. Sivan, and C. Mahendran

Abstract This chapter explores the application of design thinking principles to tackle the pressing issue of plastic waste management in India. With a focus on user-centric methodologies, the approach integrates empathy, creativity, and iterative prototyping to address the environmental challenges posed by plastic waste. By analyzing case studies and leveraging interdisciplinary research, the chapter highlights successful initiatives that reduce plastic pollution, enhance recycling processes, and promote sustainable practices. It emphasizes the need for collaborative efforts, aligning stakeholders from government, industry, academia, and civil society to foster innovation and effective waste management strategies. Through this holistic approach, design thinking not only addresses the technical aspects of waste management but also considers the socio-economic factors that influence waste practices across diverse Indian regions, contributing to the development of a resilient and sustainable waste management ecosystem.

Keywords Plastic waste management · Design thinking · Sustainable practices · Recycling innovations · Stakeholder collaboration · Environmental impact

3.1 Introduction

In India, the surge in plastic waste has emerged as a formidable challenge, intensified by its economic advantages in production and widespread consumption. The country's struggle with effective waste management exacerbates the situation,

C. Gajendran (✉) · D. Geetha
Department of Civil Engineering, SNS College of Technology, Coimbatore, Tamil Nadu, India
e-mail: cgajendran@gmail.com

P. P. Sivan
Department of Civil Engineering, Government College of Engineering, Thrissur, Kerala, India

C. Mahendran
Department of Electronics and Communication Engineering, Alagappa Chettiar Government College of Engineering and Technology, Karaikudi, Tamil Nadu, India

S. Suriyanarayanan et al. (eds.), *Combating Plastic Pollution in Terrestrial Environment*,
https://doi.org/10.1007/978-981-96-2343-3_3

necessitating innovative solutions to mitigate the environmental impact (Kumar & Agrawal, 2020; Vanapalli et al., 2021). India's position as the 12th highest in plastic waste mismanagement underscores the urgency for change, prompted by inadequate enforcement of existing policies, individual negligence, and insufficient waste management infrastructure (Banu, 2019). The core of the problem lies in the inefficacy of municipal solid waste management systems, unable to cope with the increasing volumes of plastic waste, much of which ends up in landfills, polluting the environment and squandering resources (Vimal et al., 2020; Pani & Pathak, 2021). Addressing this crisis requires a shift towards a holistic approach, incorporating design thinking to foster collaborative, innovative strategies for recycling and waste disposal (Ayyamperumal et al., 2022). Key to overcoming these challenges is the implementation of improved waste management frameworks and learning from global best practices, such as the successful single-use plastic bans in countries like Tanzania (Fig. 3.1).

These examples offer valuable insights for India to develop and enforce more effective strategies (Kumar et al., 2017; Pani & Pathak, 2021). The report of "Waste to Energy and Waste Management Market in India-2018" highlights the potential of waste-to-energy systems as a sustainable solution, emphasizing the need to divert plastic waste from landfills towards more productive and environmentally friendly uses (Lakshmikanthan, 2019; Sharma et al., 2020). By adopting such comprehensive systems, alongside enhancing recycling efforts and introducing stringent policy measures, India can significantly reduce the environmental footprint of its plastic waste. A design thinking approach, engaging stakeholders across various sectors, can catalyzes the development of effective plastic waste management strategies. This method emphasizes understanding the root causes of plastic waste generation and exploring innovative solutions through collaboration and ideation (Vanapalli et al., 2021; Vimal et al., 2020; Sharma et al., 2021; Borrelle et al., 2017; Chaudhari et al., 2021; Sarker 2011; De et al., 2023; Ganguly and Chakraborty, 2021 Bharti and Sharma 2021; Leone et al., 2023; MacDonald et al., 2023; Pandey and Manuja, 2021; Chiranjeevi et al., 2018; Mallick et al., 2021; Pearson and Khare 2022; TERI 2020; Rafey and Siddiqui, 2023; Zhang et al., 2021; Ray and Galotra, 2017; Pruneau et al., 2022; Sarkar et al., 2022; Sustainable Urban Plastic Waste Management 2023). To conclude, addressing India's plastic waste dilemma requires a multifaceted strategy that combines advanced waste management technologies, policy reform, and societal behaviour change. By drawing inspiration from global successes and leveraging innovative approaches like design thinking, India can establish a sustainable, integrated system for managing its plastic waste, aligning with both environmental sustainability and economic development goals.

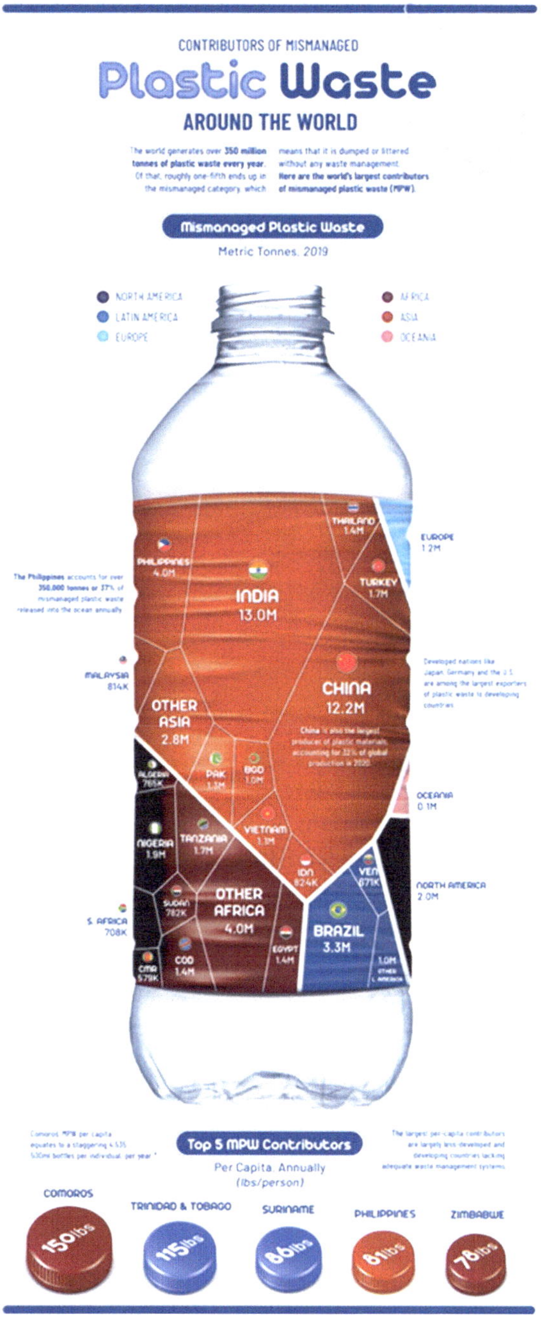

Fig. 3.1 Geographic distribution of mismanaged plastic waste and per capita contributions worldwide. *Source*. "More than 1000 rivers account for 80% of global riverine plastic emissions into the ocean". Science Advances. 718), eaaz5803

3.2 Concept of Design Thinking

Design Thinking is a problem-solving approach that prioritizes empathy, creativity, and iterative prototyping. It begins with empathizing to deeply understand user needs, followed by defining clear problem statements. Ideation then generates diverse solutions, which are quickly prototyped and tested with users. Iteration based on feedback refines ideas to better meet user needs. Its scope extends across various disciplines and industries, fostering innovation in product design, service delivery, and organizational processes. By encouraging a human-centered perspective and embracing experimentation, Design Thinking enables individuals and organizations to tackle complex challenges effectively, leading to solutions that are both innovative and user-centric.

It is a methodology that emphasizes empathy with users, collaboration, and iteration to create solutions that are viable, feasible, and desirable. Here's a simplified breakdown of the key steps in the design thinking process as per Stanford's methodology:

Empathize: Understand the perspectives and experiences of stakeholders involved in plastic waste management, including consumers, waste management workers, policymakers, and environmental activists. This involves conducting interviews, observations, and research to gain insights into their needs, challenges, and behaviours.

Define: Define the specific problem or challenge within the realm of plastic waste management. This might involve identifying key pain points, inefficiencies, or missed opportunities in the current system. For example, it could be the lack of effective recycling infrastructure, widespread single-use plastic consumption, or inadequate public awareness about plastic pollution.

Ideate: Generate a wide range of creative ideas and potential solutions to address the defined problem. Encourage brainstorming sessions and creative thinking techniques such as mind mapping, role-playing, or reverse thinking. The goal is to explore diverse possibilities without judgment and foster innovative approaches to plastic waste management.

Prototype: Develop prototypes or small-scale models of selected ideas to test their feasibility and effectiveness. Prototyping allows for quick experimentation and iteration, enabling designers to refine their concepts based on feedback and insights gained through testing. This could involve creating pilot programs, mock-ups, or simulations to simulate real-world scenarios and evaluate potential outcomes.

Test: Test the prototypes with target users and stakeholders to gather feedback and data on their performance. This phase involves observation, interviews, surveys, and other evaluation methods to assess how well the solutions address the identified problem and meet the needs of the intended users. Based on the test results, iterate on the prototypes and make necessary adjustments to improve their functionality and usability.

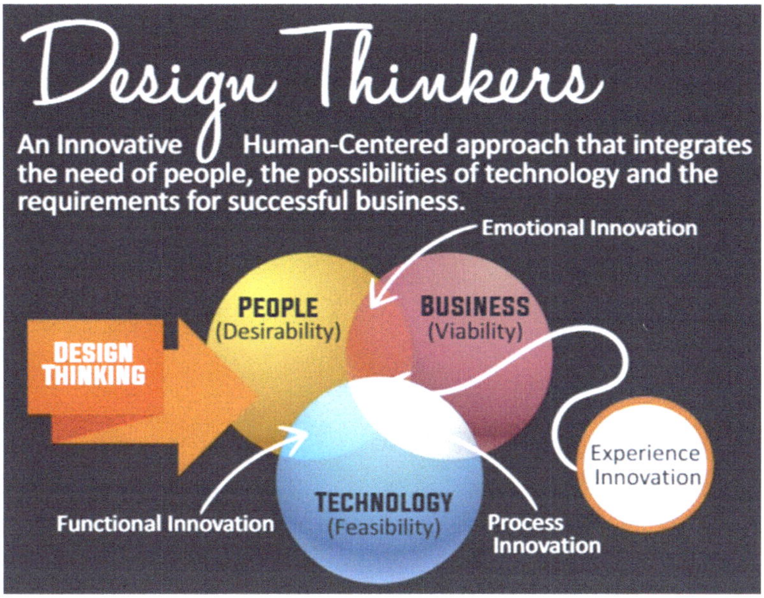

Fig. 3.2 Insight of design thinking

Throughout the design thinking process, collaboration and interdisciplinary teamwork are crucial. By bringing together diverse perspectives and skill sets, design teams can leverage the collective creativity and expertise of their members to develop innovative solutions that address the needs of users effectively (Figs. 3.2 and 3.3).

3.3 Case Studies with Statistics and Significant on Design Thinking in Plastic Waste Management

Here are a few examples:

a. The Ocean Cleanup: The Ocean Cleanup, founded by Boyan Slat, aimed to develop advanced technologies to rid the world's oceans of plastic. While not a traditional "design thinking" project, it embodies the principles of innovative problem-solving. Their design approach involved multiple iterations and prototyping to create efficient systems to collect plastic waste from the ocean's surface. Although specific statistics may vary, their efforts have gained significant attention and support worldwide, illustrating the power of innovative design in addressing complex environmental challenges.
b. IDEO's Circular Design Guide: IDEO, a global design firm, created the Circular Design Guide, which offers practical tools and resources for designing products and systems that contribute to a circular economy, aiming to eliminate waste

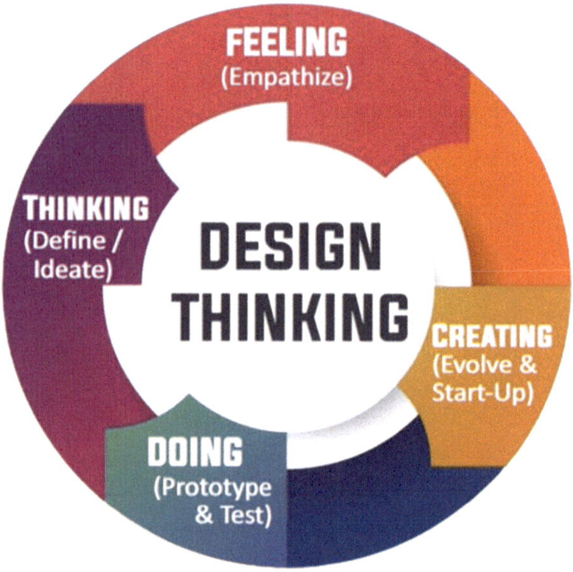

Fig. 3.3 Process of design thinking

and promote sustainability. While not focused solely on plastic waste, the guide emphasizes design thinking principles to rethink and redesign products and materials to reduce waste and encourage recycling and reuse. Though not specific to plastic waste management, the principles and methodologies outlined in the guide can be applied to various contexts, including plastic waste reduction.

c. Precious Plastic: Precious Plastic is an open-source project that empowers communities to tackle plastic pollution by providing them with the tools and knowledge to recycle plastic locally. Founded by Dave Hakkens, the project offers blueprints for building DIY recycling machines that enable individuals and small businesses to turn plastic waste into valuable products. While precise statistics may vary, Precious Plastic has inspired a global community of makers and entrepreneurs to take action against plastic waste, demonstrating the potential of grassroots initiatives fueled by design thinking principles.

Through iterative prototyping, user-centered design, and systems thinking, design thinkers can develop effective solutions that address the root causes of plastic pollution while meeting the needs of stakeholders and promoting environmental sustainability.

3.4 Understanding the Challenges of Plastic Waste in India

In India, the management of plastic waste presents a complex challenge that intersects with environmental sustainability, economic efficiency, and societal practices. Plastics, integral to modern life, pose a significant environmental management issue due to the difficulties in recycling mixed waste and the inadequate infrastructure for effective waste management. These challenges have led to an inefficient system unable to cope with the increasing flows of waste generation, resulting in substantial environmental pollution and resource wastage (Yadav et al., 2022). To combat these issues, a comprehensive approach encompassing waste-to-energy systems, enhanced recycling methods, and policies aimed at reducing plastic waste generation is essential. Implementing strategies such as the proposed ban on single-use plastics and segregation at the source can significantly improve the processing efficiency of plastic waste and mitigate the environmental impacts associated with landfill disposal. The "Waste to Energy and Waste Management Market in India-2018" report sheds light on the potential for innovative solutions within the waste and energy management sector. It highlights the importance of adopting sustainable and context-specific strategies to navigate the challenges of scaling, market dynamics, and the environmental footprint of waste management practices (Awasthi et al., 2017; Kumar et al., 2017; Shanker et al., 2022). Moreover, the proliferation of plastics and the surge in cheap packaging have altered the composition of urban waste, introducing new challenges in waste segregation, pollution, and sustainability. This scenario underscores the necessity for a paradigm shift in how we produce, consume, and dispose of plastics. Moving beyond traditional recycling methods requires a holistic strategy that leverages design thinking and innovative technologies to develop impactful solutions for plastic waste management. Adopting design-driven methodologies and cutting-edge technologies can enable India to establish a resilient waste management framework. Such a system would not only address the immediate challenges of plastic pollution but also unlock economic opportunities, ensuring a sustainable approach to environmental governance (Singh et al., 2023; Gupta, 2021; Sureshkumar & Ponnusamy, 2020). By embracing these integrated solutions, India can pave the way for a future where plastic waste management aligns with the goals of environmental preservation, economic growth, and societal well-being (Fig. 3.4).

3.5 Implementing Design Thinking in Plastic Waste Management

Implementing design thinking in plastic waste management is pivotal for crafting sustainable solutions to India's complex challenges in this domain. This approach is instrumental in enhancing the recycling processes by understanding the underlying issues leading to the deterioration of recycled plastic quality. Design thinking empowers the creation of innovative recycling methodologies that preserve the value

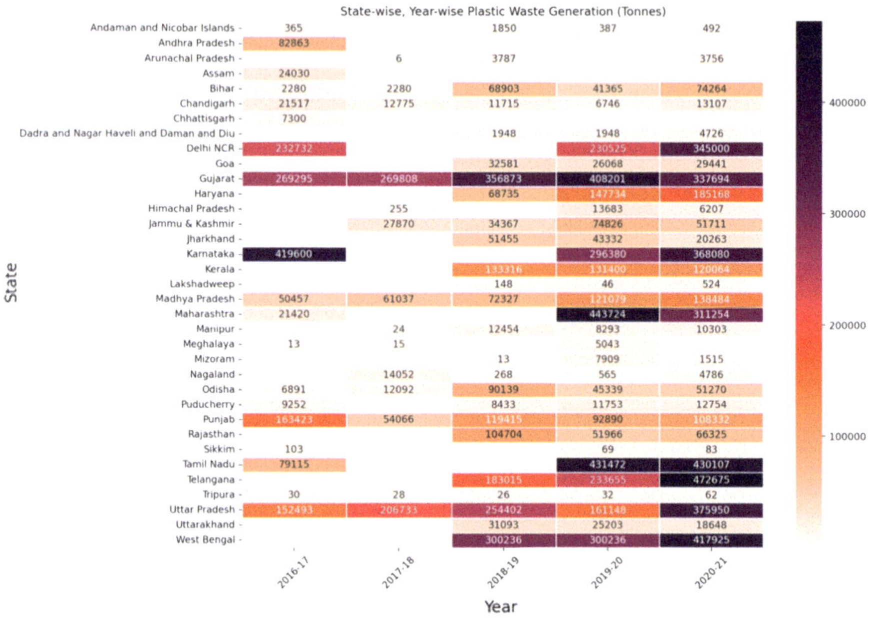

Fig. 3.4 State-wise annual plastic waste generation in India (tonnes) from 2016–17 to 2020–21

of recycled plastic, fostering a circular economy (Chauhan et al., 2021; Khandelwal & Barua, 2020; Kumar & Agrawal, 2020; Balwada et al., 2021). Moreover, design thinking facilitates the customization of waste management strategies to fit the unique requirements of different Indian regions. It considers various socio-economic and cultural factors affecting plastic waste practices, enabling the design of targeted solutions for efficient waste collection, segregation, and recycling (Awasthi et al., 2017; Shanker et al., 2022). A significant contribution of design thinking lies in transforming plastic waste into valuable energy resources.

In tackling plastic waste management, Design Thinking begins with empathizing with stakeholders affected by plastic pollution, including communities and ecosystems. Defining the problem involves identifying key challenges such as improper disposal and lack of recycling infrastructure. Ideation generates solutions like community recycling programs, biodegradable alternatives, and educational campaigns. Prototyping involves creating mock-ups of recycling systems or eco-friendly packaging. Testing involves piloting solutions in local communities to gather feedback. Iteration refines strategies based on results, ensuring effective and sustainable plastic waste management solutions that address environmental concerns and community needs.

Collaborating with stakeholders from the energy and waste sectors, it fosters the development of novel technologies and methods for plastic waste conversion into energy, offering dual benefits of waste management and sustainable energy solutions (Fiksel et al., 2021; Chauhan et al., 2021; Conlon, 2023). Community engagement and behaviour change are also crucial facets where design thinking makes a mark. It helps in crafting educational programs that promote responsible plastic use, waste segregation, and recycling, ensuring these practices resonate with local communities for lasting environmental benefits (Balwan et al., 2020; Vimal et al., 2020; Kumar et al., 2020; Somani et al., 2021). The integration of design thinking in managing India's plastic waste promises innovative, region-specific, and lasting solutions. By encouraging collaboration, creativity, and empathy, it tackles the multifaceted challenges of plastic waste, steering India towards a resilient and effective waste management system. This approach aligns with the vision of achieving sustainable plastic waste management, blending economic growth with environmental stewardship. The focus on developing efficient recycling processes, encouraging community participation, leveraging advanced technologies for waste-to-energy conversion, and fostering multi-stakeholder collaboration underscores the transformative potential of design thinking in revolutionizing plastic waste management in India (Koot, 2020; Pani & Pathak, 2021; Shanker et al., 2022; Conlon, 2023). Through such a holistic and empathetic approach, India can navigate towards a sustainable future, effectively managing plastic waste while promoting environmental conservation and energy sustainability (Fig. 3.5).

Design thinking emerges as a transformative approach in addressing the complexities of waste management, particularly in tackling the pervasive challenge of plastic waste in India. This methodology emphasizes a user-centered perspective, fostering innovation and environmental sustainability. By incorporating design thinking into waste management strategies, India stands to revolutionize its approach towards plastic waste, identifying root causes and pioneering feasible, impactful solutions (Balu et al., 2020; Balwada et al., 2021; Dora et al., 2020; Pani & Pathak, 2021; Tantiyaswasdikul, 2020). Through design thinking, alternative, eco-friendly packaging solutions can be developed, significantly reducing the environmental footprint. This approach also enhances waste management infrastructure, enabling the creation of user-centric recycling programs, efficient collection and sorting systems, and the adoption of advanced waste treatment technologies. Moreover, design thinking cultivates a culture of collaboration across various stakeholders, including waste management entities, communities, and government bodies, ensuring a comprehensive and inclusive approach to tackling plastic waste challenges. Additionally, design thinking plays a pivotal role in promoting sustainable behaviours and raising public awareness about the importance of minimizing plastic waste. This can be achieved through creative campaigns, interactive educational resources, and public initiatives that advocate for responsible waste practices. In essence, by leveraging design thinking, India can navigate the multifaceted challenges of plastic waste management, fostering innovation, stakeholder collaboration, and public engagement towards a more sustainable future.

Fig. 3.5 The innovator's vision—a conceptualization of design thinking and creativity in problem-solving

3.6 Potential Opportunities in Indian Plastic Waste Management

Leveraging design thinking principles in India's battle against plastic waste could revolutionize the development of sustainable and eco-friendly plastic materials. This approach promotes creativity and collaboration, essential for crafting new recycled materials that not only enhance industrial processes but also align with global sustainability objectives (Singh et al., 2017; Sharma et al., 2020; Sharma et al., 2020). A promising avenue is the advancement of plastic waste recycling technologies capable of producing materials with improved properties and innovative structural designs. Such advancements could broaden the applications of recycled materials, significantly diminishing the environmental toll of plastic waste and catering to the growing demand across diverse industries (Balwada et al., 2021; Shanker et al., 2022). Furthermore, the adoption of cutting-edge recycling processes can bolster the circular economy concept. This paradigm seeks to reintegrate recycled plastic materials back into production cycles, thereby curtailing the dependency on new plastic production and slashing the environmental footprint (Bhandari et al., 2021; Gupta, 2021; Pani & Pathak, 2021). Additionally, design thinking facilitates a re-evaluation of the production, usage, and disposal methods of plastic items, steering towards designs that are both functional and environmentally benevolent. Such innovations not only meet practical needs but also foster responsible consumption and waste practices (Dutta et al., 2021; Tanksale et al., 2021). The integration of design thinking into plastic waste management also uncovers opportunities for developing eco-friendly packaging solutions. This can significantly reduce the dependency on single-use plastics, improving packaging sustainability across sectors and marking a pivotal step towards environmental stewardship (Han et al., 2021; Rois et al., 2020). In essence, by applying design thinking principles, India can spearhead the creation of advanced recycling technologies, sustainable product designs, and a robust circular economy. This strategic pivot can catalyze the emergence of innovative, tailored solutions that foster a sustainable and environmentally conscious framework for managing plastic waste, positioning India as a leader in the global effort towards sustainable waste management (Fig. 3.6).

Fig. 3.6 Creative illustration of plastic waste management initiatives in India

3.7 Understanding User Needs and Behaviours

Design thinking emphasizes the importance of understanding user needs and behaviours, which is particularly relevant in the context of waste management. By conducting in-depth research and engaging with diverse stakeholders, including households, businesses, and waste management entities, India can gain valuable insights into the patterns of waste generation, disposal habits, and prevailing challenges. This user-centric approach can inform the design of targeted interventions that resonate with the specific needs of different user groups, leading to more effective waste management strategies (Singh et al., 2023; Koot, 2020). Driving Innovation in Packaging Materials and Waste Management Systems Design thinking can drive innovation in packaging materials and waste management systems by encouraging a holistic approach that considers the entire life cycle of plastic products. This includes exploring alternative materials for packaging that are eco-friendly and biodegradable, as well as developing innovative recycling technologies to efficiently process plastic waste. Design thinking can also promote the integration of circular economy principles in plastic waste management. By adopting a cyclical approach that focuses on reducing, reusing, and recycling plastic waste, India can move away from the linear model of production and dispose of waste, thereby minimizing the environmental impact of plastic waste.

3.8 Emphasizing Environmental Impact

Design thinking places a strong emphasis on considering the environmental impact of solutions throughout the ideation and implementation phases. In the context of waste management, this means prioritizing approaches that minimize environmental harm, reduce carbon footprint, and promote sustainable practices. By integrating life-cycle assessment and environmental impact analysis into the design process, India can ensure that its waste management strategies align with global sustainability goals and contribute to the preservation of natural resources.

3.9 Encouraging Cross-Sector Collaboration

Design thinking promotes a collaborative approach that brings together stakeholders from diverse sectors, including government, industry, academia, and civil society. This collaborative ethos can be instrumental in fostering cross-sectoral partnerships that drive innovative solutions and best practices in plastic waste management. Through facilitated workshops, co-creation sessions, and multi-stakeholder engagement platforms, India can encourage knowledge exchange and collective action, leading to more holistic and sustainable approaches to waste management. By embracing design thinking as a fundamental framework for waste management, India can leverage its potential to redefine plastic waste management, promote innovation, and foster a culture of environmental responsibility. This user-centered and interdisciplinary approach has the power to guide India towards a future where plastic waste is managed sustainably, in alignment with global best practices for environmental conservation and resource efficiency.

3.10 Advancing Cross-Disciplinary Research and Collaboration

Embracing design thinking in plastic waste management, India positions itself at the forefront of innovative and sustainable practices. This approach catalyzes cross-disciplinary research and collaboration, uniting scientists, engineers, social scientists, and policymakers in a common quest to devise comprehensive solutions to the plastic waste dilemma. Through such synergies, India can pioneer the development of groundbreaking technologies, effective policy frameworks, and behavioural strategies aimed at mitigating plastic waste challenges (Koot, 2020; Fiksel et al., 2021; Noklebye, 2021). This collaborative ethos is not just about creating new solutions but also about fostering a paradigm shift towards sustainability and environmental stewardship. By integrating design thinking into its waste management strategies, India nurtures a culture of innovation, empathy, and cooperation. Such a culture is crucial for transitioning to practices that ensure plastic waste is managed responsibly and in harmony with the ecosystem. India's dedication to leveraging design thinking reflects a progressive stance on environmental challenges, emphasizing innovation and inclusivity. This methodological approach transforms waste management from a series of challenges into opportunities for sustainable development, ensuring a cleaner and healthier future. Incorporating life-cycle assessments and environmental impact analyses into the design process guarantees that waste management strategies not only align with but also contribute to global sustainability goals, helping preserve natural resources. Moreover, design thinking fosters a spirit of collaboration across various sectors, promoting partnerships that bridge government, industry, academia, and civil society. Through activities like workshops, co-creation sessions, and engagement platforms, India facilitates knowledge exchange and collective action, leading to more comprehensive and sustainable waste management approaches (India community action—Appropedia, the sustainability wiki, 2019; Balu et al., 2020; atulpathaki-imnagpur.ac.in, 2021). In summary, by adopting design thinking as a fundamental framework for addressing plastic waste, India can redefine waste management practices. This interdisciplinary and user-centered approach holds the promise of guiding India towards a future where plastic waste is managed sustainably and efficiently, in line with the best global practices for environmental conservation and resource efficiency. Through innovation, empathy, and collaborative effort, design thinking enables India to navigate the complexities of plastic waste management, heralding a future of environmental responsibility and sustainable development.

3.11 Case Studies of Successful Plastic Waste Management

One example of a successful plastic waste management initiative is the city of Surat in India. Surat implemented a comprehensive waste management system, including separate collection and recycling of plastic waste. This initiative not only reduced the amount of plastic waste sent to landfills but also created employment opportunities for local communities through the establishment of recycling facilities. Another success story is the Recyclekaro initiative in Bengaluru, India. Recyclekaro is a social enterprise that focuses on reducing plastic waste by promoting recycling and upcycling practices (Vimal et al., 2020; Gopinath et al., 2020; atulpathakiimnagpur.ac.in, 2021). Through their efforts, Recyclekaro has been able to divert significant amounts of plastic waste from landfills and create sustainable livelihoods for waste pickers and recyclers. Overall, the challenges and opportunities in plastic waste management in India are significant. India has the potential to overcome these challenges and pave the way for a more effective and sustainable plastic waste management system. Continued innovation, development of solid policies, sustainable approaches, and integration of design thinking principles are crucial in this effort. Use of design thinking principles, such as empathy, ideation, and prototyping can help identify innovative and inclusive solutions for plastic waste management in India, taking into account the diverse needs and perspectives of different stakeholders.

Versova beach cleanup initiative:

Statistics: Versova Beach in Mumbai was once heavily polluted with plastic waste, with an estimated 5.3 million kilograms of trash accumulated over the years. However, a citizen-led initiative, spearheaded by lawyer Afroz Shah, mobilized volunteers to clean up the beach. Over 200 weeks, they removed around 13 million kilograms of plastic waste.

Significance: The Versova Beach cleanup initiative showcases the power of community action in addressing plastic pollution. It not only transformed a once-degraded coastline into a clean and vibrant environment but also inspired similar cleanup efforts across the country, raising awareness about the importance of responsible waste management.

Ban on single-use plastics in maharashtra:

Statistics: In 2018, Maharashtra became the first Indian state to enforce a ban on single-use plastics, including items like plastic bags, straws, and cutlery. The ban aims to reduce plastic pollution and encourage the use of alternatives. According to estimates, Maharashtra generates approximately 1,200 metric tons of plastic waste daily.

Significance: Maharashtra's ban on single-use plastics sets a precedent for other states and regions in India to adopt similar measures. By restricting the usage of disposable plastic items and promoting eco-friendly alternatives, the ban contributes to mitigating the environmental impact of plastic waste and fostering a culture of sustainability.

Integration of informal waste pickers into formal recycling systems:

Statistics: India has a vast network of informal waste pickers who play a crucial role in collecting and sorting recyclable materials, including plastic waste. It's estimated that there are millions of waste pickers across the country, contributing significantly to the recycling sector.

Significance: Integrating informal waste pickers into formal recycling systems not only improves their livelihoods but also enhances the efficiency of plastic waste management. By providing training, access to resources, and fair compensation, these initiatives empower waste pickers and contribute to the circular economy while reducing the burden on landfills.

3.12 Future Trends in Plastic Waste Management in India

3.12.1 Leveraging Technology for Effective Waste Management

As India progresses towards a more sustainable approach to plastic waste management, leveraging technology will be instrumental in optimizing waste collection, recycling processes, and monitoring environmental impact. Implementation of advanced tracking and sensor technologies can provide real-time data on waste generation and movement, enabling more efficient allocation of resources and proactive intervention to address any potential environmental risks. Additionally, the use of blockchain technology can enhance transparency and traceability in the recycling supply chain, ensuring the authenticity of recycled materials and promoting ethical practices.

3.12.2 Empowering Local Communities Through Circular Economy Models

Empowering local communities to actively participate in the circular economy model can significantly contribute to reducing plastic waste. Encouraging the establishment of community-based recycling centres and incentivizing the exchange of recyclable materials can create a self-sustaining ecosystem of waste management at the grass-roots level. This approach not only reduces the burden on formal waste management infrastructure but also promotes economic growth within communities by providing avenues for income generation through waste recovery and recycling.

3.12.3 Fostering International Collaboration for Knowledge Exchange

Collaborating with global stakeholders and experts in sustainable waste management can provide valuable insights and best practices for India's plastic waste management efforts. Knowledge exchange and partnerships with international organizations and initiatives can facilitate the adoption of innovative technologies, policy frameworks, and strategies that have proven successful in other parts of the world. Furthermore, such collaboration can also open avenues for investment and support from the global community, enhancing the scalability and impact of India's plastic waste management initiatives. As India continues to navigate the complexities of plastic waste management, incorporating these future trends into its strategies will be crucial for driving sustainable change and achieving long-term success in addressing the plastic waste challenge. By embracing technology, empowering local communities, and fostering international collaboration, India can advance towards a more resilient and environmentally conscious approach to plastic waste management, setting a positive example for the global community.

3.13 Conclusion

In conclusion, India's plastic waste management challenges require an innovative approach that embraces design thinking, leveraging advanced technologies, circular economy principles, community involvement, and international cooperation. This approach aims to develop sustainable solutions that address environmental concerns, foster economic growth, and enhance social well-being. By engaging various stakeholders and considering India's unique socioeconomic and cultural context, a holistic approach can be established to establish a self-sustaining ecosystem for plastic waste management. Through waste minimization, increased recycling, and reduced environmental impact, India can effectively address the challenges posed by plastic waste. In addition to technological and systemic solutions, raising public awareness and promoting behavioural change is essential. Educational campaigns and community-led initiatives can empower citizens to adopt responsible consumption habits and participate in waste reduction efforts. Furthermore, implementing policies that enforce extended producer responsibility will hold manufacturers accountable for managing their products' entire life cycle, encouraging sustainable practices. Waste-to-energy technologies offer a promising solution for managing non-recyclable plastics while contributing to renewable energy production. By investing in research and infrastructure development in this area, India can reduce landfill volumes and advance its sustainable energy objectives. In summary, integrating these strategies into India's plastic waste management framework will enhance the nation's efforts towards sustainability. By combining public engagement, policy reforms, and technological innovations, India can establish a resilient and environmentally conscious waste management system, setting an example for global sustainability initiatives and ensuring a cleaner future for generations to come.

References

Awasthi, A. K., Shivashankar, M., & Majumder, S. (2017). Plastic solid waste utilization technologies: A review. *IOP Conference Series: Materials Science and Engineering, 263*(2), 022024. IOP Publishing.

Ayyamperumal, R., Huang, X., Li, F., Chengjun, Z., Chellaiah, G., Gopalakrishnan, G., ... & Antony, J. K. (2022). Investigation of microplastic contamination in the sediments of Noyyal River-Southern India. Journal of Hazardous Materials Advances, 8, 100198.

Balu, S., Sampath, K. S., Hari, L. R., Ramprasath, M., Aishwarya, N. K., Aiswarya, M. S., ... &Nandanan, K. (2020, December). Assessment of waste management practices and its impact on the village ecosystem in Himachal Pradesh, India. In *2020 IEEE 8th R10 Humanitarian Technology Conference (R10-HTC)* (pp. 1–6). IEEE.

Balwada, J., Samaiya, S., & Mishra, R. P. (2021). Packaging plastic waste management for a circular economy and identifying a better waste collection system using analytical hierarchy process (AHP). *Procedia CIRP, 98*, 270–275.

Balwan, W. K., Saba, N., Singh, N., & Rasool, N. (2020). Solid waste management: first report on garbage problem in Doda Region of Jammu and Kashmir. *India. Population, 11*(17), 41.

Banu, N. (2019). Single-use plastic ban and its public health impacts: A narrative review. *Annals of SBV, 8*(1), 14.

Bhandari, N. L., Bhattarai, S., Bhandari, G., Subedi, S., & Dhakal, K. N. (2021). A review on current practices of plastics waste management and future prospects. *Journal of Institute of Science and Technology, 26*(1), 107–118.

Bharti, R., & Sharma, R. (2021). Analysis of plastic waste management: Utilization, issues & solutions. *Materials Today: Proceedings, 45*, 3625–3632.

Borrelle, S. B., Rochman, C. M., Liboiron, M., Bond, A. L., Lusher, A., Bradshaw, H., & Provencher, J. F. (2017). Why we need an international agreement on marine plastic pollution. *Proceedings of the National Academy of Sciences, 114*(38), 9994–9997.

Chaudhari, U. S., Lin, Y., Thompson, V. S., Handler, R. M., Pearce, J. M., Caneba, G., ... & Shonnard, D. R. (2021). Systems analysis approach to polyethylene terephthalate and olefin plastics supply chains in the circular economy: a review of data sets and models. *ACS Sustainable Chemistry & Engineering, 9*(22), 7403–7421.

Chauhan, A., Jakhar, S. K., & Chauhan, C. (2021). The interplay of circular economy with industry 4.0 enabled smart city drivers of healthcare waste disposal. *Journal of Cleaner Production, 279*, 123854.

Chiranjeevi, P., Dahiya, S., & Kumar, N. (2018). Waste derived bioeconomy in India: A perspective. *New Biotechnology, 40*, 60–69.

Conlon, K. (2023). Emerging transformations in material use and waste practices in the global south: Plastic-free and zero waste in India. *Urban Science, 7*(2), 47.

De, S. K., Kawdia, P., Gupta, D., & Pragya, N. (2023). Packaging plastic waste management in the cosmetic industry. *Management of Environmental Quality: An International Journal, 34*(3), 820–842.

Dora, K. N., Kumari, K., Srivatsava, M., & Dalai, N. (2020). Withdrawn: A review on various techniques for municipality waste management & product development.

Dutta, P., Talaulikar, S., Xavier, V., & Kapoor, S. (2021). Fostering reverse logistics in India by prominent barrier identification and strategy implementation to promote circular economy. *Journal of Cleaner Production, 294*, 126241.

Fiksel, J., Sanjay, P., & Raman, K. (2021). Steps toward a resilient circular economy in India. *Clean Technologies and Environmental Policy, 23*, 203–218.

Ganguly, R. K., & Chakraborty, S. K. (2021). Integrated approach in municipal solid waste management in COVID-19 pandemic: Perspectives of a developing country like India in a global scenario. *Case Studies in Chemical and Environmental Engineering, 3*, 100087.

Gopinath, K. P., Nagarajan, V. M., Krishnan, A., & Malolan, R. (2020). A critical review on the influence of energy, environmental and economic factors on various processes used to handle and recycle plastic wastes: Development of a comprehensive index. *Journal of Cleaner Production, 274*, 123031.

Gupta, P. (2021). Plastic waste management, a concern for community. *The Holistic Approach to Environment, 11*(2), 49–66.

Han, J., Gu, L., & Chen, D. (2021). Application of innovative design thinking in product design* intelligent waste paper recycling machine design case. In *E3S Web of Conferences* (Vol. 236, p. 04062). EDP Sciences.

India community action—Appropedia, the sustainability wiki. (2019, July 4). https://www.appropedia.org/India.

Khandelwal, C., & Barua, M. K. (2020). Prioritizing circular supply chain management barriers using fuzzy AHP: Case of the Indian plastic industry. *Global Business Review*, 0972150920948818.

Koot, B. (2020). Basic solutions to waste management issues that will bring clean culture in India. In *E3S Web of Conferences* (Vol. 170, p. 04004). EDP Sciences.

Kumar, A., & Agrawal, A. (2020). Recent trends in solid waste management status, challenges, and potential for the future Indian cities—A review. *Current Research in Environmental Sustainability, 2*, 100011–100011. https://doi.org/10.1016/j.crsust.2020.100011.

Kumar, S., Smith, S. R., Fowler, G., Velis, C., Kumar, S. J., Arya, S., ... & Cheeseman, C. (2017). Challenges and opportunities associated with waste management in India. *Royal Society Open Science, 4*(3), 160764.

Lakshmikanthan, P. (2019). Value creation with waste to energy: Economic considerations. In Current Developments in Biotechnology and Bioengineering (pp. 307–318). Elsevier.

Leone, G., Moulaert, I., Devriese, L. I., Sandra, M., Pauwels, I., Goethals, P. L., ... & Catarino, A. I. (2023). A comprehensive assessment of plastic remediation technologies. *Environment International*, 107854.

MacDonald, A., Allen, D., Williams, L., Flowers, P., & Walker, T. R. (2023). People, plastic, and behaviour change–a comment on drivers of plastic pollution, barriers to change and targeted behaviour change interventions. *Environmental Science: Advances, 2*(4), 551–557.

Mallick, S. K., Pramanik, M., Maity, B., Das, P., & Sahana, M. (2021). Plastic waste footprint in the context of COVID-19: Reduction challenges and policy recommendations towards sustainable development goals. *Science of the Total Environment, 796*, 148951.

Pandey, S., & Manuja, S. (2021). *Plastic waste management: turning challenges into opportunities.* The Energy and Resources Institute (TERI).

Pani, S. K., & Pathak, A. A. (2021). Managing plastic packaging waste in emerging economies: The case of EPR in India. *Journal of Environmental Management, 288*, 112405.
Pearson, M., & Khare, E. (2022). Addressing the single-use plastic proliferation problem. *MIT Science Policy Review*.
Pruneau, D., Khattabi, A., Léger, M., & Rahmani, Z. (2022). How design thinking helped craftwomen to solve a plastic pollution problem. *Environmental Sciences Proceedings, 15*(1), 40.
Rafey, A., & Siddiqui, F. Z. (2023). A review of plastic waste management in India–challenges and opportunities. *International Journal of Environmental Analytical Chemistry, 103*(16), 3971–3987.
Ray, A., & Galotra, A. (2017). A household sustainable product design for India. *International Journal of Sustainable Design, 3*(2), 79–99.
Rois, M., Mubarak, A., & Suzianti, A. (2020). Designing solution for organic waste management system with design thinking approach (case study in Depok). *IOP Conference Series: Earth and Environmental Science, 464*(1), 012002. IOP Publishing.
Sarkar, B., Dissanayake, P. D., Bolan, N. S., Dar, J. Y., Kumar, M., Haque, M. N., ... & Ok, Y. S. (2022). Challenges and opportunities in sustainable management of microplastics and nanoplastics in the environment. *Environmental Research, 207*, 112179.
Sarker, M. (2011). Converting waste plastic to hydrocarbon fuel materials. *Energy Engineering, 108*(2), 35–43.
Shanker, R., Khan, D., Hossain, R., Islam, M. T., Locock, K., Ghose, A., ... & Dhodapkar, R. (2022). Plastic waste recycling: existing Indian scenario and future opportunities. *International Journal of Environmental Science and Technology* 1–18.
Sharma, S., Basu, S., Shetti, N. P., Kamali, M., Walvekar, P., & Aminabhavi, T. M. (2020). Waste-to-energy nexus: A sustainable development. *Environmental Pollution, 267*, 115501.
Sharma, B., Goswami, Y., Sharma, S., & Shekhar, S. (2021). Inherent roadmap of conversion of plastic waste into energy and its life cycle assessment: A frontrunner compendium. *Renewable and Sustainable Energy Reviews, 146*, 111070.
Singh, N., Hui, D., Singh, R., Ahuja, I. P. S., Feo, L., & Fraternali, F. (2017). Recycling of plastic solid waste: A state of art review and future applications. *Composites Part B: Engineering, 115*, 409–422.
Singh, D., Dikshit, A. K., & Kumar, S. (2023). Quantifying the impact of municipal solid waste litter using environmental status index in urban areas. *Journal of Cleaner Production, 430*, 139653.
Somani, P., Navaneethan, R. D., & Thangaiyan, S. (2021). Integrated solid waste management in urban India: A mini review. *Journal of Physics: Conference Series, 1913*(1), 012084. IOP Publishing.
Sureshkumar, K., & Ponnusamy, V. (2020). Hybrid renewable energy systems for power flow management in smart grid using an efficient hybrid technique. *Transactions of the Institute of Measurement and Control, 42*(11), 2068–2087.
Sustainable Urban Plastic Waste Management|United Nations Development Programme. (2023, January 1). https://www.undp.org/india/publications/sustainable-urban-plastic-waste-management.
Tanksale, A. N., Das, D., Verma, P., & Tiwari, M. K. (2021). Unpacking the role of primary packaging material in designing green supply chains: An integrated approach. *International Journal of Production Economics, 236*, 108133.
Tantiyaswasdikul, K. (2020, September). how design thinking can foster environmental sustainability: integrating design thinking into circular design guide. In *Proceedings of the 6th International Conference on Industrial and Business Engineering* (pp. 157–162).
Vanapalli, K. R., Sharma, H. B., Ranjan, V. P., Samal, B., Bhattacharya, J., Dubey, B. K., & Goel, S. (2021). Challenges and strategies for effective plastic waste management during and post COVID-19 pandemic. *Science of the Total Environment, 750*, 141514.

Vimal, K. E. K., Mathiyazhagan, K., Agarwal, V., Luthra, S., & Sivakumar, K. (2020). Analysis of barriers that impede the elimination of single-use plastic in developing economy context. *Journal of Cleaner Production, 272*, 122629.
Plastic Waste Management: Turning Challenges into Opportunities|TERI. (2020, December 2). https://www.teriin.org/policy-brief/plastic-waste-management-turning-challenges-opportunities.
Yadav, V., Sherly, M. A., Ranjan, P., Prasad, V., Tinoco, R. O., & Laurent, A. (2022). Risk of plastics losses to the environment from Indian landfills. *Resources, Conservation and Recycling, 187*, 106610.
Zhang, F., Zhao, Y., Wang, D., Yan, M., Zhang, J., Zhang, P., ... & Chen, C. (2021). Current technologies for plastic waste treatment: A review. *Journal of Cleaner Production, 282*, 124523.

Chapter 4
Microplastics Screening in Tigris River of Baghdad City/Iraq

Mohammad Abdul Wahab Shaker Al-Aadhami, Maitham Abdulah Sultan, Noor Nihad Baqer, and Hassan H. Sultan

Abstract Microplastics are usually difficult to degrade or bioaccumulate, can transfer via the nutrition chain, and can at the latest be a risk to people's health, and are widely spread in aquatic environments including oceans, in lakes, in rivers, and even drinking water. This study aimed to detect microplastics in Tigris River for Baghdad city. It was collected sample in one liter from six areas of Baghdad city: Fahama, Muthana Bridge, Sarafya Bridge, Abo Noas, Doura, and Madaen. The microplastic was estimated by a fluorescent microscopic screening using Nile dye. FTIR is applied to estimate microplastic compounds. Findings appeared there were important microplastic compounds in river samples including Nylon, Polyurethane, Polypropylene, Polystyrene, Polyvinyl chloride, Polypropylene, Polyethylene terephthalate, Polycarbonate, Nitrile, Polytetrafluorethylene. The size of microplastic at 38 μm with fiber shape is the most numerous and common in most of the study areas: Muthana Bridge (71%), Sarafya Bridge area (78.7%), Madaen area (48.78%). While fragments in Fahama area (55.5%) and Abo Noas area (65.38%), also pellets in Doura area (42.85%) were the most prevalent. In conclusion, Tigris River was polluted with microplastics. The fiber microplastics are dominant in the river.

Keywords Microplastic · Tigris river · FTIR

4.1 Introduction

Microplastics are generally defined as small pieces of plastic that are less than 5 mm in diameter (Andrady, 2017; De-la-Torre, 2020), and microplastic is classified into primary and secondary. Primary microplastic is produced to be of the microscopic volume (Cole et al., 2011) involving pre-production resin spherules, beads used in cosmetics, tubes of toothpaste, and textiles coating, (Shim et al., 2018). Whilst

M. A. W. S. Al-Aadhami · M. A. Sultan · N. N. Baqer (✉) · H. H. Sultan
Ministry of Higher Education and Scientific Research, Scientific Research Commission, Research and Technology Center of Environment, Water & Renewable Energy, Baghdad, Iraq
e-mail: noornihadbaqer@gmail.com

S. Suriyanarayanan et al. (eds.), *Combating Plastic Pollution in Terrestrial Environment*,
https://doi.org/10.1007/978-981-96-2343-3_4

secondary microplastic is obtained by the large plastic materials decomposition (Gesamp, 2016), as a result of photolysis, biological degeneration, and breakdown (Browne et al., 2007).

Plastics are manufactured by long chains of polymeric molecules that are created from organic and inorganic materials, including hydrogen, oxygen, silicon, carbon, and chloride, which can usually be obtained from natural gas, oil, and carbon oil (Sánchez-Hernández et al., 2021). Many products in our daily lives consist of plastic, and with the global population steadily rising, the use and consumption of plastic products will continue to increase. Annually, 8 million tons of plastic waste are disposed of within the ocean, and experts believe that this plastic waste will double by 2030 (Zhang et al., 2021). Microplastics, which are usually difficult to degrade or bioaccumulate, can transfer through the food chain and can ultimately be hazardous to human health (Osman et al., 2023), and are widely spread in aquatic environments including oceans (Saeed et al., 2020), in lakes (Pan et al., 2023), in rivers (Tien et al., 2020), and even drinking water (Ali et al., 2023). Most studies focused on microplastics in the marine environment, with 4% of publications concerned with the freshwater ecosystem (Lambert & Wagner, 2018).

The major sources of people's exposition to microplastic are inhalation, contact with the skin, and ingestion. Inhalation of microplastic by air mainly comes from artificial textiles, dust, and crushed elastic tires (Prata, 2018).

Consuming polluted seafood, and contaminated drinking water represents the essential pathways into digestive system. Whole organisms ingestion represents a danger (Carbery et al., 2018). Despite the human dermis prohibits microplastics and pollutants from passsing, several potential entrance paths are sweat gland, skin damage, and follicles hair (Rashed et al., 2023). After microplastic ingestion, particles smaller than 150μ may travel to the lymphatic system and circulation, but less than 3% of the ingested particle is expected to be absorbed (Barboza et al., 2018).

Bisphenol A represents one of the plastic additives applied as antioxidants and also stabilizers. (Galloway et al., 2018) It is an artificial chemical vastly applied to manufacture epoxy resin, polycarbonate plastic, and polymer matter. Bisphenol A is numerous in environment, containers of foods, bottles of water, thermic papers, medical instruments and devices, children's toys, and is combined with soil and water by leaching and then transferred to the food chain (Manzoor et al., 2022; Ohore & Zhang, 2019).

Bisphenol A has a role in endocrine disruption and has the possibility of changing mechanisms in the body. Research assured its estrogen-analogous impacts, and anti-androgen activity which cause several negative effects on health, particularly on immunity, neuroendocrinal actions, and reproduction mechanism. Further, it induces mutagenesis (genetic mutation) and carcinogenic effects (Gonzalez et al., 2021; Manzoor et al., 2022).

Studies demonstrated the chemical components that attached to microplastics, such as styrene low molecular weight polyvinyl chloride monomers, polycyclic aromatic hydrocarbons, PCBs, OCPs, PBDEs, and pharmaceuticals including their metabolites, also cause mutagenic, carcinogenic, and endocrine disruption after absorption. Evidence of the influence of microplastics in freshwater organisms is restricted (Ullah et al., 2023). Microplastic can cross into the lung and the epithelial layer of the digestive system and transfer to various tissues, however, experimental evidence about this is still limited (Abad López et al., 2023).

Taken together, this study aimed to investigate microplastics in the Tigris River due to pollution with plastic material and increased pollution with wastewater in Baghdad city.

4.2 Methods

4.2.1 Sites Studied and Field Modeling

The study areas on the Tigris River in Baghdad city were chosen based on the proximity of water intake sites from the river to the drinking water treatment plants spread along the river in the center of the town. On this basis, Al-Fahama area was chosen as the first study station located in the north of the city. The second station was at the Muthana Bridge, the third area was at the Sarafya Bridge, the fourth area was at Abu Noas area, the fifth area was in the Doura area, the sixth area was in the far south of Baghdad city at Madaen area.

4.2.2 Preparation of Water Sample

The water samples were prepared based on protocol by US NOAA (Masura et al., 2015). The protocol was represented by adding a suitable quantity of H_2O_2 (30%) to 40 ml of each water sample (1000 ml). All samples were left (24–48) hours to degrade organic material. The sample was filtered by sieve (holes 38μ). It was washed the sieve via distilled water for times. Microplastics in washing water were collected in two litters of glass beakers. The washing water was filtered via filter papers (0.45μ) by filtration using vacuum pumps. Filter papers are put on glass petri dishes for microscopic examination. Steps of water sample processing as shown in Fig. 4.1.

4.2.3 Fluorescent Screening and Analysis of FTIR

The fluorescent microscope (MEIJI/ MX/Japan) examined the microplastics. The microplastics in filter paper were stained by Nile red dye (It was prepared (5 mg/ L) by dissolving the Nile red dye powder in acetone and left 24 h. before use). The filter paper after staining was left for 30 min at room temperature to dry and then examined in a fluorescent microscope at 10 X and 40X. The blue filter (425 nm) and

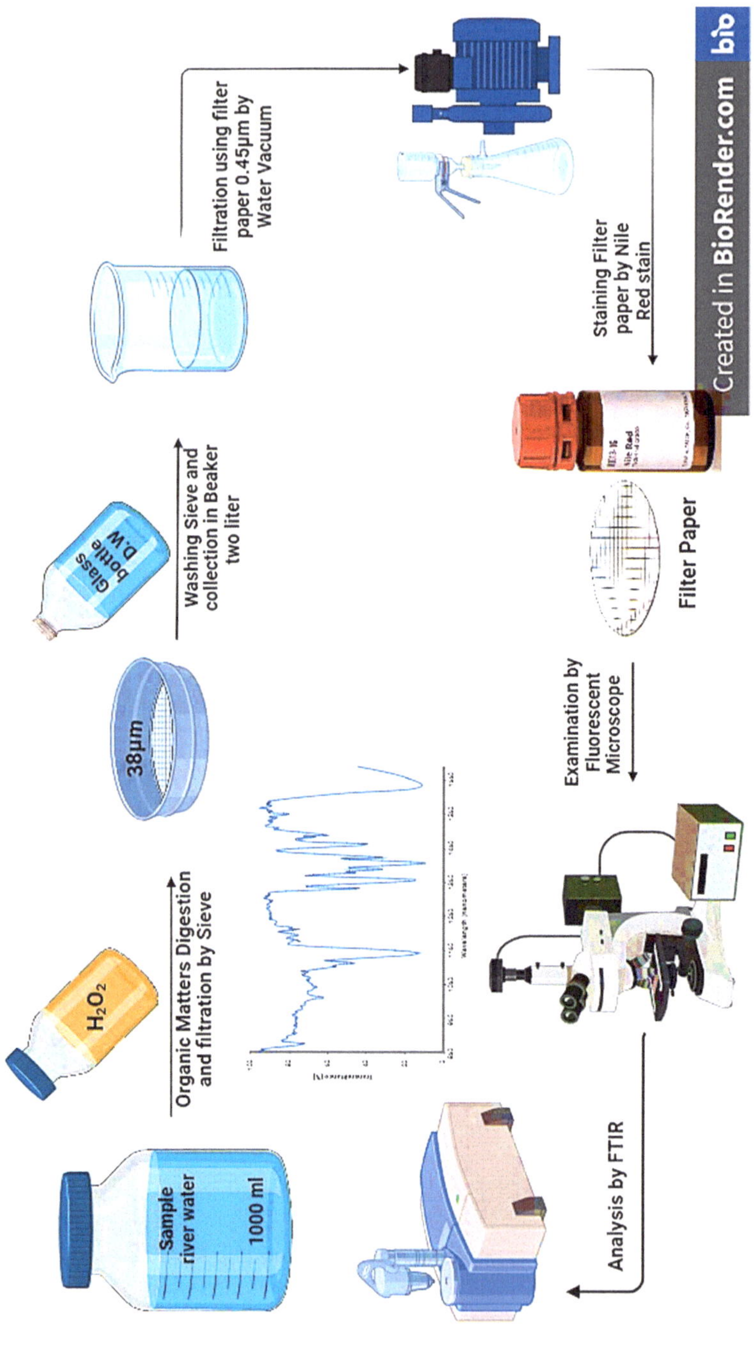

Fig. 4.1 Steps of water sample processing in this study

green filter (500 nm) were used. Microplastics are determined relying on fluorescent characteristics, their shape, and microplastic numbers.

FTIR (infrared spectroscopy) was used to distinguish and recognize microplastic polymers (Kappler et al., 2018). In this study, spectral examination were conducted by (ABB SPECTROL system, Newbury, Berkshire R G20 BBQ, UK). The measure was conducted in a wavelength range (4000–400 cm^{-1}) and transmission (zero–100).

4.3 Results and Discussion

4.3.1 Fluorescent Screening

A microscopic examination of microplastics was conducted to estimate their numbers, shapes, and colors. The outcomes shown in Table 4.1 pointed out the existence of microplastics in river water for all areas studied. The microplastic number was recorded as a high number in Madaen area (41 MPs/L) followed Sarafya Bridge area (33 MPs/L), Muthana Bridge area (31 MPs/L), Abu Noas area (26 MPs/L), Al-Fahama area (18MPs/L), and Doura area (7 MPs/L).

The results are in agreement with what was found in another study of microplastics as a pollutant in the environmental aqueous and alteration in the environmental pollutants' behavior during their reaction and absorption which cause toxic impacts, disruption, and bio-accumulation (Wang & Wang, 2018). This decline was due to the distance of Doura area from household or heavy waste disposal sites compared to the rest of the areas that were very close to these sites. Wastewater causes an increase in the amounts of microplastics released into water sources, especially rivers (Gies et al., 2018).

It is also noted in Table 4.1 and Figs. 4.2 and 4.3 that the size of microplastic at 38 μm with fiber shape is the most numerous and common in most of the study areas: Muthana Bridge (71%), Sarafya Bridge area (78.7%), Madaen area (48.78%). While fragments in Fahama area (55.5%) and Abo Noas area (65.38%), also pellets in Doura area (42.85%) were the most prevalent.

Table 4.1 Microplastic number in river water for study areas

Area name	Microplastic (particle/L^{-1})				
	Fiber	Fragment	Pellet	Total	%
Fahama	6	10	2	18	55.5% fragments
Muthana Bridge	22	7	2	31	71% fiber
Sarafya Bridge	26	3	5	33	78.7% fiber
Abo Noas	7	17	0	26	65.38% fragments
Doura	2	2	3	7	42.85% pellets
Madaen	20	10	11	41	48.78% fiber

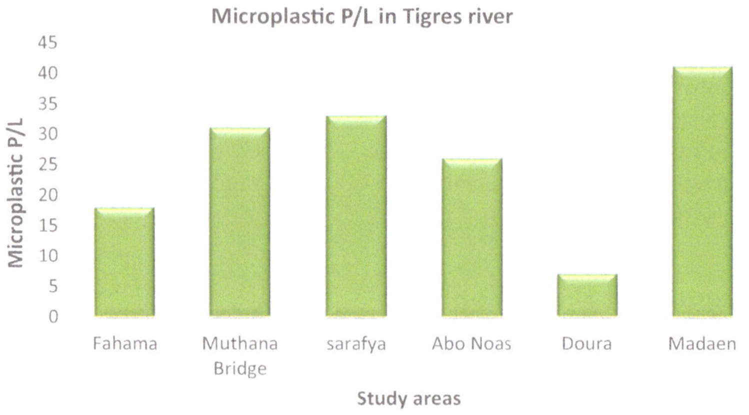

Fig. 4.2 Microplastic number in river water for study areas

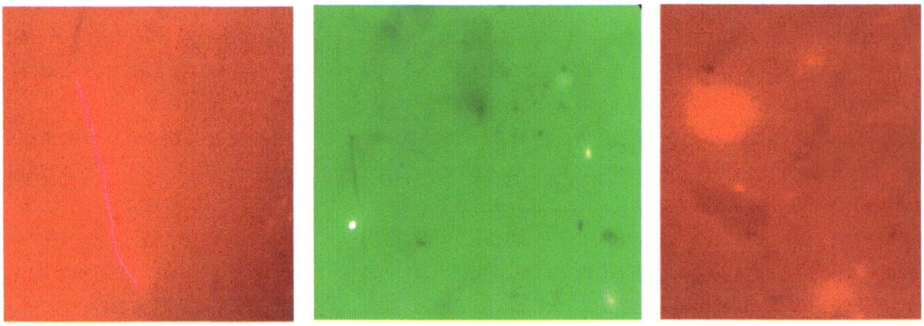

a **b** **c**

Fig. 4.3 Fluorescent microscopic examination of microplastics **a** fibre **b** pellets **c** fragments

The raised number of microplastic fibers in the river water is attributed to the fact that it is mainly produced by sewage releases (Hassan et al., 2023; Jessieleena et al., 2023). Since sewage plants are not able to remove microplastics, the effluent in sewage plants consists of a high quantity of microplastic particles (Amrutha & Warrier, 2020). The present study agreed with one study that showed fiber of microplastics was less than 1 mm which is most prevalent with (97.8%) of microplastic count in water (Lam et al., 2022).

4.3.2 FTIR Examination

The current study showed the peak spectra of compounds of microplastic for Tigris River in Fahama area (Fig. 4.4a, b), Muthana bridge area (Fig. 4.5a, b), Sarafya bridge area (Fig. 4.6a, b) AboNoas area (Fig. 4.7a, b), Doura area (Fig. 4.8a, b), and Madaen area (Fig. 4.9a, b). FTIR is an efficient procedure for detecting microplastic polymers due to absorption peaks being well-recognized and easily recognizable. FTIR supplies spectral imprints for particular polymers of microplastics and detects the existence of else compositions (additives, contaminants, polymers) (Jung et al., 2018). The FTIR was applied to confirm the presence of a pure compound and detect impurities, bounds, and side groups. FTIR method is utilized to identify functional classes in molecules and identify reactions between molecules (Bandara et al., 2023).

FTIR is used for detecting microplastics in various environments (Adhikari et al., 2022). The findings of FTIR in the current study are indicated in Figs. 4.4, 4.5, 4.6, 4.7, 4.8 and 4.9 for river water in Fahama, Muthana Bridge, Sarafya Bridge, Abo Noas, Doura, and Madaen, respectively. The FTIR for types of microplastics for each area were listed in Table 4.2.

Due to the deficiency of data about reference compositions in FTIR devices, not supplied in electronic libraries; polymer patterns are registered depending on the absorbance peaks wavelength, as identical to outcomes of past research (Manzoor et al., 2021; Veerasingam et al., 2021; Mhiret Gela and Aragaw, 2022).

Nylon represents one of the polyamides utilized in industrial and household products. FTIR of nylon characters (3300 cm^{-1}) for N-H stretch bending joined with a low peak at 93050 cm^{-1}) and allocated to fermi divided pair for N-H stretches. Besides, other bonds were established at (1630 cm^{-1}, and also 1550 cm^{-1}) identical to bonds of amide-I, and amide-II respectively. The nomenclature for Polyamides depends on the carbon atoms of number (acid and amino constitute) held to create the amide. Nylon is created via two methods: Dicarboxylic acid and di-amine (Watts et al., 2014). Through the assistance of IR detection of Nylon identical to Polyamide is significant at the same time may be difficult as a result of the polymorphism displayed in several Polyamides (Manzoor et al., 2021).

Moreover, one study revealed by FTIR technique the range of spectra of microplastic fragments (PP and PE) that increased in the density absorption of carbonyl bands C=O which located around 1700 cm^{-1}, 1027 cm^{-1} C=O, 874 cm^{-1}, and 911 cm^{-1} C=C which related to weathering method of plastics. Absorbance bands found in polymers (EVA, PEO, and EAA) identical to functional groups involving hydroxyl, hydroperoxide O-H at 3400 cm^{-1}, carbonyl C-O at1595 cm^{-1}, and ether C-O-C at 1045 cm^{-1} (Miserli et al., 2023).

Huang et al. (2023) indicated some compounds of microplastics were not detected in several samples of river water. The river water may be polluted with microplastics that have size less than (20 μm), FTIR resolution ranges 10–20 μm. Besides, FTIR process is affected by many factors including: microplastics heterogeneities, the aging of microplastic, and organic matter existence in environment. It was difficult for checking via FTIR for nonclear microplastics (Elert et al., 2017).

This current findings concur with several research that illustrated the existence of polypropylene as the microplastic compound in river (Mennekes & Nowack, 2023). One study identified some compounds of microplastics including: polypropylene, and polyethylene (Li et al., 2020). Other studies indicated the presence of different types of compound microplastics in rivers and marine organisms including polystyrene,

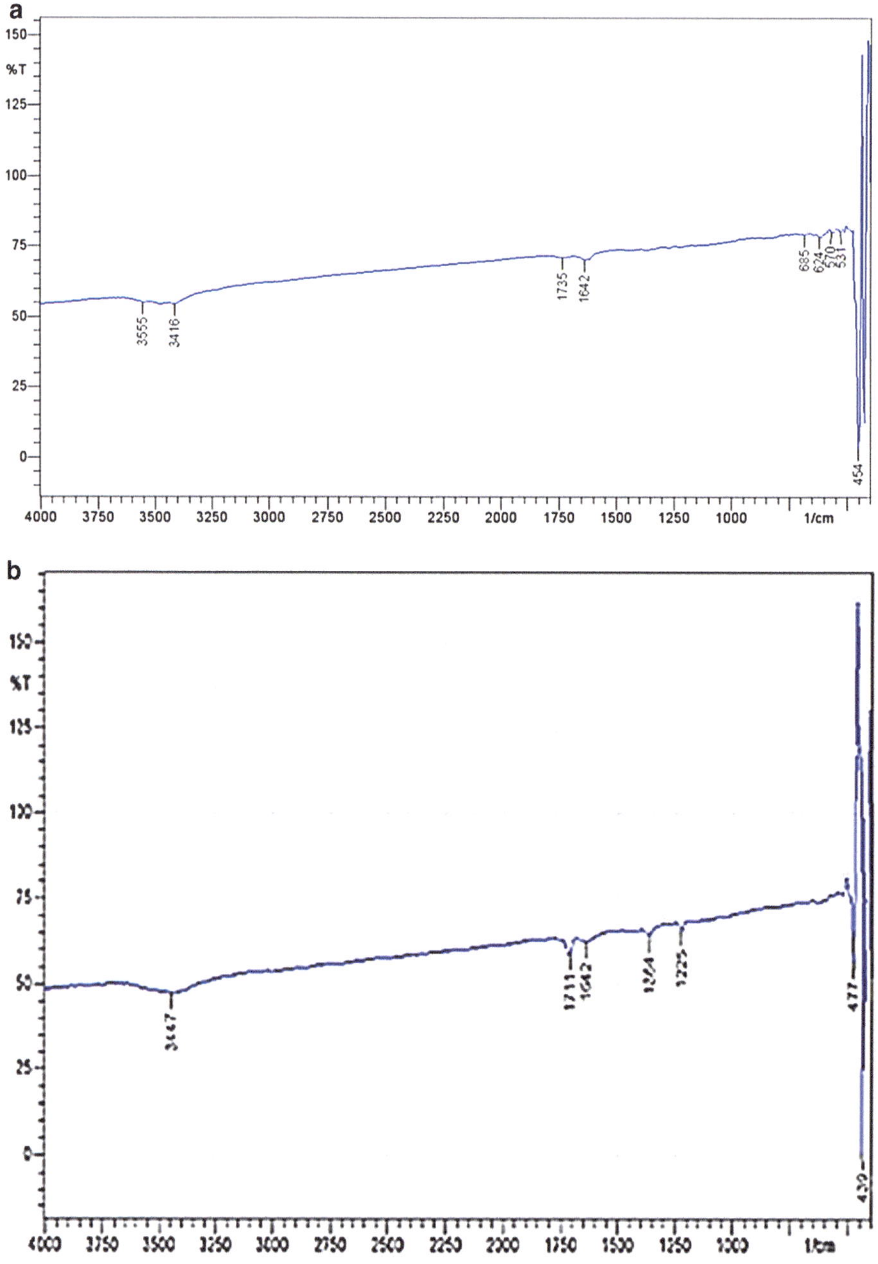

Fig. 4.4 **a** FTIR peaks spectra of microplastic for Tigris River in Fahama area. **b** FTIR peaks spectra of microplastic for Tigris River in Fahama area

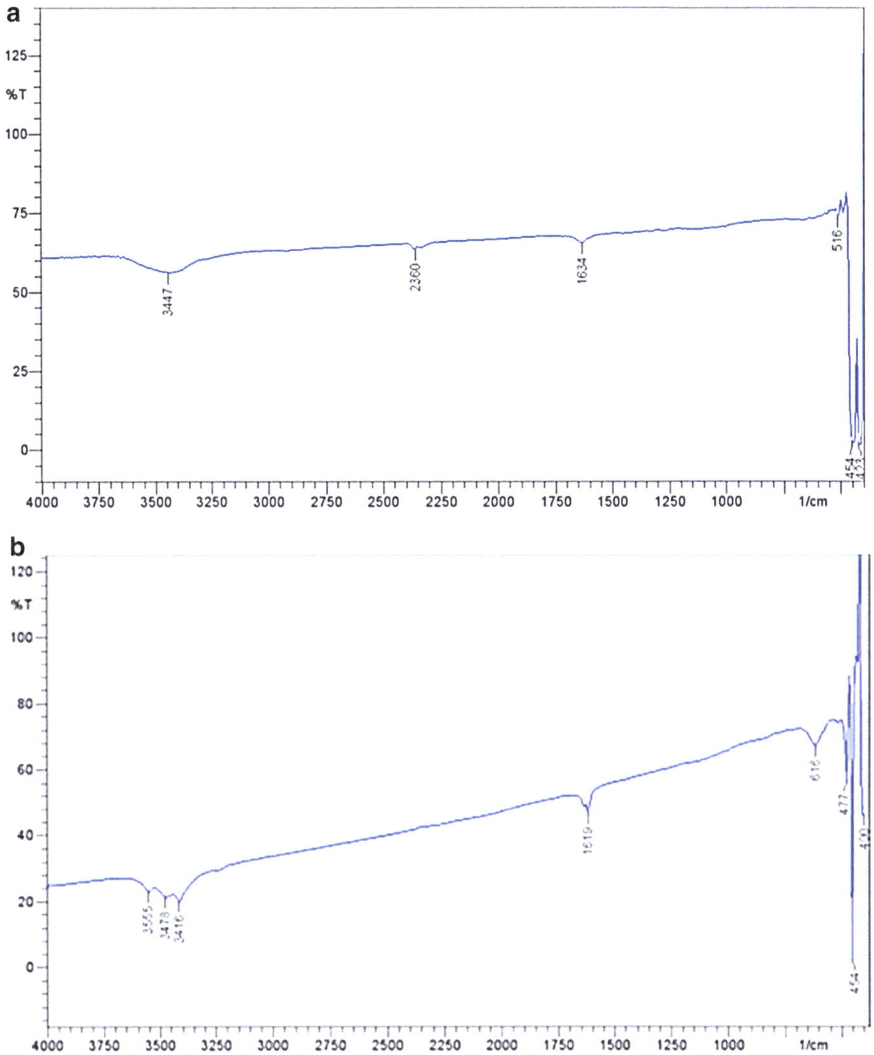

Fig. 4.5 **a** FTIR peaks spectra of microplastic for Tigris River in Muthana bridge area. **b** FTIR peaks spectra of microplastic for Tigris River in Muthana bridge area

polyamide, Polyvinyl chloride, Polypropylene, and nylon (Ismanto et al., 2023; Jung et al., 2018; Ricciardi et al., 2021).

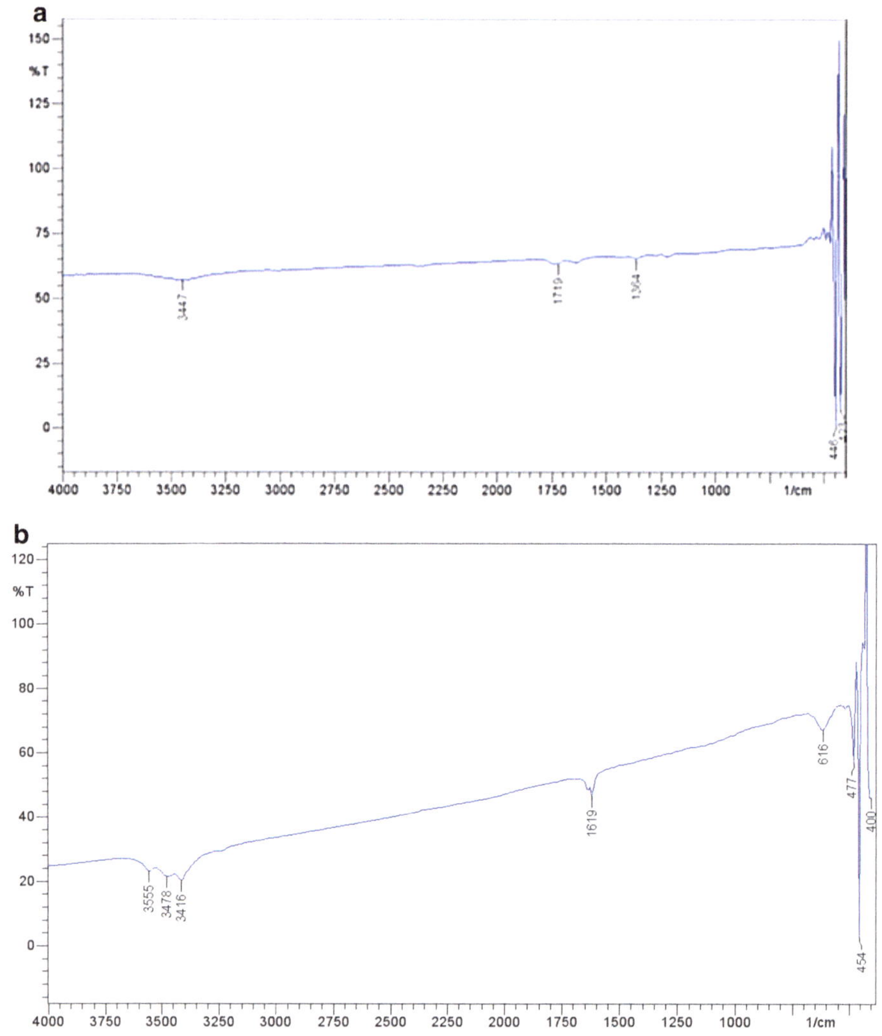

Fig. 4.6 **a** FTIR peaks spectra of microplastic for Tigris River in Sarafya bridge area. **b** FTIR peaks spectra of microplastic for Tigris River in Sarafya bridge area

4.4 Conclusions

This study showed depending on a fluorescent microscope and FTIR technique. The outcomes indicated there were important microplastic compounds in river samples including Nylon, Polyurethane, Polypropylene, Polystyrene, Polyvinyl chloride, Polypropylene, Polyethylene terephthalate, Polycarbonate, Nitrile, Polytetrafluorethylene. The fiber shape is common in most of the study areas: Muthana Bridge,

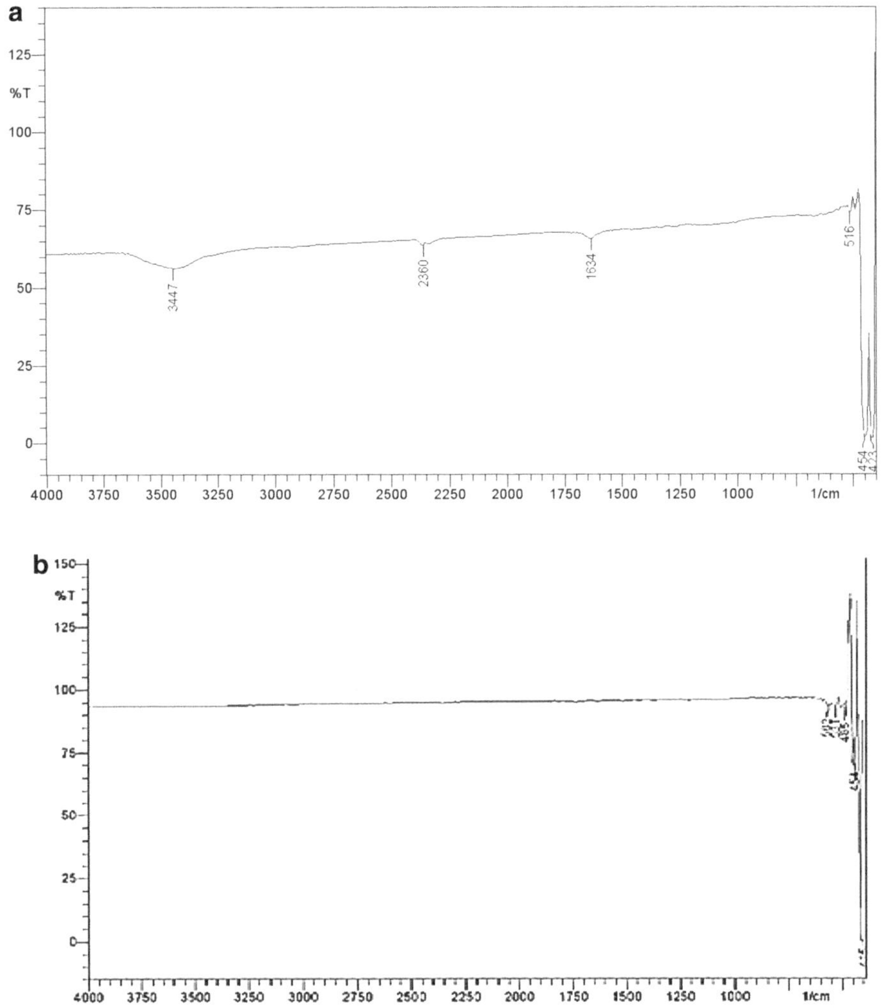

Fig. 4.7 **a** FTIR peaks spectra of microplastic for Tigris River in AboNoas area. **b** FTIR peaks spectra of microplastic for Tigris River in AboNoas area

Sarafya Bridge area, Madaen area (48.78%). While fragments from Fahama area and Abo Noas area, also pellets in Doura area were the most prevalent.

4.5 Conflict Interesting

None.

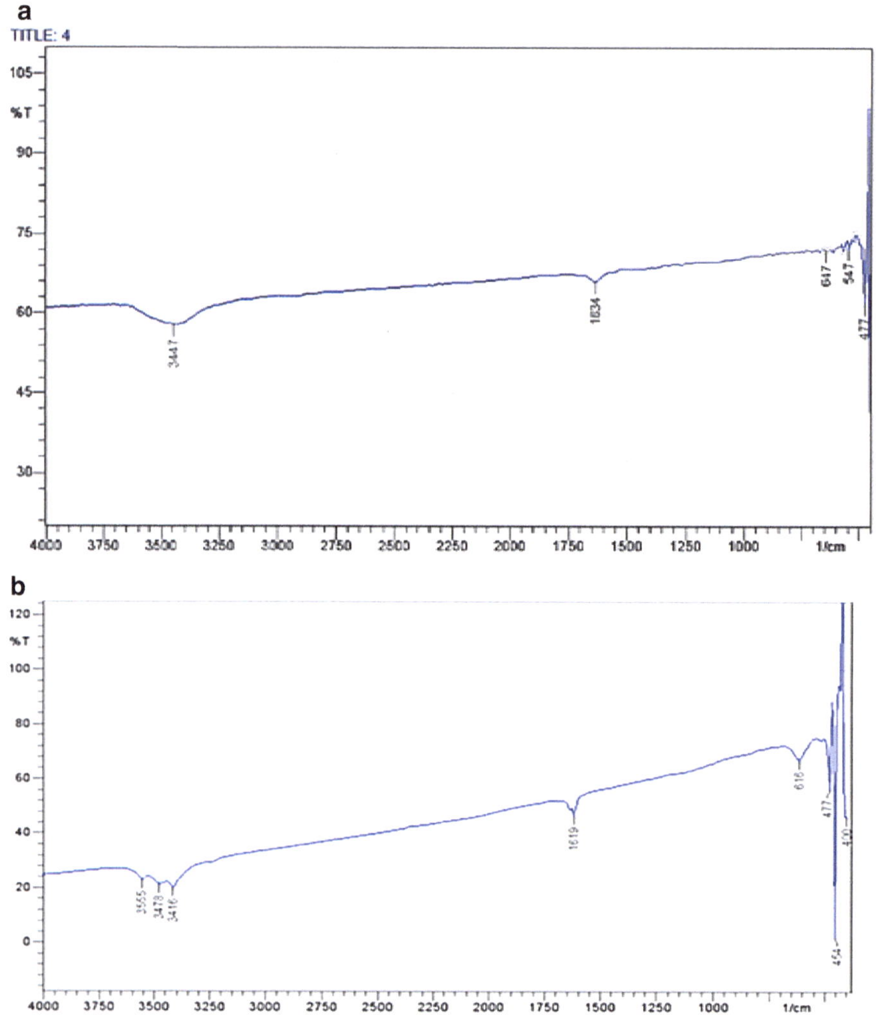

Fig. 4.8 **a** FTIR peaks spectra of microplastic for of Tigris River in Doura area. **b** FTIR peaks spectra of microplastic for of Tigris River in Doura area

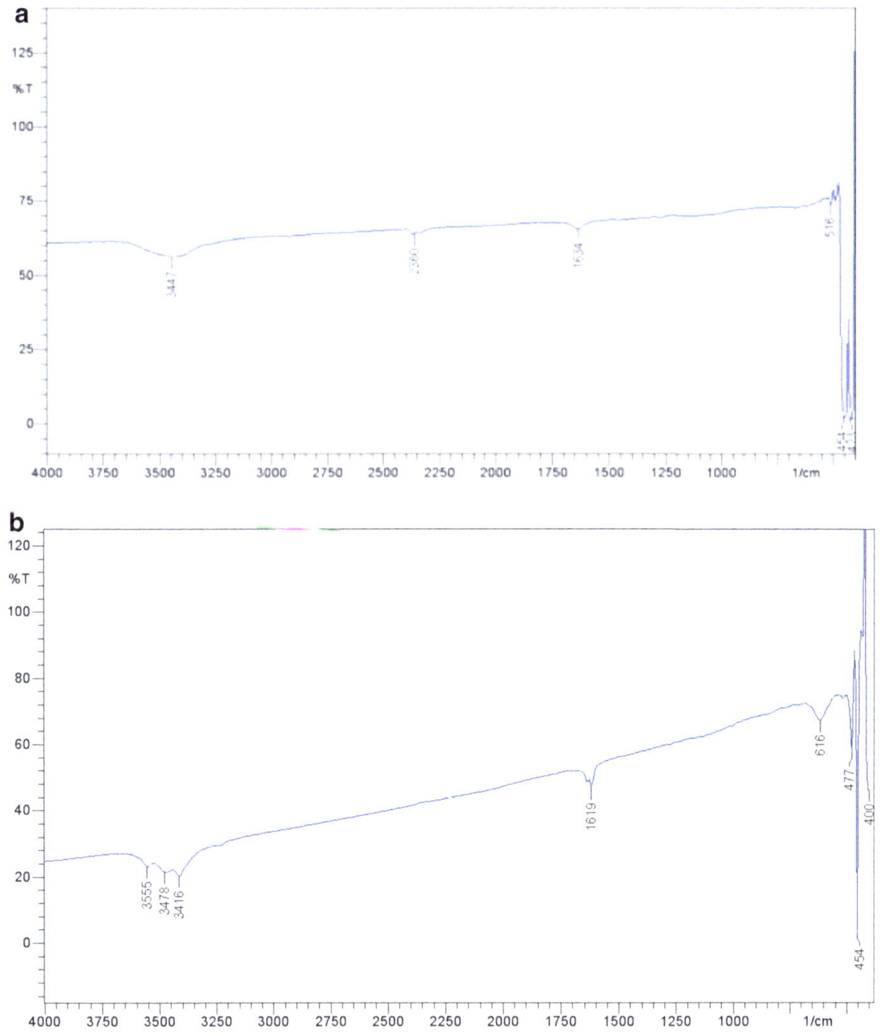

Fig. 4.9 **a** FTIR peaks spectra of microplastic for of Tigris River in Madaen area. **b** FTIR peaks spectra of microplastic for of Tigris River in Madaen area

Table 4.2 Microplastics types in river water areas

Area name	Microplastic		
	Polymer	Absorption band (cm^{-1})	Functional group
Fahama	Polyurethane (PU)	1223,1731	C=O stretch
	Nylon	1634	C=O stretch
	Polypropylene (PP)	1377	CH bending
	Polystyrene PS	694,531	CH aromatic out of the plane bend, ring aromatic out of the plane bend
Muthana Bridge	Polyvinyl chloride (PVC)	616	C-Cl stretch
	Polypropylene (PP)	1377	CH bending
	Nylon	1634	C=O stretch
Sarafya Bridge	Polyethylene terephthalate (PET)	1719	C=O stretch
	Polycarbonate (PC)	1364	CH_3 bending
	PVC	616	C-Cl stretch
Abo Noas	Nylon	1634	C=O stretch
	Polytetrafluorethylene (PTFE)	554	CF_2 bending
	Polyvinyle chloride (PVC)	616	C-Cl stretch
Doura	Nylon	1634	C=O stretch
	Polyvinyle chloride (PVC)	616	C-Cl stretch
Madaen	Nylon	1634	C=O stretch
	PVC	616	C-Cl stretch
	Polycarbonate (PC)	1364	CH_3 bending

References

Abad López, A. P., Trilleras, J., Arana, V. A., Garcia-Alzate, L. S., & Grande-Tovar, C. D. (2023). Atmospheric microplastics: Exposure, toxicity, and detrimental health effects. *RSC Advances, 13*(11), 7468–7489.

Adhikari, S., Kelkar, V., Kumar, R., & Halden, R. U. (2022). Methods and challenges in the detection of microplastics and nanoplastics: A mini-review. *Polymer International, 71*(5), 543–551.

Ali, N., Khan, M. H., Ali, M., Ahmad, S., Khan, A., Nabi, G., … & Kyzas, G. Z. (2023). Insight into microplastics in the aquatic ecosystem: Properties, sources, threats and mitigation strategies. *Science of the Total Environment, 913*, 169489.

Amrutha, K., & Warrier, A. K. (2020). The first report on the source-to-sink characterization of microplastic pollution from a riverine environment in tropical India. *Science of the Total Environment, 739*, 140377.

Andrady, A. L. (2017). The plastic in microplastics: A review. *Marine Pollution Bulletin, 119*(1), 12–22.

Bandara, T. M. W. J., Thennakoon, T. M. A. A. B., & Nishshanke, G. B. M. M. M. (2023). Preparation, characterization, and applications of graphene-based quantum dots (GQDs). In *Functional materials from carbon, inorganic, and organic sources* (pp. 21–69). Woodhead Publishing.

Barboza, L. G. A., Vethaak, A. D., Lavorante, B. R., Lundebye, A. K., & Guilhermino, L. (2018). Marine microplastic debris: An emerging issue for food security, food safety and human health. *Marine Pollution Bulletin, 133*, 336–348.

Browne, M. A., Galloway, T., & Thompson, R. (2007). Microplastic–an emerging contaminant of potential concern? *Integrated Environmental Assessment and Management, 3*(4), 559–561.

Carbery, M., O'Connor, W., & Palanisami, T. (2018). Trophic transfer of microplastics and mixed contaminants in the marine food web and implications for human health. *Environment International, 115*, 400–409.

Cole, M., Lindeque, P., Halsband, C., & Galloway, T. S. (2011). Microplastics as contaminants in the marine environment: A review. *Marine Pollution Bulletin, 62*(12), 2588–2597.

De-la-Torre, G. E. (2020). Microplastics: An emerging threat to food security and human health. *Journal of Food Science and Technology, 57*(5), 1601–1608.

Elert, A. M., Becker, R., Duemichen, E., Eisentraut, P., Falkenhagen, J., Sturm, H., & Braun, U. (2017). Comparison of different methods for MP detection: What can we learn from them, and why asking the right question before measurements matters? *Environmental Pollution, 231*, 1256–1264.

Galloway, T. S., Lee, B. P., Burić, I., Steele, A. M., Kocur, A. L., Pandeth, A. G., ... & BPA Schools Study Consortium. (2018). Plastics additives and human health: A case study of Bisphenol A (Bpa). *The Royal Society of Chemistry* 131–155.

Gesamp. (2016). Sources, fate and effects of microplastics in the marine environment: part two of a global assessment. In: P. J. Kershaw, C. M. Rochman (Eds.), IMO/FAO/UNESCO-IOC/UNIDO/WMO/IAEA/UN/ UNEP/UNDP joint group of experts on the scientific aspects of marine environmental protection. Rep. Stud. GESAMP No. 93, 220.

Gies, E. A., LeNoble, J. L., Noël, M., Etemadifar, A., Bishay, F., Hall, E. R., & Ross, P. S. (2018). Retention of microplastics in a major secondary wastewater treatment plant in Vancouver, Canada. *Marine Pollution Bulletin, 133*, 553–561.

Gonzalez, J. A., Histed, A. R., Nowak, E., Lange, D., Craig, S. E., Parker, C. G., ... & Rhodes, J. S. (2021). Impact of bisphenol-A and synthetic estradiol on brain, behavior, gonads and sex hormones in a sexually labile coral reef fish. *Hormones and Behavior, 136*, 105043.

Hassan, F., Prasetya, K. D., Hanun, J. N., Bui, H. M., Rajendran, S., Kataria, N., … & Jiang, J. J. (2023). Microplastic contamination in sewage sludge: Abundance, characteristics, and impacts on the environment and human health. *Environmental Technology & Innovation, 31*, 103176.

Huang, Z., Hu, B., & Wang, H. (2023). Analytical methods for microplastics in the environment: A review. *Environmental Chemistry Letters, 21*(1), 383–401.

Ismanto, A., Hadibarata, T., Sugianto, D. N., Zainuri, M., Kristanti, R. A., Wisha, U. J., … & Abbasi, A. M. (2023). First evidence of microplastics in the water and sediment of Surakarta city river basin, Indonesia. *Marine Pollution Bulletin, 196*, 115677.

Jessieleena, A., Rathinavelu, S., Velmaiel, K. E., John, A. A., & Nambi, I. M. (2023). Residential houses—A major point source of microplastic pollution: Insights on the various sources, their transport, transformation, and toxicity behaviour. *Environmental Science and Pollution Research, 30*, 1–22.

Jung, M. R., Horgen, F. D., Orski, S. V., Rodriguez, V., Beers, K. L., Balazs, G. H., ... & Lynch, J. M. (2018). Validation of ATR FT-IR to identify polymers of plastic marine debris, including those ingested by marine organisms. *Marine pollution bulletin, 127*, 704–716.

Kappler, A., Fischer, M., Scholz-Böttcher, B. M., Oberbeckmann, S., Labrenz, M., Fischer, D., … & Voit, B. (2018). Comparison of μ-ATR-FTIR spectroscopy and py-GCMS as identification tools for microplastic particles and fibers isolated from river sediments. *Analytical and Bioanalytical Chemistry, 410*, 5313–5327.

Lam, T. W. L., Ho, H. T., Ma, A. T., & Fok, L. (2020). Microplastic contamination of surface water-sourced tap water in Hong Kong—A preliminary study. *Applied Sciences, 10*(10), 3463.

Lambert, S., & Wagner, M. (2018). *Microplastics are contaminants of emerging concern in freshwater environments: an overview* (pp. 1–23). Springer International Publishing.

Li, Y., Lu, Z., Zheng, H., Wang, J., & Chen, C. (2020). Microplastics in surface water and sediments of Chongming Island in the Yangtze Estuary, China. *Environmental Sciences Europe, 32*(1), 1–12.

Masura, J., Baker, J., Foster, G., & Arthur, C. (2015). *Laboratory methods for the analysis of microplastics in the marine environment: Recommendations for quantifying synthetic particles in waters and sediments.*

Manzoor, S., Kaur, H., & Singh, R. (2021). Existence of microplastic as pollutant in Harike Wetland: An analysis of plastic composition and first report on Ramsar Wetland of India. *Current World Environment: An International Research Journal of Environmental Sciences, 16*, 123–133.

Manzoor, M. F., Tariq, T., Fatima, B., Sahar, A., Tariq, F., Munir, S., Khan, S., Nawaz Ranjha, M. M. A., Sameen, A., Zeng, X. A., & Ibrahim, S. A. (2022). An insight into bisphenol A, food exposure and its adverse effects on health: A review. *Frontiers in Nutrition, 9*, 1047827. https://doi.org/10.3389/fnut.2022.1047827

Mennekes, D., & Nowack, B. (2023). Predicting microplastic masses in river networks with high spatial resolution at country level. *Nature Water* 1–11.

Mhiret Gela, S., & Aragaw, T. A. (2022). Abundance and characterization of microplastics in main urban ditches across the Bahir Dar City, Ethiopia. *Frontiers in Environmental Science, 35.*

Miserli, K., Lykos, C., Kalampounias, A. G., & Konstantinou, I. (2023). Screening of microplastics in aquaculture systems (fish, mussel, and water samples) by FTIR, scanning electron microscopy-energy dispersive spectroscopy and micro-raman spectroscopies. *Applied Sciences, 13*(17), 9705.

Ohore, O. E., & Zhang, S. (2019). Endocrine disrupting effects of bisphenol A exposure and recent advances on its removal by water treatment systems. A review. *Scientific African, 5*, e00135.

Osman, A. I., Hosny, M., Eltaweil, A. S., Omar, S., Elgarahy, A. M., Farghali, M., ... & Akinyede, K. A. (2023). Microplastic sources, formation, toxicity and remediation: a review. *Environmental Chemistry Letters, 21*(4), 2129–2169.

Pan, T., Liao, H., Yang, F., Sun, F., Guo, Y., Yang, H., ... & Wang, Q. (2023). Review of microplastics in lakes: sources, distribution characteristics, and environmental effects. *Carbon Research, 2*(1), 25.

Prata, J. C. (2018). Airborne microplastics: Consequences to human health? *Environmental Pollution, 234*, 115–126.

Rashed, A. H., Yesilay, G., Hazeem, L., Rashdan, S., AlMealla, R., Kilinc, Z., ... & Kamel, A. H. (2023). Micro-and nano-plastics contaminants in the environment: Sources, fate, toxicity, detection, remediation, and sustainable perspectives. *Water, 15*(20), 3535.

Ricciardi, M., Pironti, C., Motta, O., Miele, Y., Proto, A., & Montano, L. (2021). Microplastics in the aquatic environment: Occurrence, persistence, analysis, and human exposure. *Water, 13*(7), 973.

Saeed, T., Al-Jandal, N., Al-Mutairi, A., & Taqi, H. (2020). Microplastics in Kuwait marine environment: Results of first survey. *Marine Pollution Bulletin, 152*, 110880.

Sánchez-Hernández, L. J., Ramírez-Romero, P., Rodríguez-González, F., Ramos-Sánchez, V. H., Montes, R. A. M., Rubio, H. R. P., ... & Jonathan, M. P. (2021). Seasonal evidences of microplastics in environmental matrices of a tourist dominated urban estuary in Gulf of Mexico, Mexico. *Chemosphere, 277*, 130261.

Shim, W. J., Hong, S. H., & Eo, S. (2018). Marine microplastics: abundance, distribution, and composition. In *Microplastic contamination in aquatic environments* (pp. 1–26). Elsevier.

Tien, C. J., Wang, Z. X., & Chen, C. S. (2020). Microplastics in water, sediment and fish from the Fengshan River system: Relationship to aquatic factors and accumulation of polycyclic aromatic hydrocarbons by fish. *Environmental Pollution, 265*, 114962.

Ullah, S., Ahmad, S., Guo, X., Ullah, S., Nabi, G., & Wanghe, K. (2023). A review of the endocrine disrupting effects of micro and nano plastic and their associated chemicals in mammals. *Frontiers in Endocrinology, 13*, 1084236.

Veerasingam, S., Ranjani, M., Venkatachalapathy, R., Bagaev, A., Mukhanov, V., Litvinyuk, D., … & Vethamony, P. (2021). Contributions of Fourier transform infrared spectroscopy in microplastic pollution research: A review. *Critical Reviews in Environmental Science and Technology, 51*(22), 2681–2743.

Wang, W., & Wang, J. (2018). Investigation of microplastics in aquatic environments: An overview of the methods used, from field sampling to laboratory analysis. *TrAC Trends in Analytical Chemistry, 108*, 195–202.

Watts, A. J., Lewis, C., Goodhead, R. M., Beckett, S. J., Moger, J., Tyler, C. R., & Galloway, T. S. (2014). Uptake and retention of microplastics by the shore crab Carcinus maenas. *Environmental Science & Technology, 48*(15), 8823–8830.

Zhang, F., Zhao, Y., Wang, D., Yan, M., Zhang, J., Zhang, P., ... & Chen, C. (2021). Current technologies for plastic waste treatment: A review. *Journal of Cleaner Production, 282*, 124523.

Part II
Fundamental Studies/Strategies for Combating Plastic Pollution

Chapter 5
Plastic and Polythene Waste Degradation

Amar P. Garg and Afeefa Ateeq

Abstract Plastic has become an important part of our daily life and a vital product in the economic sector because of its wide usage at large scale and is used in all sectors such as agriculture, building, construction, health and in consumer goods as transporting and packing material. A significant amount of plastic is used in television, sound systems, cell phones, computers and office materials, household products, (plastic foam in furniture, plastic chair, kitchen appliances, decoratory products), toys, acrylic composite countertop, sports safety gear, transport, clothing materials, food items, pharmaceuticals, detergents, cosmetics, sanitary wares, tiles, plastic bottles, artificial leather, and likewise. The use of plastic in every sector poses a serious threat to the environment and to all ecosystems. There is an urgent need for a method by which plastic waste can be managed and degraded. The biological techniques are environmentally safe and eco-friendly. Disposal of plastic on both land and marine causes problems for humans, animals, and different life forms. The plastic gets affected by the action of microbes in the natural environment and the microbes have the ability to degrade the plastics and polythene. It is very important to understand the interaction between microbes and plastics. Previous research have revealed that several microbes including bacteria, fungi, actinomycetes, nematodes and arthropods have the capabilities to degrade plastics and polythene. In this chapter we will focus on the basic types of plastics and polythene, their degradation and conversion by different methods. Characterization of bio-degradation is an important part of an ecosystem and circular economy and the factors affecting the bio-degradation of plastics and polythene need more investigation. The microbes possessing different sets of enzyme complexes responsible for natural and commercial degradation of plastic and polythene require their identification, design and production of genetically modified organisms and their characterization.

A. P. Garg (✉)
Dean Academics and Director Research, Swami Vivekanand Subharti University, Meerut, India
e-mail: amarprakashgarg@yahoo.com

A. Ateeq
School of Biological Engineering and Life Sciences, Shobhit Institute of Engineering and Technology (NAAC 'A' Grade Deemed-to-be-University) Modipuram, Meerut, India

S. Suriyanarayanan et al. (eds.), *Combating Plastic Pollution in Terrestrial Environment*,
https://doi.org/10.1007/978-981-96-2343-3_5

Keywords Plastic · Polythene · Microbial degradation · Waste management · Conversion of plastic into energy · Pollution

5.1 Introduction

Plastics are manmade heterochain polymers that originate from oil, coal, and natural gas (Azeko et al., 2015). It was first made by a Belgian chemist and a clever merchant Leo Baekeland who prepared the first fully synthetic plastic in 1907. The twentieth and twenty-first centuries are the age of plastics. Plastic is the loose term for chronicle materials that can be molded under heat and pressure, a process called polymerization. Plastic polymers are generally strong but non-degradable and their accumulation leads to greater environmental pollution. These are used for various purposes like food packaging, detergent, clothing, shelter, transportation, toys, household products, office material, healthcare products and machines, cell phones, electronic products, agricultural and scientific equipment including chemical substances. These plastics are of different types like polyethylene (Low density polyethylene (LDPE), Linear low density polyethylene (LLDPE), Medium density polyethylene (MDPE), and High density polyethylene (HDPE), Polypropylene (PP), Polystyrene (PS), Polyvinyl Chloride (PVC), and Polyethylene terephthalate (PET) (Montazer et al., 2020). 350–360 million metric tons of plastic is produced worldwide, mostly single-used and shockingly, this large volume of 350 million tons of plastic weight is equivalent to the whole human population of the world (Thushari et al., 2020).

The disposal of plastics is one of the biggest problems, mainly because of its resistance and complicated chemical structure, however, the plastics have a number of advantages like cost-effective, exceptionally durable, and moderately unbreakable (Isangedighi et al., 2018). But its major disadvantages are that it does not break down in the environment, has the tendency to float, long-term perseverance, and iniquitousness in the marine environment and plastic wastes pose a diversity of hazards to land and marine life (Thushari et al., 2020) including the generation of toxic gases. There is now growing public concern over environmental decline with the disposal of plastics. Discarded plastics in landfills are a big issue for a rapidly increasing percentage of solid waste that are harmful to the environment and the ecosystem as a whole (Ateeq et al., 2023; Raaman et al., 2012).

Not only landfill but the marine ecosystem is also disturbed by plastic disposal and it is listed as one of the top ten environmental problems (Muhnoja et al., 2018) which needs to be resolved globally since the United Nations Environmental Program (UNEP) issued the theme of World Environmental Day-2023, "Beat Plastic Pollution" and proposed to all countries to work together to overcome from this problem. In 2018, European Commission issued the "Plastic Strategy in European Circular Economy" which proposed at least half of the plastic waste should be recycled by 2030. After these strategies more than 15 countries/regions successively issued policies to "ban plastic" and "limit plastic" including New York State, Washington State of America, EU member states, Hainan and Jilin provinces of China and India is also part of it. The Government of India has also issued a policy to deal with plastic and e-waste from April 1, 2023 and it has been made mandatory to all manufacturers to recycle at least 60% of the plastic and e-waste in their gross production (Chaudhary and Garg, 2023). The Government of India has fixed the lifetime of all domestic products made up of plastic from 3 to 10 years depending upon its utility and durability. After the expiry of the period, the companies are required to buy back the units to recycle them. It will reduce the free disposal of the plastic to save the environment. The actual measures focus on reducing, reusing, and recycling of plastics, and its main aim is to stop the disposal of plastic on land and water before being delivered to oceans, seas, and rivers but it highly depends on the enforcement by the respective Government and raising public awareness about the environment (Muhnoja et al., 2018, Taghavi et al., 2021). Though awareness about plastic waste is increasing in most of the countries, numerous actions are being taken to tackle plastic collection by motivating the active involvement of consumers, producers, industry, and companies but some more stringent measures are required to be implemented for future and sustainable world. There are many other attempts that have been already taken to clean up the seawater but these attempts cannot clean the macro and microplastics from the sea and these should be cleaned up (Monika and Garg, 2023; Wang et al., 2020).

There is 82.4% green biomass on earth in terms of carbon and the real problem of pollution lies in accumulation of plastic and polythene on earth planet and not the scarcity of green biomass. The accumulation of plastic in nature is mainly because of its slow degradation in nature that may be due to the loss or lesser microbial diversity/activity in comparison to the amount of disposal of plastic and polythene in nature. The polythene is clogging the soil pores and reducing the percolation of rainwater resulting in the reduction of ground water level each year. Further, its disposal in the soil is emitting toxic chemicals and hazardous gases. Similarly, the accumulation of plastic and e-waste is threatening the soil fertility, microbial diversity and enhancing pollution problem. The disposal of plastic in marine water during the last two decades has greatly adversely affected the marine flora and fauna and its micro particles have entered into the human blood through food chain. Hence, the waste management of plastic and polythene is essential for sustainability of the environment which is possible either by reuse or recycle or by its conversion into energy or other utilizable products (Garg, 2023; Chaudhary and Garg, 2023).

5.2 Plastic Pollution in Marine Environment

The collected plastics from the ocean may be classified into four categories based on their sizes: mega-plastics (>1 m), macroplastics (<2.5 cm), mesoplastics (>2.5), and microplastics (<5 mm) (Loganathan and Kizhakedathil, 2023). Microplastics come from manufactured or commercial products or cosmetic products. Microplastics are present in marine and coastal systems in a wide range and polypropylene (PP) and polyethylene (PE) float on the surface of oceans due to its light weight, while polystyrene (PS), polyvinyl chloride (PVC), polyamide (PA), and polyethylene terephthalate (PET) have higher density so they do not float on the surface of the oceans. Microplastics are widely distributed in every pelagic and benthic zone of marine systems. (Thushari et al., 2020; Taghavi et al., 2021).

There is a direct and indirect source of accumulation of plastic waste in the ocean. In the coastal and marine ecosystem, plastic pollution comes from land and ocean-based sources. Mainly domestic and residential activities, tourism, and economic actions are the major land-based plastic pollution sources. In marine ecosystem, 75% of plastic pollution is caused by land sources (Cai et al., 2023; Andardy et al., 2011). Coastal zones are highly populated and industrialized or urbanized areas. Residents who use plastic products, could directly discharge into the coastal zone not only the residential waste, but sometimes the whole containers are released into the drainage systems (Browne et al., 2007; Taghavi et al., 2021). A large amount of freshwater ecosystem with direction, accumulates to the ocean through fast flow rate. The primary source of microplastics in Goiana Estuary, South America, is harmed by river basins (Devi Sangeetha et al., 2015; Lima et al., 2014).

Types of Plastics: (Isangedighi et al., 2018).

There are 7 major different types of plastic present in the environment that are used worldwide over a long period. Their chemical structures have been given by Afeefa and Garg (2023).

(A) **Polyethylene Terephthalate**: Introduced by Whinfield & Dickson in 1940. From PET plastic the beverage bottles are being made and interestingly it is produced by Coca-Cola and Pepsi (Fig. 5.1).

(B) **High-Density Polyethylene (HDPE)**: In 1953, Ziegler & Halzkamp used catalyst and low pressure to create high-density polyethylene. It is used for pipes, drains and Culverts and for a wide variety of products (Fig. 5.2)

(C) **Polyvinyl Chloride (PVC)**: It was made by a German Chemist in 1872. It is one of the most non-recycled plastic present in the environment. It is also called "Poison Plastic" (Fig. 5.3).

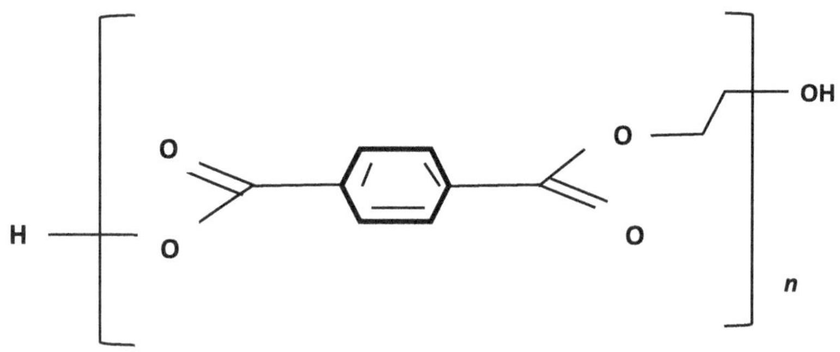

Fig. 5.1 Chemical structure of PET

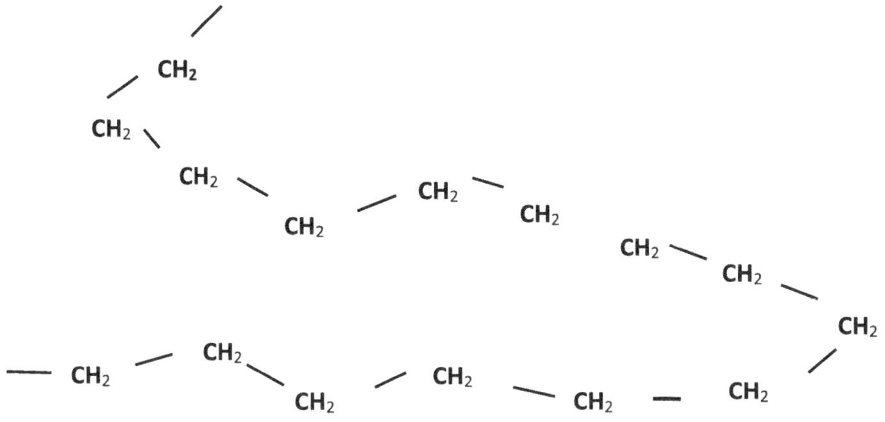

Fig. 5.2 Chemical structure of HDPE

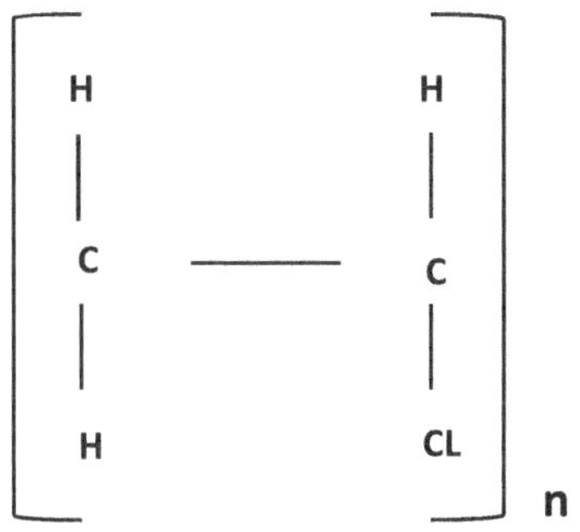

Fig. 5.3 Chemical structure of PVC

(D) **Low-Density Polyethylene (LDPE)**: It was the first Polyethylene that was produced. It has less weight than HDPE. LDPE is considered a separate recycling material. Packaging & Containers are made from LDPE and 75% of LDPE (Fig. 5.4) comes from residential households (Taghavi et al., 2021).

(E) **Polypropylene (PP)**: In 1951, Paul & Robert discovered polypropylene. It is mainly used in plastic woven bags, tarpaulins and ropes, injection molded products, film materials, and pipes, etc. (Fig. 5.5).

(F) **Polystyrene or Styrofoam (PS)**: Eduard Simon prepared medication from polystyrene in 1839. Hermann Staudinger a German chemist expanded its uses. Polystyrene (Fig. 5.6) is lightweight and easy to form into plastic materials. It is very harmful to the environment because it breaks effortlessly and is disrupted into minor and micro particles. Petri dishes are most commonly used in microbial culture in biological and medical research including diagnostics.

(G) **Miscellaneous Plastics**: Other remaining plastics are Polycarbonate, Polylactide, Acrylic, Acrylonitrile butadiene, Styrene, fiberglass, nylon, etc.

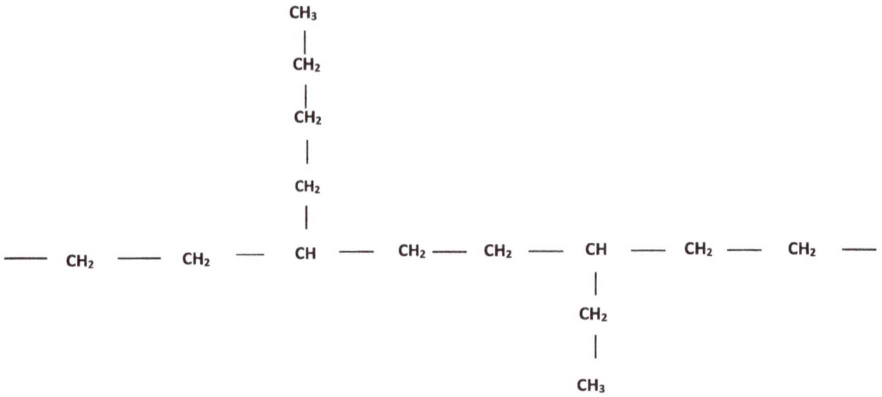

Fig. 5.4 Chemical structure of LDPE

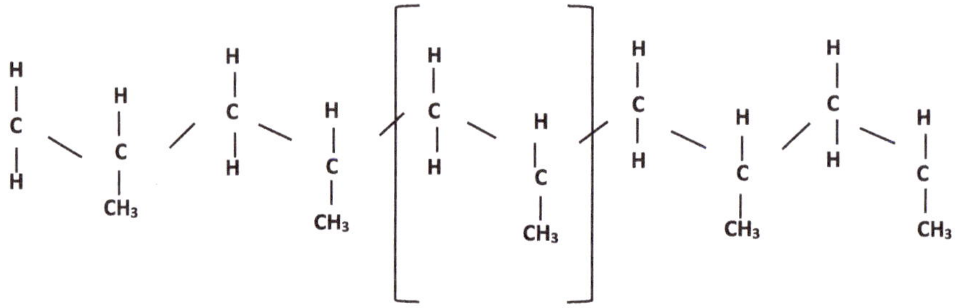

Fig. 5.5 Chemical structure of PP

Fig. 5.6 Chemical structure of PS

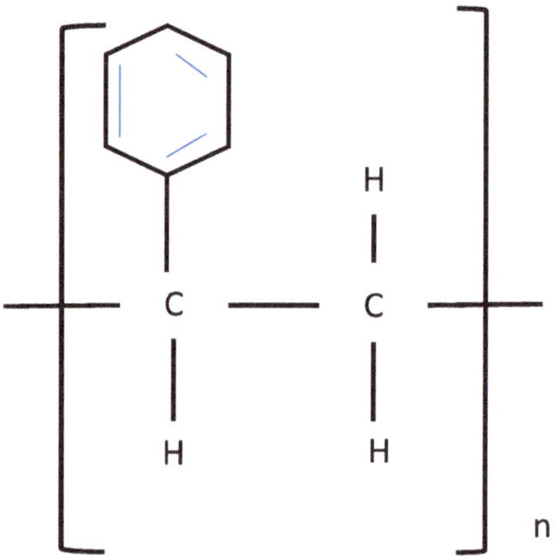

(H) **Microplastics**: Microplastic and Microlitter have been defined by Andrady (2011) as invisible particles which cannot be seen by naked eyes and the diameter of such microplastic particles ranges in between 0.06–0.5 mm (i.e. 60–500 μm). Scientifically and technically, the definition of microparticles given by Andrachy (2011) is not correct as the word "micro" is used for particles less than 15 μm, hence, it is proposed that the particles smaller than 15 μm be referred as "microparticles" while the particles / pieces larger than 15 μm but smaller than 500 μm (0.5 mm) be refereed as "milliparticles" and those bigger than 500 μm be referred as "macroparticles". These contaminated plastic particles when ingested by marine fauna, open a route by which the Persistent organic pollutants (POPs) can enter the marine food web and finally reach the humans through consumption of marine flora and fauna. (Moore, 2008; Teuten et al., 2007).

Conversion of Plastic Using Pyrolysis: Pyrolysis may be used to convert the plastic into pyrolysis oil (high density) which may be converted into energy (Fig. 5.7) In pyrolysis, the plastic is heated at high temperature in the absence of oxygen, to convert it into energy (Fig. 5.7) (Rehan et al., 2017). In this method different kinds of catalysts are used to for the improvement of pyrolysis process of plastic waste. Many types of catalysts have been used in pyrolysis process but ZSM-5 (zeolite socony mobil-5), zeolite, Y-zeolite, FCC (fluid catalytic cracking), and MCM-41 (Mobil composition of matter no. 41) are extensively used (Ratnasari et al., 2017). Energy generation from plastic waste oil, and fossil fuels using high quality burners reduce fuel costs by 30–70% and facilitate to achieve 50% reduction in air pollutants. It also facilitates reduction in carbon emissions. Conversion of plastic into energy using incineration technology coupled with pyrolysis is a new concept that produces high density pyrolysis oil which may be used in heavy industries and also to convert it into electricity. It helps in circular economy and sustainable development. It is eco-friendly and also acceptable economically. Korea and other countries are using the technology. The team India under the mentorship of Prof Amar P. Garg (one of the authors of this chapter) made a presentation at the United Nation's conference on Sustainable Development Goals (SDG) on November 13, 2023 about this technology which was highly appreciated by the experts. Efforts are being made at Swami Vivekanand Subharti University, Meerut to design a protocol for the production of pyrolysis oil from plastic waste in India as we are currently disposing of the biomedical waste using incineration technology while it may be converted into energy and finally into electricity. This process is safe, eco-friendly, less expensive when the cost of environment purification is considered, sustainable, easy to adopt at commercial level (Sánchez, 2009; Vankatesh and Vijayalakshmi, 2021).

Types of Degradable Plastics:

(A) **Photo-degradable plastics**: Photo-degradable plastic is connected to the polymer backbone which is light sensitive. By giving a long time exposure to UV radiation it can be disintegrated. When the radiation supply is stopped

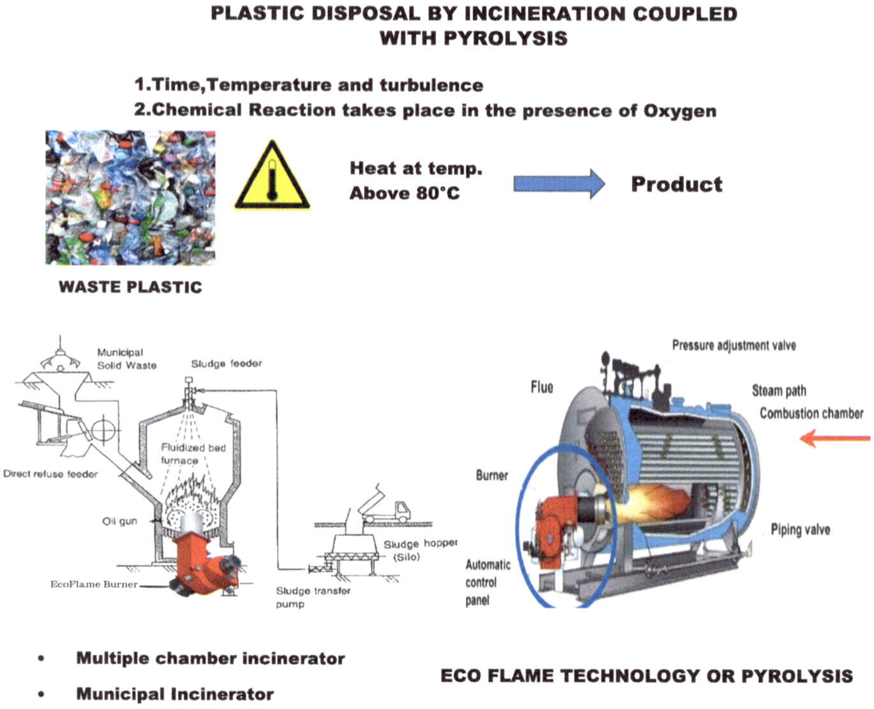

Fig. 5.7 Conversion of plastic into energy using pyrolysis

or reduced by the sunlight on the land, it is not degraded. But if artificial photo-degradation light is used it can release toxic volatile compounds (VOCs) and it can cause various problems to the environment (Lomonaco et al., 2020).

(B) **Bio-based Bio-plastic**: Bio-based bioplastic has 100% of carbon from renewal resources like forestry and agricultural resources. Starch, corn, soybean, and cellulose are examples of such resources. (Marichelvam et al., 2019). Bio-based plastics are beneficial for the environment because they can reduce fossil fuel consumption.

(C) **Compostable Bio-plastic**: Composting is a subclass of bio-degradation but every compostable material is not biodegradable in other environments such as marine, soil, and landfill. Composting can be divided into two classes home composting and industrial composting. In industrial composting conditions are controlled and home composting is slower than industrial composting (Meereboer et al., 2020). In composting the plastics are decomposed leaving no toxic compounds in the environment.

(D) **Biodegradable plastics**: Plastic that is degraded by the microorganisms in the natural environment is called as biodegradable plastic. It is disintegrated into biogases and biomass by the microorganisms (Jain et al., 2010). Various microbes use plastic as their substrate.

5.3 Degradation of Plastic

There are many ways to overcome this problem which is associated with plastic waste. It could be recycling, energy recovery, ban on specific products, and the production, use and encouragement of biodegradable plastics. Recycling is an appealing method for thermoplastics, mainly for high volumes of used plastics like PETE and Polyethylene (PE) (Taghavi et al., 2021). It can decrease the use of natural resources as well as reduce the chances of global warming. Although, it only requires a separation technique for recycling that can be incorporated into the new products from pure raw materials. From direct combustion, the recovery of energy can be achieved. Its advantages are also there since it has the capacity to treat mixed plastic. But special care is needed when dealing with PVC because incineration produces acid emissions in the environment. Waste to energy is also needed to control because it produce toxic compounds such as dioxins and furanes (Quecholac-Piña et al., 2022).

On the other hand, biodegradable plastics are also an alternative to overcome this problem due to their functional property. The brand name Eoflex® is producing complete biodegradable aromatic and aliphatic polyesters by BASF. They are making polyester by blending with bio-based polymers like starch or polylactic acid (PLA), and they are obtaining new and compelling properties of biodegradable plastics. It is also a substitute for standard plastics such as polyethylene or polystyrene. Its application range is also wide like production of waste bags to shopping bags from agriculture mulch film to knitted bags and many more items are being made by them (Siegenthaler et al., 2012).

Synthetic plastics are of two types—plastics with a carbon backbone and plastic with heteroatoms in the main chain. The plastics which have the backbone and only built carbon atoms are PE, PS, and PVC. On the other hand, the plastics which have heteroatoms in the main chain are PU and PET (Taghavi et al., 2021). There are several types of polyesters that were used as alternatives to some petro-plastics. (Mohanan et al., 2020).

Bioplastics are biodegradable because they are either manufactured from biomass or fossils. During bio-degradation, the microbes use polymers as their carbon source and some microbes use this polymer as a carbon source and energy both. Microorganisms secrete different types of enzymes to degrade the plastic, reducing polymer chain length by oxidation. Biodegradability of plastics can be detected by the production of CO_2 and increasing microbial growth (Trivedi et al., 2016).

Degradation through microorganisms is one of the easiest, most cost-effective, and eco-friendly processes, polymers like synthetic and natural both can be degraded through microbial enzymatic degradation (Taghavi et al., 2021). Microorganisms use polymers as substrates for growth when they start degrading. Polymer degradation can be seen in various types such as cracking, erosion, discoloration, and phase separation (Mukherjee & Chatterjee, 2014). Polymers basically are made up of several monomers that are mineralized into monomers and their constituent elements by the action of different sets of enzyme complexes, secreted by microbes. The monomers can pass through the cell membrane after depolymerization, enter the cell and mineralize and biodegrade the molecules inside the cell. Physical and biological forces react to breakdown the polymer. Physical forces involve heating, cooling, freezing, thawing, wetting, drying, etc. through cracking which cause mechanical damage (Muthukumar et al., 2014).

Types of Degradation:

(A) **Photo-oxidative degradation**: Light is the primary source of damaging and degrading the polymers. When the polymer absorbs light, the photodegradation process starts. UV radiation degrades synthetic polymers. Some polymeric materials are used for various applications for a lifetime. Sunlight is the source of radiation for the degradation of plastic (Fig. 5.8) (Ranby, 1989).

(B) **Thermal degradation**: Thermal and photochemical processes generally are the same because in both the processes, oxidation takes part. The main difference in both the processes is the sequence of initial steps and the site of reaction. In thermal degradation, the reaction takes part in the whole polymer but in

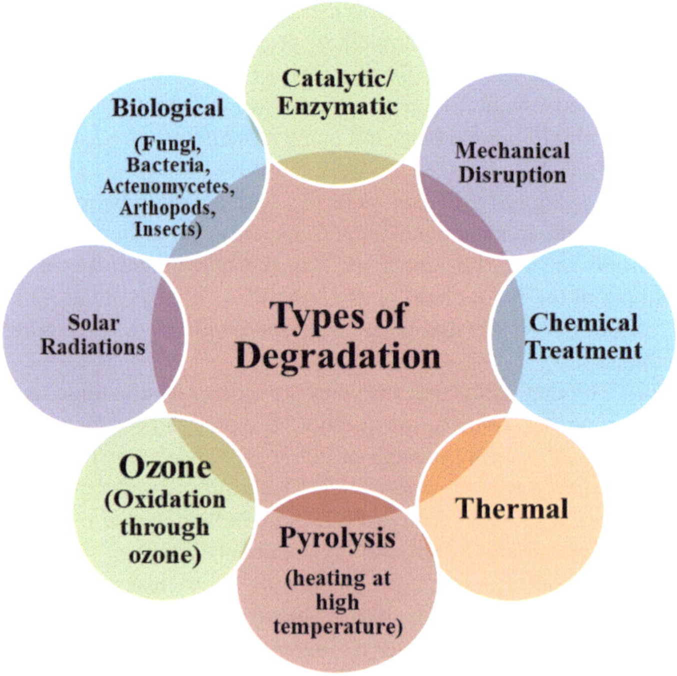

Fig. 5.8 Various types of degradation of plastic

photochemical degradation, reactions start and take part only on the surface area. For the initial stage of thermal degradation temperature and UV light are required (Fig. 5.8) (Teare et al., 2000).

(C) **Mechano-chemical Degradation**: In mechano-chemical degradation, two processes take part simultaneously *i.e.* mechanical and chemical (Fig. 5.8). In this process the polymer chain is broken down by mechanical pressure and ultrasonic irradiations. Due to chain-side reaction, the branches are increased in number, and the molecular weight is decreased. The double bond concentration is also changed. (Striegel, 2003). In the air, the molecular weight of polyvinyl chloride is also reduced by mechanic chemical dichlorination with different oxide powders like SiO_2, CaO, Al_2O_3, and Fe_2O_3 (Inoue et al., 2004)

(D) **Biological Degradation**: Biological degradation is the process of plastic being degraded by microorganisms such as Bacteria, Fungi, and Actinomycetes that actively degrade natural and synthetic plastics (Fig. 5.8). Microbes utilize plastic as an energy resource and grow on the plastic and degrade from the surface area. (Alshehrei, 2017)

(E) **Ozone Degradation**: Ozone is present in the atmosphere which causes polymer degradation. Ozone presents in the environment in small amounts but ozone has a marked great effect on polymer. Ozone formed reduction in molecular weight by a change in mechanical properties of the polymer. The exposure of the polymer to ozone then formed different types of carbonyl groups. These products are based on ketones, lactones, esters, etc. (Allen et al., 2003) (Fig. 5.8).

(F) **Catalytic/Enzymatic Degradation**: Catalytic degradation is basically a transformation of polymer into hydrocarbons which is an area of interest of microbiologists, biotechnologists and environmentalists. Catalytically degraded polymer produces oils and gases. By using this method of degradation of polymer not only the quality but the desired product can also be achieved. Different types of catalysts like Pt–Mo, and Pt–Co maintained by SiO_2 are used for degradation (Zeenat Bashir et al., 2019).

Aerobic Degradation: In aerobic biological degradation microorganisms break down the large organic compounds into smaller organic compounds in the presence of oxygen and microbes use oxygen as electron acceptors. The by-products of this reaction are water and carbon dioxide (Fig. 5.9) (Priyanka and Archana, 2011).

Anaerobic Degradation: In this process, oxygen is not required for the breakdown of organic compounds through microbial action. Oxygen is an important component for the natural elongation of contaminants at hazardous waste sites. In the anaerobic degradation process, anaerobic bacteria nitrate, iron, sulfate, manganese, and carbon dioxide an electron acceptor in place of oxygen for the breakdown of the large organic compounds into small organic compounds (Fig. 5.9) (Gu, 2003a, 2003b).

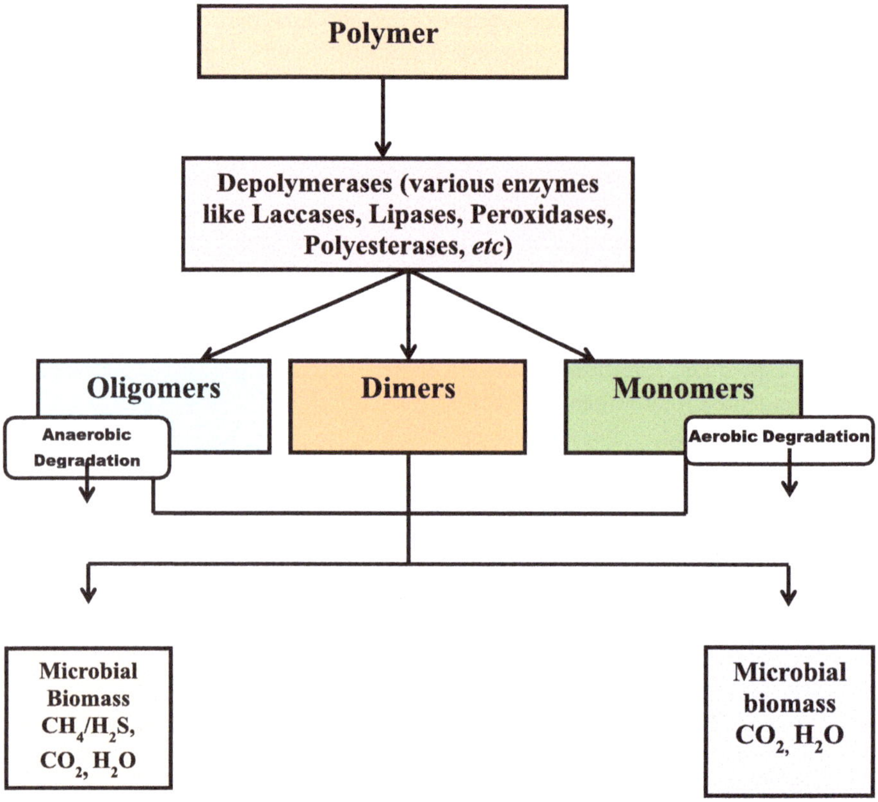

Fig. 5.9 Plastic polymer degradation under aerobic and anaerobic degradation

5.4 Carbon (Plastic) Methane + Carbon Dioxide + Water + Carbon Residual

Mechanism of plastic bio-degradation: Biological degradation of plastic starts when the microbes are attached to the surface of the plastic and utilize it as their energy source by secreting different sets of enzyme complexes that are able to cleave the respective complex bonds for the degradation of plastic (Fig. 5.10) (Danzo et al., 2018).

Polymer degradation by Microorganisms: Microbial degradation of polymers includes bio-deterioration, fragmentation, mineralization, and assimilation (Fig. 5.11) (Zeenat et al., 2021).

(A) **Bio-deterioration**: Bio-deterioration is a process in which the microbes affect the surface of plastics and change their physical, chemical, and mechanical properties. Chemical and structural changes depend upon the polymer composition. Environmental conditions also affect the changes in polymer properties.

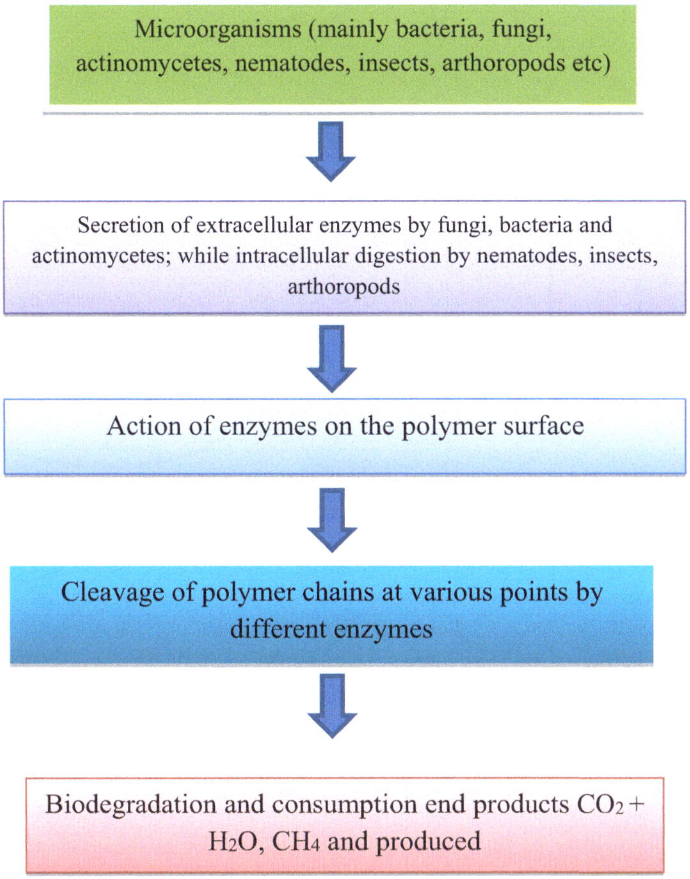

Fig. 5.10 Microbial degradation of plastic polymer

In the deterioration process substrate, biofilm formation takes place (Fig. 5.11) (Vivi et al., 2019).

(B) **Bio-fragmentation**: Bio-fragmentation is the second step after bio-deterioration where enzymatic process takes part in the plastic polymer. In bacteria, mostly oxygenase enzymes are secreted which have the ability to break down the polymer molecules to carbon chain, alcohol, and peroxyl products at the end. These products are less harmful to the environment (Fig. 5.11) (Pathak & Nichter, 2019).

(C) **Mineralization**: After the bio-fragmentation process plastic polymer enters into the microbial cells via the cell membrane where monomers that are large in size cannot reach inside the cell they stay outside. The small monomers that enter the cell membrane microbes use the monomers as the energy source and this energy is used for biomass production of microbes (Fig. 5.11) (Kale et al., 2015; Skariyachan et al., 2017).

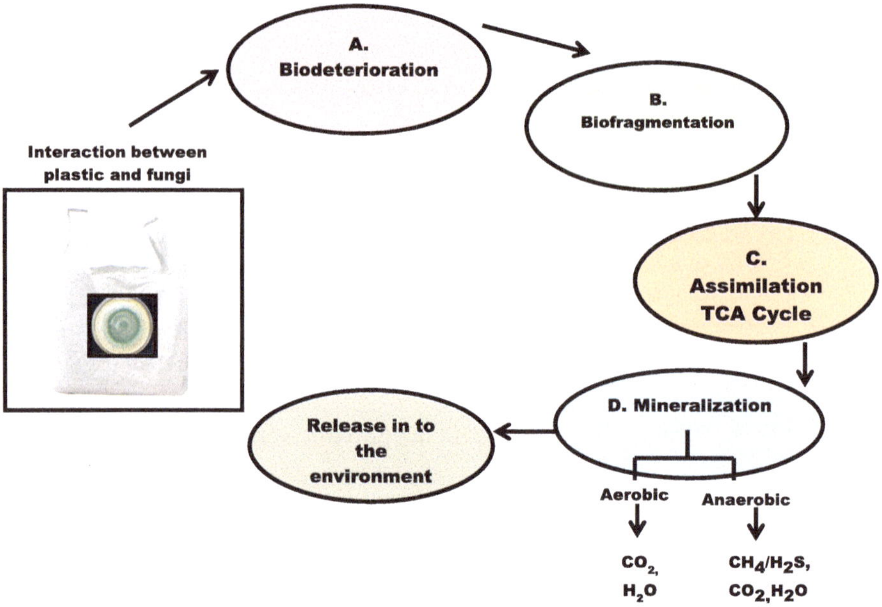

Fig. 5.11 Interaction between fungi and plastic (Adapted and modified from Varshney et al., 2023)

(D) **Assimilation**: The assimilation process is the last step for the degradation of atoms finally combined in the microbial cell for complete degradation. Now the secondary metabolites are transported either outside the cells or other microbes they further perform degradation and utilize these metabolites. In metabolite degradation the end products such as CO_2, N_2, H_2O, and CH_4 are released during metabolite degradation (Fig. 5.11) (Krzan et al., 2006). Fungi play a greater and higher role in degradation of plastic because the apical anastomoses of fungi provide additional mechanical power to the fungal tip to separate the complex plastic fibres for a forceful entry and the large surface area of mycelium enable the fungi to produce large amounts of extracellular enzymes that cleave the complex chemical bonds to release oligomers and monomers. There are fungi that are active even under anaerobic and harsh conditions, even at very high concentration of carbon dioxide.

Enzymatic Degradation of Plastic: Hydrolysable groups are absent in the carbon–carbon backbone so the degradation of plastic is a difficult task by microbial enzymes. In abiotic and biotic factors, the reduction of molecular weight is the first achievement. Microbial enzymes attack the carbonyl group of the polymer through which the degradation takes part. In enzymatic degradation, different enzymes are involved such as laccases, manganese, peroxidases, oxidases, catalases, ureases, lipases, and proteases etc. Laccases enzyme can degrade the polyethylene just within 48 h of incubation. (Jaiswal et al., 2020).

Biological Mechanisms and Enzymatic Activity of Plastic Degrading Fungi: The mechanisms and enzymatic activity of fungi involves the microbial enzyme on the surface of the plastics. Different types of microbes such as: bacteria, fungi and actinomycetes utilize the plastic surface as the source of nutrition to grow rapidly. After that the surface area of plastic gets depolymerized and degradation assembled by the mineralization process. The form of end products H_2O (water), CO_2 (carbon dioxide) and CH_2 (methane) are released (Frazer, 1994; Montazer et al., 2019). Fungi have ability to invade substrates which uses enzyme and these enzymes can degrade pollutants. Fungi have ability to produce hydrophobin proteins that are surface-active proteins and hydrophobins protein coat hyphal walls to hydrophobic substrate (Fig. 5.12). Some fungi cause bursting and swelling at small scale and make a hole in polymers (Grifn 1980). Extra- and intra-cellular enzymes of fungi take part in degradation. The intra-cellular enzymes play an important role in adaptation of fungi and they act as internal mechanisms for detoxification (Jeon et al., 2016; Olicón-Hernandez et al., 2017; Schwartz et al., 2018; Shin et al., 2018). This system is conciliated by cytochrome P450 family (CYP), in oxidation and conjugation reaction Phase I enzyme epoxidase and, Phase II enzyme transferases are involved. Cytochrome family is involved in catalyzing various enzymatic reactions because cytochrome P450 family heme-containing mono-oxygenases (Shin et al., 2018). For primary metabolisms, cytochrome P450 enzymes are very important. Cytochrome P450 family enables the protection of the hyphal wall integrity and formed spore outer wall (Crešnar and Petric 2011). Cytochrome isoforms are attached to the membrane of endoplasmic reticulum and connected to the both systolic and membrane environment after they uptake substrate from both surroundings (Šrejber et al., 2018). $NADP^+$, H^+ FAD and Heme are three cofactors CYP contains and NADPH: CYP reductase and and cytochrome P-450 hydrolase are two enzymes which CYP contains. The extracellular enzymatic system contains hydrolytic system and it produces hydrolases and they play a major role in polysaccharide degradation and in the breaking down of complex structures (lignin structures) where unspecific oxidative systems are involved. Wide range of substrates oxidized through unspecific oxidative system. Oxidoreductases formed by class II peroxidases enzyme such as: manganese peroxidases, lignin peroxidases and versatile peroxidases, laccases and peroxygenases. These enzymes play a role in transferring electron from organic substrates to molecular oxygen through oxidation and reduction using H_2O_2 as electron accepting substrate or by epoxidation, sulfonification and preoxygenation (Karich et al., 2017).

Factors affecting the degradation: Factors that affect the bio-degradation process are summarized in Fig. 5.13 as the organisms, substrate, and environmental factors (Gu, 2021). All these factors affect the degradation of the plastic individually and collectively in various ways.

Microbial degradation of low-density polyethylene (LDPE): Microbial degradation of polyethylene is the best and cheaper method to degrade polyethylene because the low-density polyethylene is a non-recyclable product that is circulated in the markets and homage and after disposal of this polythene, it causes a serious problem to the environment and the animals that feed on the material contained in

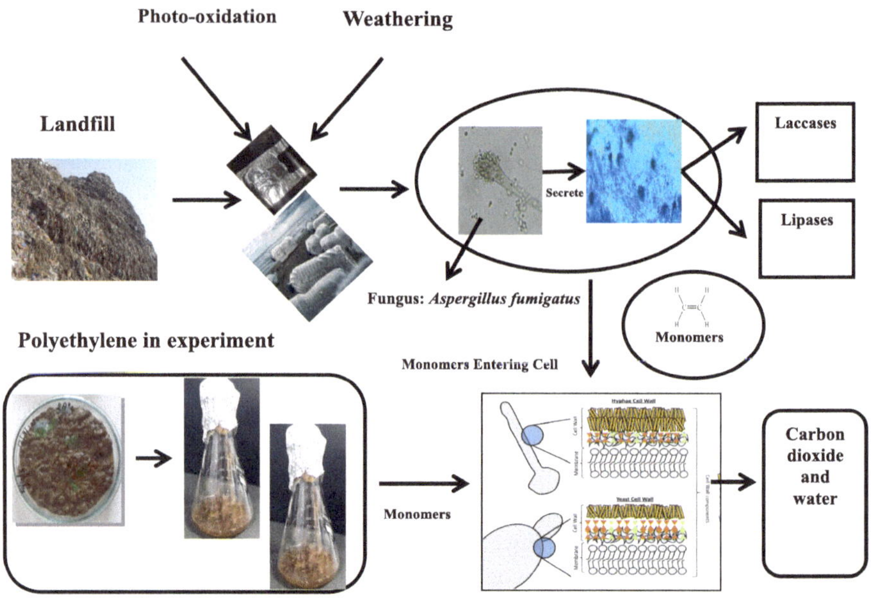

Fig. 5.12 Biological Mechanism of Plastic Degradation by Fungi

these polythene bags. When ingested by the animals, the polythene causes serious damage to the stomach and sometimes it leads to the death of the animal. The usage of polyethylene is increasing day by day in the form of polyethylene bags which is estimated to 500 billion to one trillion annually all around the globe (Madhu et al., 2014).

Polyethylene is hydrophobic in nature microorganisms do not eat the plastic and they do not have any mechanism to digest the plastic easily. Microbes used plastic as a substrate for their growth. Polyethylene indicates degradation through cracking, erosion, discoloration, and phase separation (Trivedi et al., 2016).

In previous studies and research (Table 5.1), the microbes have been isolated from the natural environment that have the ability to degrade polyethylene. Polyethylene and polypropylene were used as substrate for degradation (Park & Kim,). Microbial species such as *Streptococcus, Klebsiella, Micrococcus, Staphyloccocus*, and *Pseudomonas* were identified as the degrading microbes of polyethylene (Das & Kumar, 2015). Apart from the bacteria fungi also have the ability to degrade polyethylene. Fungal species like *Aspergillus oryza, Aspergillus nomius,* and *Aspergillus flavus* have the ability to degrade polyethylene. (Muhnoja et al., 2018, Taghavi et al., 2021). The additives which were used for the degradation of polyethylene oxidation were enhanced by which the molecular weight of polyethylene was reduced this was the reason, the microbes can easily degrade the polyethylene (Gamage and Senevirathna, 2020; Skariyachan et al., 2017).

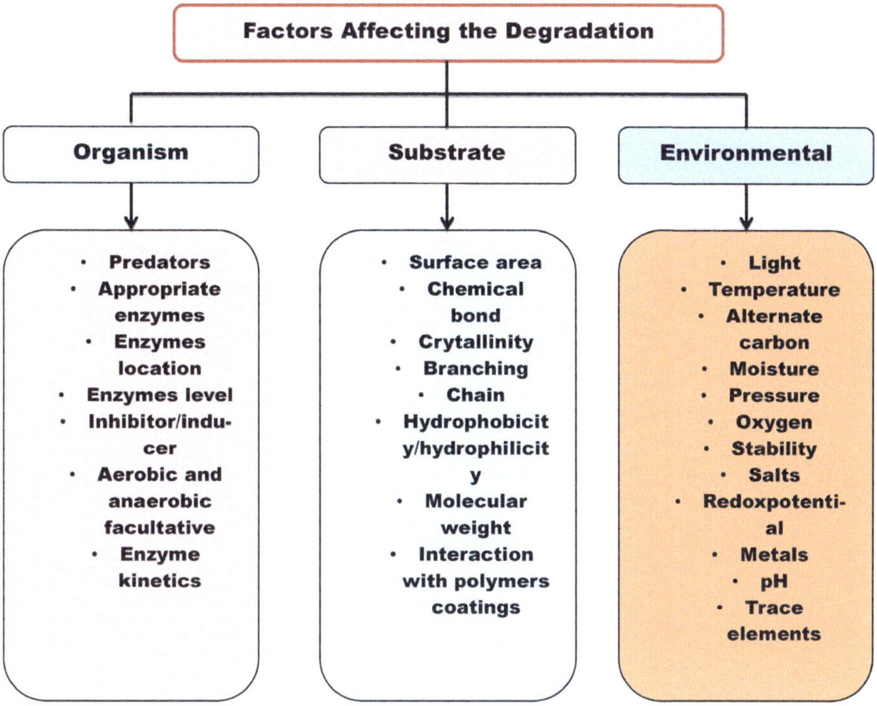

Fig. 5.13 Factors affecting the bio-degradation process (Adapted and modified from Gu, 2021)

5.5 Conclusions

The usage of plastics is spreading worldwide and it pollutes almost every ecosystem, hence, strict laws and action plan is required to restrict, reduce, recycle and re-use the plastic and plastic based products. Plastic waste is around 10% in municipal waste and 85% in marine debris. The five heaviest plastic polluter countries are China, Indonesia, the Philippines, Vietnam, and Sri Lanka. These five countries contribute at least 56% of total plastic waste in the world. Plastics are petroleum-derived polymer which is used for various purposes. Polythene bags are used all over the world. The degradation of polythene or plastic is a major concern because plastic degradation in the natural environment takes almost hundreds of years. Other than physical and chemical methods, microbial degradation of plastic is the need for sustainability of environment because these methods are tricky, time-consuming, and hazardous to the environment. The microbes have the potential to degrade polyethylene. We have isolated 10 species of Aspergillus that can degrade the polythene under laboratory conditions just within 30 days and the products of degradation may be used as biofertilizers for flowering plants, however, more researches are required to purify the enzymes involved during the degradation of plastic and polythene.

Table 5.1 Polymer degradation by microbes in previous studies

S. no	Microorganisms	Enzymes	Plastic type	Degradation time	Degradation %	Refrences
1	*Pseudomonas florescens*	Lipases esterases	PE	270	18.0	Thomas et al., (2015)
2	*Bacillus vallismoris*	Lipases	HDPE	120	75.0	Skariyachan et al., (2017)
3	*Klebsiella pneumonia*	Laccase	HDPE	60	18.4	Awasthi et al., (2017)
4	*Bacillus cereus*	Lipases	LDPE	112	36.4	Muhnoja et al. (2018)
5	*Trichoderma viridae*	Licases and peroxidases	LDPE	112	35.7	Muhnoja et al. (2018)
6	*Aspergillus nomius*	Magnese peroxidases	LDPE	45	5.13	Munir et al. (2018)
7	*Bacillus sp. & Paenibacillus*	Lipases	PE	45	6.63	Munir et al. (2018)
8	*Aspergillus flavus*	Lipases	HDPE	60	14.7	Taghavi et al., (2021)
9	*Bacillus siamensis*	Lipases	LDPE	100	5.5	Maroof et al., (2021)
10	*Fuzarium*	Cutinases	LDPE	60	8.46	Shimo (2015)
11	*Penicillium*	Lipases	LDPE	45	20.6	Tokiwa et al. (2009)
12	*Aspergillus fumigatus*	Polyesterases, esterases, lipases and hydrolases	LDPE	45	45.25	Ateeq and Garg (2023)

Acknowledgements The authors are thankful to Shobhit Institute of Engineering & Technology, Meerut for facilities.

Conflict of Interest The authors declare no conflict of interest.

References

Allen, N. S., Edge, M., Mourelatou, D., Wilkinson, A., Liauw, C. M., Parellada, M. D., Barrio, J. A., & Santa Quiteria, V. R. (2003). Influence of ozone on styrene-ethylene-butylene-styrene (SEBS) copolymer. *Polymer Degradation and Stability, 79*(2), 297–307.

Alshehrei, F. (2017). Biodegradation of synthetic plastic by microorganisms. *Journal of Applied and Environmental Microbiology, 5*(1), 8–19.

Andardy, A. L., Hamid, S. H., Hu, X., & Torikai, A. (2011). Effect of increased solar ultraviolet radiation on materials. *Journal of Photochemistry and Photobiology b: Biology, 46*(1–3), 96–103.

Andrady, A. L. (2011). Microplastics in the marine environment. *Marine Pollution Bulletin, 62*(8), 1596–1605.

Ateeq, A., & Garg, A. P. (2023). Biological degradation of low density polyethylene (LDPE). *Asian Journal of Microbiology, Biotechnology & Environmental Sciences, 25*(4), 49–53.

Ateeq, A., Neha, B., & Garg, A. P. (2023). *Plastic waste: Current Environmental Pollution, Health Hazard and Bio-degradation Strategies and its Management*. IIP Book Series (in press).

Awasthi, S., Srivastava, N., Singh, T., Tiwary, D., & Mishra, P. K. (2017). Biodegradation of thermally treated low density polyethylene by fungus *Rhizopus oryzae* NS 5. *3 Biotech, 7*(1), 73.

Azeko, S. T., Etuk-Udo, G. A., Odusanya, O. S., Malatesta, K., Anuku, N., & Soboyejo, W. O. (2015). Biodegradation of linear low density polyethylene by Serratiamarcescens subsp. Marcescens and its cell free extract. *Journal of Waste and Biomass Valorization, 6*(6), 1047–1057.

Browne, A. M., Galloway, T. S., & Thompson, R. (2007). Microplastic—An emerging contaminant of potential concern. *Integrated Environmental Assessment and Management, 3*(4), 559–561.

Cai, Z., Li, M., Zhu, Z., Wang, X., Huang, Y., Li, T., ... & Yan, M. (2023). Biological degradation of plastics and microplastics: A recent perspective on associated mechanisms and influencing factors. *Microorganisms, 11*(7), 1661.

Črešnar, B., & Petrič, Š. (2011). Cytochrome P450 enzymes in the fungal kingdom. *Bbaproteins Proteom, 1814*, 29–35.

Danzo, D., Chow, J., & Streit, W. R. (2018). Plastic: Environment and biotechnological perspectives on microbial degradation. *Applied and Environmental Microbiology, 85*(19), e01095.

Das, M. P., & Kumar, S. (2015). An approach to low-density polyethylene biodegradation by *Bacillus amyloliquefaciens*. *3 Biotech, 5*(1), 81–86.

Devi Sangeetha, R., Kannan, V. R., Nivas, D., Kannan, K., Chandru, S., & Antony R. A. (2015). Biodegradation of HDPE by *Aspergillus spp.* from marine ecosystem of gulf of Mannar, India. *Marine Pollution Bulletin*

Frazer, A. C. (1994). O-methylation and other transformations of aromatic compounds by acetogenic bacteria. In H. L. Drake (Ed.), *Acetogenes* (p. 1994). Chapman & Hall.

Gamage, T., & Senevirathna, J. D. M. (2020). Plastic pollution in the marine environment. *Heliyon, 6*(8), e04709. https://doi.org/10.1016/j.heliyon.2020.e04709.

Garg, A. P. (2023, January 3–7). What do we need to do for sustainability of environment and development" Delivered Presidential Address in 108th Indian Science Congress on "Proceedings of Section Environmental Sciences, The Indian Science Congress Association, held at RTM University, Nagpur (pp. 5–17).

Grifn, G. J. L. (1980). Synthetic polymers and the living environment. *Pure and Applied Chemistry, 52*, 399–407.

Gu, J. D. (2003a). Microbiological deterioration and degradation of synthetic polymeric materials: Recent research adances. *International Biodeterioration & Biodegradation, 52*(2), 69–91.

Gu, J.-D. (2003b). Microbiological deterioration and degradation of synthetic polymeric materials: recent research advances. *Journal of International Biodeterioration & Biodegradation, 52*, 69–91.

Gu, J. D. (2021). Biodegradability of plastics: the issues, recent advances, and future perspectives. *Environmental Science and Pollution Research, 28*(2), 1278–1282.

Inoue, T., Miyazaki, M., Kamithani, M., Kano, J., & Saito, F. (2004). Machenochemicaldechlorination of polyvinyl chloride by co-grinding with various metal oxides. *Advanced Powder Technology, 15*(2), 215–225.

Isangedighi, A. I., David, G. S., & Obot, O. I. (2018). Plastic waste in the aquatic environment: Impact and management. *Journal of Environment, 2*, 1–31.

Jain, R., Kosta, S., & Tiwari, A. (2010). Polyhydroxyalkonates: A way to sustainable development of bioplastics. *Chonicles of Young Scientist, 1*(3), 10.

Jaiswal, S., Sharma, B., & Shukla, P. (2020). Integrated approaches in microbial degradation of plastics. *Environmental Technology and Innovation, 17*, 100567.

Jeon, H., Durairaj, P., Lee, D., Ahsan, M. M., & Yun, H. (2016). Improved NADPH regeneration for fungal cytochrome P450 monooxygenase by co-expressing bacterial glucose dehydrogenase in resting-cell biotransformation of recombinant yeast. *Microbiology and Biotechnology Journal, 26*, 2076–2086.

Kale, S. K., Deshmukh, A. G., Dudhare, M. S., & Patil, V. B. (2015). Microbial degradation of plastic: A review. *Journal of Biochemical Technology, 6*(2), 952–961.

Karich, A., Ullrich, R., Scheibner, K., & Hofrichter, M. (2017). Fungal unspecifc peroxygenases oxidize the majority of organic EPA priority pollutants. *Frontiers in Microbiology, 8*, 1463.

Krzan, A., Hemjinda, S., Miertus, S., Corti, A., & Chiellini, E. (2006). Standardization and certification in the area of environmentally degradable plastics. *Polymer Degradation and Stability, 91*(12), 2819–2833.

Lima, A. R. A., Barletta, M., Costa, M. F., Ramos, J. A. A., Dantas, D. V., Melo, P. A. M. C., Justino, A. K. S., & Ferrierab, G. V. B. (2014). Changes in the composition of ichthyoplankton assemblage and plastic debris in mangrove creeks relative to moon phases. *Journal of Fish Biology, 89*(1), 619–640.

Loganathan, Y., & Kizhakedathil, M. P. J. (2023). A review on microplastics-an indelible ubiquitous pollutant. *Biointerface Research in Applied Chemistry, 13*(2), 126.

Lomonaco, T., Manco, E., Corti, A., La Nasa, J., Ghimenti, S., Biagini, D., Di Francesco, F., Modugno, F., Ceccarini, A., Fuoco, R., & Castelvetro, V. (2020). Release of harmful volatile organic compounds (VOCs) from photo-degraded plastic debris: A neglected source of environmental pollution. *Journal of Hazardous Matter, 394*, 122596.

Madhu, G., Bhunia, H., & Bajpai, P. K. (2014). Investigation of physico-mechanical properties and biodegradability of high density polyethylene/polylactide blends. *Journal of Polymer Materials, 31*(4), 381.

Marichelvam, M. K., Jawaid, M., & Asim, M. (2019). Corn rice starch-based bioplastic as alternative packaging materials. *Journal of Fibers, 7*(4), 32.

Maroof, L., Khan, I., Yoo, H. S., Kim, S., Park, H., Ahmed, B., & Azam, S. (2021). Identification and characterization of low density polyethylene- degradation bacteria isolated from soils of waste disposal sites. *Environmental Engineering Research, 26*(3), 200167.

Meereboer, K. W., Misra, M., & Mohanty, A. K. (2020). Review of recent advances in the biodegradability of polyhydroxyalkonoate (PHA) bioplastics and their composites. *Green Chemistry., 22*, 5519.

Mohanan, N., Montazer, Z., Sharma, P., & Levin, D. B. (2020). Microbial and enzymatic degradation of synthetic plastics. *Frontiers in Microbiology, 11*, 580709.

Monika, C., & Garg, A. P. (2023). Global impact of plastic pollution and its management for sustainable development (inpress).

Montazer, Z., Habibi Najaf, M. B., & Levin, D. B. (2019). Microbial degradation of lowdensity polyethylene and synthesis of polyhydroxyalkanoate polymers. *Canadian Journal of Microbiology, 65*, 1–11. https://doi.org/10.1139/cjm-2018-033.

Montazer, Z., Habibi Najafi, M. B., & Levin, D. B. (2020). Challenges with verifying microbial degradation of polyethylene. *Polymers, 12*(1), 123. https://doi.org/10.3390/polym12010123.

Moore, C. J. (2008). Synthetic polymers in the marine environment: A rapidly increasing, long-term threat. *Environmental Research, 108*(2), 131–139.

Muhnoja, C. N., Makonde, H., Magoma, G., & Lmbuga, M. (2018). Biodegradability of polyethylene by bacteria and fungi from Dandora dumpsite Nairobi-Kenya. *PLoS ONE, 13*(7), e0198446.

Mukharjee, S., & Chatterjee, S. (2014). A comparative study of commercially available plastic carry bag biodegradation by microorganism isolated from hydrocarbon effluent enriched soil. *International Journal of Current Microbiology and Applied Science, 3*(5), 318–325.

Munir, E., Harefa, R. S. M., Priyani, N., & Suryantoi, D. (2018). Plastic degrading fungi *Trichodermaviride and Aspergillusnomius* isolated from local landfill soil in Medan. *IOP Conference Series: Earth and Environmental Science, 126*, 012145.
Muthukumar, T., Arvinthana, A., Dineshram, R., Venkatesan, R., & Doble, M. (2014). Biodegradation of starch blended high density polyethylene using marine bacteria associated with biofilm formation and its isolation characterization. *Microbial & Biochemical Technology, 6*, 3.
Olicón-Hernández, D. R., González-López, J., & Aranda, E. (2017). Overview on the biochemical potential of flamentous fungi to degrade pharmaceutical compounds. *Frontiers in Microbiology, 8*, 17.
Park, S. K., & Kim, C. G. (2019a). Biodegradation of micro-polyethylene particles by bacterial colonization of a mixed microbial consortium isolated from landfills site. *Chemosphere, 222*, 527–553.
Park, S. Y., & Kim, C. G. (2019b). Biodegradation of micro-polyethylene particles by bacterial colonization of a mixed consortium isolated from landfill site. *Chemosphere, 222*, 527–533.
Pathak, G., & Nichter, M. (2019). The anthropology of plastics: An Agenda for local studies of a global matter of concern. *Medical Anthropology Quarterly, 33*(3), 307–326.
Priyanka, N., & Archana, T. (2011). Biodegradability of polythene and plastic by help of microorganism: A way for Brighter future. *Journal of Environmental and Analytical Toxicology, 1*, 1000111.
Quecholac-Piña, X., Hernández-Berriel, M. D., Mañón-Salas, M. D., Espinosa-Valdemar, R. M., & Vázquez-Morillas, A. (2022). Degradation of plastics under Anaerobic conditions: A short Review. *Polymers, 12*, 109; 10.3390.
Raaman, N., Rajitha, N., Jayshree, A., & Jegadeesh, R. (2012). Biodegradation of plastic by Aspergillus spp. isolated from polythene polluted sites around Chennai. *Journal of Academic, Industrial Research, 1*(6), 2278–5213.
Ranby, B. (1989). Photodegradation and photooxidation of synthetic polymers. *Journal of Analytical and Applied Pyrolysis, 15*, 237–247.
Ratnasari, D. K., Nahil, M. A., & Williams, P. T. (2017). Catalytic pyrolysis of waste plastics using staged catalysis for production of gasoline range hydrocarbon oils. *Journal of Analytical and Applied Pyrolysis, 124*, 631–637.
Rehan, M., Miandad, R., Barakat, M. A., Ismail, I. M. I., Almeelbi, T., Gardy, J., ... & Nizami, A. S. (2017). Effect of zeolite catalysts on pyrolysis liquid oil. *International Biodeterioration & Biodegradation, 119*, 162–175.
Sánchez, C. (2009). Lignocellulosic residues: Biodegradation and bioconversion by fungi. *Biotechnology Advances, 27*, 85–194.
Schwartz, M., Perrot, T., Aubert, E., Dumarçay, S., Favier, F., Gérardin, P., Morel-Rouhier, M., Mulliert, G., Saiag, F., Didierjean, C., & Gelhaye, E. (2018). Molecular recognition of wood polyphenols by phase II detoxifcation enzymes of the white rot trametes versicolor. *Science and Reports, 8*, 8472.
Shimo, M. H. U. (2015). *Plastic Recycling in Bangladesh, What needs to be done?*
Shin, J., Kim, J. E., Lee, Y. W., & Son, H. (2018). Fungal cytochrome P450s and the P450 complement (CYPome) of fusarium graminearum. *Toxins, 10*, 112.
Siegenthaler, K.O., Kunkel, A., Skupin, G., & Yamamto, M. (2012). Ecoflex and ecovio: Biodegradable performance-enabling plastics. In *Advance in Polymer Science, Synthetic Biodegradable Polymer*.
Skariyachan, S., Setlur, A. S., Naik, S. Y., Naik, A. A., Usharani, M., & Vasist, K. S. (2017). Enhanced biodegradation of low and high density polyethylene by novel bacteria consortia formulated from plastic-contaminated cow dung under thermophilic conditions. *Environmental Science and Pollution Research, 24*, 8443–8457.
Šrejber, M., Navrátilová, V., Paloncýová, M., Bazgier, V., Berka, K., Anzenbacher, P., & Otyepka, M. (2018). Membrane-attached mammalian cytochromes P450: An overview of the membrane's efects on structure, drug binding, and interactions with redox partners. *Journal of Inorganic Biochemistry, 183*, 117–136.

Striegel, A. M. (2003). Influence of chain architecture on the machenochemical degradation of macromolecules. *Journal of Biochemical and Biophysical Methods, 56*(1–3), 117–139.

Taghavi, N., Singhal, N., Zhuang, W., & Baroutian, S. (2021). Degradation of plastic waste using stimulate and naturally occurring microbial strains. *Chemosphere, 263*, 127975.

Teare, D., Emmison, N., Ton-That, C., & Bradley, R. (2000). Cellular attachment to ultraviolet ozone modified polystyrene surfaces. *Langmuir, 16*(6), 2818–2824.

Teuten, E. L., Rowland, S. J., Galloway, T. S., & Thompson, R. C. (2007). Potential for plastics to transport hydrophobic contaminants. *Environmental Science & Technology, 41*(22), 7759–7764.

Thomas, B., Olanrewaju-Kehinde, D., Popoola, O., & James, E. (2015). Degradation of plastic and polythene materials by some selected microorganisms isolated from soil. *World Applied Sciences Journal, 33*(12), 1888–1891.

Thushari, G. G. N., & Senevirathna, J. D. M. (2020). Plastic pollution in the marine environment. *Heliyon, 6*(8).

Tokiwa, Y., Calabia, B. P., Ugwu, C. U., & Aiba, S. (2009). Biodegradability of plastics. *International Journal of Molecular Sciences, 10*(9), 3722–3742.

Trivedi, P., Hasan, A., Akhtar, S., Siddiqui, H. M., Sayeed, U., & Khan, M. (2016). Role of microbes in degradation of synthetic plastics and manufacture of bioplastics. *Journal of Chemical and Pharmaceutical Research, 8*(3), 211–216.

Vankatesh, S., & Vijayalakshmi, S. (2021). Microbial degradation of plastics: sustainable approach to tackling environmental threats facing big cities of the future. *Journal of King Saud University, 33*(3), 101362.

Varshney, S., Gupta, V., Yadav, A. N., Rahi, R. K., & Neelam, D. K. (2023). An overview on role of fungi in systematic plastic degradation. *Journal of Applied Biology and Biotechnology, 11*(3), 61–69.

Vivi, V. K., Franchetti, S. M., & Angelis, D. A. (2019) Biodegradation of PCL and PVC: *Chaetomiumglobosom* (ATCC 16021) activity *64*(1–7).

Wang, J., Zhao, X., Wu, A., Tang, Z., & Niu, L. (2020). Aggregation and stability of sulfate-modified polystyrene nanoplastics in synthetic and natural waters. *Environmental Pollution, 268*, 114240. https://doi.org/10.1016/j.envpol.2020.114240.

Zeenat Bashir, Z. B., Sheng LiLi, S. L., Annamma Anil, A. A., Arvind Lali, A. L., Minton, N. P., & Zhang Ying, Z. Y. (2019). *Engineering Geobacillus thermoglucosidasius for direct utilisation of holocellulose from wheat straw.*

Zeenat, Elahi, A., Bukhari, A.D., Shamim, S., & Rehyman, A. (2021). Plastic degradation by microbes: A sustainable approach. *Journal of King Saud University, 33*, 101538.

Chapter 6
Chemical Characterization of Blends of Recycled Polymethyl Methacrylate-Based Denture Base Resin

Divya Krishnamoorthi, V. Anbazhagan, and R. Ganesamoorthy

Abstract The current research was carried out to characterize the chemical properties of polymeric blends of recycled polymethyl methacrylate-based denture base resin used in dentistry. Polymeric blends of recycled denture base resins (r-PMMA) were fabricated at w/w concentrations of 10%, 20%, 30%, 40% and 50% with commercially available denture base resin. Processed disk-shaped specimens ($n = 10$) of unmodified commercially available denture base resin were considered as the control group (R_0). Similarly, experimental group specimens were fabricated using 5 polymeric blends (R_{10}, R_{20}, R_{30}, R_{40} and R_{50}). Representative cuboidal specimens (5 mm × 5 mm × 3 mm) of all 6 groups were fabricated for surface characterization by Field-emission scanning electron microscopy (FESEM) and compositional analysis was carried out using energy dispersive X-ray (EDX). The groups were studied by nuclear magnetic resonance (NMR) spectroscopy. The DC of the polymers (powdered form) was analyzed by Fourier transform infrared spectroscopy. One-way analysis of variance was used to compare the mean values of DC% among the groups followed by post hoc *Bonferroni* test. Results show that the addition of r-PMMA at the ratios of 10, 20, 30, 40, and 50% increased the weight % of C from 62.79 to 66.06%. The addition of r- PMMA improved the DC and the experimental group R_{30} showed the highest DC. Findings conclude that the increase of C % can be attributed to the addition of r-PMMA which due to polymerization has enhanced % of C. FTIR spectra revealed that no new compound was formed, eliminating the likelihood of contamination and adulteration due to mechanical recycling. The addition of r-PMMA improved the DC. The addition of r-PMMA improved the chemical properties of the Polymethyl methacrylate denture base acrylic resin could significantly impact the denture quality and durability of life for the denture wearer.

D. Krishnamoorthi (✉)
Department of Prosthodontics and Crown and Bridge, Vinayaka Mission's Sankarachariyar Dental College, A Constituent Institute of Vinayaka Mission's Research Foundation (DU), Salem, Tamil Nadu, India
e-mail: drdivyakrishna@vmsdc.edu.in

V. Anbazhagan · R. Ganesamoorthy
Vinayaka Mission's Kirupananda Variyar Arts and Science College, A Constituent Institute of Vinayaka Mission's Research Foundation (DU), Salem, Tamil Nadu, India

S. Suriyanarayanan et al. (eds.), *Combating Plastic Pollution in Terrestrial Environment*,
https://doi.org/10.1007/978-981-96-2343-3_6

Keywords Fourier transform infrared · Magnetic resonance imaging · Polymethyl methacrylate · Recycle · Spectroscopy

6.1 Introduction

Use of removable prosthesis for the rehabilitation of missing dentition and associated structures is a time-tested treatment protocol. As it's rightly compared, just as a sturdy building requires a solid foundation, creating durable and visually pleasing dentures necessitates an optimal denture base. With the progression of civilization there has been a leap in the materials available for dental professionals to choose from. The perfect dental material should be biocompatible, easily accessible, cost-effective, and simple to handle with precise technical processes. This is essential in creating prosthetics that function well and look appealing. Denture based materials have been a focal point in dental material studies (Phillips, 2005).

Glossary of Prosthodontic Terms, ed. 9, defines "DENTURE BASE MATERIAL as any substance of which a denture base may be made" and "DENTURE BASE as the part of a denture that rests on the foundation tissues and to which teeth are attached" (Phillips, 2005). Various materials have been tested as denture base materials such as wood, ivory, porcelain, rubbers and metals, but none have stood the test of time like Polymethyl methacrylate (PMMA) (Phillips, 2005). Otto Rohm produced a solid transparent sheet of acrylic acid as part of a Ph.D. thesis, along with Hass, he introduced Poly (methyl methacrylate) (PMMA) as transparent sheets. In 1937 Du Dout De Nemours introduced the powder form of PMMA and by 1946, approximately 95% of all dentures were fabricated using PMMA, A clinical evaluation of removable prosthesis was conducted by Wright and he found that PMMA fulfilled most requirements of ideal denture base material (Avinash et al., 2013; Eu, 2000; Thopegowda et al., 2013).

Vinyl acrylic, Nylon, Polyethers, Polystyrene, Polypropylene, Polycarbonate, Epoxy resins and various other high impact methacrylates were introduced and used as denture base material in 1950s and 1960s, but none would ever absolutely replace PMMA (Avinash et al., 2013; Eu, 2000; Thopegowda et al., 2013). Most ongoing research aims at improving the mechanical strength of PMMA-based denture base materials by incorporating various reinforcements/ modifications to the polymer and/ or monomer component. Even though it fulfills most of the aforementioned ideal requirements, PMMA's environmental impact cannot be overlooked.

A 2019 report of the European Commission's research and innovation quoted the global consumption of MMA as 4.2 M tonne/year approximately and global PMMA waste to be approximately 800,000–1,000,000 tonne/year (20–25% MMA consumption). European consumption alone was around 1Mtonne/year. The highlight of the report was mentioned that 200,000 tonne/year waste could be recovered by creating an economy of 1400 M€ and 1700 new jobs (Eu, 2000).

It is forecast that 100% increase in PMMA consumption by 2030 is expected resulting in catastrophic waste problems globally. China the major importer of

PMMA-based waste has imposed a ban on its imports on 01.01.2018. This may affect the developing countries more, so the development of a process to recover the monomer Methyl methacrylate (MMA) is of paramount importance (Eu, 2000).

It was noted by the "Eco-profiles and Environmental Product Declarations of the European Plastics Manufacturers", that around 116 MJ of gross energy/ primary fuel,76 l of water and more than 20 different raw materials are required to produce 1 kg beads of PMMA. In addition, considerable quantity and number of harmful by-products like carbon monoxide, carbon dioxide, methane, sulfur dioxide and mercaptan products are emitted into the environment. The gross solid waste pollutants were chalked upto 140.765 g for the production of 1 kg PMMA beads (Charmondusit et al., 2006). The depletion of natural resources to produce the virgin material for use leaves a long trail of carbon footprint (Eu, 2000). It has become a necessity to become more environmentally conscious and move towards eco-friendly ways in dentistry (Eu, 2000).

The present study characterizes denture base processed by using blends of recycled PMMA (r-PMMA) at 10, 20, 30, 40 and 50% (w/w) concentration using FESEM (Field emission scanning electron microscopy), and EDAX (Energy Dispersive Spectroscopy), NMR spectroscopy (Nuclear Magnetic Resonance spectrometry) and FTIR (Fourier transform infrared Spectroscopy).

6.1.1 Materials and Methods

The testing was carried out at Central Instrumentation Centre, Madurai Kamaraj University, Madurai, Tamil Nadu, India. Heat-activated denture base acrylic resin (DPI® Heat Cure; DentalProducts of India, Mumbai, India) was modified by adding r-PMMA at 10, 20, 30, 40 and 50% (w/w) concentration (R_{10}, R_{20}, R_{30}, R_{40} and R_{50}). Unmodified heat cure denture base resin (DPI® Heat Cure; DentalProducts of India, Mumbai, India) served as the control group (R_0) (Ashton et al., 2016; Charmondusit et al., 2006; Krishnamoorthi et al., 2022; Krishnamoorthi et al., 2022; Makenji & Savage, 2012; Maris et al., 2018; Suresh et al., 2017, 2018). Triple blinding was followed to avoid expectation bias (Fig. 6.1).

6.1.1.1 FESEM and EDAX

A single cuboidal sample (measuring 5 mm $\times$ 5 mm $\times$ 3 mm) was produced for each group. The surface morphology of these specimens was analyzed using a field-emission scanning electron microscope (TESCAN VEGA3 SBH, manufactured by Tescan Orsay Holding, a.s., Czech Republic) (Gaabour, 2021). This equipment captured high-resolution images. The unfinished cuboidal samples were positioned on labeled specimen mounts, maintaining a beam-to-specimen surface distance of 10 mm. The surfaces were scanned at 10 kV with a magnification of 3,000 $\times$ magnification (Fig. 6.2).

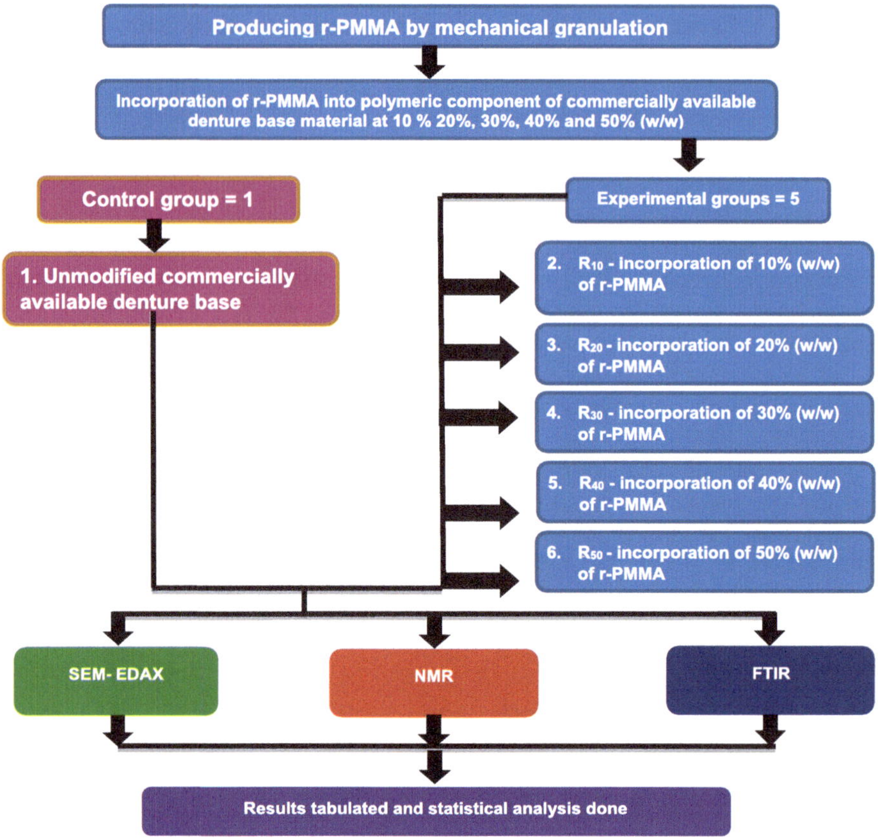

Fig. 6.1 Flow chart of the study methodology

The chemical composition of the specimens was determined using EDAX spectroscopy (utilizing TESCAN's Essence™ software). X-ray incident photons interacted with the electrons of atoms, and the resulting distribution of impacted particles was detected through EDAX (Gaabour, 2021).

6.1.1.2 NMR Spectroscopy

Disk-shaped specimens measuring 25 mm × 2 mm (n = 10) were created for each group as per the manufacturer's guidelines (Table 6.1) and utilizing stainless steel dies of matching dimensions to create the necessary mold space. For analysis purposes, the polymerized disk specimens from all groups were ground into fine powder particles using a clean tungsten carbide bur (Ajay et al., 2020). The resulting powder underwent sieving to ensure uniform particle size. Utilizing a digital NMR spectrometer (JNM-ECZS NMR spectrometer, Jeol Ltd., Japan), both ^{1}H- and ^{13}C-NMR spectra were

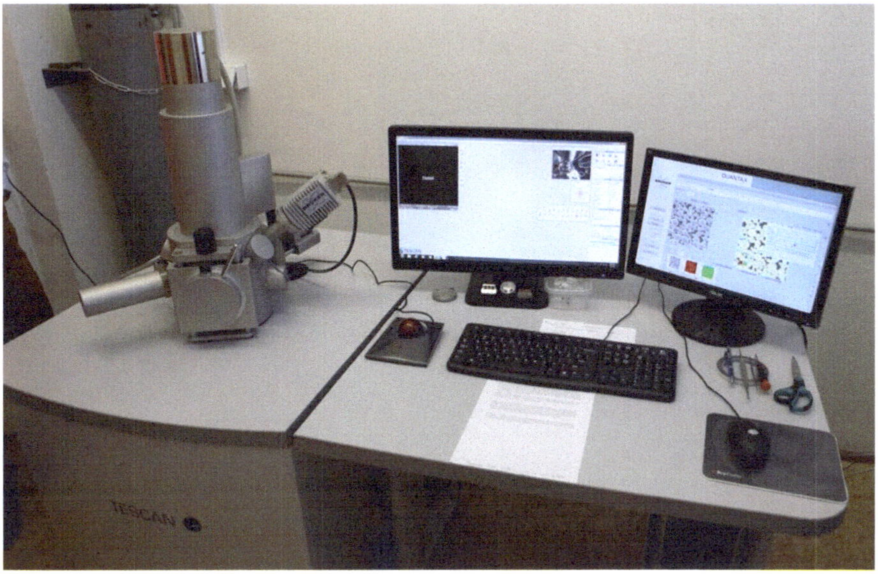

Fig. 6.2 Scanning electron microscope

Table 6.1 Summary of all the experiments along with the sample size and dimension for each

Experiment	Sample size/group	Total no. of specimens tested	Dimension
SEM-EDAX	1	6	5 mm × 5 mm × 3 mm
FTIR	10	70	Powdered specimen
NMR	10	60	Powdered specimen

obtained for the groups (Fig. 6.3). For ^{1}H-NMR, 25 mg of powdered resin specimen was used, while 50 mg was employed for ^{13}C-NMR, each placed in a thin glass tube. Deuterated chloroform ($CDCl_3$), measuring one mL, was added to the tube and thoroughly shaken to completely dissolve the powdered resin. Tetramethylsilane served as the preferred internal standard (Ajay et al., 2020).

6.1.1.3 FTIR

To conduct the analyses, the polymerized disk specimens fabricated as mentioned above of all groups were ground into fine powder particles using a clean tungsten carbide bur. A quantity of 2 mg of specimen powder was then combined with 100 mg of dry potassium bromide (KBr) powder. This mixture was compressed into a 0.5 mm thick disk using a specialized mold and hydraulic pellet press. Placing this pellet vertically in the specimen compartment, an infrared beam was transmitted through it. Specimens of uncured denture base resin were prepared by heating at 70°C for

Fig. 6.3 NMR spectrometer

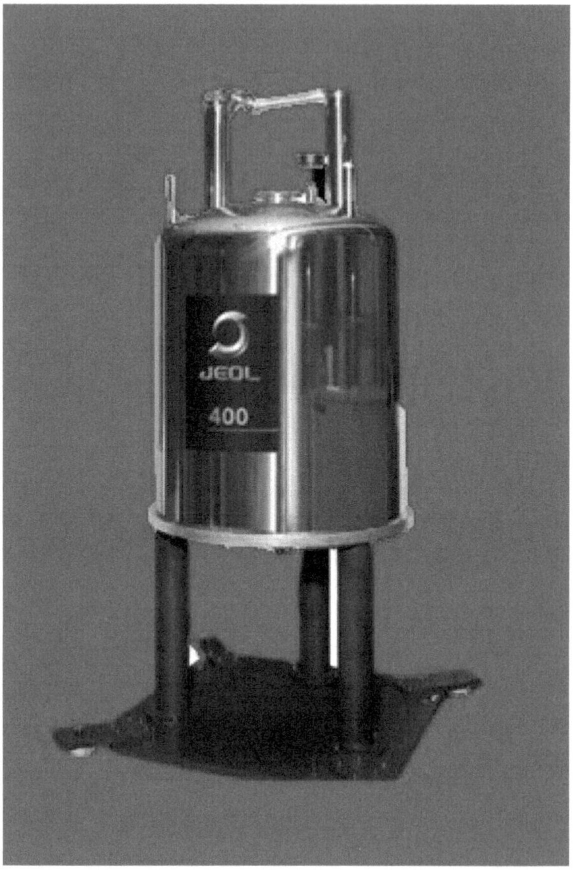

30 h, causing the initiator, benzoyl peroxide, to degrade. This led to the degradation of the initiator, benzoyl peroxide. Once the monomer was mixed with this treated powder it eschewed the polymerization reactions producing a resin paste. This paste was placed between two polyethylene films, compressed to form an extremely thin film, and subjected to the infrared beam (Ajay et al., 2019).

FTIR analysis (IRSpirit ™, Shimadzu, Kyoto, Japan) was used to study the Degree of conversion of monomer to polymer (Fig. 6.4). The analysis was conducted with a resolution of 4 cm^{-1} and 16 scans in absorbance (Abs) mode. The degree of conversion (DC) was assessed by comparing the Abs intensities of aliphatic C=C and carbonyl bond C=O at wavenumbers 1,638 and 1,720 cm^{-1}, respectively. These specific wavenumbers serve as common reference points for assessing the polymerized material's internal standards, applicable to both cured and uncured materials. The DC was computed using the following equation (Ajay et al., 2019):

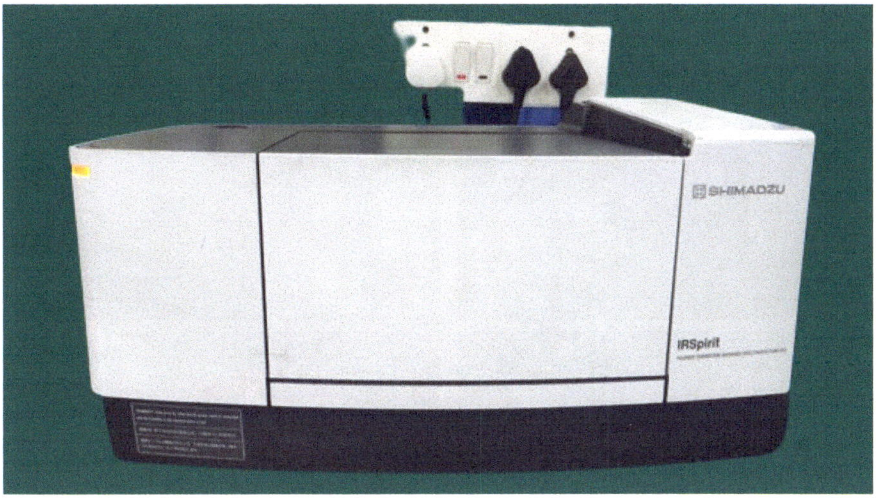

Fig. 6.4 FTIR spectrometer

$$DC\% = \left[1 - \frac{\left(\frac{Abs\,1559}{Abs\,1721}\right) Cured}{\left(\frac{Abs\,1559}{Abs\,1721}\right) Uncured}\right] \times 100 \quad \text{(Ajay et al.2019)}$$

where DC is Degree of conversion.

Abs is Absorbance intensity at given wavelength.

To minimize bias, all polymeric specimens (whether in powder or solid form) were prepared by a single investigator.

6.2 Statistical Analysis

The mean and standard deviation (SD) were computed for each group. Data analysis involved utilizing SPSS (Statistical Package for Social Sciences) Version 26.0 software for both descriptive and analytical statistics. A parametric One-Way ANOVA test was employed to examine variations in mean scores among the groups, followed by pairwise comparisons conducted using Bonferroni's post hoc test.

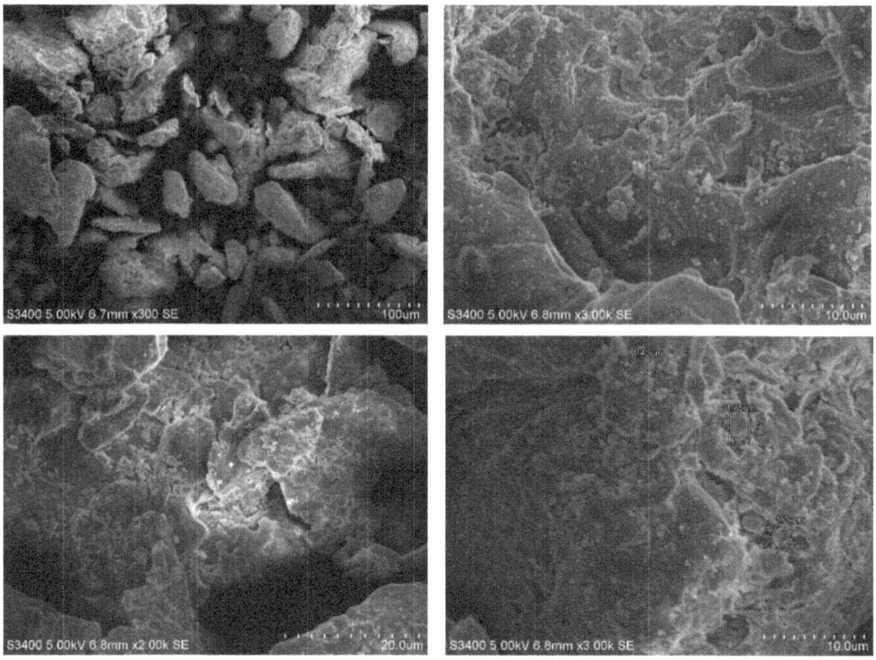

Fig. 6.5 SEM of r-PMMA

6.2.1 *Results*

6.2.1.1 FESEM and EDAX

The SEM analysis of r-PMMA revealed an amorphous complex structure (Fig. 6.5). The SEM analysis revealed that the unmodified commercially available polymeric powder (control group) revealed smooth spherical beads (Fig. 6.6).

The EDAX analysis showed the weight % of C and O as 60.53 and 38.60% respectively in the unmodified commercially available polymeric powder (control group) (Fig. 6.7). The modification by addition of r-PMMA at the ratios of 10, 20, 30, 40, and 50% resulted in an increase of the weight % of C from 62.79 to 66.06% (Figs. 6.8, 6.9, 6.10, 6.11 and 6.12).

6.2.1.2 FTIR

The Fourier-transform infrared spectroscopic analysis was performed to analyze the functional groups present in both control and experimental groups. Cured and uncured polymeric specimens were analyzed and spectrum were recorded from 4000 to 500 cm^{-1}. All the groups displayed several peaks.

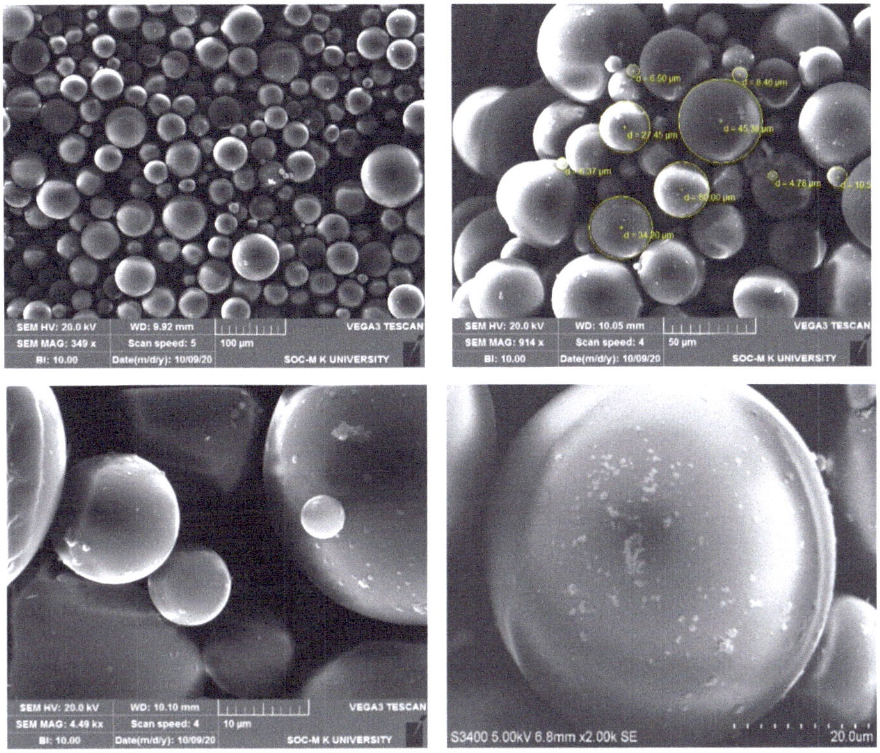

Fig. 6.6 SEM of control group

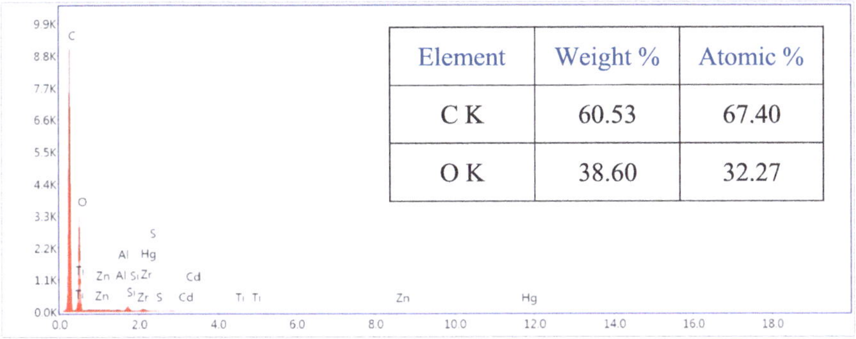

Element	Weight %	Atomic %
C K	60.53	67.40
O K	38.60	32.27

Fig. 6.7 EDAX of Control group

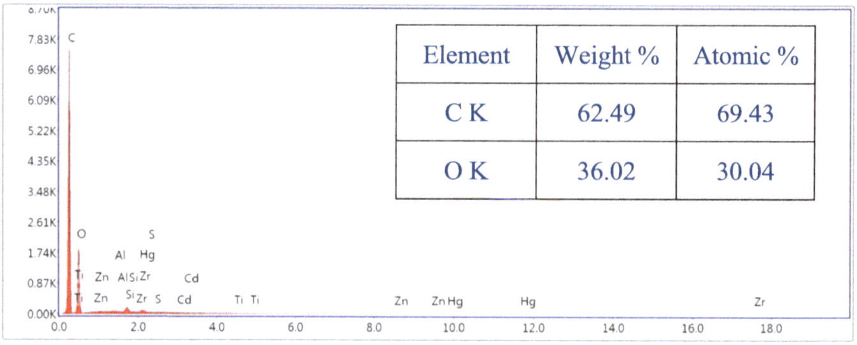

Element	Weight %	Atomic %
C K	62.49	69.43
O K	36.02	30.04

Fig. 6.8 EDAX of R_{10}

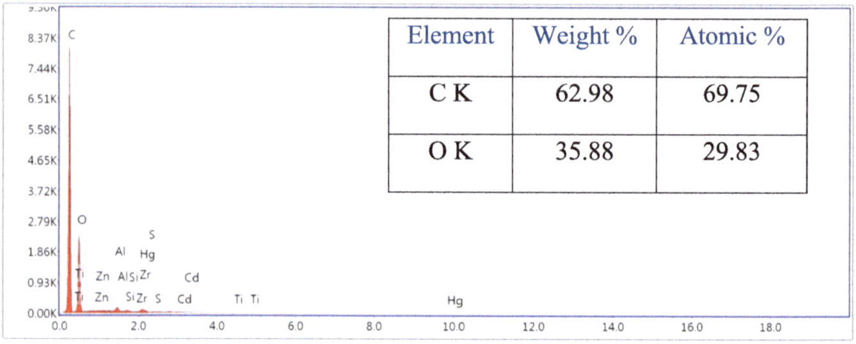

Element	Weight %	Atomic %
C K	62.98	69.75
O K	35.88	29.83

Fig. 6.9 EDAX of R_{20}

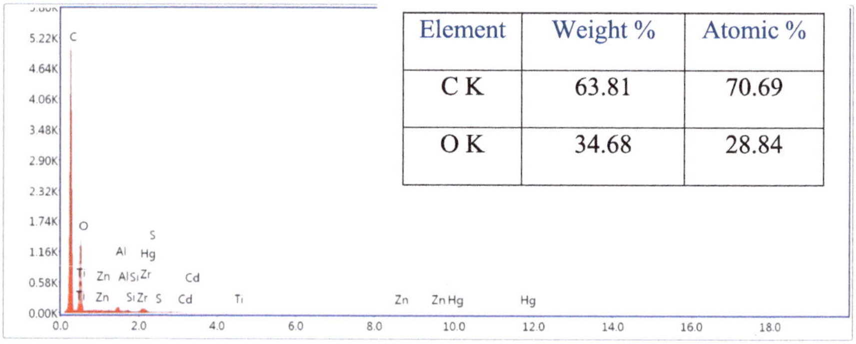

Element	Weight %	Atomic %
C K	63.81	70.69
O K	34.68	28.84

Fig. 6.10 EDAX of R_{30}

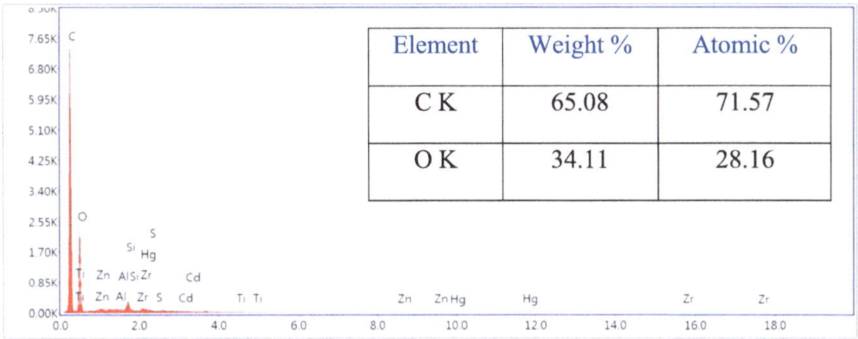

Element	Weight %	Atomic %
C K	65.08	71.57
O K	34.11	28.16

Fig. 6.11 EDAX of R_{40}

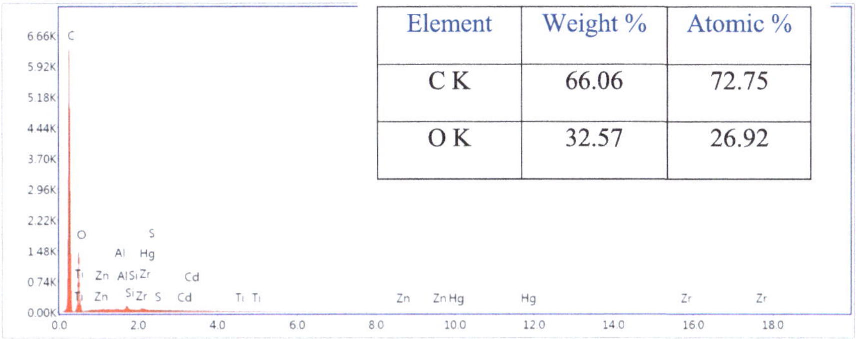

Element	Weight %	Atomic %
C K	66.06	72.75
O K	32.57	26.92

Fig. 6.12 EDAX of R_{50}

Figure 6.13 shows the FTIR spectral observations of unpolymerized polymeric component of the Control group R_0. The spectrum depicted a weak peak at 2995.02 cm^{-1} owing -CH_2 ester methylene group. At 2952.18 cm^{-1} the peak depicts the presence of -CH_3 stretching vibrations. The series of weak clustered peaks from 2192.91 to 1957.29 cm^{-1} can be attributed to the presence of the C≡N, from the initiator di-isobutylazonitrile ($((CH_3)_2C(CN)]_2N_2$). The strong and broad peak at 1721.83 cm^{-1} is owing to C=O ester carbonyl stretching. The moderate peak at 1434.12 cm^{-1} is attributed to C=C symmetrical alkenyl stretch. Peaks at 1385.10 and 1189.25 cm^{-1} indicate the presence of C–C symmetrical stretching. Peak at 1238.59 cm^{-1} depicts the presence of C-O ester band stretch. The strong and broad peak at 1140.38 cm^{-1} is owing to C–O–C asymmetrical stretch vibrations. The weak peak the 1061.74 cm^{-1} is owing to C=C asymmetrical alkenyl stretching. Moderate peaks at 985.42 and 837.94 cm^{-1} can be credited to C-H out of plane bending. The sharp and moderate peak at 748.56 cm^{-1} is owed to *cis*- C-H out of plane bending.

Figure 6.14 shows the FTIR spectral observations of polymerized specimen of the control group R_0. Multiple peaks were observed, these were compared and

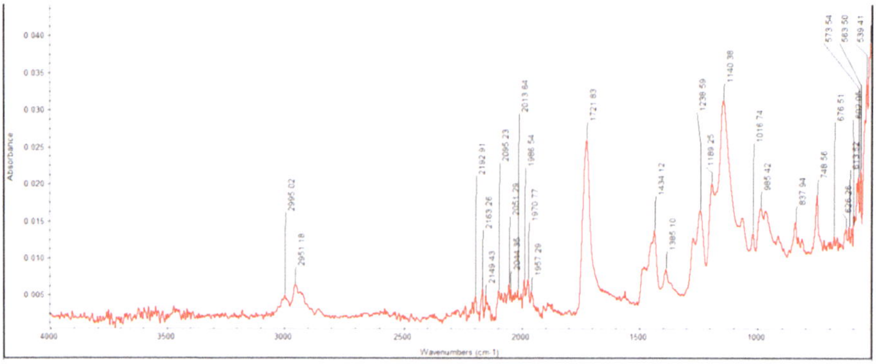

Fig. 6.13 FTIR spectrum of unpolymerized polymeric component

contrasted to the spectral observations of the unpolymerized polymeric component of the Control group FUR_0. The spectrum depicted the shift of the weak peak at 2995.02 cm^{-1}, owing to -CH_2 ester methylene group, to 2992.24 cm^{-1}. The shift of peak at 2952.18 cm^{-1} depicting the presence of -CH_3 stretching vibrations to 2948.49 cm^{-1} was also observed. The diminishing of the series of weak clustered peaks from 2192.91 to 1957.29 cm^{-1} attributed to the presence of the C≡N, from the initiator di-isobutylazonitrile ($((CH_3)_2C(CN)]_2N_2$) to weaker and fewer peaks at 2026.63 cm^{-1} was observed. The strong and broad peak at 1721.83 cm^{-1} owing to C=O ester carbonyl stretching had shifted to 1722.24 cm^{-1}. The moderate peak at 1434.12 cm^{-1} attributed to C=C symmetrical alkenyl stretch had shifted to 1433.81 cm^{-1}. Peak at 1385.10 cm^{-1} indicating the presence of C–C symmetrical stretching, shifted to 1387.08 cm^{-1}. Peak at 1238.59 cm^{-1} depicts the presence of C-O ester band stretch. The strong and broad peak at 1140.38 cm^{-1} owing to C–O–C asymmetrical stretch vibrations, shifted to 1142.73 cm^{-1} with increase in sharpness and intensity. The weak peak at 1061.74 cm^{-1} owing to C=C asymmetrical alkenyl stretching had further weakened. Moderate peaks at 985.42 and 837.94 cm^{-1} which can be credited to C-H out of plane bending have shifted to 984.07 and 839.29 cm^{-1} respectively.

Figures 6.15, 6.16, 6.17, 6.18 and 6.19 show the FTIR spectral observations of polymerized specimens of experimental group R_{10}–R_{50}. Multiple peaks were observed, these were compared and contrasted to the spectral observations of the previous groups. The peak at 2952.18 cm^{-1} depicting the presence of -CH_3 stretching vibrations which made a negative shift to 2990.20 cm^{-1}, was seen to diminish considerably from Experimental group R_{10}–R_{50} evocative of further lower degree of polymerization.

6.2.1.3 Degree of Conversion (Dc)

The FTIR spectra of all the tested uncured specimens showcase two absorbance (Abs) peaks at 1559 and 1721 cm^{-1}. This represented the unreacted C=C in the uncured

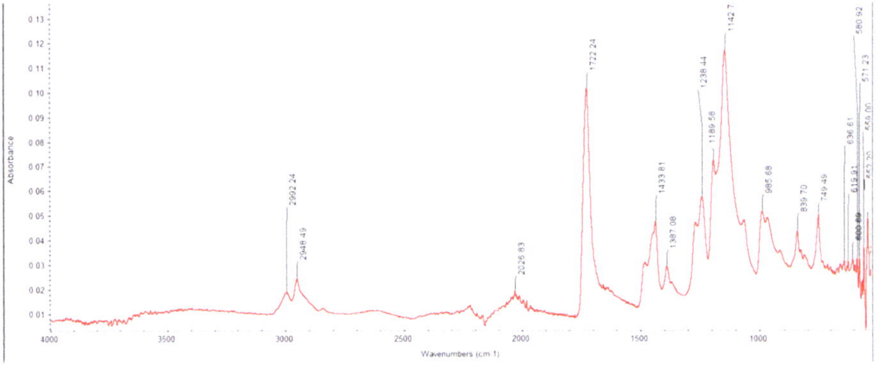

Fig. 6.14 FTIR spectrum of polymerized control group R_0

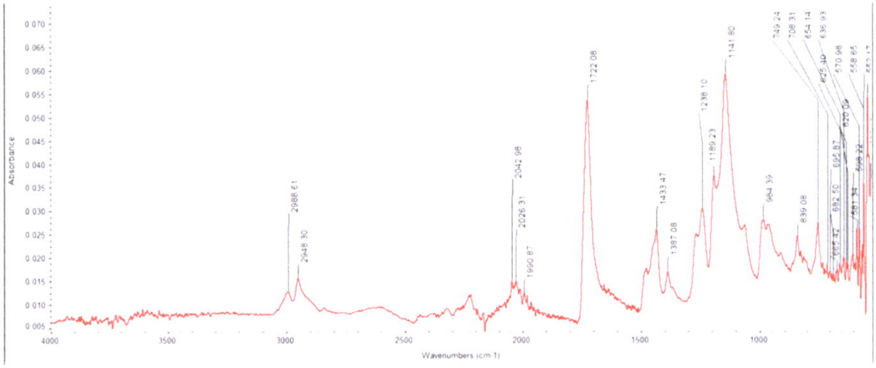

Fig. 6.15 FTIR spectrum of R_{10}

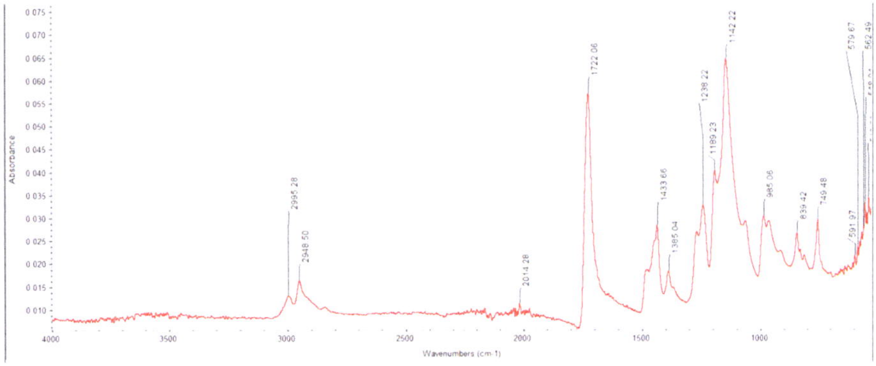

Fig. 6.16 FTIR spectrum of R_{20}

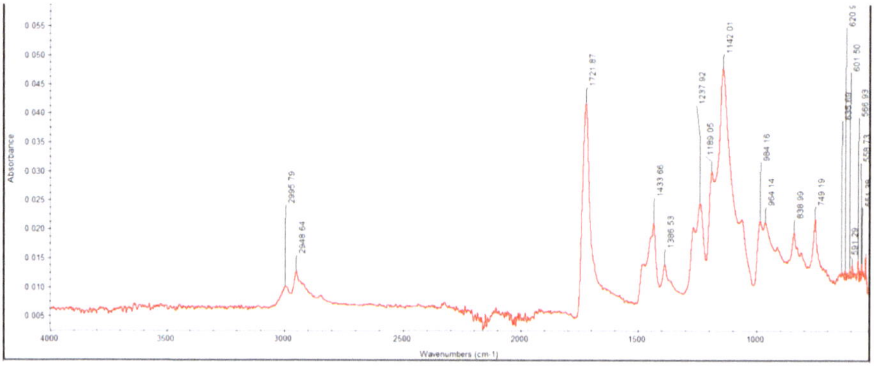

Fig. 6.17 FTIR spectrum of R_{30}

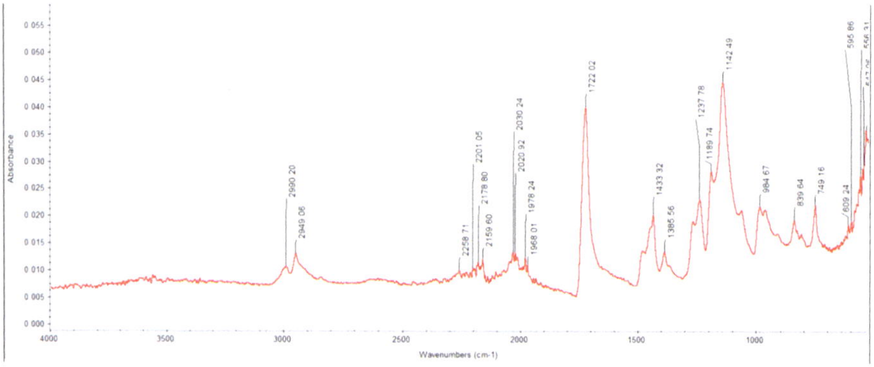

Fig. 6.18 FTIR spectrum of R_{40}

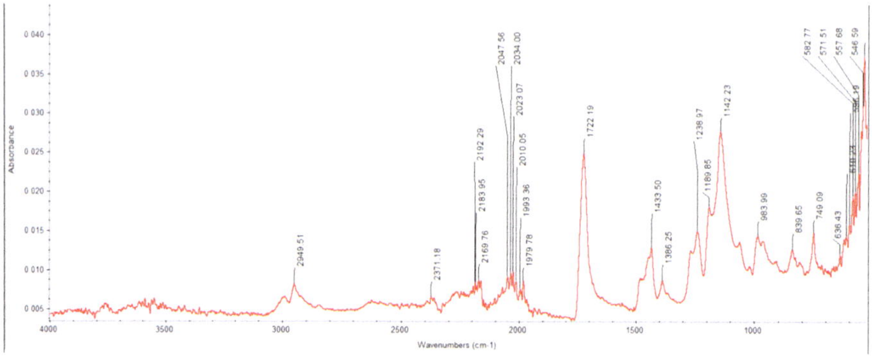

Fig. 6.19 FTIR spectrum of R_{50}

specimens. It was noted that the peak at 1559 cm^{-1} was weaker in the spectra of all 5 of the experimental group uncured specimens compared to the control group uncured specimen. This confirms the presence of higher degree of conversion of C=C to C–C due to the addition of the r-PMMA, which was already polymerized. This peak was seen to dramatically diminish in all the cured specimens. Table 6.2 represents the mean DC % and Standard deviation (SD) of control and experimental groups. One-way ANOVA showed statistically significant difference between the groups. Table 6.3 showed the Post hoc Scheffe test to compare the DC % between the groups. It was noted that there was a significant difference between control group and all experimental groups, as well as between R_{40} and R_{50}.

Table 6.2 One-way ANOVA of mean DC%

Group		Mean DC (%)	SD	F-Value	p value
Control group R_0	R_0	79.03	0.74	12.359	> 0.05
Experimental group R_{10}	R_{10}	81.01	0.83		
Experimental group R_{20}	R_{20}	81.38	0.71		
Experimental group R_{30}	R_{30}	81.51	0.73		
Experimental group R_{40}	R_{40}	81.27	0.91		
Experimental group R_{50}	R_{50}	80.52	1.02		

Table 6.3 Post hoc Scheffe tests of DC %

Group	Comparative group	Mean difference	Significance
R_0	R_{10}	−1.98*	0.00
	R_{20}	−2.35*	0.00
	R_{30}	−2.48*	0.00
	R_{40}	−2.24*	0.00
	R_{50}	−1.49*	0.01
R_{10}	R_{20}	−0.37	0.96
	R_{30}	−0.50	0.87
	R_{40}	−0.26	0.99
	R_{50}	0.49	0.88
R_{20}	R_{30}	−0.13	1.00
	R_{40}	0.11	1.00
	R_{50}	0.24	0.99
R_{30}	R_{40}	0.99	0.23
	R_{50}	0.75	0.55
R_{40}	R_{50}	−1.98*	0.00

6.2.1.4 NMR Spectroscopy

Nuclear magnetic resonance (NMR) spectroscopy was done to characterize and study the polymer structure–property relationships. The peak signals observed were assigned to specific atoms along the polymer backbone and side chains.

Figure 6.20a shows the peak signals of ^{1}H-NMR of the control group polymerized specimen (R_0). After polymerization, the disappearance of doublet and triplet splitting of the $-CH_2-$ (methylene group) between 1.741–1.925 ppm was observed. In addition, other peaks appeared to be broadened. The experimental groups (R_{10}, R_{20}, R_{30}, R_{40} and R_{50}) showcased similar patterns (Figs. 6.21a, 6.22a, 6.23a, 6.24a, and 6.25a). The slight variation in the peak signals is highlighted in the following Table 6.4.

Figure 6.20b shows the ^{13}C-NMR peak signals of the control group polymerized specimen (R_0). The doublet peaks between 18.641 and 16.351 ppm which were attributed to—CH_3 (methyl) carbon show a diminished intensity. Similar pattern was observed in the peak signals of ^{13}C-NMR of all the unpolymerized specimens of the 5 experimental groups (R_{10}, R_{20}, R_{30}, R_{40} and R_{50}) as depicted in Figs. 6.21b, 6.22b, 6.23b, 6.24b, and 6.25b. The slight variation in the peak signals is highlighted in Table 6.5.

6.2.2 Discussion

Denture base resins have proved themselves to be the front runners for ideal denture base materials. This product took dentistry by storm and by 1946, more than 90% of the dentures were fabricated using PMMA (Phillips, 2005).

This research aimed to address the problem of NON-BIODEGRADABLE DENTAL HEALTH CARE WASTE, in the form of discarded Heat cure denture base resin by attempting to adapt recycling treatments and methods used in the plastic industries to recycle PMMA-based denture base resin in-house, in the dental institutes, so as to be reused in dentistry. The research also aimed to evaluate the effect of substitution of this recycled denture base resin (r-PMMA) at the concentration of 10%, 20%, 30%, 40% and 50% (w/w) to commercially available heat cure denture base resin on the chemical characteristics of the cured denture base resin. The null hypothesis of the present research was that the incorporation of recycled heat cure denture base polymeric powder at various concentrations (10–50% (w/w)) would not affect the chemical characteristics, of PMMA-based heat cure denture base resin.

An extensive review of literature showed that initiatives and legislative regulations, have been implemented to keep dental waste production and management in check. But most of these regulations have been put in place with regard to disposal of DENTAL AMALGAM, of which MERCURY is a major component (Avinash et al., 2013; Chopra & Gupta, 2014; Mohelay et al., 2016). Attempts to include other metal alloys and even dental waxes in the recycling system to prolong its lifecycle in dental use have been noted (Avinash et al., 2013; Chopra & Gupta, 2014; Mohelay

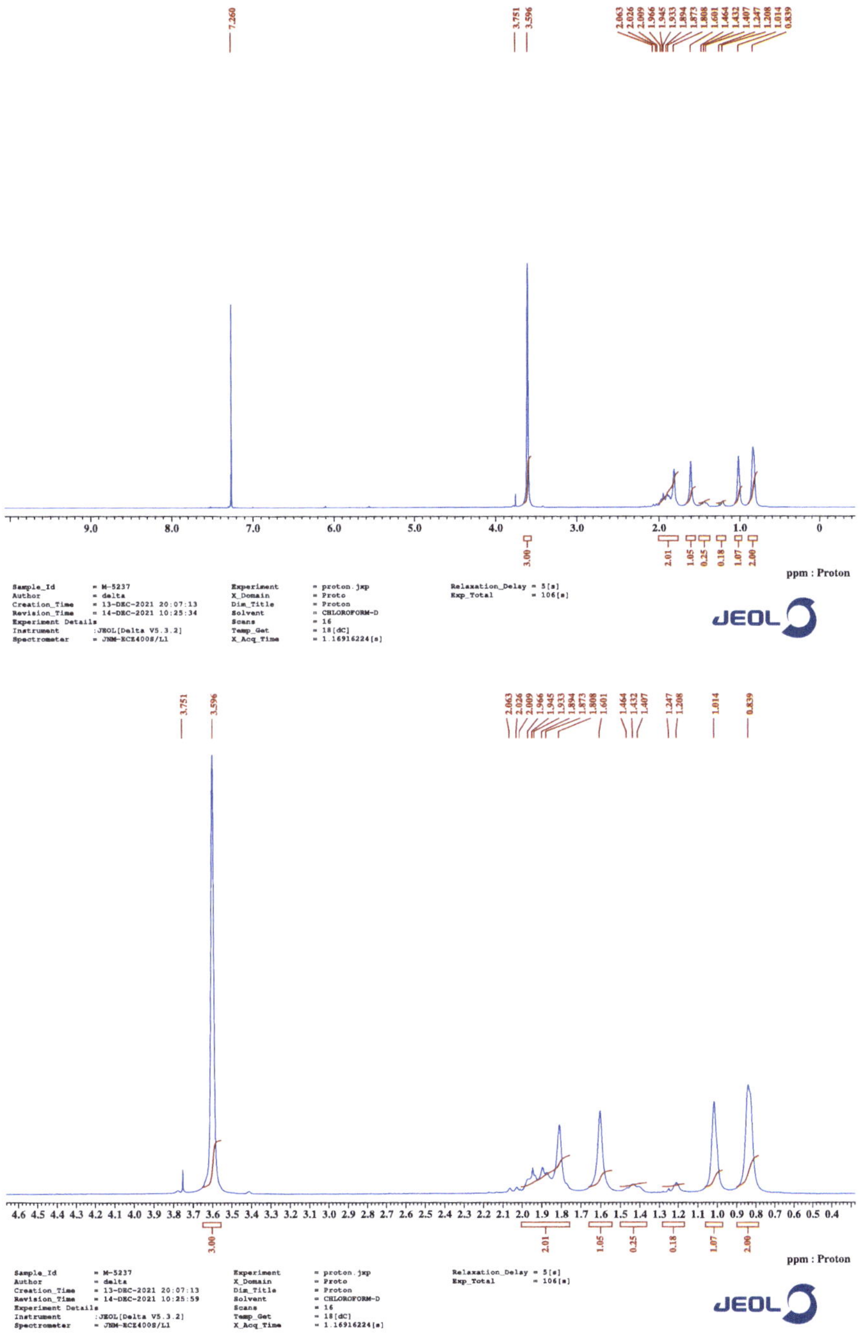

Fig. 6.20a ^1H-NMR spectrum of R_0

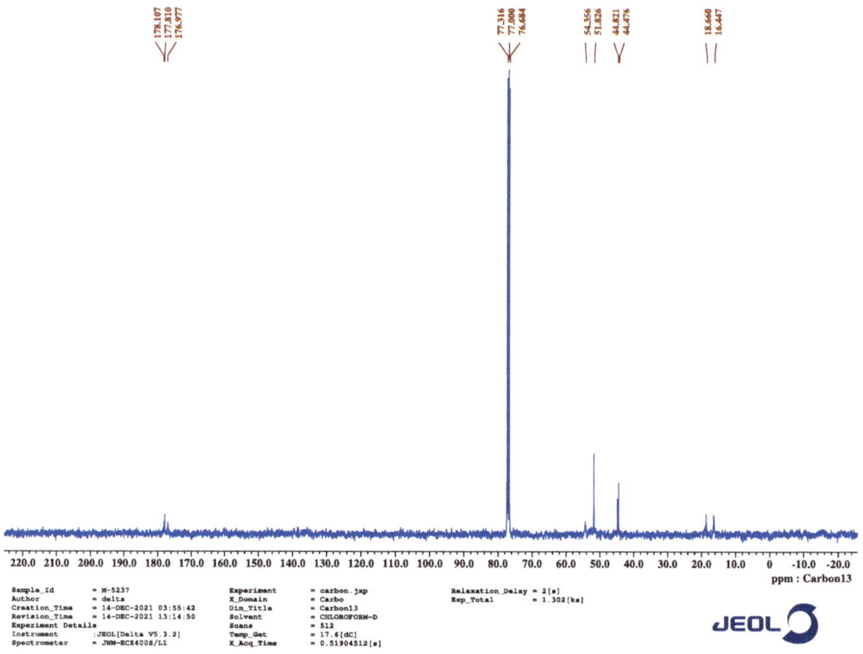

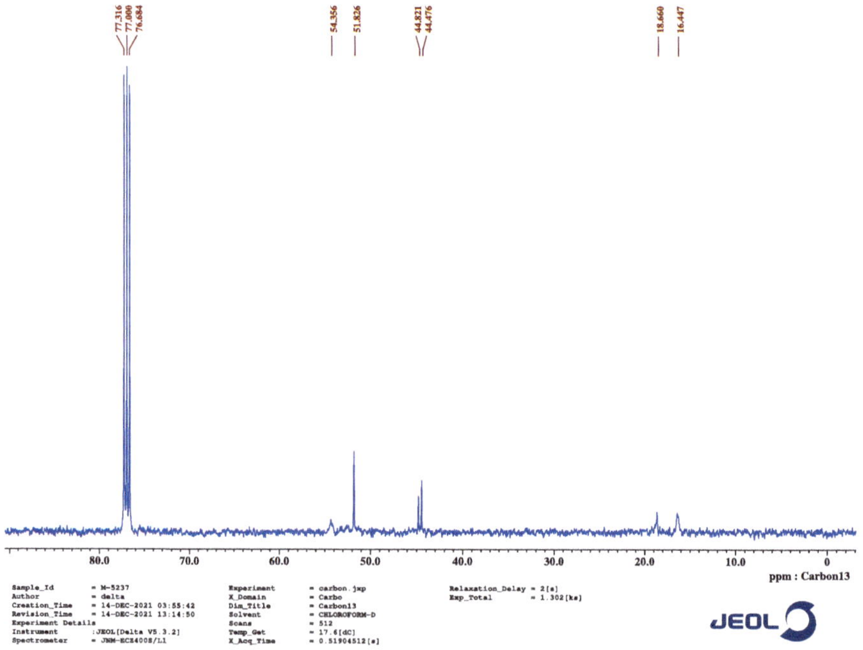

Fig. 6.20b ^{13}C-NMR spectrum of R_0

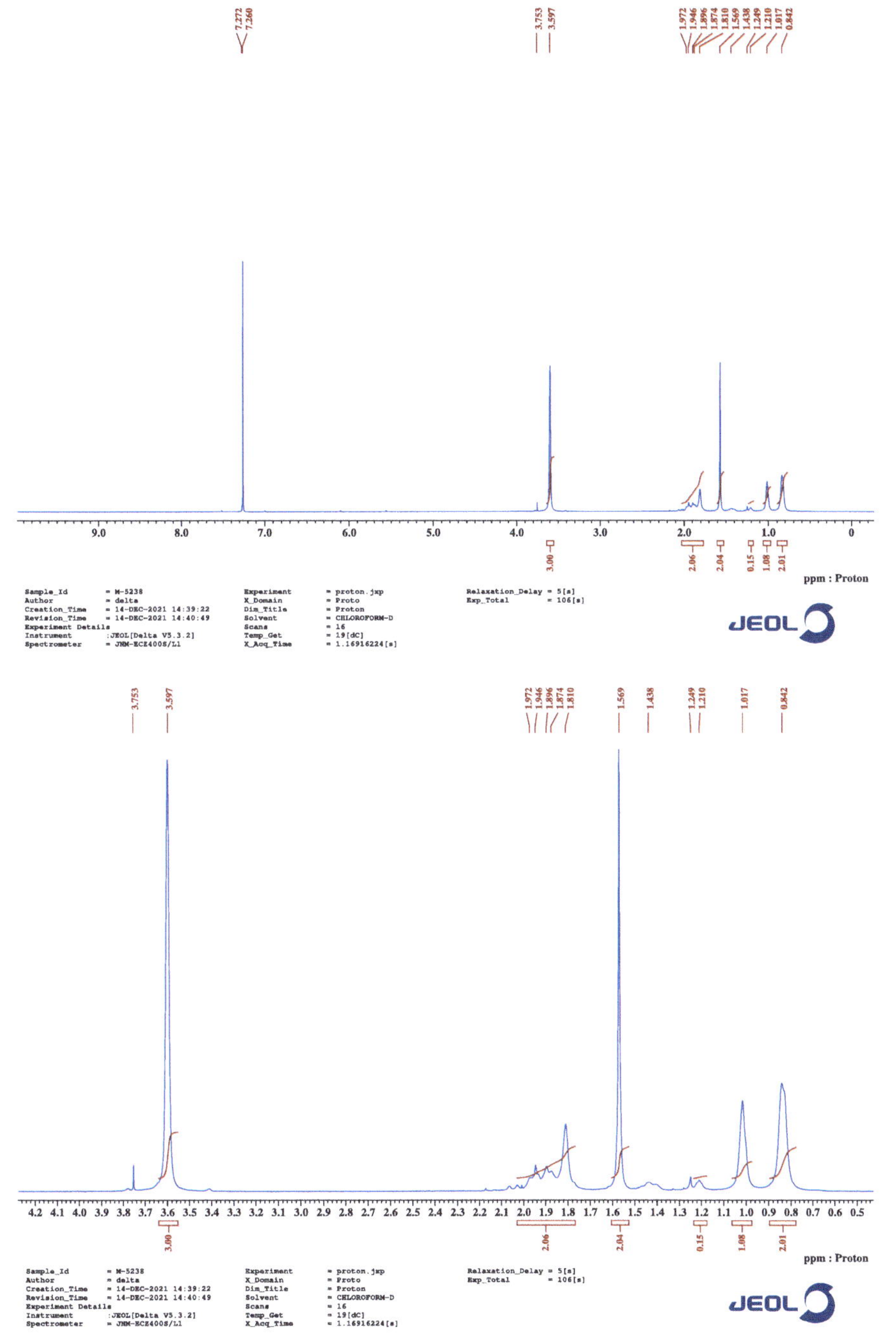

Fig. 6.21a ^{1}H-NMR spectrum of R_{10}

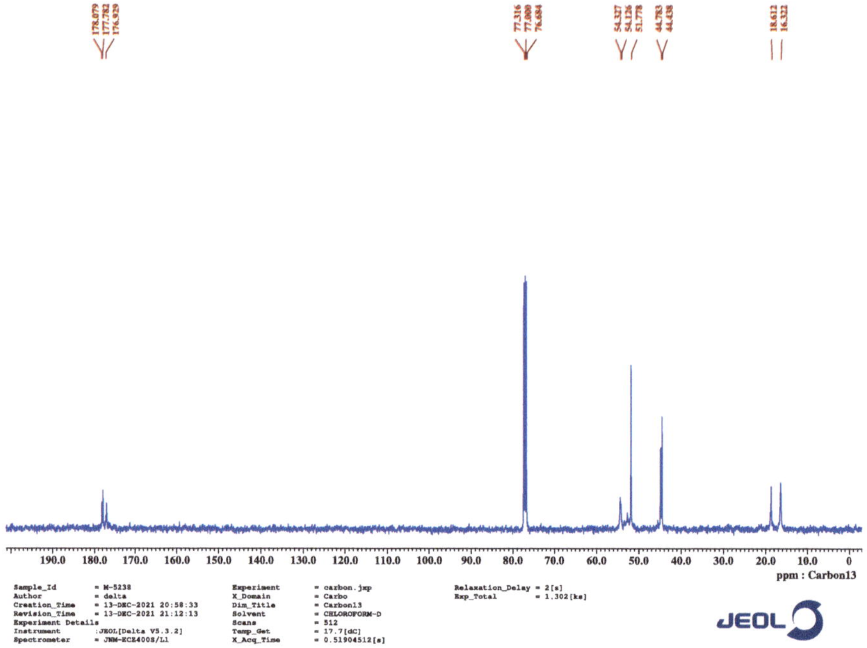

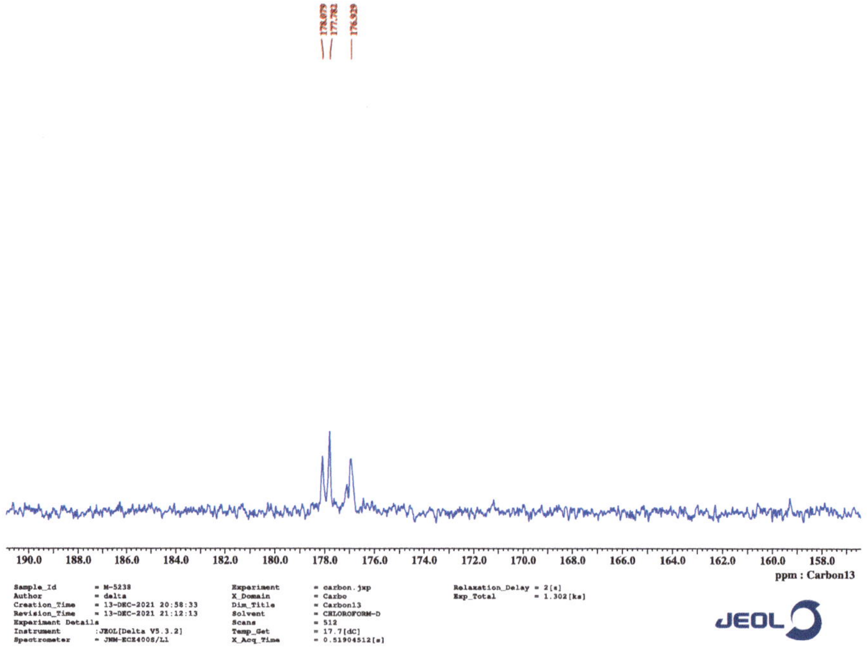

Fig. 6.21b ^{13}C-NMR spectrum of R_{10}

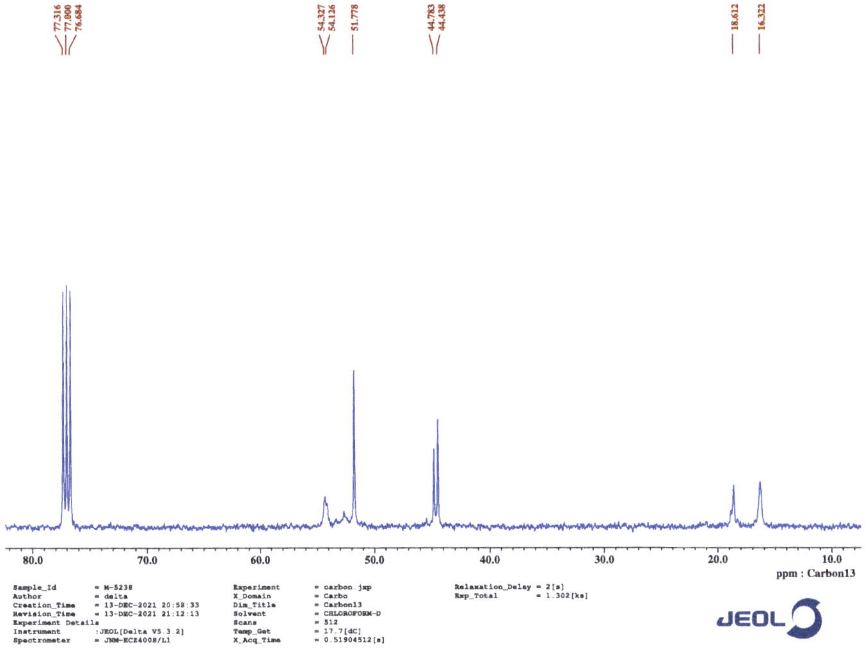

Fig. 6.21b (continued)

et al., 2016; Thopegowda et al., 2013). Yet no attempt has been made with respect to recycling and reusing DENTAL POLYMER RESINS.

Due to the lack of literature with regard to denture base resin recycling in dentistry, this research turned to literature on plastic recycling and attempted to adapt these methods to recycle denture base resin.

The SEM analysis of r-PMMA revealed amorphous complex structure. The SEM analysis revealed that the unmodified commercially available polymeric powder (control group) revealed smooth spherical beads. The amorphous complex structure increases the surface area of the individual particle facilitating better wettability.

The EDAX analysis showed the weight % of C and O as 60.53% and 38.60% respectively in the unmodified commercially available polymeric powder (control group). The modification by addition of r-PMMA at the ratios of 10, 20, 30, 40, and 50% resulted in increase of the weight % of C from 62.79 to 66.06%. This increase can be attributed to the addition of r-PMMA which due to polymerization has enhanced % of C.

FTIR represents a form of infrared spectroscopy that gathers detailed data across a broad spectrum, making it the favored technique for chemically characterizing denture base resin. Operating within the infrared segment of the electromagnetic spectrum, FTIR measures the absorption of light by vibrating molecular bonds, offering a distinctive molecular fingerprint.

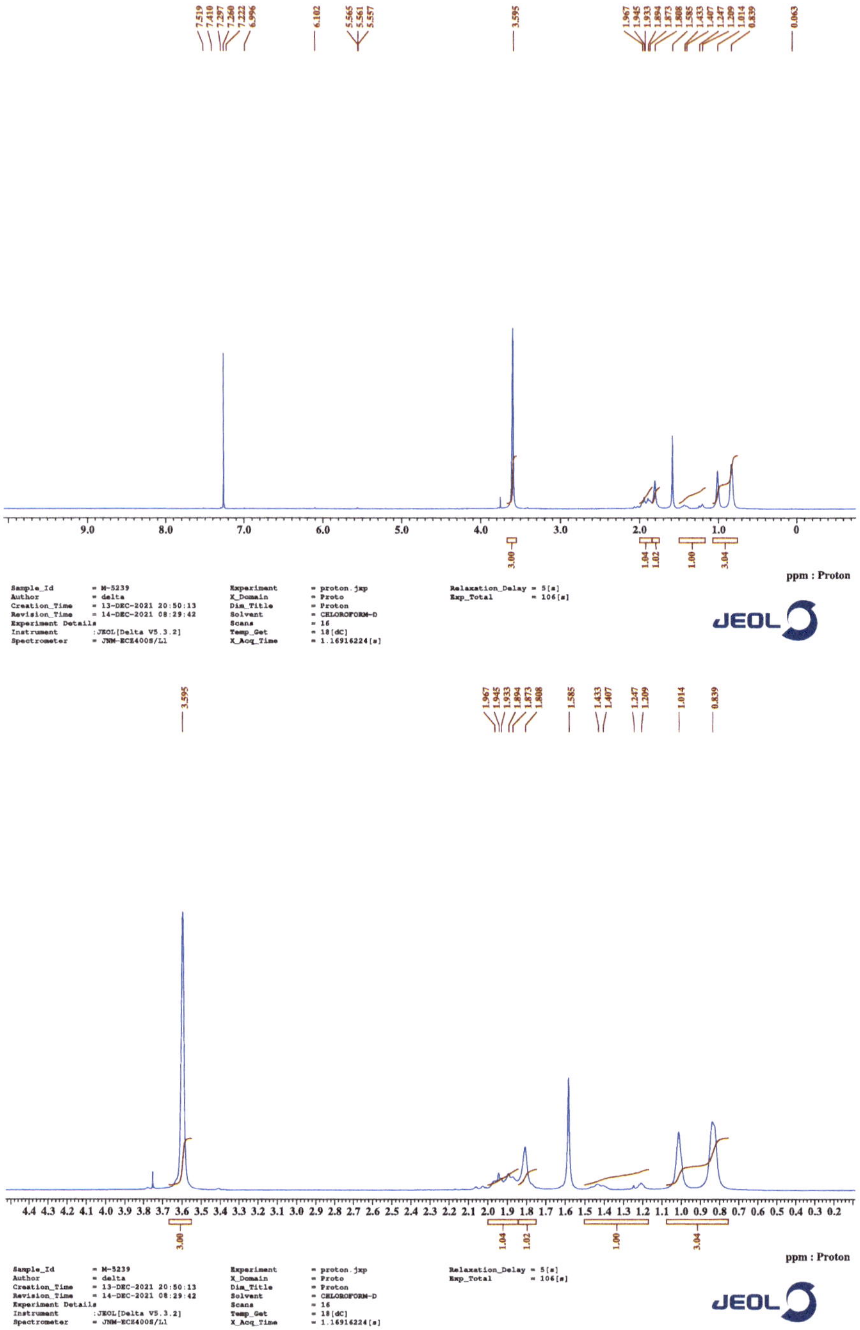

Fig. 6.22a ^{1}H-NMR spectrum of R_{20}

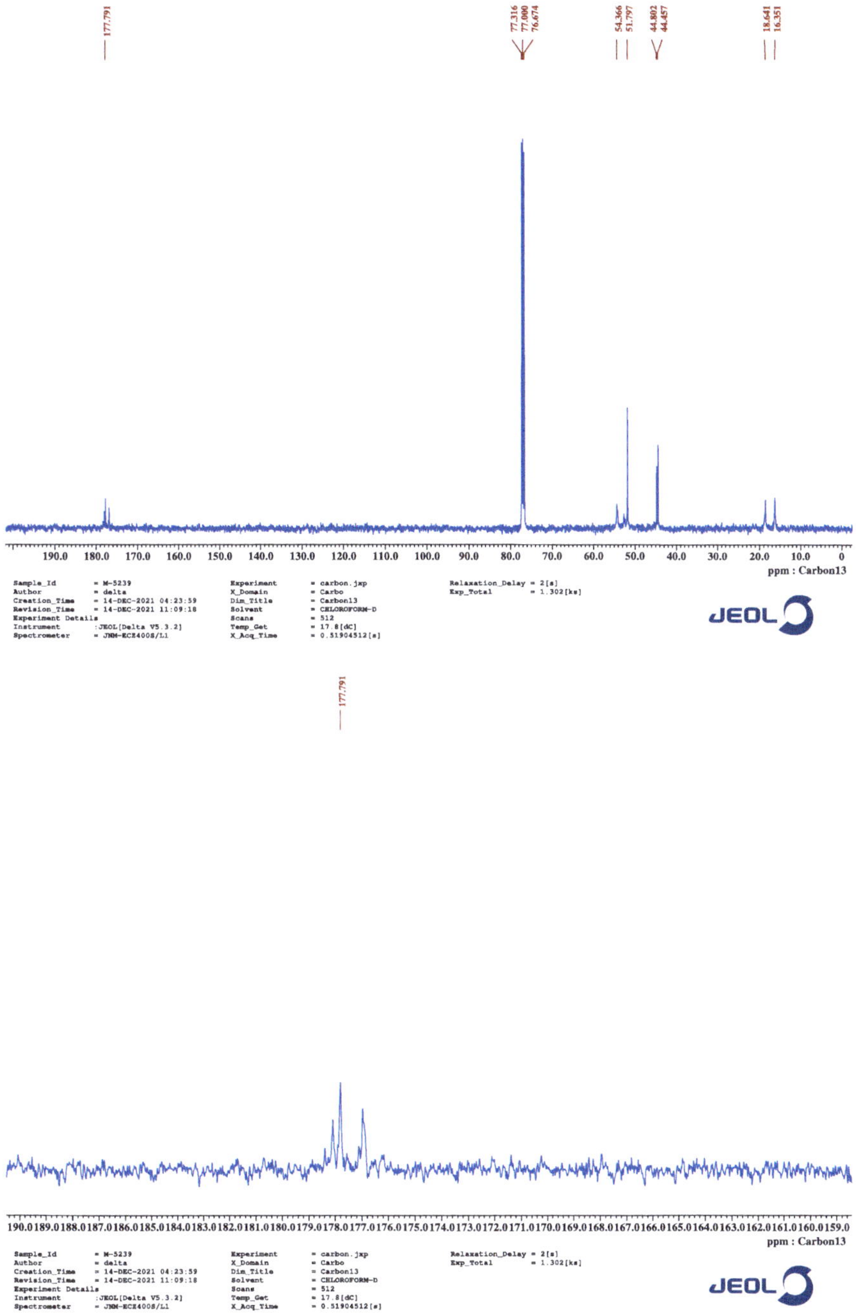

Fig. 6.22b ^{13}C-NMR spectrum of R_{20}

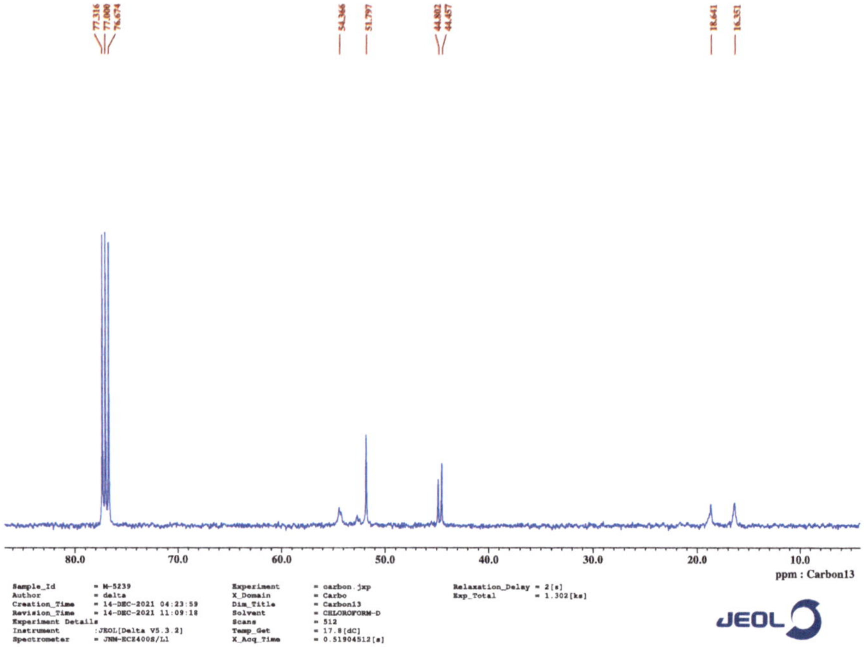

Fig. 6.22b (continued)

The FTIR spectral observations of the control group in this research were similar to those reported in previous studies by Vuluga et al., Ajay et al., Sunil et al., and Mahfooz et al. (Ajay et al., 2019; Al-Ali & Kassab-Bashi, 2015; Asim et al., 2015; Bin et al., 2018; Kowalczuk & Pitucha, 2019). The spectrum observations of experimental groups containing various concentrations of r-PMMA fabricated in the research depicted similar peaks to that of the control, with exception of the peak at 1559.86 cm^{-1}, which corresponds to the aliphatic carbon–carbon double bond (C=C). The absence of any new peaks indicates the absence of formation of any new chemical structure during the mechanical recycling process. This confirms that only mechanical size reduction has been carried out without alternation and adulteration to the chemical structure of PMMA.

The diminishing of the series of weak clustered peaks from 2192.91 to 1957.29 cm^{-1} attributed to the presence of the C≡N, from the initiator di-isobutylazonitrile ($((CH_3)_2C(CN)]_2N_2$) to weaker and fewer peak 2026.63 cm^{-1} indicated the depletion of the initiator, pointing to the active role of the initiator in the polymerization process both in the control and experimental groups.

The degree of conversion (DC) was determined through FTIR spectral analysis of both uncured and cured specimens. The carbonyl group (C=O) at 1721 cm^{-1} served as the internal standard peak, commonly utilized due to its stability before and after polymerization. The aliphatic group (C=C) at 1559 cm^{-1} was used in the DC calculation (Ajay et al., 2019; Al-Ali & Kassab-Bashi, 2015; Asim et al., 2015;

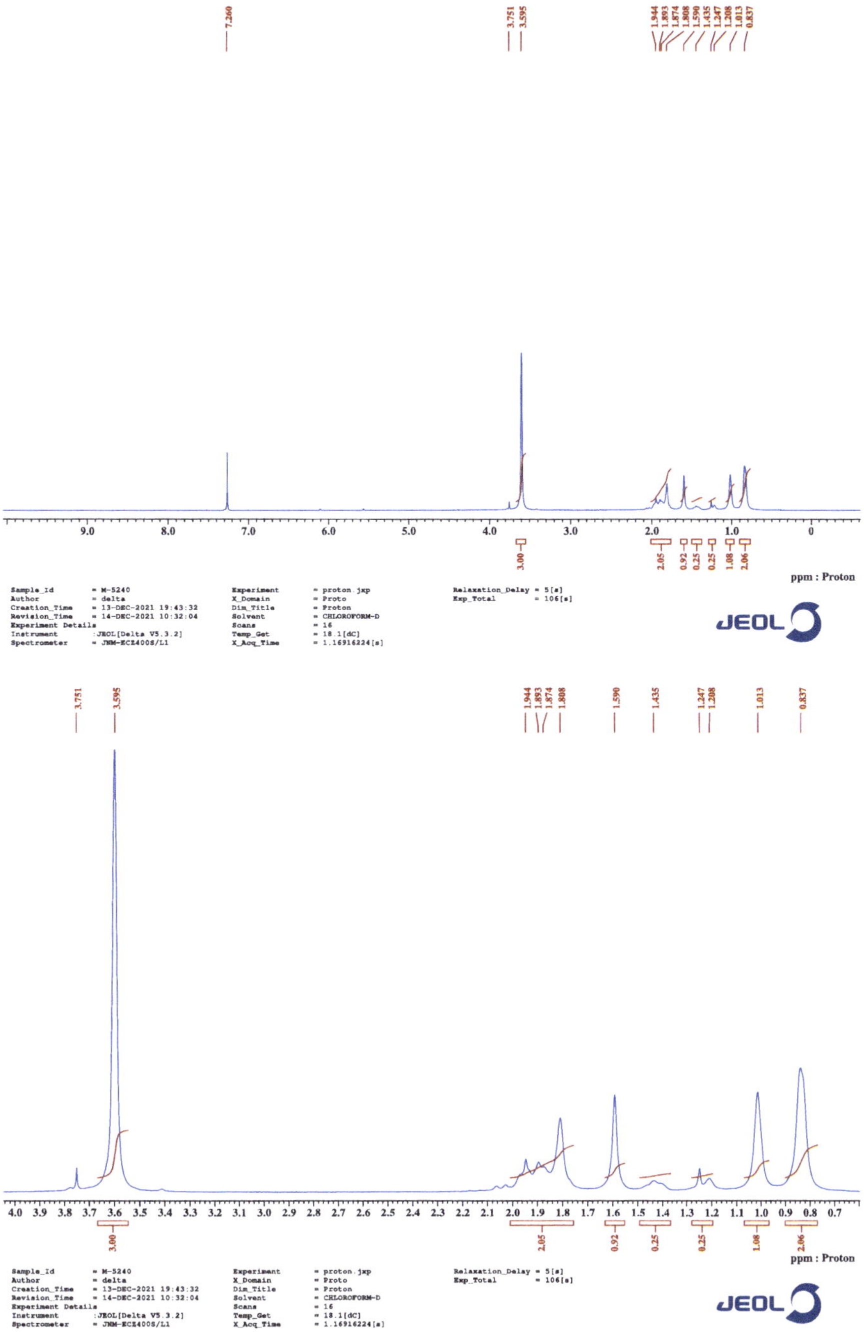

Fig. 6.23a ^{1}H-NMR spectrum of R_{30}

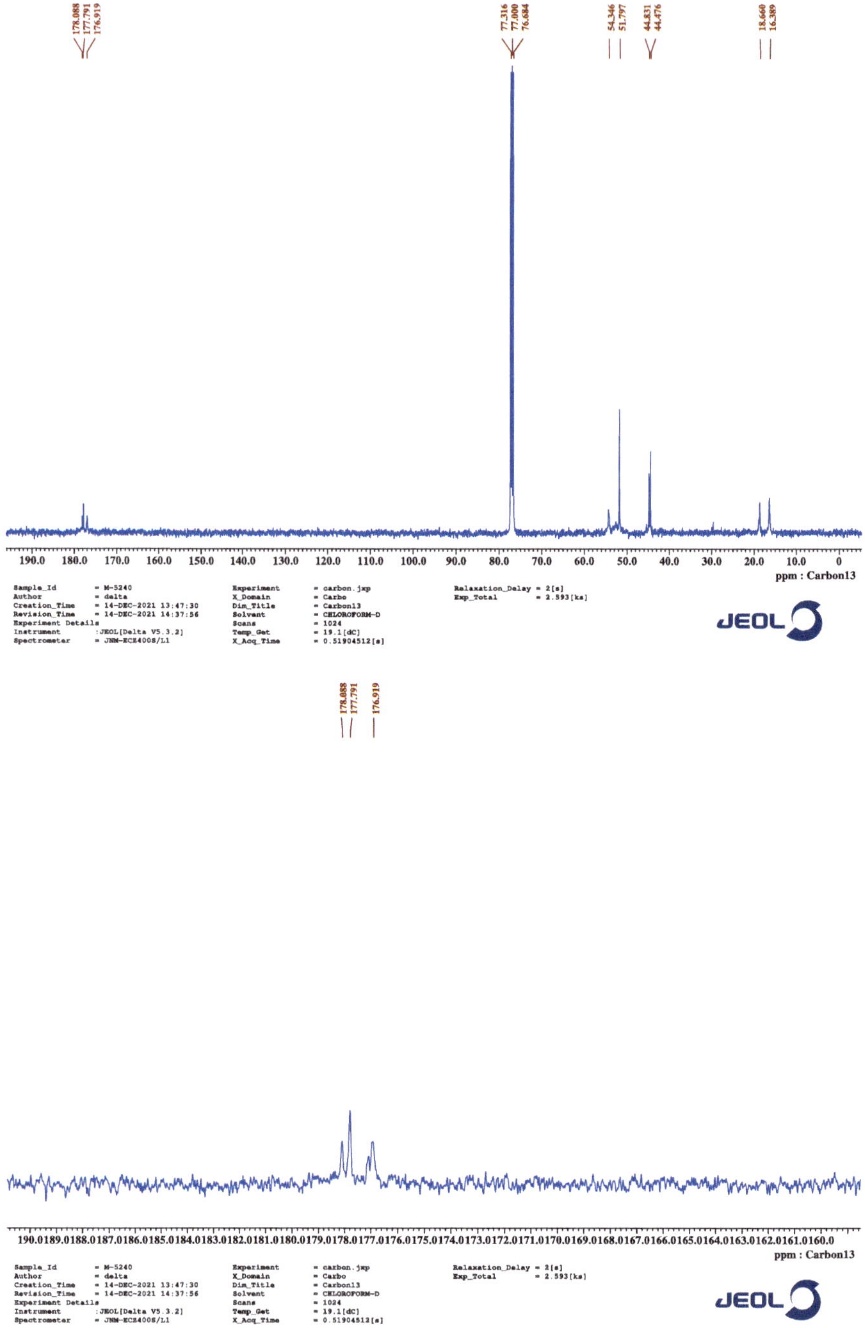

Fig. 6.23b ^{13}C-NMR spectrum of R_{30}

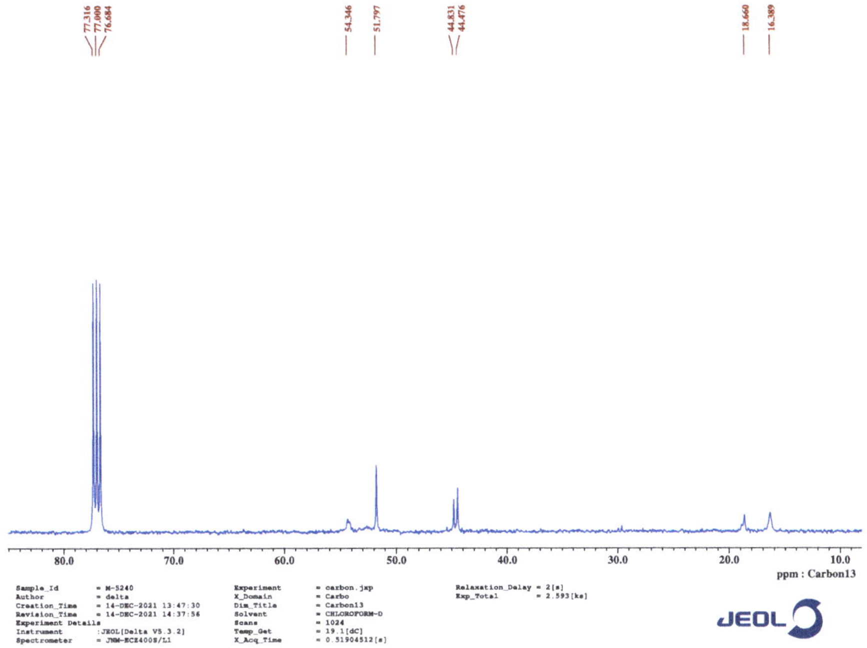

Fig. 6.23b (continued)

Bin et al., 2018; Kowalczuk & Pitucha, 2019). In resin-based denture materials, the DC holds significant clinical relevance as it correlates closely with the material's physical and mechanical properties. A higher DC implies fewer remaining unreacted monomers outside the system, reducing material cytotoxicity and allergic reactions, and thereby enhancing clinical safety (Al-Ali & Kassab-Bashi, 2015).In this study, the DC exhibited an increasing trend with rising concentrations of r-PMMA. Notably, the selected r-PMMA concentrations (10–50 wt %) did not compromise the DC. This suggests that incorporating r-PMMA up to 50 wt % minimally impacted the polymerization reaction of the heat-cured denture base resin. Consequently, the physical and mechanical characteristics of the denture base resin might also be minimally affected.

NMR is the foremost method for discerning the structure of organic compounds. Among spectroscopic techniques, NMR uniquely demands a comprehensive analysis and interpretation of the entire spectrum. While it requires a larger sample quantity than mass spectroscopy, NMR is non-destructive. Modern instruments can yield reliable data from samples weighing less than a milligram. Successful utilization of NMR as an analytical tool hinges on comprehending the underlying physical principles (NMR Spectroscopy [Internet], 2022).

In contrast to infrared and UV–visible spectroscopy, where absorption peaks locate themselves solely by frequency or wavelength, NMR resonance signals' positioning relies on both external magnetic field strength and rf frequency. One solution

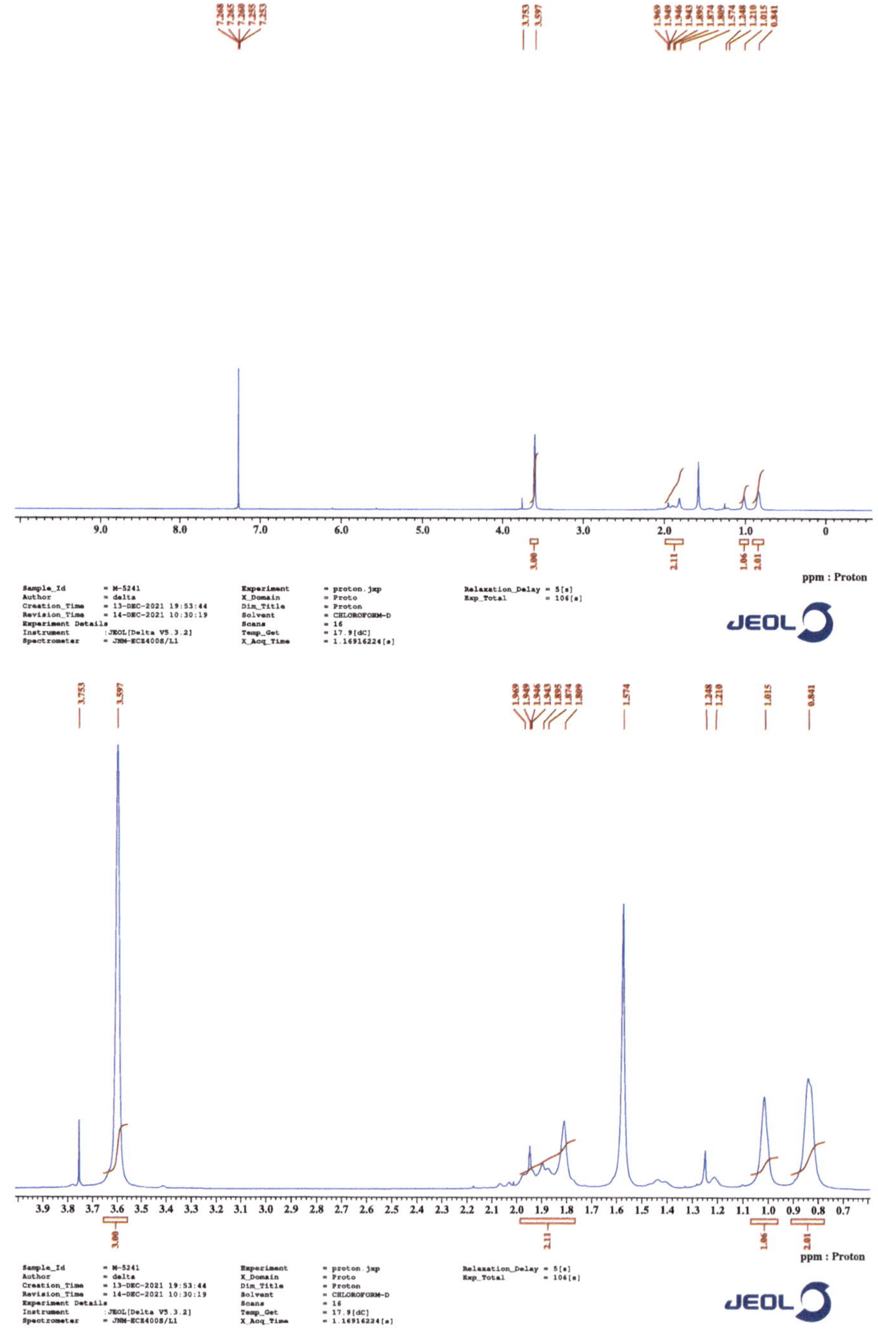

Fig. 6.24a ^{1}H-NMR spectrum of R_{40}

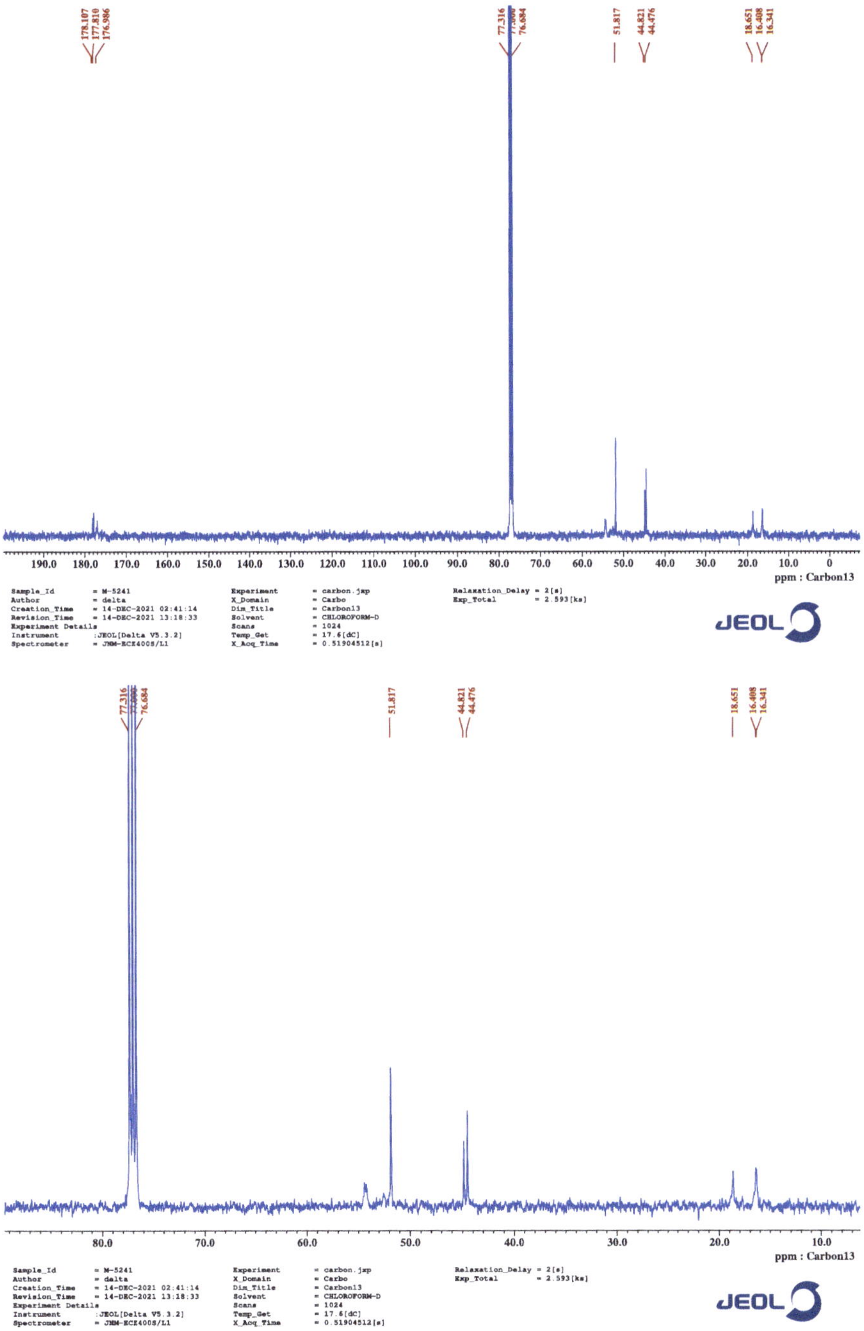

Fig. 6.24b ^{13}C-NMR spectrum of R_{40}

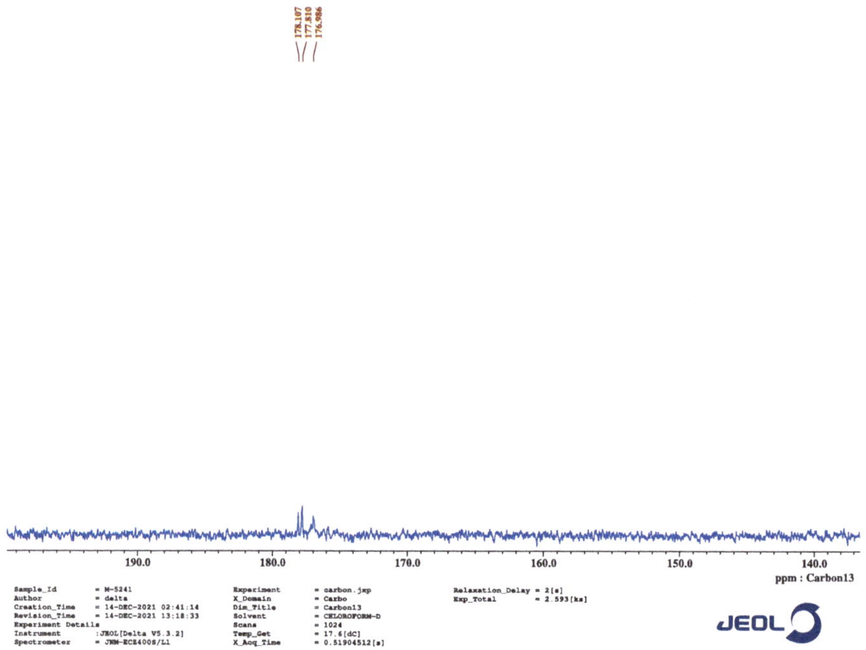

Fig. 6.24b (continued)

involves reporting an NMR signal's location relative to a reference signal from a standard compound added to the sample. This reference standard should be chemically inert, easily removable from the sample post-measurement, and yield a single sharp NMR signal that doesn't disrupt organic compound resonances. Tetramethylsilane, (CH3)4Si, known as TMS, fulfills these criteria and has become the preferred reference compound for proton and carbon NMR (NMR Spectroscopy [Internet], 2022; Spěváček et al., 1990). This study also employed this internal reference compound.

The control group showed ^{1}H-NMR and ^{13}C-NMR peaks with appropriate intensities and splitting to previous articles (Ajay et al., 2020). The NMR peaks of r-PMMA fabricated for the research exhibited the disappearance of doublet and triplet splitting of the –CH_2- (methylene group) between 1.741 and 1.925 ppm. In addition, other peaks appeared to be broadened. This confirmed the thermal induced cross linking between the monomers and the polymeric granules. In the ^{13}C-NMR peaks of r-PMMA demonstration the doublet peaks between 18.641 and 16.351 ppm which were attributed to—CH_3 (methyl) carbon at diminished intensity. This indicates the involvement of the alkylic carbon in the polymerization process. Similar pattern was observed in the peak signals of ^{13}C-NMR of all the 5 experimental groups.

The absence of the peak singlet around 5.90 ppm and 5.40 ppm confirms the presence of higher degree of PMMA. This indicates that the PMMA chains were able to act as macro-RAFT (Reversible addition – fragmentation chain-transfer)

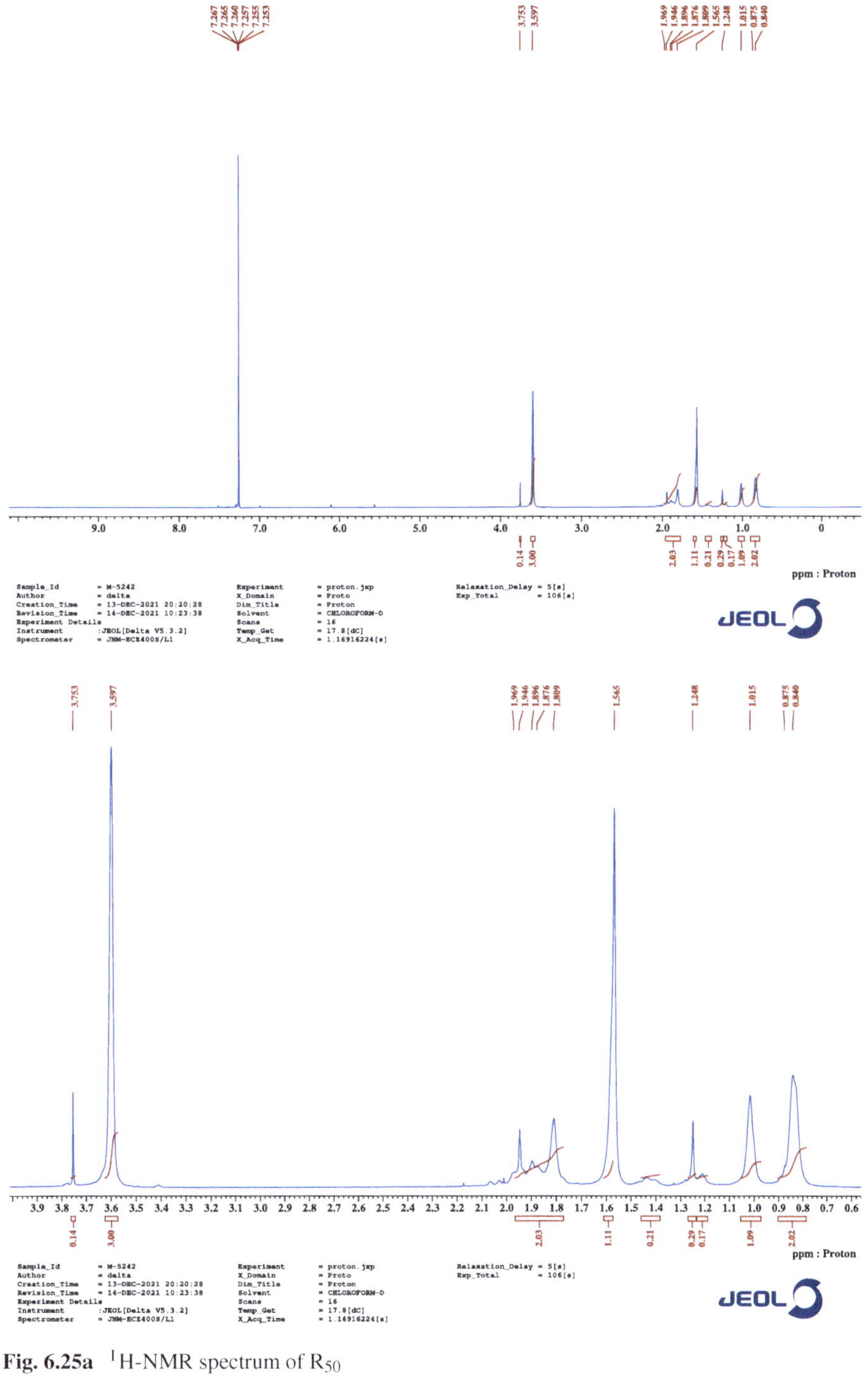

Fig. 6.25a ^{1}H-NMR spectrum of R_{50}

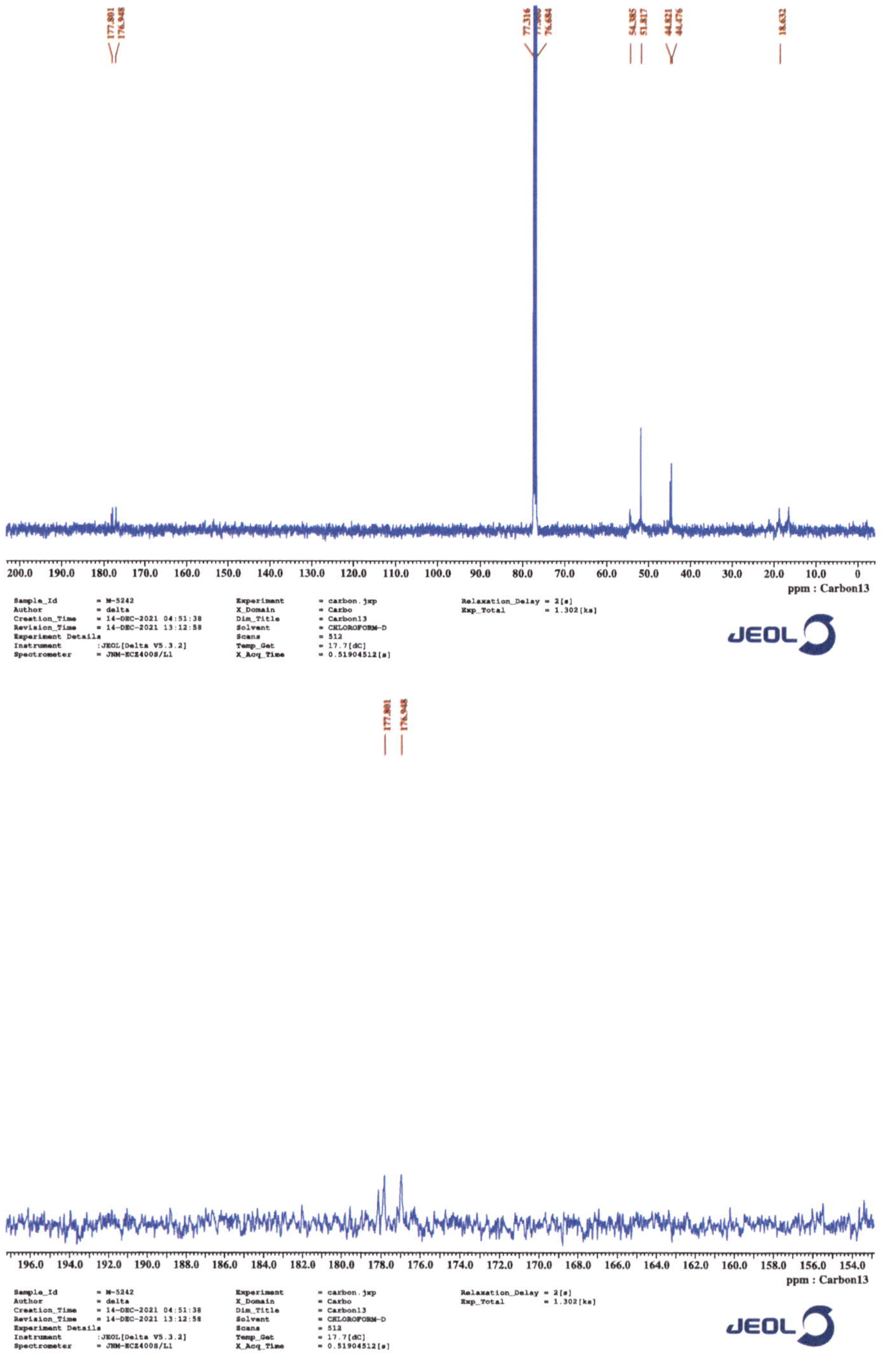

Fig. 6.25b ^{13}C-NMR spectrum of R_{50}

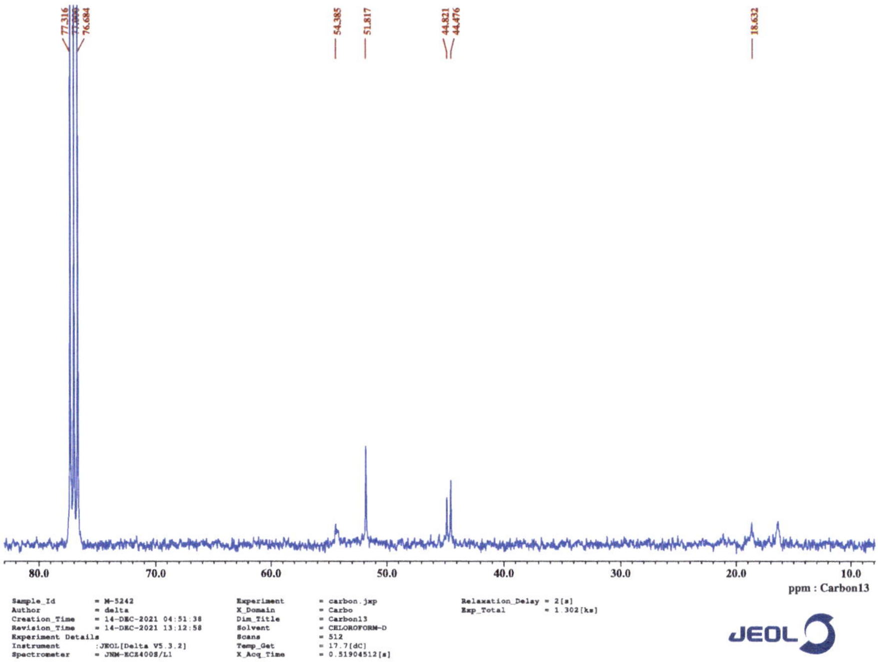

Fig. 6.25b (continued)

Table 6.4 ^{1}H-NMR peak signals

Group	–CH_3	–O-CH_3	–CH_2–
R_0	0.84–1.01	3.59	1.80–1.96
R_{10}	0.84–1.01	3.59	1.80–1.96
R_{20}	0.83–1.01	3.59	1.80–2.06
R_{30}	0.82–0.99	3.58	1.72–1.92
R_{40}	0.83–1.01	3.59	1.80–2.02
R_{50}	0.83–1.01	3.59	1.80–2.02

Table 6.5 ^{13}C-NMR peak signals

Group	–CH_3	–CH_2–	Quaternary carbon	–O-CH_3	-C=O
R_0	18.63	44.47–44.82	51.81	54.38	176.94–177.80
R_{10}	16.37–18.64	44.46–44.82	51.79	54.15–54.35	176.95–178.08
R_{20}	16.34–18.64	44.45–44.80	51.79	54.34	176.95–178.08
R_{30}	16.36–18.65	44.45–44.81	51.79	54.14–54.36	176.96–178.08
R_{40}	16.35–18.65	44.47–44.82	51.81	54.36	176.96–178.08
R_{50}	16.37–18.66	44.47–44.82	51.81	54.16–54.37	176.97–178.10

agent and were used in further polymerization process. This indicates the multiple time reusability of the r-PMMA with the control monomer units.

In the present research modification by substitution with 10, 20, 30, 40, and 50% (w/w) of recycled denture base polymeric powder (r- PMMA) were assessed. Future studies may be carried out by increasing the substitution concentration (w/w %) of recycled denture base polymeric powder (r- PMMA) and assessing its effect on the chemical, physio-mechanical and biological properties. The number of times the denture base resin is subjected to mechanical recycling processes could also be subjected to future research. r-PMMA fabricated with dentures fabricated at the preclinical level with highly cross-linked denture teeth, which may alter the material properties may be investigated. This novel research that employed mechanical recycling methods to recycle denture base acrylic resin and create blends of the recycled denture base polymeric powder (r- PMMA) with virgin commercially available material, hence the results should be interpreted with caution and carefully extrapolated. The effect of various mechanical and chemical denture cleaning processes in use may also affect the surface roughness and hardness of the material. Future inquiries may be carried out to evaluate other dimensions of the polymeric modification, to better understand the scope of the modification, before real-world application.

6.2.3 Social Implications

The recycle and reuse of dental materials is an area of focus for the sustainable growth of dentistry in future. The ability to reuse a dental material will greatly reduce the burden on the environment and the non-renewable natural resources which are used to produce these materials. The long-term well-being and health of the global population is determined by our attitude toward our environment. It has become inevitable for each and every individual to be conscious of his/ her carbon footprint and make appropriate lifestyle changes to diminish the environmental burden caused by them. The need for dentists to be environmentally conscious and practice eco-dentistry has become the inexorable need of the hour. A pressing matter in dentistry as we move forward is the management of solid waste. Environmental consciousness and accountability should be our foremost responsibility as individuals and community. Energy efficiency, renewability, non-toxicity, sustainability, low carbon foot print, minimal invasiveness and reduced carbon emissions are all practices which come under the canopy of broader terms as "GOING GREEN" AND "ECO-FRIENDLY".

6.2.4 Conclusion

Within the limitations of the research, it can be concluded that heat cure denture base resin can be effectively recycled and reused to create blends of recycled heat cure denture base polymeric powder (r-PMMA) at the concentration of 10%, 20%,

30%, 40% and 50% (w/w) (R_{10}, R_{20}, R_{30}, R_{40}, and R_{50}) to commercially available PMMA- based heat cure denture base resin. Through FTIR analysis it was confirmed that no new compound was formed, eliminating the likelihood of contamination and adulteration due to the mechanical recycling of the denture base resin. FTIR spectra also showcased that the blends fabricated had a better degree of conversion than the control. Incorporation of 30% r-PMMA showcased the highest DC of all the groups evaluated. NMR analysis indicated that the PMMA chains were able to act as macro-RAFT (Reversible addition – fragmentation chain-transfer) agent and were used in further polymerization process. This indicated the multiple time reusability of the r-PMMA with the control monomer units. The blends also demonstrated reduced residual monomer content owing to better degree of conversion.

References

Ajay, R., Suma, K., JayaKrishnaKumar, S., Rajkumar, G., Kumar, S. A., & Geethakumari, R. (2019). Chemical characterization of denture base resin with a novel cycloaliphatic monomer. *The Journal of Contemporary Dental Practice, 20*(8), 940–946.

Ajay, R., Suma, K., Sasikala, R., Sajidabegum, S., Vignesh, V., & Bhuvaneshkumar, D. (2020). Chemical structure and physical properties of heat-cured poly(methyl methacrylate) resin processed with cycloaliphatic comonomer: An in vitro study. *The Journal of Contemporary Dental Practice, 21*(3), 286–290.

Al-Ali, A. A. S., & Kassab-Bashi, T. Y. (2015). Fourier Transform Infra Red (FTIR) spectroscopy of new copolymers of acrylic resin denture base materials *4*(4), 172–180.

Ashton, E. G., Kindlein, W., Demori, R., Cândido, L. H. A., & Mauler, R. (2016). Recycling polymeric multi-material products through micronization. *Journal of Cleaner Production, 116*, 268–278.

Asim, A., Al-Ali, S., Kassab-Bashi, T. Y. (2015). Fourier Transform Infra Red (FTIR) spectroscopy of new copolymers of acrylic resin denture base materials. *Science Technology and Engineering [Internet], 4*(4), 172–180. www.erpublications.com.

Avinash, B., Avinash, B. S., Shivalinga, B. M., Jyothikiran, S., & Padmini, M. N. (2013). Going green with eco-friendly dentistry. *The Journal of Contemporary Dental Practice, 14*(4), 766–769.

Charmondusit, K., Arleewong, P., Studies, R., Campus, S., & Section D. (2006). The application of cleaner technology in the acrylic cast sheet production: Case study of acrylic cast sheet company in Thailand *01*(3), 515–523.

Chopra, A., Gupta, N., Rao, N. C. V. S. (2014). Eco-dentistry: The environment-friendly dentistry. *Saudi Journal of Health and Science* 361–365.

Eu. (2000). Draft proposal for a European Parliament and Council directive on Waste Electric and Electronic Equipment, Brussels, Belgium. [Internet]. Publications Office of the European Union, 2000 [cited 2022 Jan 2]. https://eur-lex.europa.eu/legal-content/EN/TXT/?uri=CELEX:52023PC0063.

Gaabour, L. H. (2021). Effect of addition of TiO 2 nanoparticles on structural and dielectric properties of polystyrene/polyvinyl chloride polymer blend. *AIP Advances, 11*, 105120. https://doi.org/10.1063/5.0062445.

Kowalczuk, D., & Pitucha, M. (2019) Application of FTIR method for the assessment of immobilization of active substances in the matrix of biomedical materials. *Materials (Basel), 12*(18).

Krishnamoorthi, D., Karthigeyan, S., Ali, S. A., Rajajayam, S., Gajendran, R., & Rajendran, M. (2022). Evaluation of the in-vitro cytotoxicity of heat cure denture base resin modified with recycled PMMA-based denture base resin. *Journal of Pharmacy and Bioallied Science, 14*(Suppl 1), S719–S725.

Krishnamoorthi, D., Suma, K., Ali, S. A., et al. (2022). Effect of recycled denture base polymeric powder incorporation on the surface properties of heat-cured PMMA denture base acrylic resin: An In vitro study. *World Journal of Dentistry, 13*(5), 527–533.

Mahfooz, A. M. B., & Alammari, M. R. (2018). The use of Fourier Transform Infra-Red (FTIR) Spectroscopic analysis and cell viability assay to assess pre-polymerized CAD\CAM acrylic resin denture base materials. *International Journal of Pharmaceutical Research & Allied Sciences, 7*(2), 111–118.

Makenji, K., & Savage, M. (2012). Mechanical methods of recycling plastics from WEEE. In *Waste Electrical and Electronic Equipment Handbook* (pp. 212–238).

Maris, J., Bourdon, S., Brossard, J. M., Cauret, L., Fontaine, L., & Montembault, V. (2018). Mechanical recycling: Compatibilization of mixed thermoplastic wastes. *Polymer Degradation and Stability, 147*, 245–266.

Mohelay, N., Deolia, S. G., Jagyasi, D., Lakhwani, R., Sen, S., & Chapekar, J. (2016). Eco-friendly dentistry: A green business with teeth. *International Journal of Oral Health and Medical Research, 3*(2), 66–70.

NMR Spectroscopy [Internet]. Michigan, USA: Michigan state university [cited 2022 June 20]. https://www2.chemistry.msu.edu/faculty/reusch/virttxtjml/spectrpy/nmr/nmr1.htm.

Phillips, R. W. (2005). *Skinner's Science of Dental Materials* (11th ed., pp. 162–169). W.B Saunders.

Spěváček, J., Schneider, A., & Straka, J. (1990). High-resolution carbon-13 nuclear magnetic resonance spectra and structure of amorphous and crystalline forms of stereoregular poly(Methyl methacrylate)s+. *Macromolecules, 23*(12), 3042–3051.

Suresh, S. S., Mohanty, S., & Nayak, S. K. (2017). Preparation and characterization of recycled blends using poly (vinyl chloride) and poly(methyl methacrylate) recovered from waste electrical and electronic equipments. *Journal of Cleaner Production, 149*, 863–873.

Suresh, S. S., Mohanty, S., & Nayak, S. K. (2018). Preparation of poly (vinyl chloride)/poly(methyl methacrylate) recycled blends: Effect of varied concentration of PVC and PMMA in stability of PVC phase on the recycled blends. *Materials Today Proceedings, 5*(2), 8899–8907.

Thopegowda, N. B., Shenoy, K., Shankarnarayana, R. K., Kukkila, J., Vaddya, S. B., & Gingipalli, K. (2013). Recycling of materials used in dentistry with reference to its economical and environmental aspects. *International Journal of Health and Rehabilitation Sciences, 2*(3), 140–145.

Chapter 7
Plastic Waste Management: Microplastic Phytoextraction from Textile Waste Sludge Using *Lemna Minor*

Clemant Jayasuriya, Harshani Munasingha, Nimila Dushyantha, and Titus Cooray

Abstract Microplastics (MPs) have gained significant global concern due to their widespread presence, potential harm to ecosystems, and ability to enter biological systems. Wastewater treatment plant (WWTP) sludge, rich in fibrous-shaped MPs, is a primary source of MPs in terrestrial environments. Managing MPs in soil from WWTP sludge presents significant challenges. Therefore, it is necessary to develop effective management solutions for this waste stream. This study aimed to assess the potential of duckweed (*Lemna minor*), an aquatic macrophyte, in phytoextracting MPs from textile waste sludge. Preparation of dry samples and wet peroxide oxidation processes were done for the identification of fiber-shaped MPs in selected sludge. Microscopic analysis revealed an average MP concentration of 2.2 particles per gram in the selected textile waste sludge, primarily in a fibrous form. *Lemna minor* was cultivated in the sludge samples. As the plants matured, a steady increase in fibrous-shaped MPs phytoextraction was evident, with concentrations reaching 1.3 MPs per plant after 7 days, 5.6 MPs per plant after 21 days, and 9.2 MPs per plant after 45 days. The visual sorting method was carried out to count MPs using an optical microscope. These results highlighted the proficiency of *Lemna minor* in MP phytoextraction through an eco-friendly and economically viable process to mitigate MP contamination from textile waste sludge. Once these macrophytes reach their maximum biomass yield, it is recommended to harvest and burn them in available burning facilities. The resulting remediated sludge can be used as a fertilizer. In addition, attention should be directed toward industrial plants with potential for textile fiber production, such as corn, wheat, hemp, and bamboo, to offer a more comprehensive solution to MP phytoextraction aligned with circular economy principles. This study demonstrates the potential of using *Lemna minor* for effective MP remediation from textile waste, contributing to sustainable waste management practices.

C. Jayasuriya · H. Munasingha · N. Dushyantha (✉) · T. Cooray
Department of Applied Earth Sciences, Faculty of Applied Sciences, Uva Wellassa, University, Badulla, Sri Lanka
e-mail: nimila.dush@uwu.ac.lk

S. Suriyanarayanan et al. (eds.), *Combating Plastic Pollution in Terrestrial Environment*,
https://doi.org/10.1007/978-981-96-2343-3_7

Keywords Phytoremediation · Aquatic macrophyte · Microplastic pollution · Textile waste

7.1 Introduction

Microplastics (MPs) are small particles of plastic (with a size less than 5 mm), which have emerged as a significant global concern as an environmental pollutant due to their diminutive size, widespread presence, ability to enter biological systems, and detrimental effects on ecosystems (Chamanee et al., 2023; Rafiq & Xu, 2023). They have accumulated in the environment through direct release or the fragmentation of larger plastic debris, termed primary and secondary MPs, respectively. Because of their small size, MPs are pervasive in the environment, causing adverse effects (Hettiarachchi & Meegoda, 2023). Consequently, the presence of MPs and their impacts on both marine and terrestrial environments have created substantial attention in recent years. MPs are known to be bioavailable to marine organisms throughout the food chain, leading to reduced reproduction and development, thereby challenging their long-term survival (Sharifinia et al., 2020). The composition and larger surface area of MPs make them prone to attracting organic pollutants and leaching harmful plasticizers. Furthermore, aside from their impact on marine ecosystems, MPs have affected terrestrial environments for decades due to the increasing abundance of plastic waste (Kim & Hwang, 2023). Current research indicates that MPs are causing adverse effects on humans through their presence in the food chain and the environment. Thus, the need for a remedial approach to address the abundance of MPs in the environment is strongly indicated. Microfibers originating from synthetic textiles are a prominent contributor to the presence of MPs in both marine and terrestrial ecosystems (Xu et al., 2020). Fibers released during the manufacturing process of synthetic textiles are identified as a primary source of MPs, while those released during washing and disposal constitute a secondary source (Acharya et al., 2021; Gao et al., 2023).

The majority of microfibers found in the environment are believed to stem from wastewater treatment plants (WWTPs) handling textile waste. While effluent water is the primary source for introducing microfibers into marine and aquatic environments, residual sludge is responsible for the presence of MPs in terrestrial settings (De Falco et al., 2019; Ziajahromi et al., 2016). WWTPs play a major role in MP contamination in the aqueous environment. Despite claims that over 90% of MPs are removed from the effluent water by WWTPs (Bläsing & Amelung, 2018), they remain a major contributor to marine MP pollution. It has been reported that 35% of all MPs found in oceans worldwide originate from WWTP effluents (Boucher & Friot, 2017). Despite the substantial removal efforts, a considerable volume of effluent continues to be discharged into aquatic systems globally through various types of WWTPs, causing pollution. Textile wastewater contains a diverse array of MPs, textile dyes, inorganic salts, and organic additives, often accompanied by elevated levels of heavy metals (Aslam et al., 2004). Within synthetic textile mills, the concentrations of these components are notably higher, owing to processes like raw fabric scouring and dyeing. In addition, the concentration of airborne MPs is relatively high in these industrial zones, which further contributes to MPs in wastewater, albeit receiving less attention (Prata, 2018). In the WWTP, treated water is released into the environment, while abandoning all of these toxic elements and chemicals including a high number of MPs in the sludge. However, managing MPs in terrestrial settings has proven challenging, primarily due to difficulties in identification and quantification. Given that WWTP sludge stands as a primary source of MPs in soil, there is a pressing need to devise a solution for its management.

Phytoextraction is a phytoremediation technique employed for the removal or regulation of pollutants from water, soil, and air (Bhat et al., 2022). Plants used in this technique can uptake pollutants from the soil or water by roots and bring them into harvestable parts of plants such as shoots. Previous studies have illustrated that plants can accumulate MPs from soil through phytoremediation (Capozzi et al., 2018; Ebere et al., 2019; Koongolla et al., 2020; Li et al., 2020; Mateos-Cárdenas et al., 2019; Rozman & Kalčíková, 2022). However, the efficiency of various phytoremediation techniques in MP removal varies, and there is limited research specifically comparing these efficiencies. This study aims to evaluate the phytoextraction potential of duckweed (Lemna minor), an aquatic macrophyte, as an ecologically and economically viable means of removing MPs from the sludge generated by textile WWTPs. This area of research remains relatively unexplored, with specific research gaps in understanding the efficiency of *Lemna minor* in comparison to other phytoremediation techniques. Furthermore, this research addresses the identified gap in the dearth of studies focusing on the extraction of MPs from textile waste sludge.

7.2 Materials and Methods

7.2.1 Sludge Sample Collection

Textile wastewater sludge is accessible at textile mills in Sri Lanka that employ synthetic yarns in their production processes. The Western Province of Sri Lanka was selected as the sample collection site due to its significance in the country's textile industry. This region contains a high concentration of textile mills, making it a major hub for textile manufacturing in Sri Lanka. Consequently, the WWTPsin this area manage substantial volumes of textile effluent, which is likely to contain high concentrations of MPs. Furthermore, the WWTP selected for this study employs advanced treatment processes representative of the region's typical practices, ensuring the results are applicable to other similar facilities. For this study, textile waste sludge samples were specifically gathered from the WWTP located in the Western Province of Sri Lanka.

7.2.2 Extraction of MPs from the Sludge

A total of 400 g of textile waste sludge was carefully weighed and subsequently dried completely at 90 °C for 24 h to ensure thorough dehydration. Before the sieving procedure, the dried sludge was crushed using a mortar and pestle. The crushed sludge sample was then sieved through sets of sieves with openings of 2 mm, 500 µm, and 125 µm, utilizing an automatic sieve shaker for approximately 2 h. The sieved sludge from the 500 µm and 125 µm sets was collected, and 50 g of sludge was separated using an analytical balance and the cone and quartering method.

To prepare a 0.05 M Fe^{2+}solution, 7.5 g of $FeSO_4.7H_2O$ was dissolved in distilled water within a volumetric flask, and the volume was adjusted to 500 mL. In addition, 3 mL of concentrated. H_2SO_4 acid was added to the prepared solution. Then, 120 12 mL of 50% H_2O_2 was used for 50 g of sludge sample (Masura et al., 2015). A total of 50 g of dry sieved sludge was transferred into a 600 mL glass beaker. Subsequently, a mixture of 0.05 M Fe^{2+} in an aqueous solution with Con.H_2SO_4 was added to the beaker, followed by the addition of 12 mL of 50% H_2O_2. The entire process was carried out in a fume hood, and the sample was left inside the fume hood until the reaction was complete. Later, an additional 20 mL of Fe^{2+}aqueous solution and 12 mL of H_2O_2 were introduced into the same beaker. After the reaction, the solution was transferred to a hot plate with a magnetic stir. Then, 6 g of NaCl was added to each 20 mL of the final volume of the sample at 70 °C and stirred at 200 rpm for 2 h. After the sample reached room temperature, the settled sludge was then separated by using an Imhoff cone (Fig. 7.1a), and effluent was then separated and transferred to the separatory funnel. The effluent was filtered through a 0.45 µm Whatman Glass Microfiber Filter paper (Model no. VE115) (Fig. 7.1b).

7.2.3 Experiment Set-Up

The duckweed (*Lemna minor*) was selected as a representative aquatic macrophyte and planted in the sludge samples. Once the macrophyte was successfully grown (Fig. 7.2a) under the temperature range of 20–27 °C with adequate light conditions and no additional nutrient, plant samples were collected after 7, 21 and 45 days. Each collected sample was air-dried and then oven-dried at 105 °C (Fig. 7.2b). Subsequently, the dried plant leaves were finely ground using a mortar and pestle in preparation for subsequent analysis. During the analysis, 3 replicates were used to assess variability.

Fig. 7.1 **a** Sludge separated by Imhoff cones and **b** separate microplastics in separatory funnels with filter papers

Fig. 7.2 Duckweed samples **a** after the growth of 45 days and **b** after oven-dried

7.2.4 Observation and Identification of MPs

MPs present in the initial sludge (n = 10) and plant samples (n = 30) were observed and identified using an Olympus SZX16 dissection microscope, equipped with a camera attached to the microscope. The MPs collected on each filter paper from the initial sludge and in plant samples were quantified, providing values for the concentration of MP particles per gram of sludge and per plant, respectively. Moreover, Fourier transform infrared spectroscopy (FTIR) analysis was carried out on both initial and 45-day-old plant samples, employing a Bruker Alpha spectrophotometer to identify functional groups within the range of 500–4500 cm^{-1}. Background corrections were applied for the absorption of atmospheric water and CO_2.

7.3 Results and Discussion

7.3.1 Identification and Quantification of MPs in Textile Waste Sludge

Microscopic analysis of the sludge revealed the presence of fibrous-shaped MPs among the separated MPs on each filter paper from the textile waste sludge (Fig. 7.3). After counting the number of fibrous-shaped MPs in the filter papers of the initial sludge, the average concentration of MPs in the textile waste sludge was determined to be 2.2 MPs per gram.

Therefore, the presence of fibrous-shaped MPs in textile waste sludge could potentially lead to various environmental and health impacts if not managed properly. Once the fibrous-shaped MPs enter to the environment, they can persist in the environment for long periods of time due to their resistance to degradation. Consequently, they have the potential to contaminate both aquatic and terrestrial ecosystems. Moreover, these MPs can adsorb other emerging contaminants from the surrounding water and soil systems, effectively acting as carriers for these substances. This results in increased mobility and persistence of these emerging contaminants (Zhou et al.,

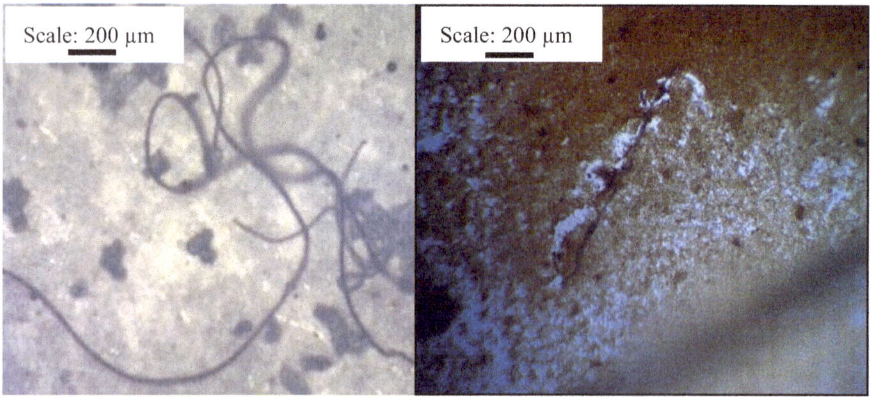

Fig. 7.3 Fibrous-shaped MPs in textile waste sludge

2019). Therefore, it is necessary to develop a sustainable remedial approach to effectively remove MPs from the textile waste sludge.

7.3.2 *Phytoextraction of MPs Using* Lemna Minor

Initial *Lemna minor* leave samples were analyzed for the presence of available MPs in their biomass. However, MP types present in the waste sludge were not observed in the optical microscope images of the initial *Lemna minor* leave samples. This indicated that the plant leaves did not contain MPs prior to their growth in the textile waste sludge. After 7 days of growth, a few fibrous-shaped MPs were observed in the *Lemna minor* leave samples (Fig. 7.4a and b). Interestingly, the number of fibrous-shaped MPs observed in the *Lemna minor* leaves samples was steadily increased as the growth of the samples reached 21 days (Fig. 7.4c and d) and 45 days (Fig. 7.4e and f), respectively. After counting the number of fibrous-shaped MPs in the leaves of the plants, the average MP concentrations of the plant leaves were calculated as 1.3 MPs per plant after 7 days, 5.6 MPs per plant after 21 days and 9.2 MPs per plant after 45 days. Figure 7.5a and b present the FTIR spectra of the *Lemna minor* leave samples prior to the growth in the textile waste sludge and after the growth of 45 days in the textile waste sludge, respectively. Therefore, it is revealed that *Lemna minor* plants continuously absorbed fibrous-shaped MPs into their leaves from the textile waste sludge, showing their potential for phytoextraction of MPs from the textile waste sludge. As the phytoextraction process is entirely natural, devoid of chemical or synthetic inputs, it stands as a non-invasive method that preserves the ecological equilibrium of the aquatic environment. Moreover, in comparison to other remediation techniques for MPs, the utilization of *Lemna minor* plants in phytoextraction proves to be a relatively cost-effective approach.

7.4 Limitations of This Study

Due to the lack of practical analytical methods and time limitations, optimizing this process became challenging, particularly when considering other aspects like utilization of diverse textile waste sludge streams and various types of aquatic macrophytes with high affinity for MPs. However, by optimizing growth conditions, it is possible to enhance the accumulation of MPs in these plants even further.

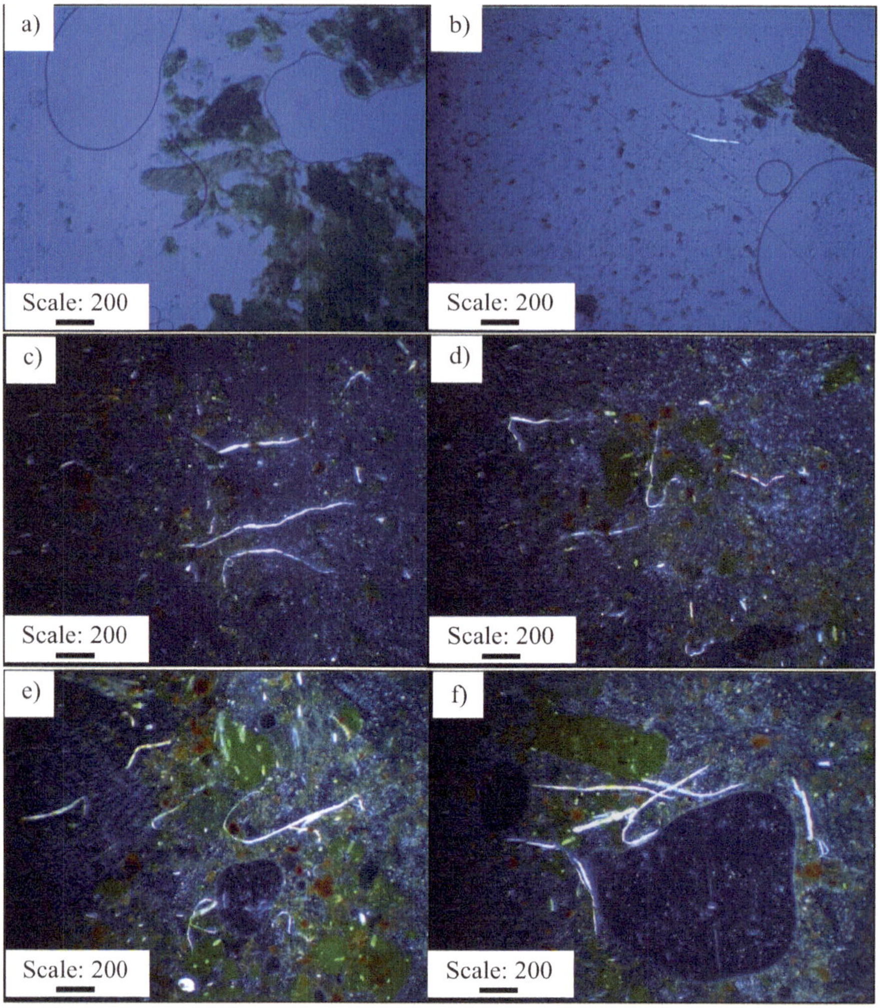

Fig. 7.4 Fibrous-type MPs present in *Lemna minor* leave samples **a** and **b** after 7 days, **c** and **d** after 21 days and **e** and **f** after 45 days

7.5 Conclusion

This study was conducted to evaluate textile waste sludge as a significant source of MP release into the environment and to develop a sustainable phytoremediation approach for its mitigation. According to the microscopic analysis after the extraction of MPs from the textile waste sludge, it was determined that the average MP concentration in the textile waste sludge was 2.2 MPs per gram, predominantly in fibrous form. The phytoextraction method employed, using *Lemna minor* as a model aquatic macrophyte, proved highly effective. As the plants matured, there was a

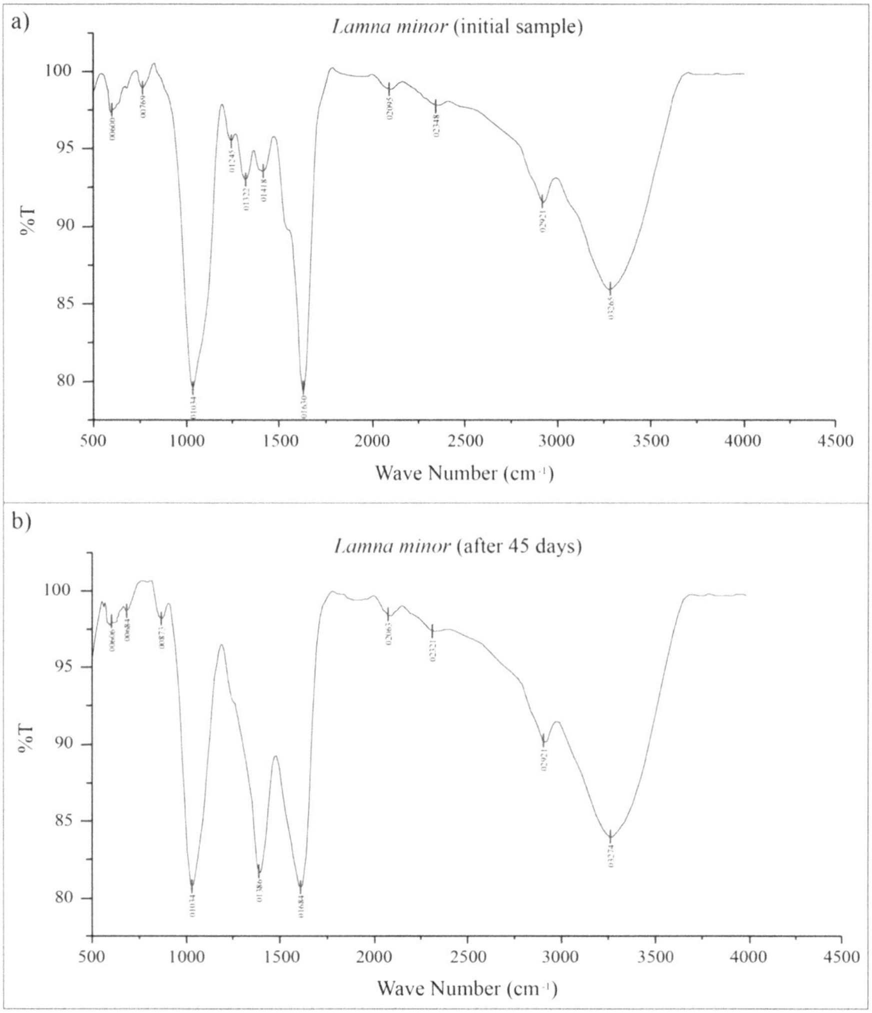

Fig. 7.5 FTIR Transmittance % results of **a** initial plant leaves and **b** plant leaves after the growth of 45 days

steady increase in fibrous-shaped MPs, with concentrations reaching 1.3 MPs per plant after 7 days, 5.6 MPs per plant after 21 days, and 9.2 MPs per plant after 45 days. These compelling results demonstrated the high affinity of *Lemna minor* plants to actively absorb fibrous-shaped MPs from textile waste sludge, showing their potential as agents for phytoextraction of MPs from such waste streams. Once these plants reach their maximum biomass yield, the biomass can be sent to cement kilns within a cement clinker production facility in Sri Lanka for high-temperature incineration. Four aquatic plants were used at the initial stage to identify the most effective plant for the particular sludge, then *Lemma minor* was selected to continue the study because it shows a rapid growth rate, easy harvesting process and it shows huge biomass in a short time duration. The resulting remediated sludge can then be utilized as fertilizer. Furthermore, this cost-effective and environmentally friendly process can be applied to remove MPs from other waste streams too.

7.6 Future Perspectives

It is necessary to assess the hyperaccumulation potential of different aquatic macrophytes for MPs. It holds great significance in addressing environmental contamination and exploring sustainable phytoremediation methods. Understanding the mechanisms underlying hyperaccumulation in aquatic macrophytes will increase their ecological importance and support the development of innovative strategies for the remediation of MP-contaminated environments. Therefore, further research is needed to assess the efficiency of phytoextraction and its viability for large-scale applications. In addition, the appropriate disposal of these macrophytes or the advanced processing of harvested biomass to extract MPs is a consideration for future endeavors. In this context, special attention should be given to industrial plants that can be used for textile fiber manufacturing. Plants such as Corn, Wheat, Hemp and Bamboo have been identified in the literature for their potential phytoremediation capabilities for MPs. Since these plants are directly involved in extracting textile fibers, evaluating and enhancing the accumulation of MPs in these plants can offer a more sustainable solution to combat MP contamination arising from the textile industry, aligning with the principles of a circular economy. This is only a qualitative analysis and this study was performed to ensure the *Lemma minor* could be used as an indicator plant for the phytoremediation of textile MPs.

7.7 Declaration of Competing Interest

The authors declare that they have no known competing financial interests or personal relationships that could have appeared to influence the work reported in this paper.

Acknowledgements The author is grateful to the Department of Applied Earth Sciences, Faculty of Applied Sciences, Uva Wellassa University for providing lab facilities for this study.

Data Availability Data will be made available on request.

References

Acharya, S., Rumi, S. S., Hu, Y., & Abidi, N. (2021). Microfibers from synthetic textiles as a major source of microplastics in the environment: A review. *Textile Research Journal, 91*, 2136–2156.

Aslam, M. M., Baig, M. A., Hassan, I., Qazi, I. A., Malik, M., & Saeed, H. (2004). Textile wastewater characterization and reduction of its COD and BOD by oxidation. *Electronic Journal of Environmental, Agricultural and Food Chemistry, 3*, 804–811.

Bhat, S. A., Bashir, O., Haq, S. A. U., Amin, T., Rafiq, A., Ali, M., Américo-Pinheiro, J. H. P., & Sher, F. (2022). Phytoremediation of heavy metals in soil and water: An eco-friendly, sustainable and multidisciplinary approach. *Chemosphere, 303*, 134788.

Bläsing, M., & Amelung, W. (2018). Plastics in soil: Analytical methods and possible sources. *Science of the Total Environment, 612*, 422–435. https://doi.org/10.1016/j.scitotenv.2017.08.086.

Boucher, J., & Friot, D. (2017). *Primary microplastics in the Oceans: A global evaluation of sources.*

Capozzi, F., Carotenuto, R., Giordano, S., & Spagnuolo, V. (2018). Evidence on the effectiveness of mosses for biomonitoring of microplastics in fresh water environment. *Chemosphere, 205*, 1–7.

Chamanee, G., Sewwandi, M., Wijesekara, H., & Vithanage, M. (2023). Global perspective on microplastics in landfill leachate; Occurrence, abundance, characteristics, and environmental impact. *Waste Management, 171*, 10–25.

De Falco, F., Di Pace, E., Cocca, M., & Avella, M. (2019). The contribution of washing processes of synthetic clothes to microplastic pollution. *Science and Reports, 9*, 6633.

Ebere, E. C., Wirnkor, V. A., & Ngozi, V. E. (2019). Uptake of microplastics by plant: A reason to worry or to be happy? *World Science News* 256–267.

Gao, Z., Cizdziel, J. V., & Chen, L. (2023). Microplastics profile in sludge from a university wastewater treatment plant and the influence of chemical digestions on Nile red stained microplastics. *Journal of Environmental Chemical Engineering, 11*, 109671.

Hettiarachchi, H., & Meegoda, J. N. (2023). Microplastic pollution prevention: The need for robust policy interventions to close the loopholes in current waste management practices. *International Journal of Environmental Research and Public Health, 20*, 6434.

Kim, J., Hwang, Y.S., 2023. Facile microplastic recovery in soil using temperature-driven phase transition of saturated fat: Optimization and evaluation. *Journal of Environmental Chemical Engineering* 111101.

Koongolla, J. B., Lin, L., Pan, Y., Yang, C., Sun, D., Liu, S., Xu, X., Maharana, D., Huang, J., & Li, H. (2020). Occurrence of microplastics in gastrointestinal tracts and gills of fish from Beibu Gulf, South China Sea. *Environmental Pollution, 258*, 113734. https://doi.org/10.1016/j.envpol.2019.113734.

Li, L., Luo, Y., Peijnenburg, W. J. G. M., Li, R., Yang, J., & Zhou, Q. (2020). Confocal measurement of microplastics uptake by plants. *MethodsX, 7*, 100750. https://doi.org/10.1016/j.mex.2019.11.023.

Masura, J., Baker, J., Foster, G., & Arthur, C. (2015). Laboratory methods for the analysis of microplastics in the marine environment: Recommendations for quantifying synthetic particles in waters and sediments.

Mateos-Cárdenas, A., Scott, D. T., Seitmaganbetova, G., van Pelt Frank, N. A. M., & Jansen Marcel, A. K. (2019). Polyethylene microplastics adhere to Lemna minor (L.), yet have no effects on plant growth or feeding by Gammarus duebeni (Lillj.). *Science of the Total Environment, 689*, 413–421.

Prata, J. C. (2018). Airborne microplastics: Consequences to human health? *Environmental Pollution, 234*, 115–126.

Rafiq, A., & Xu, J.-L. (2023). Microplastics in waste management systems: A review of analytical methods, challenges and prospects. *Waste Management, 171*, 54–70.

Rozman, U., & Kalčíková, G. (2022). The response of duckweed Lemna minor to microplastics and its potential use as a bioindicator of microplastic pollution. *Plants, 11*, 2953.

Sharifinia, M., Bahmanbeigloo, Z. A., Keshavarzifard, M., Khanjani, M. H., & Lyons, B. P. (2020). Microplastic pollution as a grand challenge in marine research: A closer look at their adverse impacts on the immune and reproductive systems. *Ecotoxicology and Environmental Safety, 204*, 111109.

Xu, C., Zhang, B., Gu, C., Shen, C., Yin, S., Aamir, M., & Li, F. (2020). Are we underestimating the sources of microplastic pollution in terrestrial environment? *Journal of Hazardous Materials, 400*, 123228.

Zhou, Y., Liu, X., & Wang, J. (2019). Characterization of microplastics and the association of heavy metals with microplastics in suburban soil of central China. *Science of the Total Environment, 694*, 133798.

Ziajahromi, S., Neale, P. A., & Leusch, F. D. L. (2016). Wastewater treatment plant effluent as a source of microplastics: Review of the fate, chemical interactions and potential risks to aquatic organisms. *Water Science and Technology, 74*, 2253–2269.

Chapter 8
Biodegradation of Polyethylene Plastic Waste by Bacterial Isolates from Earthworm Larvae

Ansam Subhi Sabei, Iman Hendi Gatea, Adnan Hannon Abbas, Qusay A. Abdulamer, Rnafawzy Tawfeeq, and Ameena Gazy Abid

Abstract For decades, polyethylene (PE) has been regarded as non-biodegradable. The literature has provided only a limited amount of valid evidence regarding the biodegradation of PE by bacterial cultures, despite the occasional description of this procedure. Two bacterial Bacilli spp. isolates were recovered from the intestine of earthworms and were found to be capable of degrading PE. In addition to microscopic examination, the bacterial isolates were characterized using selective media and a few biochemical assays. The maximal growth rates for the two isolates B1 and B2 were estimated to be 1.8 and 1.3, respectively, three weeks after the experiment commenced, using a Spectrophotometer at a wavelength of 600 nm. During the twenty-one-day incubation period of the two isolates on PE films, viable biofilms were formed, and the hydrophobicity of the PE films decreased. Subsequently, atomic force microscopy (AFM) was employed to detect evident damage, such as fissures and cavities (18.5–21.1 nm in depth), on the surfaces of the PE films for B1 and B2. Fourier transform infrared (FTIR) was employed to confirm the formation of carbonyl groups. The findings indicate that PE-degrading bacteria are present in the guts of earthworms, and they offer optimistic evidence for the biodegradation of PE in the environment.

Keywords Earthworm larvae · *Bacillus* · PE · Biofilm

8.1 Introduction

One of the most significant challenges that the environment is currently facing is the disposal of plastic, as large quantities of synthetic plastic do not biodegrade. A diverse array of microorganisms has the capacity to degrade various types of plastic

A. S. Sabei · I. H. Gatea (✉) · A. H. Abbas · Q. A. Abdulamer · R. Tawfeeq · A. G. Abid
Environment, Water and Renewable Centre, Scientific Research Commission, Ministry of Higher Education and Scientific Research, Baghdad, Iraq
e-mail: eman77aa@yahoo.com

S. Suriyanarayanan et al. (eds.), *Combating Plastic Pollution in Terrestrial Environment*,
https://doi.org/10.1007/978-981-96-2343-3_8

under the appropriate circumstances. Nevertheless, biological decomposition is a protracted process due to the fact that these polymers are difficult to degrade and are not soluble in water. The native state of the substance renders it incapable of being deconstructed into simpler components. Consequently, the utilization of these materials in the development of packaging or waste disposal products leads to adverse environmental consequences (Kwpp & Jewell, 1992; Zeb et al., 2020).

The degradation of the majority of synthetic plastics in natural environments is a relatively gradual process that is initiated by the influence of the surrounding environment and subsequently facilitated by the activity of free-living microorganisms. In general, the process of microbes breaking down plastic into smaller components is extremely slow, and certain microorganisms are incapable of breaking down specific varieties of plastic (Albertsson et al., 1994).

Chemical, thermal, photooxidation, and biodegradation are all processes that can decompose plastic. The duration of these processes is dependent on the molecular weight of the polymer, with some forms of plastics taking up to a thousand years to decompose (Mahdiyah & Mukti, 2013; Pramila & Ramesh, 2011). Biodegradable plastics are materials that are intended to decompose naturally or in biological waste treatment facilities for municipal or industrial waste (Singh & Gupta, 2014). Consequently, biodegradable plastics enable the implementation of novel waste management strategies (Gouda et al., 2012).

Carbon dioxide and water are produced during aerobic biodegradation, while carbon dioxide, water, and methane are produced during anaerobic biodegradation (Ishigaki et al., 2004). Virtually all types of tests can be conducted through visual observations, which involve the examination of observable changes in plastics. Some of the effects that are used to define degradation include the formation of biofilms on the surface, the defragmentation of the surface, the production of cavities or fissures, and changes in color. Visual alterations may serve as an early indicator of microbial invasion; however, they do not establish the existence of a biodegradation process in terms of metabolism. Advanced microscopy techniques, including atomic force microscopy (AFM), transmission optical microscopy (TOM), and scanning electron microscopy (SEM), can be employed to conduct more detailed observations, which can subsequently be employed to gain a more comprehensive understanding of the degradation mechanism. (Ikada, 1999). In addition, recent research has demonstrated that the body can effectively eliminate toxins by ingesting them through the digestive tract, as enzymes and microorganisms in the stomach collaborate (Zeb et al., 2020). Earthworm larvae are nocturnal organisms that exist in the humus-rich, humid, and moist soil of outdoor spaces such as gardens and lawns. They dig deeper into the earth to prevent drying out when conditions are dry. Their contribution to the food web for avian species is substantial due to their roles as herbivores and large-scale decomposers. Additionally, it promotes soil aeration and the enhancement of soil fertility. A complete digestive tract, which includes an esophagus and intestine, is present in earthworm larvae. The alimentary canal that commences at the mouth and concludes at the exit is its own, complete canal (Fig. 8.1).

The earthworm larvae typically consume organic decomposing materials in the soil, which are absorbed by the pharynx and brought into the crop. The gizzard,

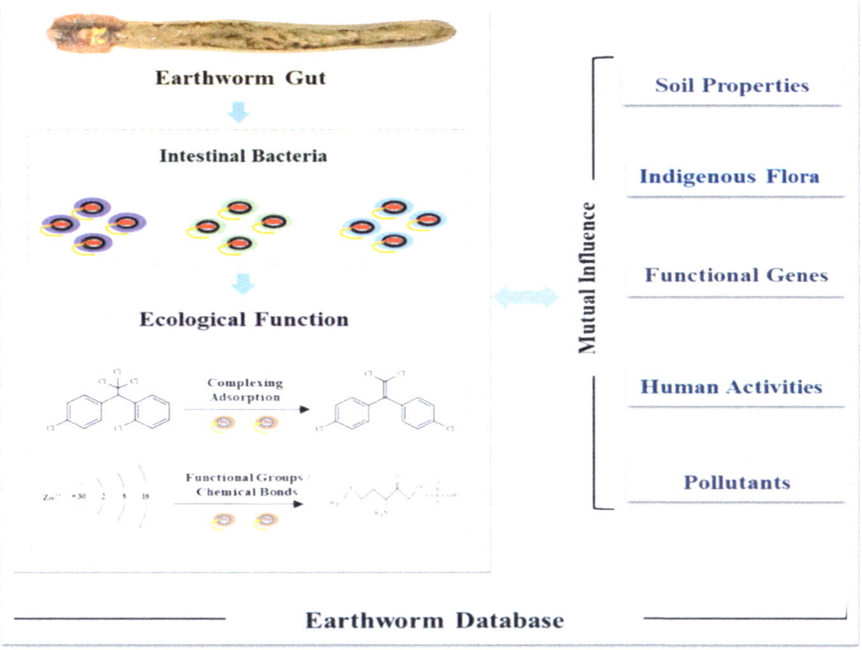

Fig. 8.1 Earthworm database Magnin et al. (2020)

which has thick muscular walls that crush and grind food with sand grains, is the next location where the food is processed. Digestion is typically external in the intestinal tract, and the absorption capacity of food particles increases as a result of the increased surface area of the alimentary canal, which is facilitated by the presence of a fold in the intestinal wall. Undigested food is expelled to the outside through the anus. Earthworm larvae, which are named after their "gut of the earth," are soil bio-organisms that support the recycling of minerals and the degradation of plant residues, including vegetation. According to Hortensia and David (2009), they contribute to the degradation and conditioning of plant detritus in their intestines.

The microbial communities that are present in the digestive systems of earthworms are influenced by the climate, soil variety, and organic material composition. They are capable of producing biogenic structures, intervening in the biological regulating systems of the soil, and removing soil particles (Rossi et al., 2006). In addition, they contribute to the regulation of the availability of nutrients that can be assimilated by plants, the infiltration of water, and the preservation of the soil structure. These nutrients consist of nitrogen in the form of nitrates and ammonia (Desjardins et al., 2003).

The significance of the earthworm's larvae intestinal microbiota in the maintenance of the earthworm's metabolism and the transformation of nutrients and pollutants in the environment has been increasingly demonstrated by evidence. The earthworm's larvae alimentary canal is conducive to the colonization of bacteria, which connects the canal to the surrounding environment, depending on its nutrient content. Consequently, this investigation examined the capacity of bacteria in the alimentary canal of earthworms to decompose plastic waste, which is presently one of the most significant environmental pollutants.

8.2 Material and Methods

8.2.1 Prepare the Larvae for the Experiments

Adult earthworms weighing 50 g were obtained from a local animal supply store, selecting only those in optimal health. To obtain the stomach contents of the worm larvae for the enrichment of PE-degrading bacteria, the surfaces of the larvae were sterilized by submerging them in 75% ethanol for one minute, followed by two rinses with sterile saline solution. In a tissue homogenizer, fifty earthworm larvae were desiccated and ground into a powder. One gram of dry powder from earthworm larvae was transferred into a small glass vial, followed by a series of dilutions.

8.2.2 Serial Dilution and Screening

The specimen underwent several dilutions; initially, 9 ml of sterile distilled water was incorporated into the desiccated weight derived from the worms; subsequently, 1 ml was extracted from the fourth dilution; this aliquot was then disseminated onto the nutrient agar medium and incubated at 30° C for 48 h (Fig. 8.2).

8.2.3 **Purification of Bacterial Isolates*:***

Before being used in further studies, bacterial isolates had been purified on nutrient agar media and stored in the fridge.

Fig. 8.2 Adult earthworms serial dilution

8.2.4 Measuring the Efficiency of Bacterial Isolates on Plastic Degradation

The liquid basal medium (LBM), was made with deionized water and contained 1.0 g of NH4NO3, 0.005 g of NaCl, 0.002 g of FeSO4. 7H2O, 0.002 g of ZnSO4. 7H2O, and 0.001 g of Mn per 1000 ml, was filled with the bacterial inoculum from purified colonies. The Erlenmeyer flask held the LBM (G22-76, 1996). After autoclaving the medium for twenty minutes at 121 degrees Celsius, 0.25 g of the tiny PE fragments were added (the PE film was chopped into squares that measured ten millimeters by ten millimeters). The components of polyethylene (PE) were weighed, cleaned in 75% ethanol, and allowed to air dry in a sterile bench with laminar flow before being used. The flasks were incubated at 30 °C for 3, 7, 14, 21, and 28 days, respectively, on a rotary shaker set at 120 rpm.

8.3 Measure Growth and Prepare Samples to Detect Dissociation Efficiency

A spectrophotometer was used to measure growth at various incubation times. The measurements were noted. Biodegradation Assays Following a 28-day incubation period, the PE biodegradation process was detailed. A 4-h washing procedure with a 2% w/v aqueous sodium dodecyl sulfate (SDS) solution completely removed the biofilm that was on the PE sheet (Gilan et al., 2004; Sivan et al., 2006). The PE sheets that were utilized in the sterile control group were also treated using the identical procedure. Consequently, each and every PE sheet's surface topography was characterized using the atomic force microscope (AFM) (Gilan et al., 2004; Sudhakar et al. 2008). FTIR imaging was used to look at the changes that occurred in the chemical components and function groups of the surface.

8.4 Results and Discussion

Serial dilutions of worm larvae samples were prepared at the very beginning of the experiments. These samples were procured from local markets for the purpose of preparing sustenance for fish and other animals. Subsequently, they were cultured by spreading them on nutrient agar medium and cultivated at a temperature of 30 °C. The growth was monitored for 48 h, and the bacterial isolates were subsequently purified on the same medium. They were obtained 18 pure bacterial isolates, which were stored in the refrigerator until they were needed. The isolates were cultivated on a liquid medium that contained a group of mineral salts and were equipped with plastic pieces of polyethylene that were prepared as described in the working methods. The medium was incubated at a temperature of 30 °C, and growth readings were obtained using a spectrophotometer to assess absorbance. The results indicated that two bacterial isolates from the genus Bacillus were present. The microscopic examination and certain biochemical experiments (Fig. 8.3) and (Table 8.1) have determined that it is capable of degrading plastic and promoting growth by utilizing it as the sole carbon source.

The best growth was 1.8 and 1.4 for the two isolates B1, and B2, respectively, after 21 days of growth on plastic slides (Fig. 8.4).

The fundamental mechanism for the biodegradation of this type of polymer is the oxidation or hydrolysis of a high molecular weight polymer by an enzyme, which leads to the formation of functional groups and an increase in the polymer's hydrophilicity. We employed FTIR analysis to ascertain whether the worm homogenate caused the PE polymer to endure chemical degradation upon contact. After 28 days of incubation in LBM at 30 °C, PET sheets were subjected to FTIR spectroscopy in the wave number range of 450–4000 cm^{-1}. Using Fig. 8.2, we can illustrate that PET exhibited the formation of three distinct absorption peaks at 1472, 1462, and 729 cm^{-1} in response to C-H bond elongation (Methylene group) under laboratory conditions.

The FTIR spectra indicate that the emergence of new absorption peaks at 2920 and 2850 cm^{-1} indicates C-H bond stretching (CH2). The 2669 cm^{-1} peak is attributed to C=O bond stretching (Carbonyl group), and the 2341 and 874 cm^{-1} peaks are assigned to C=C bond stretching (Benzene ring). The spectroscopic data corroborated the identity of the PE film, as it exhibited peaks at 2,920 and 2,850 cm^{-1}, which are the

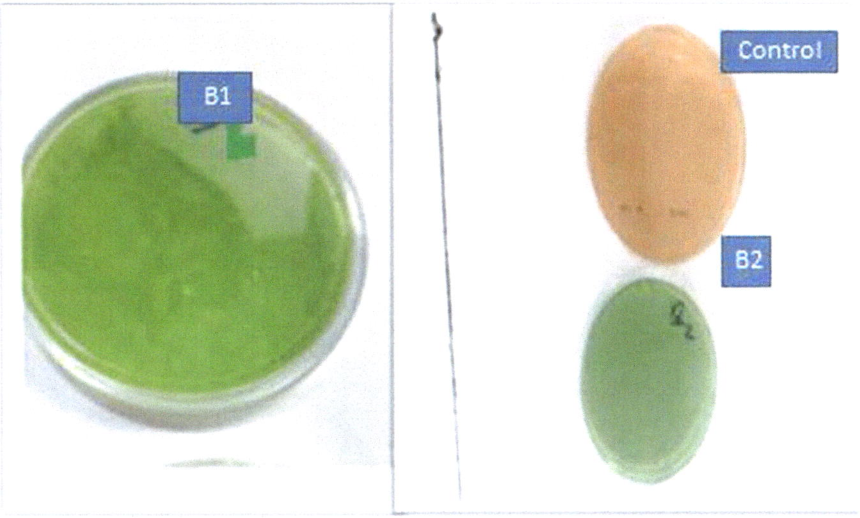

Fig. 8.3 Growth pf bacterial isolates on selective medium

Table 8.1 Characterization of active isolates

Test		Bacterial isolate	
		Bacillus spp B1	*Bacillus* spp B2
Gram-stain reaction		Gram-positive endospore-forming rod-shaped	Gram-positive endospore-forming rod-shaped
Colonial appearance		Rough circular colony , fuzzy yellowish white with jagged edges	Spreading, and irregularly shaped round whitish
Sugar fermentation	Mannitol	+	-
	Glucose	+	+
	Lactose	-	-
	Maltose	-	-
Oxidase		+	+
Catalase		-	-
Indole and methyl red tests		-	+
Starch hydrolysis		+	+
Penicillin susceptibility		R	R
Lecithinase		-	+

traditional hallmarks of PE, when the FTIR probe was directed at untreated samples (Fig. 8.5).

Variations in the Surface's Physicochemical Properties The physicochemical surface properties of the incubated PE film sheets on the LBM were observed to alter after the bacterial growth was completely removed from the PE samples, following a

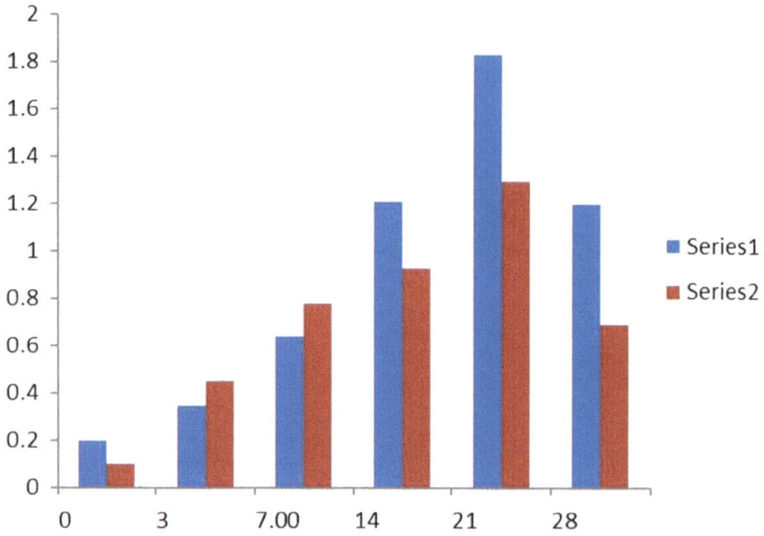

Fig. 8.4 Growth of B1 (series 1), B2 (series 2) isolates on PE at 30 °C

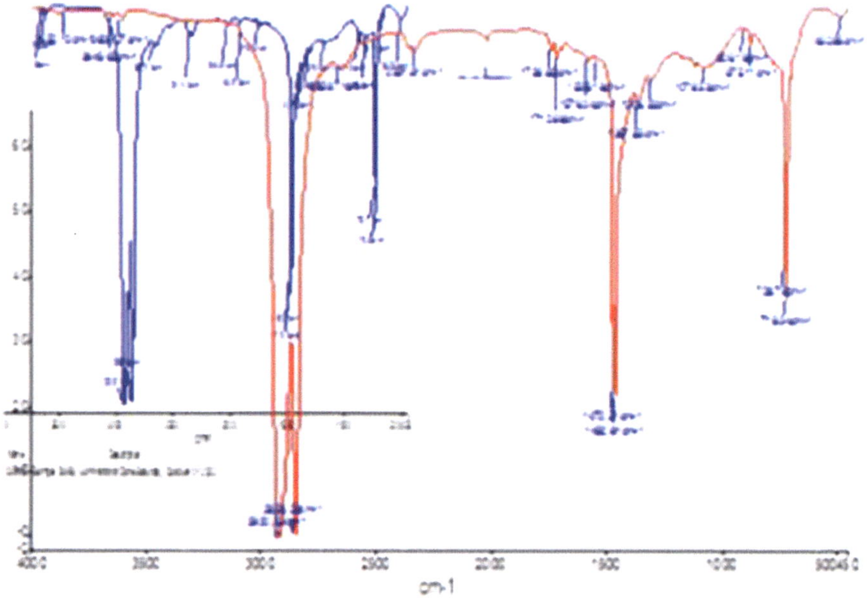

Fig. 8.5 FTIR spectra of PE samples after the 28-day incubation with strains B1 (Red) and B2 (Blue)

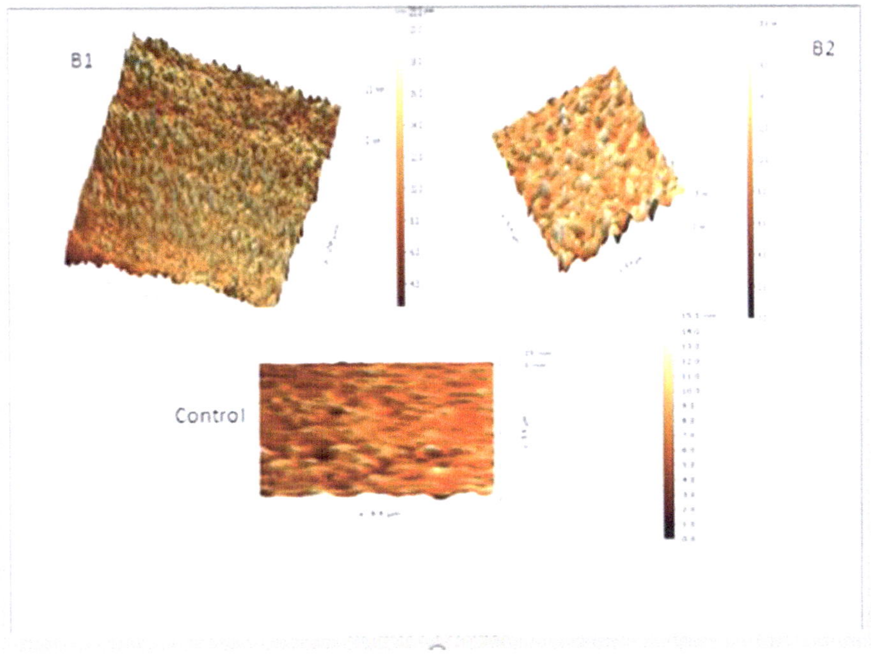

Fig. 8.6 AFM observations of the physical surface topography of the sterile control versus the PE sheets incubated with strains B1 and B2 after 28 days

28-day incubation period. This was performed subsequent to the complete removal of bacterial proliferation from the PE samples. The surface deterioration was determined by analyzing the change in surface topography using the AFM, and the appearance of fissures and cavities on the surfaces of the PE samples that had been inoculated with bacteria (Fig. 8.6).

Smooth surfaces, devoid of any imperfections, were preserved by the controls. The cavities on the surfaces of the PE sheets attained their maximum depths of approximately 21.1 and 18.6 nm after being inoculated with strains B1 and B2, respectively (Fig. 8.6).

Sudhakar et al. (2008) and Tribedi and Sil (2013) have not reported the discovery of this specific form of deep cavity by other researchers who are investigating the characterization of microorganisms that are capable of digesting polyethylene. The two strains did, in fact, impair the physical integrity of the PE material, as indicated by the microscale surface topography findings.

8.5 Conclusion

The investigation involved the evaluation of bacterial isolates from the intestines of worm larvae to determine their efficacy in disassembling polyethylene, a type of plastic. The isolates demonstrated their ability to disassemble through the use of contemporary techniques, including AFM and FTIR technology. Consequently, it is strongly advised that this biological method be implemented as a primary strategy for the reduction of plastic pollution in both the marine and terrestrial environments. This will be achieved through partnerships between local governments and corporations. Despite the necessity of remedial measures, including recycling and cleansing, to eliminate these pollutants.

References

Albertsson, A. C., Barenstedt, C., & Karlsson, S. (1994). Abiotic degradation products from enhanced environmentally degradable polyethylene. *Acta Polymerica, 45*, 97–103.

ASTM G 22-76. (1996). Standard practice for determining resistance of plastics to bacteria.

Desjardins, T., Charpentier, F., Pashanasi, B., Pando-Bahuon, A., Lavelle, P., & Mariotti, A. (2003). Effects of earthworm inoculation on soil organic matter dynamics of cultivated ultisol. *Pedobiologia, 47*, 835–841.

Gilan, I., Hadar, Y., & Sivan, A. (2004). Colonization, biofilm formation and biodegradation of polyethylene by a strain of Rhodococcusruber. *Applied Microbiology and Biotechnology, 65*(1), 97–104.

Gouda, M. K., Swellam, A. E., & Omar, S. H. (2012). Biodegradation of synthetic polyesters (BTA and PCL) with natural flora in soil burial and pure cultures under ambient temperature. *Research Journal of Environmental and Earth Science, 4*, 325–333.

Hortensia, B.-V., & David, E.-V. (2009). Bacterial diversity in the digestive tract of earthworms (Oligochaeta). *Journal of Biological Sciences, 9*, 192–199.

Ikada, E. (1999). Electron microscope observation of biodegradation of polymers. *Journal of Environmental Polymer Degradation, 7*, 197–201.

Ishigaki, T., Sugano, W., Nakanishi, A., Tateda, M., Ike, M., & Fujita, M. (2004). The degradability of biodegradable plastics in aerobic and anaerobic waste landfill model reactors. *Chemosphere, 54*, 225–233.

Kwpp, L. R. & Jewell, W. J. (1992). Biodegradability of modified plastic films in controlled biological environments. *Environmental Science and Technology, 26*, 193–198; Lee, B., Pometto, A. L., Fratzke, A., & Bailey, T. B. (1991). *Applied and Environmental Microbiology, 57*, 678–685.

Magnin, A., Pollet, E., Phalip, V., & Avérous, L. (2020). Evaluation of biological degradation of polyurethanes. *Biotechnology Advances, 39*, 107457.

Mahdiyah, D. & Mukti, B. H. (2013). Isolation of polyethylene plastic degrading-bacteria. *Bioscience International, 2*, 29–32.

Pramila, R., & Ramesh, K. V. (2011). Biodegradation of low density polyethylene (LDPE) by fungi isolated from municipal landfill area. *Journal of Microbiology and Biotechnology, 1*, 131–136.

Rossi, J. P., Huerta, E., Fragoso, C., & Lavelle, P. (2006). Soil properties inside earthworm patches and gaps in a tropical grassland (la Mancha, Veracruz, Mexico). *European Journal of Soil Biology, 42*, S284–S288.

Singh, J., & Gupta, K. (2014). Screening and identification of Low density Polyethylene (LDPE) degrading soil fungi isolated from polythene polluted sites around Gwalior city (MP). *International Journal of Current Microbiology and Applied Science, 3*, 443–448.

Sivan, A., Szanto, M., & Pavlov, V. (2006). Biofilm development of the polyethylene-degrading bacterium Rhodococcusruber. *Applied Microbiology and Biotechnology, 72*(2), 346–352.

Sudhakar, M., Doble, M., Sriyutha, M. P., & Venkatesan, R. (2008). Marine microbe-mediated biodegradation of low-and high-density polyethylenes. *International Biodeterioration and Biodegradation, 61*(3), 203–213. https://doi.org/10.1016/j.ibiod.2007.07

Tribedi, P., & Sil, A. K. (2013). Low-density polyethylene degradation by Pseudomonas sp. AKS2 biofilm. *Environmental Science and Pollution Research, 20*(6), 4146–4153.

Zeb, A., Li, S., Wu, J., Lian, J., Liu, W., Sun Y. (2020). Insights into the mechanisms underlying the remediation potential of earthworms in contaminated soil: A critical review of research progress and prospects. *Science of the Total Environment, 740*, Article 140145.

Chapter 9
Community-Based Plastic Waste Management Initiatives: A Key to Combating Plastic Pollution

Arockia E J Ferdin, Jia Wei Chook, and Patricia Manzano-Fischer

Abstract The advent of plastic in the Anthropocene is transforming our blue planet into a 'plastic planet.' While key technologies exist to reduce, recycle, and repurpose plastic waste, it remains a human-induced hazard threatening life on Earth. This chapter underscores the importance of establishing community-based plastic waste management (CBPWM) initiatives to combat plastic pollution. It provides a definition of CBPWM and elaborates on the essential components necessary for implementation. Through a series of case studies spanning diverse global contexts, including urban and rural settings, this chapter illustrates how communities have responded to their unique needs. It also highlights the challenges faced by these communities in establishing effective CBPWM systems. Drawing on these case studies and identified challenges, the authors offer recommendations that may assist individuals, communities, local organizations, government authorities, environmental educators, and policymakers in formulating locally adapted policies to address community-specific needs. These recommendations are a vital contribution to the global fight against plastic pollution, demonstrating that community engagement at the local level can have far-reaching impacts on environmental protection and the well-being of all living things on Earth.

Keywords Beat plastic pollution · Bottom-up approach · Community-based plastic waste management · Community empowerment · Community engagement

---

A. E J Ferdin (✉)
Department of Natural Resources and Environmental Studies, College of Environmental Studies and Oceanography, National Dong Hwa University, Hualien, Taiwan
e-mail: ferdinenv@gmail.com

J. W. Chook
DOMI Earth, Taipei, Taiwan
e-mail: christina.c25@outlook.com

P. Manzano-Fischer
Instituto de Geografía, Universidad Nacional Autónoma de México, Mexico City, Mexico
e-mail: pmanzano@comunidad.unam.mx

S. Suriyanarayanan et al. (eds.), *Combating Plastic Pollution in Terrestrial Environment*,
https://doi.org/10.1007/978-981-96-2343-3_9

9.1 Introduction

Plastics, though inexpensive to produce, present a significant challenge due to their non-biodegradable nature, despite not being a major component of solid waste (Kaza et al., 2018). Plastics can break down into macro, micro and nanoplastics, existing in substantial quantities and various forms across every continent, causing harm to the environment and all living organisms (Lim, 2021). Plastics are also integrated into non-plastic manufactured products, making household-level separation challenging. For instance, Tetra Pak cartons contain paper, aluminium, and low-density polyethylene (LDPE), a type of plastic that poses challenges in post-consumer recovery and recycling (Ma, 2018). While seven common types of plastics are produced and used globally, segregating these types in practice is challenging. It is essential to segregate plastic waste at its source (Dharia et al., 2023). Plastic is a human-induced hazard, and managing plastic waste takes place in the context of human society (Chook et al., 2024). Implementing management strategies tailored to local contexts plays a pivotal role in the global fight against plastic pollution (Willis et al., 2022). The extensive evidence demonstrates that community engagement, in collaboration with relevant stakeholders, effectively addresses issues related to plastic waste management (Kaza et al., 2018; Mutemani et al., 2022). Moreover, focusing on plastic waste management at the local level can complement existing solid waste management systems worldwide.

In this context, community-based plastic waste management (CBPWM) is defined as an alternative, bottom-up approach to solid waste management. It is a form of community-based waste management that primarily focuses on the reduction, reuse, and recycling of plastic waste. CBPWM initiatives are specifically designed, co-managed, and collaboratively implemented with key local stakeholders to meet the needs of communities in handling their plastic waste.

These CBPWM initiatives often emerge in response to inefficiencies in local authorities' efforts to manage plastic waste within their neighborhoods. For instance, community-based waste management programs were established due to the local government's inability to offer efficient waste collection services and the presence of outdated waste collection vehicles (Colon & Fawcett, 2006; Mutemani et al., 2022). Recognizing the imperative need to maintain clean and healthy communities (Kaza et al., 2018), these initiatives empowered residents to manage their own waste at the household level (Ambastha & Aich, 2020; Wijayanti & Suryani, 2015). By encouraging waste separation at its source (Colon & Fawcett, 2006), CBPWM initiatives aimed at fostering a sense of environmental responsibility among community members, emphasizing the 3R approach—Reduce, Reuse, Recycle (Yandri et al., 2023). Moreover, these efforts not only focused on efficient waste management but also aimed to create local job opportunities and additional income (Fahmi, 2005), making waste management economically advantageous to cover their living expenses (Rathi, 2006; Wijayanti & Suryani, 2015), while addressing local environmental concerns.

9.1.1 Impacts of Plastics on Rural and Urban Communities

We live in a plastic society, where plastics are intertwined with our daily lives. While companies primarily bear responsibility for creating plastic products, it's the consumers who contribute significantly to the high consumption and disposal of plastic waste. However, the use of plastic, disposal, and its impacts vary between rural and urban communities, depending on the lifestyle and applications. For example, rural communities dependence on plastics introduces problems, including the breakdown of plastics into micro and nanoplastics during and after the use of agricultural plastic products, which pollute and affect soil health, exhibit root uptake by plants reaching edible vegetables and fruits (Oliveri Conti et al., 2020) and entering our food chain, and are transported to off-sites via wind and water runoff (FAO, 2021). A new research conducted in the agricultural lands in Bhopal, India identified five kinds of microplastics present in all their gathered soil samples (Singh et al., 2023). Microplastics have even been reportedly found in earthworms and being transported by them to the different soil layers and could potentially reach groundwater (FAO, 2021; Rillig et al., 2017). This shows the heightened interaction between fragmented plastics with vertebrates and invertebrates on soil and animals and birds may consume them on the ground.

Furthermore, the widespread illegal dumping of plastic waste in rural area poses significant challenges (Fig. 9.1), attracting a multitude of birds, domesticated and non-domesticated animals often assume them as a food or food source. Numerous studies have reported instances of ingestion of plastic by birds (Zhao et al., 2016) and the use of plastics in bird nests (FAO, 2021), ingestion of plastics by domesticated animals was found on cattle (Turnbull & Barua, 2023), sheep and goats (Otsyina et al., 2014), and wildlife such as Arabian camels (*Camelus dromedarius*) (Eriksen et al., 2021) Asian elephants (*Elephas maximus*) (Katlam et al., 2022) and the Indian rhinoceros (*Rhinoceros unicornis*) (Awasthi et al., 2023). This not only results in increased mortalities due to mismanaged plastic waste, but particularly endangers threatened species, opening a new arena for conservation biologists. Further, plastics in its various forms deteriorate the wildlife habitats, disrupting ecosystem functioning and its ability to adapt to climate change.

Various sources, such as households, commercial establishments, and industries generate plastic waste in urban areas. Improperly disposed of plastic waste can clog drainage systems (Fig. 9.2), leading to flooding and the spread of waterborne diseases, thus impacting the quality of life of residents. Additionally, high urban density can make it more difficult to implement effective waste separation and recycling programs, as there may be limited space for waste storage and processing facilities. There exists a trade-off between saving space and converting plastic waste into energy or fuels, leading to the increased release of toxic substances into the air. This can harm not only the urban community through air pollution and respiratory illness but also affect the urban environment's ability to adapt to climate change. Furthermore, countries in the Global North have been using developing countries as a dumping ground for decades by sending their plastic waste because it is cheaper than processing it

Fig. 9.1 Plastic bags scattered from a landfill near the Sierra Gorda Biosphere Reserve in Queretaro, Mexico (Photo credit: Rurik List)

domestically. This influx of surplus and sometimes illegal unwanted plastic wastes adds further strain to the urban waste management system, putting a heavy burden on urban communities in the Global South and leading to social and environmental injustice. Consumers are unknowingly eating plastics of an equivalent size of a debit card or five grams per week through food and beverages (Senathirajah et al., 2021). This could disrupt the functioning of entire human systems, leading to an array of health problems (Consumer Reports, 2024).

The role of plastic waste and its impacts on rural and urban communities has highlighted the urgency and complexity of the plastic waste problem. There is no one-size-fits-all solution for managing plastic waste in these different contexts. Empowering communities and relevant stakeholders to reduce, reuse, and recycle plastic waste in their localities within these settings through CBPWM is one possible way to achieve this. The aim of this paper is to outline the key components of CBPWM, case studies from urban and rural settings, the challenges and opportunities to different contexts, and recommendations for reducing plastic waste in the environment.

Fig. 9.2 Trash from upstream villages flows through canals into the Lerma River, Mexico's second-longest river (Photo credit: Rurik List)

9.2 Key Components of CBPWM

CBPWM is a project driven by local communities and groups of people who are aware of the excessive plastic waste problem and want to take action. The key components required for effective CBPWM can be metaphorically explained as the human heart, brain, muscles, and bones, meaning that anyone who has these components can participate in the initiatives.

9.2.1 Heart: Perception and Awareness

The heart symbolizes the perceptions and awareness of individuals regarding plastic waste. As previously mentioned, consumers play a significant role in plastic consumption and its resulting waste generation. Many hold the misconception that plastic packaging, such as single-use plastics (SUPs) and food wrapped in plastic bags, is more hygienic compared to alternatives. Therefore, gaining insights into how communities perceive plastic consumption, the impacts of plastic waste, and its management is crucial. This collective understanding of community perceptions can greatly assist local authorities, policymakers, and relevant stakeholders in developing effective solutions and strategies for reducing, reusing, and recycling plastic waste.

These strategies should be tailored to address the specific concerns, needs, and demands of diverse communities. Community perceptions are complex to understand, yet they represent a fundamental component at the core of CBPWM. Since different communities have varying consumption preferences based on their lifestyles, considering these perceptions can significantly enhance the success of implementing effective CBPWM initiatives.

Another crucial and inseparable component of CBPWM is awareness, closely intertwined with consumers' perceptions. However, there is a segment of consumers, whose awareness regarding the impact of plastics on their health and the environment remains limited, leading them to continue their participation in the disposable plastic culture. For instance, a study conducted by Van Rensburg et al. (2020) in Dublin, South Africa found that 69% of their respondents perceived the severity of SUPs on the environment. Another study conducted by Chook et al. (2024) in Kuala Lumpur, Malaysia found that 95% of the respondents are aware of the environmental impacts of plastics. It has become evident from these findings that, for many people, plastic is convenient, lightweight, and protects their food and products. It is easy to carry, and there is often a lack of viable alternative products, making the use of plastics a habitual choice.

Therefore, awareness becomes paramount in the battle against plastic waste, as it plays a central role in mitigating the impact on both individuals and the environment. Understanding community perceptions and awareness is crucial to CBPWM, as it can drive positive behavioral changes in efforts to combat plastic pollution.

9.2.2 Brain: Knowledge

Brain represents the educational knowledge of tackling plastic waste and collaborative skills with other entities in CBPWM. Environmental education stands as a crucial element in CBPWM. Research has shown that a mere 1% increase in individual environmental knowledge leads to a 0.4% improvement in environmental behavior among university students (Vicente-Molina et al., 2013). Additionally, individuals with higher levels of education tend to exhibit greater concerns regarding the plastic waste issue (Chook et al., 2024). Early exposure to environmental education is particularly effective in helping children comprehend these issues and motivating them to take action.

Environmental education provides individuals with the ability to grasp the interconnectedness between waste, human health, and the environment, along with their respective impacts. In a broader context, there are four primary educational methodologies: education driven by community initiatives, education steered by governmental bodies, education fostered by business entities, and education conducted within academic institutions (Chow et al., 2017). Therefore, integrating a waste management module into academic curricula is essential for equipping elementary, junior, and senior high school students with the knowledge needed to manage waste responsibly. Likewise, empowering communities with the necessary knowledge to refuse, reduce, and manage plastic waste, along with community driven education initiatives, plays a crucial role in fostering responsible practices.

In the realm of plastic waste management, an array of stakeholders, beyond civic authorities, plays a significant role. These stakeholders encompass but not limited to households, community-based organizations, small businesses, waste pickers, waste collectors, and non-governmental organizations (NGOs) (Muller et al., 2001). Collaboration among these diverse entities necessitates both human capital and social capital, working collectively to devise organized solutions for improving plastic waste management within the community. The multi-layered collaboration among stakeholders proves invaluable for CBPWM, as it fosters a comprehensive understanding of the issue. Each stakeholder within the team contributes unique resources across various aspects, thereby enhancing the efficiency of plastic waste management. Most importantly, such collaboration encourages the exchange of knowledge, best practices, and innovative ideas among stakeholders.

9.2.3 Muscles: Participation and Engagement

Community participation and community engagement, metaphorically represented as the muscles of CBPWM, are integral components that embody the active involvement and commitment of individuals within a community. To establish CBPWM, it necessitates the active participation and engagement of community members. The primary distinction between participation and engagement lies in the fact that community engagement entails a proactive and purposeful exchange of ideas between citizens and public decision-makers, while participation can solely originate from community members.

Community engagement plays a pivotal role in promoting behavioral change among community members, fostering awareness, and emphasizing the significance of this bottom-up approach. It encourages individuals to develop environmental awareness and implement practices such as the 3Rs. Additionally, community engagement enhances communication among individuals within a community, addressing their concerns, offering guidance, and promoting shared responsibility approaches. This enables community members to engage in meaningful conversations and participate in the decision-making process. For instance, the formation of voluntary and community sector organizations focused on waste management in the UK demonstrates their dual role in environmental preservation by collecting, reusing, and minimizing waste, while also providing social support to vulnerable households (Forrest & Wiek, 2014). Engaging communities in waste management structures not only promotes sustainability but also fosters the capacity for change within communities and society at large (Collier et al., 2013; Tudor et al., 2011).

Households' participation constitutes the largest category of stakeholders in plastic waste management. However, challenges persist, such as a lack of willingness or sense of responsibility toward plastic waste segregation and management, a low inclination to maintain public areas clean, and a scarcity of financial incentives for participation, among others. There is often a common assumption that municipal authorities are solely responsible for plastic waste management (Zurbrugg & Ahmed, 1999). Achieving effective plastic waste management necessitates top-down communication about the consequences of waste mismanagement among community members and the public, fostering consensus. This can be accomplished through various means, including household plastic waste segregation and the establishment of neighborhood-level recycling centres.

9.2.4 Bones: Infrastructure and Laws

Finally, to achieve solutions to the plastic waste problem, we need the infrastructure and legislation that allows strategies to take place, laying the foundation upon which effective strategies can be implemented. Legislative measures can span a range from incentivizing responsible business practices to imposing penalties for non-compliance entities. From having Extended Producers Responsibility (EPR) where manufacturing corporations have to find a way to recycle their packaging and waste, thus ensuring they take responsibility in the full lifecycle of their products. Incentives, like the refund or reward points given back when you return a crystal glass bottle or aluminium can. Last but not least, fines and penalties for not separating waste to enforce compliance.

The infrastructure component is equally important in CBPWM, which includes having places for the community to separate and recycle waste, and transport to the companies that conduct the recycling of the different types of plastic, facilitating the transformation of plastic waste into valuable resources.

In the end, to develop effective and sustainable strategies for CBPWM, it is essential that all components work together to reduce, reuse, and recycle much like a body functioning as a whole. Local solutions that address the plastic waste problem should consider the needs and requirements of all 'body parts' to achieve consensus-driven solutions.

9.3 Case Studies of Successful Community-Based Initiatives in Rural and Urban Settings

Communities worldwide are taking matters into their own hands by proposing alternatives to reduce and recycle plastic. Various strategies are being implemented by communities globally. We have selected case studies to illustrate how both urban and rural settings are tackling to address the plastic waste problem.

9.3.1 Community-Led Initiatives to Reduce Plastic Waste

1. Mexico

   (a) San Juan Quiahije

Rural communities lacking waste collection services face significant challenges in waste management. San Juan Quiahije, an indigenous town of 4,000 people in the state of Oaxaca, Mexico, encountered solid waste issues. Recycling companies charged fees for polyethylene terephthalate (PET), aluminium, or cardboard, necessitating the town to address plastic waste management. In 2020, the community took action by banning SUP bags and containers, plastic bottles under two litres, snack packaging and fast food services (Cruz Soto, 2023).

Collaborating with local authorities, the community implemented various strategies to reduce plastic waste. They stopped purchasing and selling beverages in plastic bottles under two litres, opting for washable containers and leaves for wrapping food. This initiative yielded success, reducing waste by a third from 32 tons per month in 2019 to 10 tons in 2022 (NVI Noticias, 2022).

Furthermore, the town launched an educational campaign promoting healthier habits and a diet free of fast or highly processed food. They also published a book featuring healthy, accessible recipes based on local ingredients. To further reduce solid waste, they initiated material collection for recycling, with local authorities receiving waste management training from federal and state governments (Gobierno del Estado de Oaxaca, 2020).

(b) Yalálag

In 2017, the community of Yalálag, a small town in Oaxaca, Mexico, grew weary of the sight of plastic waste spread across its streets. Concerned about health, food, waste, and the environment, the town took action. In a town assembly where 600 out of the 2,500 inhabitants participated, they collectively decided to ban potato chips and instant noodle soups due to the excessive plastic waste they generated. Additionally, the town implemented a plastic waste management program and began composting organic waste (Ferri, 2020).

2. San Pedro La Laguna, Guatemala

In 2016, San Pedro La Laguna, a Guatemalan Mayan village with 14,000 inhabitants, implemented a ban on the sale and use of SUP bags, straws, and styrofoam food containers (United Nations Environment Programme, 2018). This action positioned the village as a pioneer in combating plastic pollution, marking it as the first in Guatemala to do so. Prior to this ban, San Pedro La Laguna was overwhelmed with SUPs, resulting in overflowing landfills and severe pollution of Lake Atitlán, which is vital for the community's livelihood (United Nations Environment Programme, 2018; World Economic Forum, 2021).

Recognizing the gravity of the situation, the local authorities, community members, and religious leaders endorsed the SUPs ban initiative. Initially, there was resistance from the residents, and the village faced lawsuits from plastic companies attempting to halt the ban. Nevertheless, the national court upheld the prohibition, siding against the companies seeking to overturn it (The GAIA Alliance, 2020). Subsequently, 80% of the villagers ceased using SUPs, switching from plastic bags to palm baskets and adopting banana leaves for wrapping food (United Nations Environment Programme, 2018; World Economic Forum, 2021). This ban gave the Tz'utujil Mayan community the opportunity to return to their traditional way of life that has been practiced by their ancestors for centuries (The GAIA Alliance, 2020). The initiative exemplifies a robust community-driven approach, supported by the municipal authorities, to curb the generation and consumption of plastic waste.

3. New South Wales, Australia

The protection of a community's natural resources can be a powerful motivator for organization and action. When a bottling company proposed to set up in Bundanoon, a small community in New South Wales was displeased. The town decided to halt the construction of the water bottling plant and banned the sale of bottled water in 2009. The community began purchasing reusable water bottles and refilling them at free public water fountains throughout the town. Actions taken by the community to enforce the water bottle ban included installing drinking water fountains, providing refillable bottles to local businesses, promoting tourism, and encouraging support for local enterprises (JoAnna, 2018). Bundanoon is recognized as the world's first town to be free from bottled water and has continued to be free from plastic bottles to this day (Jain et al., 2019).

4. Leeds University, United Kingdom

Universities can serve as a prime example of the power of young people working toward a common goal. In 2018, students at Leeds University were the first in the country to ban SUPs from university shops, demonstrating that a small movement can set a precedent for towns and cities. They have successfully eliminated over a million SUP items and are actively addressing the challenges of replacing SUPs, particularly in complex environments such as laboratories (Our Plastics Programme, n.d.). To further support the initiative, the students charted the locations of water dispensers across the campus, enabling everyone to easily locate refill stations and thus discouraging the purchase of bottled water on campus.

9.3.2 Innovative Recycling and Up-Cycling Initiatives

1. Malaysia

In a multi-ethnic country, religion can play a significant role in community-based efforts to address conservation and sustainability issues. The Taiwan-based Buddhist Tzu-Chi Foundation, established in 1966, has been present in Malaysia since 1989 (Mohamad et al., 2012). Recycling forms a crucial part of their service to the community, not only in Malaysia but around the world (Mohamad et al., 2012). In Malaysia, the organization operates over 1,000 community recycling stations through four regional headquarters, serving high-density residential areas. The initiative engages a diverse range of community members, including housewives, retirees, workers, and families. Currently, the Tzu-Chi Foundation Malaysia boasts a network of more than 18,000 volunteers nationwide (Branda, 2022).

The foundation also runs thrift shops where items in good condition are sold at affordable prices. The recycling stations, strategically located within community areas, provide infrastructure for local residents to segregate waste at home and deposit recyclables at the nearest Tzu Chi stations. For instance, approximately 480 tonnes of recyclables were collected in the Klang Valley alone (Branda, 2022). These stations not only facilitate recycling but also encourage individuals to volunteer. The work of Tzu Chi not only advances recycling practices among individuals and families but also fosters social learning, potentially leading to further improvements in plastic waste management in the future.

2. Nairobi, Kenya

The transformation of plastic waste into new materials is a common practice in many countries, adding value to what would otherwise be discarded. Gjenge Makers, a community-focused Kenyan enterprise, upcycles plastic waste into alternative construction materials, specifically paving bricks used in houses, commercial buildings, and roads. Gjenge Makers specializes in processing three types of plastics: PET, high density polyethylene (HDPE), and polypropylene (PP) to create paving bricks of varying loads and thicknesses. It works closely with independent waste pickers and operates numerous collection and sorting facilities across the nation, with its main processing facility in Nairobi (Gjenge Makers, n.d.).

These bricks are waterproof, more affordable, and twice as strong as traditional concrete blocks. For instance, Gjenge Makers repurposed 14.9 tonnes of plastic waste to construct pavers within their country. According to their website, they have upcycled over 200,000 kg of plastic waste into bricks or pavers. The initiative has so far provided employment for over 600 individuals, including waste pickers, women, and members of various organizations. Their aim is to recycle plastic waste, promote a culture of recycling and upcycling, provide sustainable building materials, and create job opportunities for women, all while helping to combat plastic pollution (Gjenge Makers, n.d.).

3. Kamikatsu, Japan

Kamikatsu is globally recognized for its pioneering community-based zero waste initiative. Although waste segregation began in 1990, the open burning of household products was prevalent in the town. The short-lived operation of incinerators and the absence of waste collection facilities led the town to seek alternative waste management solutions. In 2003, the Kamikatsu Town Council passed a 'zero waste declaration', and since then, Kamikatsu has been at the forefront of Japan's zero waste movement. The first and second authors visited Kamikatsu in October 2022, located on Shikoku Island, which boasts a unique approach to waste compared to other Japanese municipalities. For example, while many municipalities segregate waste into combustible and non-combustible categories, Kamikatsu's system is based on recyclability. This approach has enabled Kamikatsu to achieve a recycling rate of over 80% and is on course to meet its zero waste target by 2030.

At the Zero Waste Centre, the town's approximately 1,500 residents can exchange their waste for eco-friendly products. The town exemplifies how citizen cooperation and appropriate policies can meet community needs. The recycling center, constructed from recycled materials, showcases the principle of reuse. Waste is segregated into more than 40 categories, with each category assigned a contractor for disposal once capacity is reached. The center also houses a thrift shop, inviting visitors to take items for free to promote reuse. Adjacent to the center is a hotel that offers travelers an experience in sustainable tourism, constructed entirely from recycled materials. Additionally, the town has implemented a food waste reduction program and a ride-share system, demonstrating a commitment to a circular economy and a sustainable lifestyle.

9.4 Challenges in Establishing CBPWM in Rural and Urban Settings

Many communities lack adequate facilities and infrastructure for plastic waste management due to factors such as remoteness, cost constraints, limited awareness of the impacts of plastics, and local governments' inability to provide essential services.

As a result, these communities may resort to unsustainable practices like openly dumping household plastic waste into the environment. Furthermore, the absence of proper waste collection and disposal services often leads to the open burning of plastics, exacerbating environmental pollution and affecting nearby communities, their pets, livestock, and wildlife.

In Malaysia, some communities have the opportunity to recycle their old products at dedicated sustainable corners (Fig. 9.3). However, many face financial constraints and budgetary limitations that prevent the construction of such recycling stations in their neighborhoods. The lack of these facilities means that communities are often unable to segregate waste, are unaware of how to do so, or do not know where to dispose of their recyclable items. Additionally, the installation and maintenance of these stations require ongoing care and resources, presenting another challenge.

Approximately two billion people worldwide lack access to basic municipal solid waste services, and a further three billion are without proper solid waste disposal facilities, including for plastics. This leads to the mismanagement of plastic waste, which particularly affects urban communities in low- and middle-income countries (UN-Habitat, 2021).

Rural communities are heavily dependent on subsistence agriculture for their livelihoods, with plastics playing an integral role in various agricultural activities (Fig. 9.4). The versatile use of plastics includes irrigation tubes, greenhouse nets, drip lines, seedling bags, fertilizer and pesticide containers, sacks, bottles, trays, protective

Fig. 9.3 Sustainable corner in a Kuala Lumpur shopping mall parking lot

layers for fruit, pipes, and films, among others (FAO, 2021). The open burning of these agricultural plastics, a common practice among farmers after the harvest season, releases toxic substances harmful to the environment, humans, and wildlife adapted to human-dominated landscapes. Often, these plastics are left behind, buried, or landfilled without proper collection and disposal. Over time, they break down into micro and nano plastics, adversely affecting soil and plant health. Moreover, mulch films that accumulate can reduce crop yields (FAO, 2021). Due to their size, large plastics pose a serious threat to wildlife and birds through entanglement and ingestion, leading to the potential dysfunction of entire ecosystems.

The entanglement and ingestion of plastic waste by marine mammals are well-documented phenomena. However, terrestrial mammals and birds are increasingly encountering and interacting with plastics. This issue is particularly acute in the fringes of protected areas, where agricultural plastics, open dumping, and the unchecked influx of tourists due to wildlife tourism contribute to the mismanagement of plastic waste. Such practices exert additional pressure on these protected areas, leading to biodiversity loss, ecosystem dysfunction, and a reduced capacity to withstand climate change. Tourists, favoring convenience, often purchase food, snacks, and water in plastic packaging, which undermines local efforts to manage plastic waste and poses further challenges to the communities and wildlife including their habitats.

Fig. 9.4 Application of mulch films in watermelon fields of Datong Township, Yilan County, Taiwan

Supermarkets and restaurants are major contributors to plastic waste generation in urban settings. Supermarkets depend on plastic packaging to handle products, extend shelf life, protect goods, enhance visual appeal, and convey hygiene. A Greenpeace report states that supermarkets in the United Kingdom generate over 800,000 tonnes of plastic waste annually (Environmental Investigation Agency & Greenpeace, 2021). Restaurants use plastics to reduce labor costs associated with washing and cleaning kitchenware, offer hygiene and convenience to customers, and facilitate takeaways.

Urban communities encounter numerous obstacles in managing plastic waste, exacerbated by rapid urbanization, high population density, and elevated consumption rates. Other factors include mass migration and the proliferation of commercial entities and large industries, such as hotels and hospitals. Additionally, a lack of awareness and poor waste segregation practices at the household level aggravate the situation. These elements strain urban waste management systems and infrastructure, resulting in issues like overflowing landfills, illegal dumping, littering, and pollution.

However, the unsustainable use of SUPs in both sectors is influencing consumer behavior, attitudes, and preferences toward plastics, leading to a heavy reliance on SUPs as part of a 'plastic society'. This reliance comes at a significant cost to consumers, with the presence of plasticizers—commonly known as phthalates—found in humans, alongside other chemicals used to produce and enhance the flexibility and durability of plastics (Consumer Reports, 2024). These substances are found in all the plastic packaging products that are widely available at restaurants and supermarkets (Fig. 9.5). These chemicals pose a risk of entering the food chain and being ingested by humans.

This presents additional challenges for local authorities, as the majority of urban residents, whose daily lives are inextricably linked with the use of plastics, are unaware of the negative impacts of littering and the open dumping of plastic waste into the environment. Moreover, they lack sufficient awareness and knowledge about how to segregate plastic waste, handle it properly, or where to seek help, resulting in non-compliance with waste management practices. For instance, a case study from Kuala Lumpur revealed that while 95% of urban dwellers are aware of the environmental impacts of plastics, only 17% actively segregate waste at their households (Chook et al., 2024).

Another challenge is the limited access to environmental education among rural populations. These communities often struggle to learn about the best practices for tackling plastic pollution. For instance, rural areas typically lack the billboards, large display screens, and advertisements that inform urban dwellers about consumption behaviors, plastic waste, and the impacts of plastic pollution. Furthermore, language barriers can impede the learning process, as most information is published in a few widely spoken languages, whereas smaller communities may speak their own native languages. This can make it difficult for community members to grasp the information, leading to disinterest and disengagement. The language barrier also prevents people from acquiring new and updated knowledge promptly. For example, Taiwan, an East Asian country aiming to become bilingual by 2030, still lacks comprehensive bilingual content for educating non-Chinese speaking foreigners. Similarly, rural

Fig. 9.5 Excessive plastic packaging of vegetables at supermarket chain in Taiwan

communities in developing countries often lack effective channels to receive current information on CBPWM initiatives or to learn from other successful examples.

9.5 Recommendations and Conclusion

Based on the findings of this study on CBPWM, it is clear that addressing the complex challenges of managing plastic waste effectively requires a coordinated, multidisciplinary approach involving various stakeholders at the community to policy level. The ongoing issue of plastic waste in both urban and rural environments not only compromises environmental sustainability but also presents considerable health risks and socio-economic challenges. Consequently, this section leverages the case studies and identified challenges to propose a set of recommendations designed to enhance the long-term viability of CBPWM practices. These recommendations aim to support the establishment of CBPWM initiatives and provide a comprehensive framework for communities, local authorities, policymakers, environmental practitioners, and other stakeholders to improve CBPWM and combat plastic pollution.

Recommended actions:

1. Refuse, reduce, replace: Move beyond recycling; prioritize refusing, reducing, and replacing the use of plastics, particularly SUPs, where feasible.
2. Citizen science: Employ a citizen science approach to monitor segregation, the rate of plastic waste generation, collection, and reduction within the community.
3. NGO consultation: Encourage the community to consult local NGOs for strategies to reduce or avoid plastic use.
4. Zero waste committee: Establish a zero waste committee in the locality to explore ways to reduce plastic waste generation and encourage bulk purchasing or wholesale procurement of essential supplies for the community.
5. Infrastructure demand: NGOs, recycling companies, and communities should collaborate and advocate for the necessary infrastructure, such as household-level segregation and plastic recycling stations, in their locality.
6. Collaboration with recycling companies: In areas with existing infrastructure, partner with recycling companies to negotiate for the collection of plastic waste.
7. Plastic classification awareness: Promote a clear understanding of plastic classification to enhance recycling rates, prevent the use of non-recyclable plastics, and foster an eco-conscious community.
8. Phase out toxic chemicals: Advocate for the banning or phasing out of per- and polyfluoroalkyl substances (PFAS) and harmful plasticizers in products and manufacturing processes due to their persistent environmental impact and potential health risks.
9. Demand labeling for transparency: Demand that producers implement mandatory labeling to indicate the presence of microplastics, PFAS, and plasticizers in all food and drink products, thus allowing consumers to make informed decisions and avoid products containing these substances.

Recommendations for the communities, local authorities and policymakers:
For communities:

1. Advocate for environmental education at educational institutions: Encourage educational institutions to include the impacts of plastics on human health, the environment, and ecosystems in their curriculum.
2. Community education programs: Start with small-scale initiatives like neighborhood recycling stations and clean-up programs to foster a culture of recycling, waste reduction, and proper waste segregation at the household and community levels.
3. Adopt a zero waste lifestyle: Embrace and revert to traditional methods, such as using banana plant leaves for food wrapping, and encourage the use of cloth bags for grocery shopping, akin to the widely promoted manjappai bags (yellow cloth bags) in Tamil Nadu, South India.
4. Engage in citizen science: Monitor local waste management efforts and contribute to data collection to improve practice.
5. Voice in policy formulation: Actively participate in discussions with local NGOs and government bodies to shape policies that affect plastic waste management.

6. Sustainable procurement of products: Advocate for community-wide policies that prioritize the procurement of minimal-packaging products in wholesale to meet the needs of the community.

For local authorities:

1. Provide incentives: Offer rewards for responsible practices, drawing inspiration from models like the Zero Baht Shop in Thailand.
2. Support community education: Collaborate with NGOs to educate local communities about waste segregation and reduction.
3. Public–private partnerships: Foster partnerships with private recycling companies to develop new recycling technologies and infrastructure that align with community needs.
4. Develop infrastructure: Work with NGOs and recycling companies to build the necessary infrastructure for effective plastic waste management.
5. Sustainable public procurement: Implement procurement policies that prioritize products with minimal environmental impact and encourage suppliers to adopt sustainable practices.
6. Promote plastic classification awareness: Enhance recycling rates by educating the public about different types of plastics and their recyclability.
7. Faith-based collaboration: Local authorities can partner with religious organizations such as the Tzu Chi Foundation, which has significant influence among communities of Chinese ethnicity, to launch joint initiatives.
8. Community grants: Offer grants to communities and NGOs for projects that align with zero waste goals and plastic reduction approaches.
9. Legislative support for citizen science: Support legislative frameworks that encourage citizen science initiatives related to plastic waste management.
10. Zero waste certification: Develop a zero waste certification program for communities that meet certain waste reduction criteria.

For policymakers:

1. Establish strategic alliances: Form partnerships with health organizations, environmental NGOs, religious groups, and recycling companies to address plastic pollution comprehensively.
2. Resource allocation: Provide resources and develop comprehensive action plans to support communities in transitioning to zero waste lifestyles.
3. Empower communities: Ensure that community leaders have a say in establishing CBPWM and in advocating for supportive laws and regulations.
4. Recycling infrastructure: Invest in recycling infrastructure to ensure that plastics already in circulation are managed responsibly and do not end up as pollutants.
5. Ban on manufacturing and importing SUPs: Policymakers could introduce legislation to ban the manufacturing and importing of SUPs, especially those that are not essential and have readily available alternatives.

6. Phasing out toxic plastics: Implement a phased approach to ban plastics with extreme levels of toxicity. This could involve creating a list of prohibited substances such as PFAS and plasticizers and setting deadlines for their elimination from production processes.
7. Labeling requirements: Mandate warning labels on products that contain micro or nano plastic particles. These labels should inform consumers about the concentration of such particles and their potential impact on health, wildlife, and the environment, enabling informed purchasing decisions.
8. Incentives for alternatives: Provide incentives for companies that develop and use alternative materials to plastics.
9. Producer responsibility: Enforce EPR regulations that hold producers accountable for the entire lifecycle of their products, encouraging them to design products with less environmental impact.

Throughout this study, we have examined the impact of plastic in rural and urban settings and identified the essential components for implementing successful CBPWM. We have offered recommended actions and tailored recommendations to communities, local authorities, and policymakers to ensure the long-term success and effectiveness of CBPWM initiatives. The pervasive presence of plastic in our daily lives has transformed our society into one heavily reliant on plastic. We stand at a pivotal moment, risking the transformation of our blue planet into a 'plastic planet' with dire consequences for all life on Earth. These issues cannot be resolved overnight. The case studies presented are just a few examples of how communities are dedicated to combating plastic pollution locally. Plastic waste is a human-induced wicked problem that requires a collective effort to gradually eliminate plastic products from our lives. Numerous CBPWM initiatives remain undocumented; their innovative approaches to overcoming plastic pollution must be recognized, replicated, and scaled up globally to drive transformative change in plastic waste management.

References

Ambastha, R., & Aich, S. (2020). *Decentralized Community-Led Solid Waste Management. Sustainable Waste Management: Policies and Case Studies.*

Awasthi, B., Lamichhane, B. R., & McConkey, K. R. (2023). Plastic ingestion by greater one-horned rhinos in Nepal: An emerging conservation threat. *Global Ecology and Conservation, 46*, e02549. https://doi.org/10.1016/j.gecco.2023.e02549.

Branda, N. (2022). *Seniors set a high bar for a low-carbon lifestyle in Malaysia.* Retrieved November 09, 2023, from https://tzuchi.us/blog/low-carbon-lifestyle-malaysia.

Chook, J. W., Ferdin, A. E. J., & Lee, C.-H. (2024). Unmasking Malaysian preferences for effective plastic waste management: A choice experiment approach. *Journal of Material Cycles and Waste Management, 26*(1), 512–528. https://doi.org/10.1007/s10163-023-01850-1.

Chow, C.-F., So, W.-M. W., Cheung, T.-Y., & Yeung, S.-K. D. (2017). Plastic waste problem and education for plastic waste management. In S. C. Kong, T. L. Wong, M. Yang, C. F. Chow, & K. H. Tse (Eds.), *Emerging Practices in Scholarship of Learning and Teaching in a Digital Era* (pp. 125–140). Springer Singapore. https://doi.org/10.1007/978-981-10-3344-5_8.

Collier, M. J., Nedović-Budić, Z., Aerts, J., Connop, S., Foley, D., Foley, K., Newport, D., McQuaid, S., Slaev, A., & Verburg, P. (2013). Transitioning to resilience and sustainability in urban communities. *Cities, 32*, S21–S28. https://doi.org/10.1016/j.cities.2013.03.010.

Colon, M., & Fawcett, B. (2006). Community-based household waste management: Lessons learnt from EXNORA's 'zero waste management' scheme in two South Indian cities. *Habitat International, 30*(4), 916–931. https://doi.org/10.1016/j.habitatint.2005.04.006.

Consumer Reports. (2024). *The plastic chemicals hiding in your food*. Retrieved November 03, 2023, from https://www.consumerreports.org/magazine/2024/02/.

Cruz Soto, A. (2023). *San Juan Quiahije, el pueblo que le dijo no al plástico*. Retrieved January 10, 2024, from https://www.diariodemexico.com/mi-nacion/san-juan-quiahije-el-pueblo-que-le-dijo-no-al-plastico.

Dharia, S. V., Khushwah, A. J., Choudhari, C. M., & Kavre, M. S. (2023). Optimum source segregation bin for household solid waste and waste plastic recycling. *Journal of Material Cycles and Waste Management, 25*(4), 2312–2325. https://doi.org/10.1007/s10163-023-01689-6.

Environmental Investigation Agency and Greenpeace. (2021). *Checking out on plastics III*. Retrieved December 28, 2023, from https://eia-international.org/wp-content/uploads/Checking-Out-on-Plastics-III.pdf.

Eriksen, M., Lusher, A., Nixon, M., & Wernery, U. (2021). The plight of camels eating plastic waste. *Journal of Arid Environments, 185*, 104374. https://doi.org/10.1016/j.jaridenv.2020.104374.

Fahmi, W. S. (2005). The impact of privatization of solid waste management on the Zabaleen garbage collectors of Cairo. *Environment and Urbanization, 17*(2), 155–170. https://doi.org/10.1177/095624780501700212.

FAO. (2021). *Assessment of Agricultural Plastics and Their Sustainability—A Call for Action*. Rome. https://doi.org/10.4060/cb7856en.

Ferri, P. (2020). *Yalálag, en la Sierra Norte, prohibió comida chatarra desde hace 3 años*. Retrieved December 20, 2023, from https://imparcialoaxaca.mx/oaxaca/462745/yalalag-en-la-sierra-norte-prohibio-comida-chatarra-desde-hace-3-anos/.

Forrest, N., & Wiek, A. (2014). Learning from success—Toward evidence-informed sustainability transitions in communities. *Environmental Innovation and Societal Transitions, 12*, 66–88. https://doi.org/10.1016/j.eist.2014.01.003.

Gjenge Makers. (n.d.). *Our story*. Retrieved December 06, 2023, from https://www.gjenge.co.ke/about.

Gobierno del Estado de Oaxaca. (2020). *Capacita SEMAEDESO a autoridades de San Juan Quiahije en el manejo integral de residuos sólidos urbanos*. Retrieved February 10, 2024, from https://www.oaxaca.gob.mx/comunicacion/capacita-semaedeso-a-autoridades-de-san-juan-quiahije-en-el-manejo-integral-de-residuos-solidos-urbanos/.

Jain, B., Singh, A. K., & Susan, M. A. B. H. (2019). The world around bottled water (Chap. 2). In A. M. Grumezescu & A. M. Holban (Eds.), *Bottled and Packaged Water* (pp. 39–61). Woodhead Publishing. https://doi.org/10.1016/B978-0-12-815272-0.00002-7.

JoAnna, H. (2018). *ATTA member leads plastic water bottle ban, offers tips for other industry leaders*. Retrieved December 07, 2023, from https://www.adventuretravelnews.com/atta-member-leads-plastic-water-bottle-ban-offers-tips-for-other-industry-leaders.

Katlam, G., Prasad, S., Pande, A., & Ramchiary, N. (2022). Plastic ingestion in Asian elephants in the forested landscapes of Uttarakhand, India. *Journal for Nature Conservation, 68*, 126196. https://doi.org/10.1016/j.jnc.2022.126196.

Kaza, S., Yao, L., Bhada-Tata, P., & Van Woerden, F. (2018). *What a waste 2.0: A global snapshot of solid waste management to 2050. Urban development*; © Washington, DC: World Bank. http://hdl.handle.net/10986/30317. License: CC BY 3.0 IGO.

Lim, X. (2021). Microplastics are everywhere—But are they harmful. *Nature, 593*(7857), 22–25. https://doi.org/10.1038/d41586-021-01143-3.

Ma, Y. (2018). Changing Tetra Pak: From waste to resource. *Science Progress, 101*(2), 161–170. https://doi.org/10.3184/003685018x15215434299329.

Mohamad, Z. F., Idris, N., Baharuddin, A., Muhammad, A., & Nik Sulaiman, N. M. (2012). The role of religious community in recycling: Empirical insights from Malaysia. *Resources, Conservation and Recycling, 58*, 143–151. https://doi.org/10.1016/j.resconrec.2011.09.020.
Muller, M., Hoffman, L., & Scheinberg, A. (2001). *Community partnerships in integrated sustainable waste management*. WASTE, CW Gouda, The Netherlands. www.waste.nl.
Mutemani, J., Chinyama, A., Mohsin, M., & Kativhu, T. (2022). Evaluation of the community participation in solid waste management: Case of the city of Bulawayo, Zimbabwe. *Arabian Journal of Geosciences, 15*(10), 969. https://doi.org/10.1007/s12517-022-10228-w.
NVI Noticias. (2022). Lucha contra basura: Así han reducido sus desechos en Quiaje, Oaxaca. Retrieved January 22, 2024, from https://www.nvinoticias.com/oaxaca/general/lucha-contra-basura-asi-han-reducido-sus-desechos-en-quiahije-oaxaca/138949.
Oliveri Conti, G., Ferrante, M., Banni, M., Favara, C., Nicolosi, I., Cristaldi, A., Fiore, M., & Zuccarello, P. (2020). Micro- and nano-plastics in edible fruit and vegetables. The first diet risks assessment for the general population. *Environmental Research, 187*, 109677. https://doi.org/10.1016/j.envres.2020.109677.
Otsyina, H., Nguhiu-Mwangi, J., Mogoa, E., Mbuthia, P., & Ogara, W. (2014). A retrospective study on the prevalence of plastic materials in the rumen of sheep and goats in Nairobi, Kenya. *Bulletin of Animal Health and Production in Africa, 62*(3), 197–205.
Our Plastics Programme. (n.d.). *Sustainability at Leeds*. Retrieved February 10, 2024, from https://sustainability.leeds.ac.uk/our-work/single-use-plastics-programme/.
Rathi, S. (2006). Alternative approaches for better municipal solid waste management in Mumbai, India. *Waste Management, 26*(10), 1192–1200. https://doi.org/10.1016/j.wasman.2005.09.006.
Rillig, M. C., Ziersch, L., & Hempel, S. (2017). Microplastic transport in soil by earthworms. *Scientific Reports, 7*(1), 1362. https://doi.org/10.1038/s41598-017-01594-7.
Senathirajah, K., Attwood, S., Bhagwat, G., Carbery, M., Wilson, S., & Palanisami, T. (2021). Estimation of the mass of microplastics ingested—A pivotal first step towards human health risk assessment. *Journal of Hazardous Materials, 404*, 124004. https://doi.org/10.1016/j.jhazmat.2020.124004.
Singh, S., Chakma, S., Alawa, B., Kalyanasundaram, M., & Diwan, V. (2023). Assessment of microplastic pollution in agricultural soil of Bhopal, Central India. *Journal of Material Cycles and Waste Management*. https://doi.org/10.1007/s10163-023-01805-6.
The GAIA Alliance. (2020). *Reducing the use of disposable plastic: San Pedro La Laguna, Sololá, Guatemala*. Retrieved January 11, 2024, from https://www.no-burn.org/wp-content/uploads/06-San-Pedro-La-Laguna_Guatemala_English.pdf.
Tudor, T., Robinson, G. M., Riley, M., Guilbert, S., & Barr, S. W. (2011). Challenges facing the sustainable consumption and waste management agendas: Perspectives on UK households. *Local Environment, 16*(1), 51–66. https://doi.org/10.1080/13549839.2010.548372.
Turnbull, J., & Barua, M. (2023). Living waste, living on waste: A bioeconomy of urban cows in Delhi. *Transactions of the Institute of British Geographers, 48*(3), 474–490. https://doi.org/10.1111/tran.12573.
UN-Habitat. (2021). *Waste wise cities tool*. Retrieved February 08, 2024, from https://unhabitat.org/wwc-tool.
United Nations Environment Programme. (2018). *This Mayan village is winning the fight against plastic pollution*. Retrieved December 26, 2023, from https://www.unep.org/news-and-stories/story/mayan-village-winning-fight-against-plastic-pollution.
Van Rensburg, M. L., S'phumelele, L. N., & Dube, T. (2020). The 'plastic waste era'; social perceptions towards single-use plastic consumption and impacts on the marine environment in Durban, South Africa. *Applied Geography, 114*, 102132. https://doi.org/10.1016/j.apgeog.2019.102132.
Vicente-Molina, M. A., Fernández-Sáinz, A., & Izagirre-Olaizola, J. (2013). Environmental knowledge and other variables affecting pro-environmental behaviour: Comparison of university students from emerging and advanced countries. *Journal of Cleaner Production, 61*, 130–138. https://doi.org/10.1016/j.jclepro.2013.05.015.

Wijayanti, D. R., & Suryani, S. (2015). Waste bank as community-based environmental governance: A lesson learned from Surabaya. *Procedia—Social and Behavioral Sciences, 184*, 171–179. https://doi.org/10.1016/j.sbspro.2015.05.077.

Willis, K., Hardesty, B. D., Vince, J., & Wilcox, C. (2022). Local waste management successfully reduces coastal plastic pollution. *One Earth, 5*(6), 666–676. https://doi.org/10.1016/j.oneear.2022.05.008.

World Economic Forum. (2021). *Here's what happened when one Guatemalan town went to war on plastic waste*. Retrieved November 18, 2023, from https://www.weforum.org/agenda/2019/02/this-guatemalan-town-went-to-war-on-plastic-waste/.

Yandri, P., Budi, S., & Putri, I. A. P. (2023). Waste sadaqah: A new community-based waste-management practice in Java, Indonesia. *Sustainability: Science, Practice and Policy, 19*(1), 2212510. https://doi.org/10.1080/15487733.2023.2212510.

Zhao, S., Zhu, L., & Li, D. (2016). Microscopic anthropogenic litter in terrestrial birds from Shanghai, China: Not only plastics but also natural fibers. *Science of the Total Environment, 550*, 1110–1115. https://doi.org/10.1016/j.scitotenv.2016.01.112.

Zurbrugg, C., & Ahmed, R. (1999). Enhancing community motivation and participation in solid waste management. *Sandec News, 4*, 2–6.

Chapter 10
Efficient Solid Waste Classification Using Computer Vision and Deep Learning: A Sustainable Waste Management Approach

Geerish Suddul and Nadish Ramsurrun

Abstract Plastic waste pollution poses a significant environmental threat, harming ecosystems and wildlife. While recycling offers a solution, sorting different types of waste remains a challenge. This work proposes a Deep Learning (DL) approach using computer vision to automate solid waste identification and classification within a multi-category recycling system, aiming to improve sorting efficiency and reduce the amount of plastic and other solid waste ending up in landfills. We present the development a Convolutional Neural Network (CNN) designed for accurate waste classification from image data, thereby enabling fully automated smart bin operation. We applied different optimization techniques on several CNN architectures, including transfer learning and data augmentation, to improve the classification accuracy and computational efficiency. Our experimental results demonstrate that the proposed model achieves high accuracy in classifying plastic waste but also metal, paper, cardboard and glass. This research demonstrates that Artificial Intelligence (AI) in the form of Deep Learning has the potential to significantly contribute to a more sustainable waste management system.

Keywords Deep learning · Solid waste classification · Convolutional neural network

G. Suddul (✉) · N. Ramsurrun
University of Technology, Mauritius (UTM), Port Louis, Mauritius
e-mail: g.suddul@utm.ac.mu

N. Ramsurrun
e-mail: nadish.ramsurrun@umail.utm.ac.mu

S. Suriyanarayanan et al. (eds.), *Combating Plastic Pollution in Terrestrial Environment*,
https://doi.org/10.1007/978-981-96-2343-3_10

10.1 Introduction

Solid waste management is a growing challenge faced by every country worldwide. Improper waste disposal creates a ripple effect of problems, including contaminated water, increased disease transmission, and flooding from clogged drains. As per the report entitled "What a Waste 2.0: A Global Snapshot of Solid Waste Management to 2050"(Kaza et al., 2018) it is estimated that the world is producing a staggering 2.01 billion tonnes of solid waste annually, with a significant portion not being responsibly managed. The average person produced 0.74 kg of waste daily, but this varies greatly across regions, from 0.11 to 4.54 kg. This problem is projected to worsen, with global waste expected to reach 3.40 billion tonnes by 2050. Plastic waste is a major contributor to this issue, accounting for a huge 44% of global waste composition in 2016. Despite efforts, the world still produces a concerning 430 million tonnes of plastic annually, much of which is single-use. Failing to take decisive action, plastic pollution is predicted to triple by 2060, posing a severe threat to our environment, health, and economy, especially for developing countries. (United Nations, 2023).

Landfills, the most common waste disposal method globally (Kaza et al., 2018) pose a significant environmental threat. Leachate, toxins, and greenhouse gas emissions pollute the surrounding environment. Fortunately, recycling and composting offer a more sustainable approach. While recycling effectively reduces reliance on landfills, sorting various materials like metal, glass, plastic, and paper can be complex and expensive. Existing sorting solutions include mechanical systems, manual sorting, tagged waste with automated identification, and computer vision technologies.

While manual sorting, prevalent in developing countries, is efficient (Saar & Thomas, 2004), it exposes workers to hazardous materials and health risks. Similarly, tagged waste using barcodes or Radio Frequency Identification (RFID) streamlines sorting but raises environmental concerns as mentioned by several authors (Saar & Thomas, 2002; Tsao et al., 2017; Abdoli, 2019). This work elaborates on the methodology of our earlier work (Ramsurrun et al., 2021), providing a more comprehensive understanding of our approach to automate waste sorting through Artificial Intelligence (AI) techniques. Our initial approach consists of simulating a smart recycling bin equipped with a camera, which captures the image of the waste material, predicts the type of waste and opens the corresponding lid. This expanded work presents a more comprehensive analysis and further elaboration of the results, including the machine model evaluation with a new set of images.

10.2 Background

Traditional computer vision approaches are based on logic programming to identify and interpret visual information. Such algorithms are quite complex and require significant manual effort to define features and rules for object recognition. Machine

Learning (ML), more specifically Deep Learning (DL) has enhanced the field of computer vision. This section serves as a foundation, introducing the core concepts of Deep Learning for Computer Vision, and is based on an adaptation of Suddul and Seguin (2023). It helps to establish a basic understanding of these principles.

10.2.1 Convolutional Neural Network

A Convolutional Neural Network (CNN) is a Deep Neural Network (DNN) architecture particularly designed to work with visual type of data, such as photos and videos. It can handle a very large number of parameters to learn features from an image. A coloured image, for example, plastic waste, is recorded in a computer system with three different colours, Red, Green and Blue (RGB), known as channels (Yamashita et al., 2018). In a traditional conventional neural network perspective, using a 3 channel image of size 500 pixels in width and height, results in 750,000 individual pixels which are fed into the network. If this network has 1000 neurons in its first layer, the resulting number of input parameters is 750 million. This has a huge impact on the computation requirements of the computer. However, using a Convolutional Neural Network (CNN) significantly reduces the number of parameters through the use of specialized operations. For example, with the use of filters, a CNN can scan an image, extracting important features like edges and shapes. This reduces the number of parameters needed significantly, making CNNs much more efficient at processing visual data (LeCun et al., 1998).

A CNN can be viewed as a series of filters that analyse an input image piece by piece, as illustrated in Fig. 10.1. Here, we assume a small image with 4 pixels height, 4 pixels width, and 3 color channels (red, green, blue). The CNN uses filters to scan the image, looking for patterns and edges. These patterns become like building blocks, helping the network understand the content of the image. These building blocks are termed as features, and the collection of features forms a feature map. After applying filters, the CNN combines the features into a single stream of information. Finally, the CNN uses regular neural network layers termed as fully connected layers to analyse the features and make a final prediction about the type of waste (plastic, metal, paper, glass).

10.2.2 Convolutions

Consider an image as a grid of numbers, where each number represents the brightness or colour of a pixel. In Fig. 10.2, the image is a small 4×4 grid with just one channel that is a grayscale image. Convolution is the technique used by CNNs to analyse these grids. We use a small filter, also known as a kernel, that slides across the image, like a window. This filter also has numbers, and the technique is to multiply the corresponding numbers between the image and the filter, then add them all up.

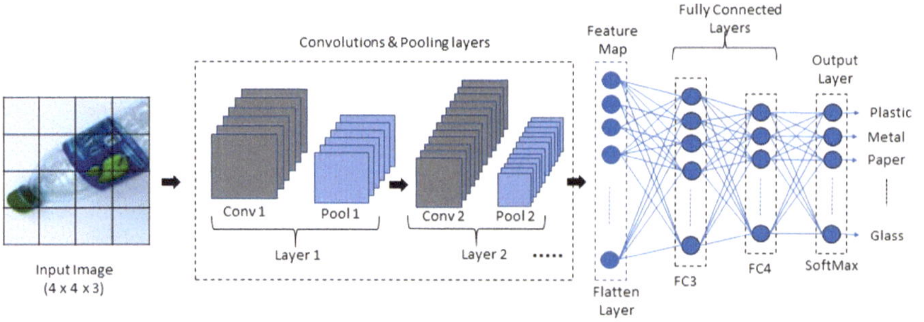

Fig. 10.1 Architecture of convolutional neural network

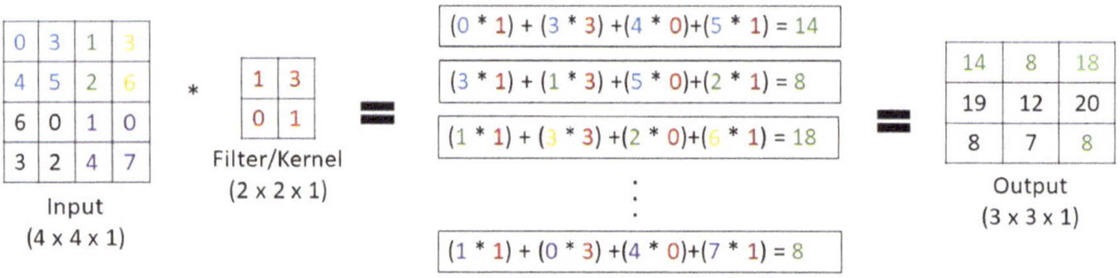

Fig. 10.2 Convolutions in action with 2 × 2 filter

The result of this multiplication and addition becomes a single number in a new grid called the activation map. We slide the filter one step at a time (stride of 1) across the entire image, performing the multiplication and addition at each spot. This creates a smaller output grid (3 × 3 in our example) that captures important features from the original image. For example, the first value (14) in the output grid is calculated by multiplying the top-left corner of the image with the corresponding values in the filter and adding them all up.

The example in Fig. 10.2 uses a single kernel and as such the output does not consists of multiple channels. Equation 10.1, describes the relationship between number of kernels and channels. The channels are represented by $m_{c'}$, with stride s and padding p. The $m \times m$ reflects the size of the output matrix and $k \times k$ the size of the kernel, having the same number of channels.

Dimension of Input matrix: $(m \times m \times m_c)$

Dimension of Filter matric: $(k \times k \times m_c)$

Dimension Output matrix based on convolutions:

$$\frac{(m-k)+2p}{s} + 1 \times \frac{(m-k)+2p}{s} + 1 \times m_c' \tag{10.1}$$

Convolution in a CNN is like breaking down an image into its building blocks which represent a simplified version of the real word. To capture the real-world complexities, CNNs use a special function called the ReLU (Rectified Linear Unit)

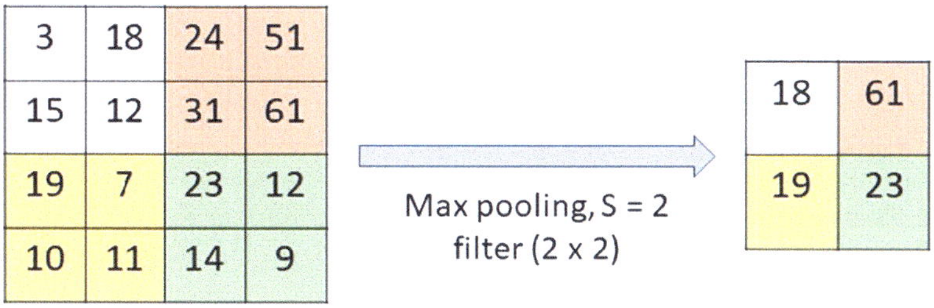

Fig. 10.3 Application of max pooling with stride of 2

after the convolution stage. This function basically acts like a switch that turns negative numbers to zero and keeps positive numbers unchanged. This adds a layer of non-linearity to the process, allowing the network to learn more intricate patterns in the image data.

10.2.3 Pooling

After the convolution process extracts features from the image, pooling layers come into play. The main role of pooling is to summarize the huge amount of information generated from the convolution step, making it smaller and less computer intensive for processing. There are different ways to perform pooling. One such approach is to use max pooling, as shown in Fig. 10.3. Here, a small filter slides across the data from the convolution layer. Within each window, max pooling picks the most significant (maximum) value. This reduces the size of the data significantly, making it easier to process. This pooling step helps to reduce computing efforts and makes the features more robust.

The function represented by Eq. 10.2 determines the dimension of the pooling output matrix.

$$\frac{(m-k)}{s} + 1 \times \frac{(m-k)}{s} + 1 \times m_c' \tag{10.2}$$

10.2.4 Pre-configured CNN Models & Transfer Learning

Creating a CNN entirely from scratch can be time consuming. It involves defining the appropriate architecture: numerous layers, settings (like stride and padding), and tweaking various parameters to find the optimal configuration. This process requires a lot of experimentation. Instead, pre-configured existing models, which have proven effective for image classification, can be used, as listed below:

- The LeNet-5 model from LeCun et al. (1998) consists of 7 layers with 4 convolutions and pooling, two fully connected layers and one output layer. This model has approximately 60,000 parameters.
- Krizhevsky, et al. (2012) created the AlexNet with a total of 8 layers. Out of which five are convolutions and pooling layers, two fully connected layers and 1 output layer. This network consists of around 60 million parameters.
- The Inception network created by Szegedy et al. (2016) has around 27 layers and 4 million parameters.
- Simonyan and Zisserman (2015) created the VGG network, with 16 layers and above 138 million parameters.
- He et al. (2016) is at the origin of the Residual Network (ResNets) with 152 layers and around 60 million parameters.
- Tan and Le (2021) built the EfficientNet algorithm with 237 layers and above 5 million parameters.

Furthermore, instead of training a complex image recognition model from scratch, transfer learning leverages pre-trained models. These models are like pre-made components, trained on massive datasets and ready to use. They come with built-in knowledge in the form of weights from their previous training. Transfer learning additionally contributes to saving time and computing resources.

10.3 Related Work

Plastic pollution is a growing environmental concern. Automating plastic waste detection using machine learning (ML) offers a promising solution for improving recycling and waste management practices. This section examines recent research, which apply machine learning for waste detection including plastic, focusing on Convolutional Neural Networks (CNNs) and their performance in waste classification tasks.

In a previous study by Yang and Thung (2016), researchers compared two machine learning techniques for classifying waste into different categories. They used the TrashNet dataset (Thung, 2016) and evaluated the performance of Support Vector Machine (SVM) and a Convolutional Neural Network (CNN) called AlexNet. While the SVM achieved an accuracy of around 63%, the CNN only reached 22% accuracy. This suggests that the CNN model may not have been optimized for this specific task. The study also identified the need for future research on classifying multiple waste objects within a single image. Building upon previous research, (Adedeji & Wang, 2019) proposed a promising approach for waste classification. They combined a powerful CNN architecture (ResNet50) with an SVM classifier to identify four waste categories (glass, paper, plastic, metal) and achieved an accuracy of around 87%. This significant improvement over prior studies such as Yang and Thung (2016) can likely be attributed to their use of data augmentation techniques to optimize their dataset size. However, a limitation of their work was the reliance on a relatively small dataset of waste images. Future research efforts could focus on acquiring and utilizing even larger datasets to potentially achieve even higher classification accuracy. Several researchers are exploring computer vision to automate recycling bins and reduce reliance on manual sorting. Desai et al. (2018) proposed a system that uses a CNN to classify waste as degradable or non-degradable. This allows a robotic arm controlled by a Raspberry Pi to open the appropriate bin lid. Building on this concept, Aral et al. compared different CNN architectures (Xception, MobileNet, DenseNet, InceptionResNetV2) on the TrashNet dataset (Thung, 2016). By optimizing the dataset size using various techniques (data augmentation), they achieved impressive accuracy rates, with DenseNet121 reaching 95%. Further research (Ziouzios & Dasygenis, 2019) explored a cost-effective smart bin system. Their system utilizes a CNN (ResNet34) to classify six waste categories (cardboard, glass, metal, paper, plastic, trash) using the TrashNet dataset. Achieving a 92% accuracy, their system also leverages a Raspberry Pi for bin control and LoRaWAN connectivity for image transmission.

Umut and Seyfi (2018) investigated various combinations of Convolutional Neural Networks (CNNs) and classifiers to improve waste classification accuracy. They tested different CNN architectures (like InceptionV3 and VGG-16) along with two classifier types (SoftMax and SVM) on the TrashNet dataset (Saar & Thomas, 2002), which includes six waste categories (glass, paper, cardboard, plastic, metal, trash). Interestingly, they found that using SVM as a classifier resulted in higher accuracy compared to SoftMax. The best performing combination achieved an impressive accuracy of nearly 98% (InceptionV3 + SVM). Convolutional Neural Networks (CNNs) have emerged as a dominant approach for image-based classification tasks, including waste detection. Bircanoglu et al. (2018) introduced RecycleNet, a CNN architecture specifically designed for waste classification. While their model achieved an accuracy of 81% using data augmentation techniques, established CNNs like DenseNet121 reached even higher accuracy (95%) on the TrashNet dataset. This highlights the potential of CNNs for plastic waste detection, but also emphasizes the importance of model optimization techniques like data augmentation. Costa et al. (2018) evaluated different algorithms for classifying waste into four categories (glass, metal, paper, plastic) using the TrashNet dataset (Thung, 2016). They compared Convolutional Neural Networks (CNNs) like VGG16 and AlexNet, along with Support Vector Machines (SVM), K-Nearest Neighbors (KNN), and Random Forests (RF). VGG16 emerged as the most effective model, achieving an accuracy of 93%.

Machine learning, particularly CNNs, has demonstrated significant potential for automating plastic waste detection. By continuously improving model performance through optimization techniques and addressing real-world challenges, this technology can play a vital role in mitigating plastic pollution and fostering a more sustainable future.

10.4 Materials and Methods

10.4.1 Experimental Approach

This study focuses on identifying the most appropriate Deep Learning model, which can accurately identify different types of waste, including plastic. This model forms the basis of a simulated system, which consists of a camera mounted on a recycling bin. The camera captures an image of the waste item which is fed into the DL model which, in turn, predicts the kind of recyclable found in the image. Finally, based on this classification, the bin lid for the specific recyclable material automatically opens up. Figure 10.4 depicts our experimental steps in our approach.

Step 1 & 2: These steps represent the images within the dataset that are used to train and test different models. The images are classified as cardboards, papers, plastics, glass and metals.

Step 3: This step represents the Convolutional Neural Networks with different model configurations selected from a preliminary experimentation. Examples of the

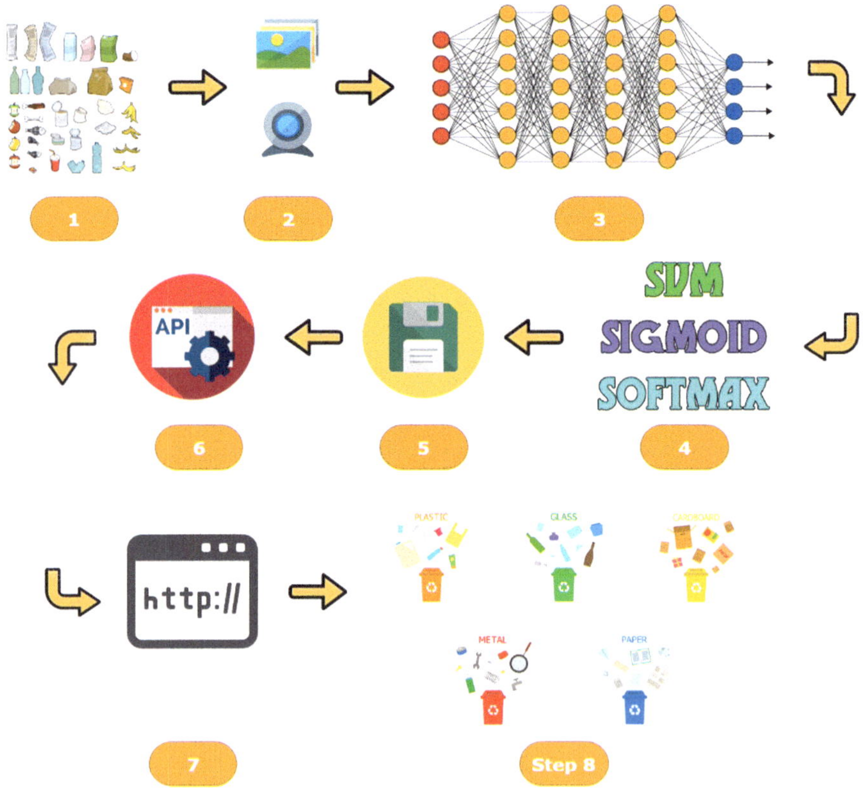

Fig. 10.4 Experimental approach

Fig. 10.5 Model selection approach

CNN are VGG16, InceptionResNetV2, Xception. These CNNs vary based on their architecture (size/layers).

Step 4: This step represents the classifiers. For this study, three different classifiers are being evaluated: Sigmoid, SoftMax and Support Vector Machines (SVM).

Part 3 and Part 4 together represent a model which is used to predict a waste category from an input image. Different experiments have been carried out to identify the best performing model based on training and testing accuracy. The experiments are:

1. Experiment 1: CNN + SVM classifier
2. Experiment 2: CNN + SoftMax classifier
3. Experiment 3: CNN + Sigmoid classifier
4. Experiment 4: Improving the models.

Experiments 1, 2 and 3 consist of 36 CNN models trained with different classifiers and after comparison, 15 models emerged, from which 5 have been further considered, based on their performance. In experiment 4, these 5 models have been exposed to longer training epochs with the integration of call-backs, and tested on 10 randomly selected images from the test dataset (Fig. 10.5).

Step 5: This part consists of saving the different models so as to be able to import the selected one to make predictions. This is done to avoid the training phase everytime the model has to be used to make predictions. The model is saved in the HDF5/H5 format.

Step 6: Once the model is chosen, it is made accessible as a reusable software component, technically known as a Rest API. As such, it can be used to make predictions from different platforms and programming languages. This is accomplished using the Flask web application framework in python. The API returns the response in the JSON format.

Step 7: This part consists of creating the client application which is used to upload images or take a picture from the webcam and send it for prediction through the API. The predicted class and the accuracy of prediction is returned. This client-side application takes the form of a simple webpage written in HTML, CSS and JS.

Step 8: This part is a simulation of the lid opening so as to dispose the trash in the appropriate bin.

With the aim of optimizing waste classification using deep learning technologies, three main experiments denoted as experiments 1, 2, and 3 are designed to assess the effectiveness of combining different classifiers with Convolutional Neural Networks. Each experiment utilizes a similar and consistent setup involving twelve

CNN models each, exploring the dynamics between the neural network architectures and three types of classifiers: Support Vector Machine (SVM), SoftMax, and Sigmoid. For all experiments we use a computer equipped with an Intel i7-9750H processor with 16.0 GB for main memory, 256 GB Solid-State Drive (SSD) for storage and additional power was provided by a NVIDIA GeForce GTX 1660 Ti Graphical Processing Unit (GPU).

10.4.2 Dataset

For this study we used the TrashNet dataset (Saar & Thomas, 2002; Simonyan & Zisserman, 2015) which consists of 2527 images divided into 6 classes: paper, glass, plastic, metal, cardboard and trash. Images representing the different classes are shown in Fig. 10.6.

The dataset demonstrates a generally balanced distribution across its classes, except for trash images as can be depicted in Fig. 10.7. In order to avoid any bias in the machine learning model during training, we decided to remove the trash category and further consolidate the balanced aspect of the dataset classes.

Furthermore, data augmentation is employed to synthetically increase the number of images and thereby further diversifying the dataset and enhancing the model's ability to generalize. Data augmentation consists of applying techniques such as rotation, shifting, shearing, zooming, and flipping on each image in the dataset. Therefore, the augmented dataset captures a broader range of variations as shown below in Fig. 10.8 for a plastic container. This augmentation strategy not only enriches the dataset but also helps in mitigating overfitting by exposing the model to a more comprehensive representation of the underlying data distribution.

Figure 10.9 shows the new distribution of classes following data augmentation. To streamline the training process and reduce computation time, the images were reduced to 224×224 pixels and the white background was maintained. Finally, the dataset was randomly divided into two sets, with 70% of the images used for training the CNN model, and the remaining 30% used for testing its accuracy. In the context of this study, validation accuracy and loss is being termed as testing accuracy and loss.

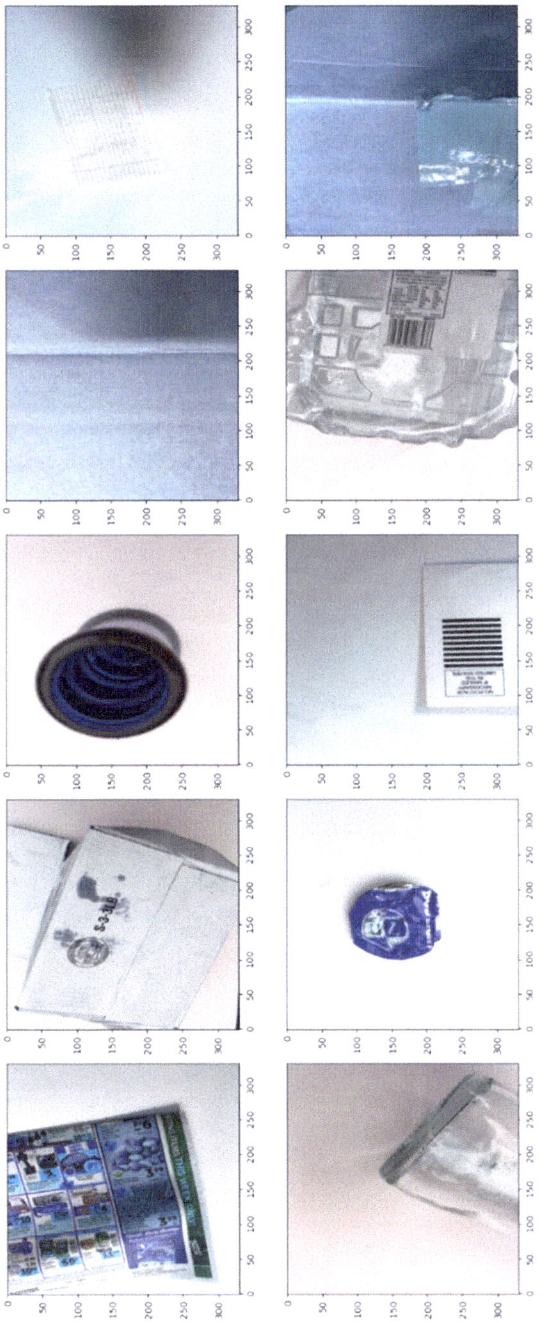

Fig. 10.6 Sample images from dataset

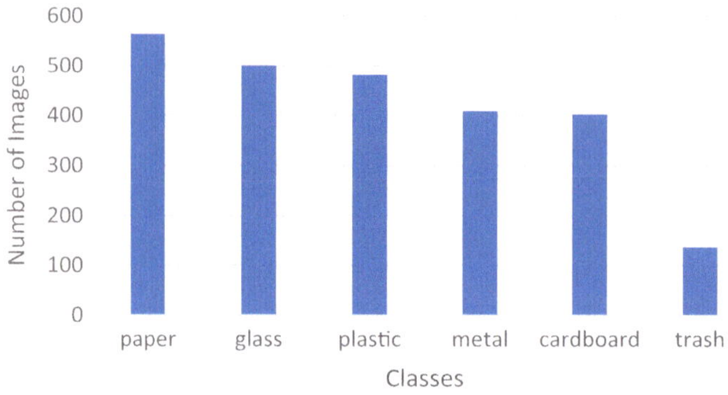

Fig. 10.7 Class distribution for TrashNet dataset

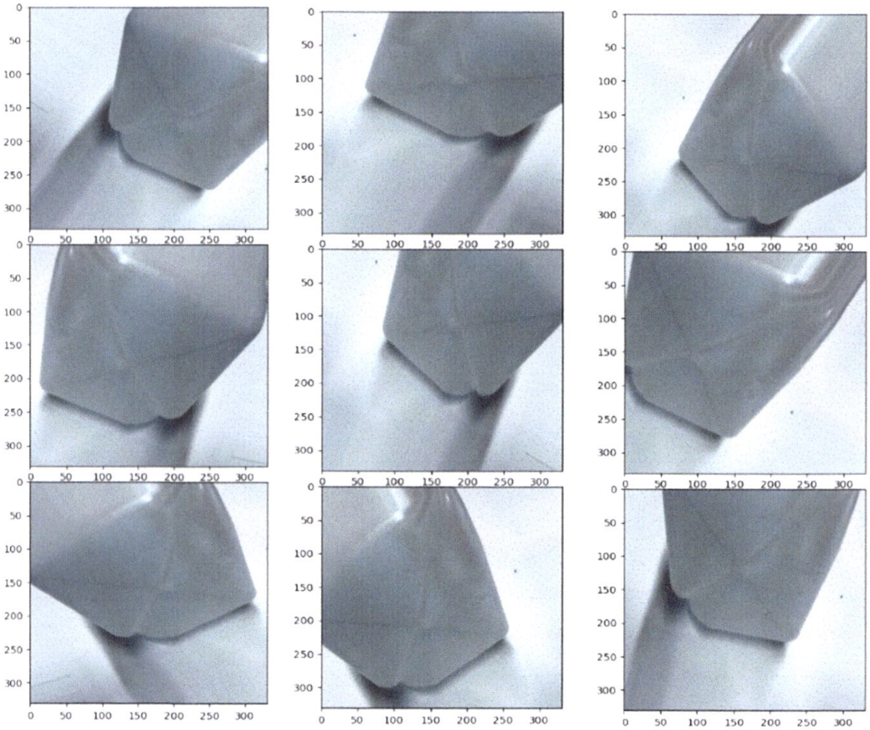

Fig. 10.8 Data augmentation

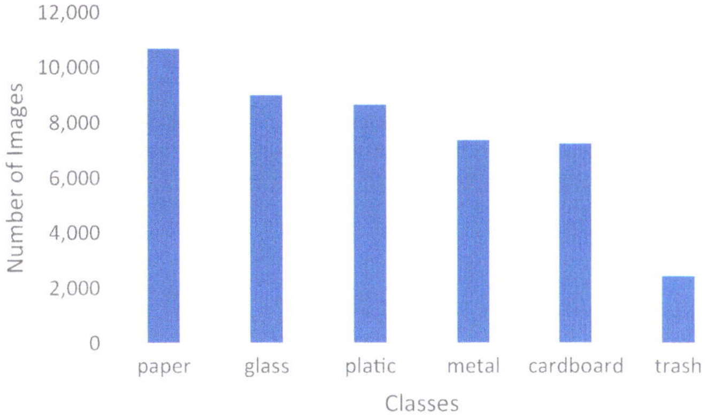

Fig. 10.9 Class distribution after data augmentation

10.5 Results and Discussion

10.5.1 Results from Experiment 1: CNN + SVM Classifier

In Experiment 1, we evaluated the performance of various Convolutional Neural Network (CNN) architectures combined with a Support Vector Machine (SVM) classifier for waste classification. The five best performing results are summarized in Table 10.1.

It can be noted that VGG19 and VGG16 architectures exhibited remarkable performance, achieving testing accuracies of 84.3% and 82.3%, respectively. Despite a slightly lower training accuracy of 95%, the VGG19 model demonstrated superior generalization ability, maintaining a high testing accuracy on unseen data, which is a critical aspect for practical deployment in waste classification systems.

While the NASNetMobile model achieved a high training accuracy of 96.3%, its testing accuracy of 76.5% suggests some degree of overfitting or limited generalization to unseen data. Similarly, the Xception and InceptionV3 models recorded high

Table 10.1 Summary of results for experiment 1

Model	Training accuracy (%)	Test accuracy (%)	Time taken (min)	Size (MB)
1. VGG19	95	84.3	34	23
2. VGG16	96	82.3	36	62
3. NASNetMobile	96.3	76.5	31	26
4. Xception	97.7	77.9	71	95
5. InceptionV3	95.2	77.5	53	99

training accuracies of 97.7% and 95.2% respectively, but their testing accuracies were relatively lower at 77.9% and 77.5%.

In terms of computational efficiency, the VGG19 model stood out with a training time of 34 minutes and a compact model size of 23 MB, making it suitable for deployment in resource-constrained environments. The VGG16 model also demonstrated efficient training time of 36 minutes, albeit with a larger model size of 62 MB.

10.5.2 Results from Experiment 2: CNN + SoftMax Classifier

Building upon the insights gained from Experiment 1, this second experiment investigates the performance of Convolutional Neural Networks (CNNs) coupled with a SoftMax classifier for waste classification. The five best performing models from this experiment are shown in Table 10.2.

The VGG19 architecture, again, demonstrated strong performance, achieving a testing accuracy of 83.9%. Despite a lower training accuracy of 93.4%, the VGG19 model exhibited robust generalization ability. The VGG16 model also delivered commendable results, with a testing accuracy of 82.7% and a training accuracy of 96.5%. With a training time of 2 hours and 40 minutes, coupled with a manageable model size of 62 MB, further enforces its suitability for practical applications in waste classification systems.

Furthermore, the ResNet101V2 model showcased an interesting training accuracy of 99.2%, but coupled with its testing accuracy of 80.5%, suggests a high degree of overfitting. Also, its relatively larger model size of 181 MB may limit its applicability in resource-constrained environments. Both the Xception and NASNetLarge models demonstrated competitive training accuracies above 98%, but their testing accuracies were comparatively lower at 78.5% and 76.4% respectively. Additionally, the NASNetLarge model required a significantly longer training time of around 5 hours, making it less efficient for deployment in real-time waste classification systems.

Table 10.2 Summary of results for experiment 2

Model	Training accuracy (%)	Test accuracy (%)	Time taken	Size (MB)
1. VGG19	93.4	83.9	3 h 7 min	23
2. VGG16	96.5	82.7	2 h 40 min	62
3. ResNet101V2	99.2	80.5	2 h 42 min	181
4. Xception	98.4	78.5	2 h 21 min	95
5. NASNetLarge	98.9	76.4	5 h	360

10.5.3 Results from Experiment 3: CNN + Sigmoid Classifier

Experiment 3 assesses the performance of Convolutional Neural Networks (CNNs) combined with a Sigmoid classifier for waste classification. The findings can be depicted in Table 10.3.

In this experiment, the VGG16 model emerged as the top performer, achieving a testing accuracy of 84.3% with a training accuracy of 96.4%. The relatively efficient training time of 69 minutes and a moderate model size of 64 MB make it a compelling choice for waste classification applications. Following closely, the ResNet101V2 model demonstrated strong performance with a testing accuracy of 83.8% and a training accuracy of 99.1%. Despite its relatively larger model size of 181 MB, its efficient training time of 76 minutes highlights its potential suitability for practical deployment, however signs of overfitting has been noted.

The VGG19 model also delivered competitive results, achieving a testing accuracy of 83.4% with a training accuracy of 94.8%. Its compact model size of 26 MB and efficient training time of 66 minutes further enhance its appeal for deployment in resource-constrained environments.

However, the NASNetLarge model exhibited a longer training time of 318 minutes, significantly higher than other models in the experiment. While it achieves a resonable training accuracy of 98.8%, its testing accuracy of 73.9% suggests limitations in generalization on unseen data. Similarly, the ResNet152V2 model demonstrated efficient training with a time of 84 minutes but exhibited a lower testing accuracy of 75.3% compared to other top-performing models.

Comparing the three experiments on CNN architectures with different classifiers for waste classification, VGG19 demonstrated a consistent and robust generalization capabilities with efficient testing accuracies ranging from 83.9% to 84.3%. The second-best performing model was VGG16 with a range from 82.3% to 84.3% accuracy and efficient training times. ResNet101V2 also demonstrated some strong results despite occasional signs of overfitting, making it suitable for less resource-constrained environments. However, NASNetLarge exhibits limitations in generalization and longer training times, hindering its practical deployment. Considering both performance and efficiency, the top five models chosen for further improvements were as follows:

1. VGG16 + SVM

Table 10.3 Summary of results for experiment 3

Model	Training accuracy (%)	Test accuracy (%)	Time taken (min)	Size (MB)
1. VGG16	96.4	84.3	69	64
2. ResNet101V2	99.1	83.8	76	181
3. VGG19	94.8	83.4	66	26
4. NASNetLarge	98.8	73.9	318	360
5. ResNet152V2	99	75.3	84	181

2. VGG19 + SVM
3. VGG16 + Sigmoid
4. VGG19 + SoftMax
5. VGG16 + SoftMax.

10.5.4 Experiment 4

This experiment builds upon the findings from experiments 1, 2, and 3 by specifically targeting the fine-tuning of the top five performing models identified in these initial trials. This focus on fine-tuning is a strategic approach to further optimize model performance, refining and adjusting the models based on their previously demonstrated capabilities. Fine-tuning in this context involves the extension of the training period and the implementation of more sophisticated training techniques and hyperparameters adjustments. By increasing the number of training epochs to 100, the models were given more exposure to the training data, which is a critical factor in deep learning models capturing more complex patterns and reducing error rates on new, unseen data.

Furthermore, experiment 4 introduces additional hyperparameter tuning, such as adjustments to SVM regularize rates. Regularization is a technique used to prevent overfitting, ensuring that the model does not just memorize the training data but also generalizes well to new data. The inclusion of a call-back mechanism to save the best-performing model during the training process is another aspect of fine-tuning that ensures the retention of the optimal model state, regardless of subsequent performance drops in later epochs. The results are shown in Fig. 10.10.

The VGG16 architecture with an SVM classifier demonstrated an exceptionally high training accuracy of 98.54%, which is indicative of the model's capacity to fit

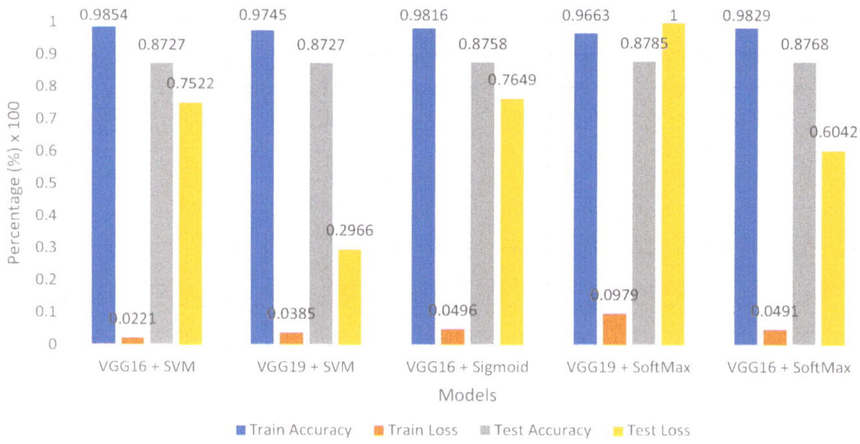

Fig. 10.10 Model accuracy and loss comparison

the training data nearly perfectly. However, this high training accuracy comes with a training loss of 2.21% and a test loss of 75.22%. While the test accuracy was strong at 87.27%, the gap between training and test loss suggests that the model is overfitting on the training data. This overfitting is further highlighted by the discrepancy between training and test accuracy, which is symptomatic of the model's limited ability to generalize to unseen data. The VGG19 model paired with an SVM classifier achieved a training accuracy slightly lower at 97.45%, showcasing a lower test loss of 29.66%. A smaller test loss value, and comparable test accuracies shows the generalization capabilities of the model.

When the VGG16 architecture was combined with a Sigmoid classifier, the experiment noted a slight decrease in training accuracy to 98.16% and an increase in training loss to 4.96%. Although a high test loss, the model achieves a test accuracy of around 87.58%. These figures suggest that the Sigmoid classifier contributs to almost the same generalization capabilities with the SVM in the case of the VGG16 model. The VGG19 + SoftMax configuration emerged as a strong contender in terms of generalization, with the highest test accuracy among the group at 87.85%. Despite a higher training loss than VGG-16 with SoftMax classifier, VGG-19 + SoftMax shows better training and test accuracy ratio, avoiding signs of overfitting. We used the cross-entropy loss as metric, and there may be the need to run the experiments further to confirm the exact loss values for VGG-19 with SoftMax. However, given the current computing resources, running longer epochs is not feasible. The VGG16 + SoftMax is showing signs of overfitting in this setting.

As for the training time, with reference to Fig. 10.11, the VGG16 + SVM model trained in just under 127 minutes while the VGG19 + SVM showed a slight improvement, requiring about 122 minutes. This suggests some efficiency despite its more complex architecture. The VGG16 + Sigmoid further reduced the training time to approximately 120 minutes. The VGG19 + SoftMax's training time increased to around 131 minutes, while the VGG16 + SoftMax took the longest at approximately 136 minutes. This progression illustrates a trade-off between model complexity and training efficiency, with the more accurate VGG16 + SoftMax configuration demanding more computational time.

Experiment 4 returned valuable insights into the performance of the selected models but, it was not conclusive enough to determine the optimal model for waste classification. While VGG-19 + SVM demonstrated robust performance, VGG-19 + SoftMax classifier exhibited better results. Due to computational resource constraints, further training and testing epochs were not feasible to validate these findings, especially for VGG-19 + SoftMax. Therefore, to assess the generalization capabilities of each model under real-world conditions, an additional evaluation was conducted. A set of 15 previously unseen images, comprising of 3 randomly selected images from each of the 5 waste classes, was used. This evaluation aimed to evaluate the models robustness and predictive accuracy when presented with data variations not encountered during training. The results of this evaluation are presented in Table 10.4.

The VGG16 + SVM model, while efficient in training time, misclassified a plastic bottle as glass showing variability in its confidence scores, already reflected in

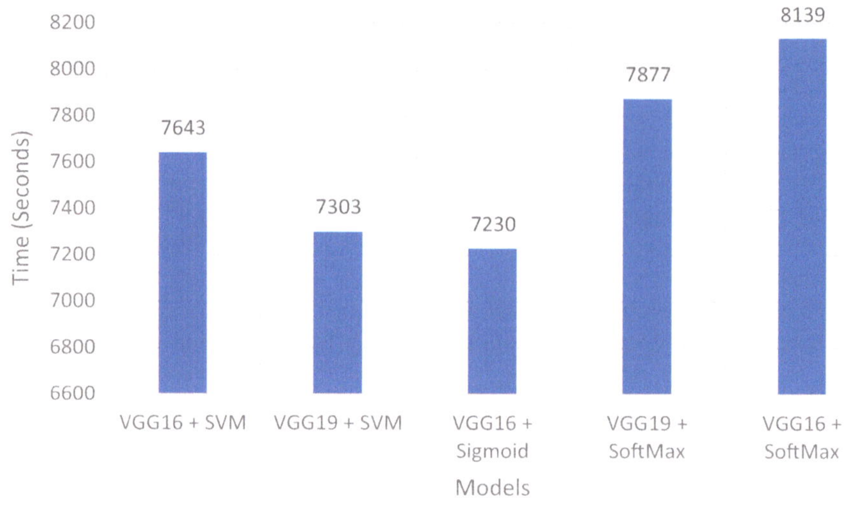

Fig. 10.11 Training time (in seconds)

the prediction probabilities. The VGG19 + SVM showed similar inconsistencies, notably misidentifying a metal can as cardboard. These misclassifications, particularly in material categories with distinct characteristics, highlight potential limitations in the model's feature extraction capabilities.

The VGG16 + Sigmoid, demonstrated strong capabilities, with accurate predictions across most classes and high confidence scores, emphasizing the benefits of its fine-tuning process. The VGG19 + SoftMax model, consistent with its high testing accuracy from the training phase, performed very well, making correct predictions with high confidence scores, affirming its robustness in handling unseen data, scoring the highest with 14 out of a possible 15 points.

The VGG16 + SoftMax configuration, despite its extended training time, did encounter some issues. It correctly identified all the materials but showed a significant confidence variation, particularly in its misclassification of a glass bottle as metal. The scores assigned to each prediction shed light on the certainty with which each model made its classifications, an essential factor in practical scenarios where decisions based on these predictions may have significant repercussions.

Among the models evaluated, the VGG19 + SoftMax configuration emerges as the most robust and reliable choice. It not only demonstrated a balanced training efficiency with commendable computational speed but also showcased superior generalization capabilities in practical testing. Its consistency in accurately predicting unseen images, along with high confidence scores, underlines its potential for real-world applications. This model strikes an optimal balance between accuracy, efficiency, and generalizability, making it the preferred candidate for further development and application in tasks requiring reliable material classification.

Table 10.4 Model evaluation with completely new images

Image	Actual	VGG16 + SVM	VGG19 + SVM	VGG16 + Sigmoid	VGG19 + SoftMax	VGG16 + SoftMax
	Cardboard	Cardboard 1.19	Cardboard 0.58	Cardboard 0.29	Cardboard 0.94	Cardboard 1.00
	Cardboard	Cardboard 0.32	Cardboard 0.50	Cardboard 0.0024	Cardboard 1.00	Cardboard 0.76
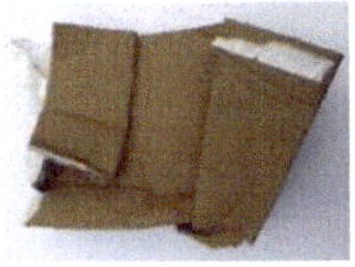	Cardboard	Cardboard 1.78	Cardboard 4.56	Cardboard 0.0044	Cardboard 0.93	Cardboard 0.97
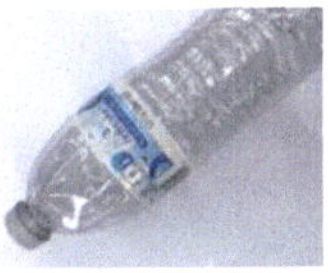	Plastic	Plastic 0.98	Plastic 1.38	Plastic 0.97	Plastic 1.00	Plastic 0.86
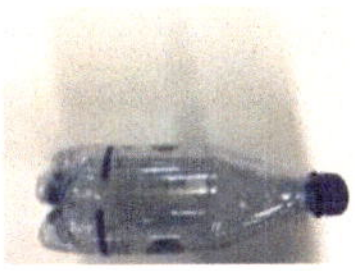	Plastic	Glass 1.18	Glass 3.36	Plastic 0.45	Glass 0.99	Plastic 0.97
	Plastic	Plastic 2.36	Glass 0.17	Plastic 1.00	Plastic 0.55	Plastic 1.00
	Paper	Paper 2.42	Paper 2.26	Paper 0.94	Paper 1.00	Paper 1.00

(continued)

Table 10.4 (continued)

Image	Actual	VGG16 + SVM	VGG19 + SVM	VGG16 + Sigmoid	VGG19 + SoftMax	VGG16 + SoftMax
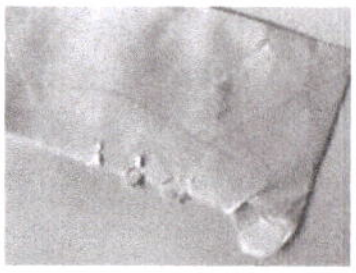	Paper	Paper	Paper	Paper	Paper	Paper
		2.34	1.18	0.98	0.98	0.99
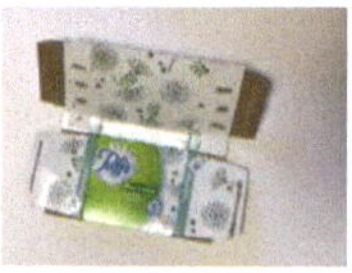	Paper	Paper	Paper	Paper	Paper	Paper
		2.22	2.41	1.00	1.00	1.00
	Metal	Cardboard	Paper	Cardboard	Metal	Paper
		0.94	0.52	0.95	0.87	0.56
	Metal	Metal	Metal	Metal	Metal	Metal
		2.91	2.72	1.00	1.00	1.00
	Metal	Metal	Metal	Metal	Metal	Metal
		1.30	0.21	0.99	0.95	0.99
	Glass	Glass	Glass	Glass	Glass	Plastic
		0.91	0.44	0.010	0.80	0.88

(continued)

Table 10.4 (continued)

Image	Actual	VGG16 + SVM	VGG19 + SVM	VGG16 + Sigmoid	VGG19 + SoftMax	VGG16 + SoftMax
	Glass	Metal	Glass	Metal	Glass	Glass
		2.71	0.083	0.16	1.00	1.00
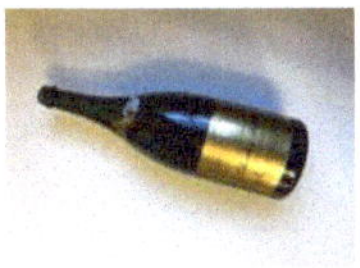	Glass	Glass	Glass	Glass	Glass	Metal
		2.13	1.87	0.88	1.00	1.00
Score (Max: 15)		12	12	13	14	13

10.6 Conclusion

This research thorougly compared a combination of 36 variations of Convolutional Neural Networks (CNNs) across different classifiers. The VGG19 CNN architecture with a SoftMax classifier emerged as the most effective model, achieving highest accuracy in classifying waste into five categories. While this result demonstrates a notable improvement, real-world deployment necessitates further refinement. Future research will prioritize expanding the dataset to encompass a wider range of waste materials, enhancing model robustness. Additional optimization techniques will also be explored, including further hyperparameter tuning, and the application Nature Inspired Algorithms (NIAs) to improve on the model minimization capabilities, reducing classification errors. Image segmentation can also be considered for more complex scenarios to identify and separate different waste materials in a single image.

References

Abdoli, S. (2019). RFID application in municipal solid waste management system. *International Journal of Environmental Research, 3*(3), 447–454.

Adedeji, O., & Wang, Z., 2019. Intelligent waste classification system using deep learning convolutional neural network. *Manufacturing* 607–612.

Bircanoglu, C., et al. (2018). RecycleNet: Intelligent waste sorting using deep neural networks. In *Innovations in Intelligent Systems and Applications (INISTA).*

Costa, B. S., et al. (2018). Artificial intelligence in automated sorting in trash recycling. In *Anais do XV Encontro Nacional de Inteligência Artificial e Computacional (ENIAC 2018).*

Desai, Y., Dalvi, A., Jadhav, P., & Baphna, A. (2018). Waste segregation using machine learning. *International Journal for Research in Applied Science & Engineering Technology (IJRASET), 6*(3).

He, K., Zhang, X., Ren, S., & Sun J. (2016). Deep residual learning for image recognition. Las Vegas, NV, USA. In *2016 IEEE Conference on Computer Vision and Pattern Recognition (CVPR)*.

Kaza, S., Yao, L. C., Bhada-Tata, P., & Van Woerden, F. (2018). *What a waste 2.0 : A global snapshot of solid waste management to 2050.* Urban Development Series. World Bank.

Krizhevsky, A., Sutskever, I., & Hinton, G. (2012). ImageNet classification with deep convolutional neural networks. *Neural Information Processing Systems.*

LeCun, Y., Bottou, L., Bengio, Y., & Haffner, P. (1998). Gradient-based learning applied to document recognition. *Proceedings of IEEE, 86*(11), 2278–2324.

Ramsurrun, N., Suddul, G., Armoogum S., & Foogooa, R. (2021). Recyclable waste classification using computer vision and deep learning. In *2021 zooming innovation in consumer technologies conference (ZINC)*, Novi Sad, Serbia (pp. 11–15). https://doi.org/10.1109/ZINC52049.2021.9499291.

Saar, S. S. M., & Thomas, V. M. (2004). Towards intelligent recycling: A proposal to link bar codes to recycling information. *Resources, Conservation & Recycling, 41*(1), 15–22.

Saar, S., & Thomas, V. (2002). Toward trash that thinks: Product tags for environmental management. *Journal of Industrial Ecology, 6*(2), 133–146.

Simonyan, K., & Zisserman, A. (2015). *Very deep convolutional networks for large-scale image recognition.* s.l., arXiv:1409.1556.

Suddul, G., & Seguin, J. F. L. (2023). A comparative study of deep learning methods for food classification with images. *Food and Humanity, 1*, 800–808. https://doi.org/10.1016/j.foohum.2023.07.018, ISSN 2949-8244.

Szegedy, C., et al. (2016). Rethinking the inception architecture for computer vision. In *Proceedings of 2016 IEEE Conference on Computer Vision and Pattern Recognition* (p. 2818).

Tan, M., & Le, Q. (2021). EfficientNet: Rethinking model scaling for convolutional neural networks. In *Proceedings of the 36th International Conference on Machine Learning, Long Beach, California.*

Thung, G. (2016). *Github.* https://github.com/garythung/trashnet.

Torres-Garcia, A., et al. (2015). Intelligent waste separator. *Computacion Y Sistemas, 19*(3), 487–500.

Tsao, Y. C., Linh, V. T., & Lu, J. C. (2017). Closed-loop supply chain network designs considering RFID adoption. *Computers & Industrial Engineering, 113*, 716–726.

Umut, O., & Seyfi, L. (2018). Fine-tuning models comparisons on garbage classification for recyclability. In *2nd International Symposium on Innovative Approaches in Scientific Studies.*

United Nations, S. D. G. (2023). *Fast facts-what is plastic pollution.*. Retrieved April, 2024, from, https://www.un.org/sustainabledevelopment/blog/2023/08/explainer-what-is-plastic-pollution/.

Yamashita, R., Nishio, M., Do, R., & Togashi, K. (2018). Convolutional neural networks: an overview and application in radiology. *Insights into Imaging, 9*, 611–629.

Yang, M. & Thung, G. (2016). *Classification of trash for recyclability status.* http://cs229.stanford.edu/proj2016/report/ThungYang-ClassificationOfTrashForRecyclabilityStatus-report.pdf.

Ziouzios, D., & Dasygenis, M. (2019). *A Smart Recycling Bin for Waste Classification. s.l., Panhellenic Conference on Electronics & Telecommunications (PACET).*

Chapter 11
Plastic Pollution, Treatment and Management Strategies in Pakistan: A Review

Iftikhar Ahmad, Aqdus Ali, and Khadija Shabbir

Abstract Plastic, a versatile and affordable polymer has been used in daily life since 1950. The rate of plastic production is rapidly increasing, leading to plastic pollution worldwide. Pakistan accounts for 15% annual increase in plastic waste due to high rate of mismanagement and improper waste disposal. Many case studies show that microplastic pollution is a growing environmental threat in Pakistan, with the Rawal Lake study revealing the presence of fibres and fragments as the most dominant types. Sustainable alternatives and recycling procedures are being developed to address the environmental effects of plastic waste. Governments worldwide are addressing plastic pollution at local, national, and international levels by prohibiting single-use plastics through reduce and reuse campaigns by turning plastic waste into asphalt or bricks. Innovative techniques like recycling, landfilling, and waste-to-energy are being explored to control plastic pollution throughout the world. Biodegradation by using plastic eating bacteria is another good option for complete plastic waste removal. Pakistan has started an initiative to control plastic pollution through Oxo-biodegradable technology to address plastic pollution in 2019, banned polythene bags in Islamabad, construction of plastic roads, and the Clean Green Pakistan Index aims to improve sanitation services through competition among cities and citizens. Waste-to-energy strategies such as pyrolysis and gasification, are also being used to recover energy and overcome recycling limitations but these strategies need effective monitoring and economic incentives in Pakistan. In addition, private organizations such as the World Wildlife Fund have launched a pilot project in Lahore to collect and recycle plastic. The policy statement prohibits the production, import, distribution, sale, and use of single-use plastic items in Islamabad and beverage containers made entirely from virgin plastic. Plastic pollution is a significant environmental threat not only in Pakistan but also throughout the world that require urgent actions through reduction, recycling, global cooperation, sustainable alternatives, strict regulation, and public awareness to protect the planet for future generation.

I. Ahmad (✉) · A. Ali · K. Shabbir
Department of Environmental Sciences, COMSATS University Islamabad, Vehari Campus, Vehari, Pakistan
e-mail: iftikharahmad@cuivehari.edu.pk

S. Suriyanarayanan et al. (eds.), *Combating Plastic Pollution in Terrestrial Environment*,
https://doi.org/10.1007/978-981-96-2343-3_11

11.1 Introduction

Plastic is a versatile and ubiquitous substance that has revolutionized modern living. Even though plastics have been used in daily life for a very long time, mass production of plastic products started all the way back to 1950 (Urbanek et al., 2018). In essence, the word "plastic" comes from the Greek word "plastikos," which means "able to be moulded into various size and shapes" (Joel, 1995). This extremely useful and affordable polymer may be moulded into a variety of shapes and expelled into a variety of practical goods, such as moulded plastic water bottles, food containers, packaging, and several other everyday plastic products (Andrady & Neal, 2009). Today, silicon, carbon, hydrogen, nitrogen, and oxygen are frequently used as raw materials in the production of plastics; nevertheless, the primary raw materials required to produce plastics are extracted from oil, coal, and natural gas (Seymour, 1989). Studies show that plastics may be categorized into four types depending on their sizes: nano (less than 100 nm), micro (less than 5 mm), meso (less than 25 mm), and macro (more than 25 mm). Each size group is composed of a distinct polymer, such as polyethylene, polypropylene, polystyrene, or polyvinyl chloride (Windsor et al., 2019). Due to the wide range of plastic items being used for both industrial and domestic applications, the rate of plastic production is rapidly rising these days (Vijayakumar & Sebastian, 2018). Pollution is a persistent issue on a worldwide scale with cross-institutional jurisdiction, transdisciplinary scope for solutions, and international roots (Iroegbu et al., 2021). Nowadays, plastic waste is present everywhere in the planet, including deep oceans, mountain summits, deserts, tropical landfills, and Arctic snow (MacLeod et al., 2021). Plastic pollution can manifest in several ways, such as the build-up of wastes, marine litter, plastic pieces or microparticles, and non-biodegradable fishing nets that trap garbage and wildlife or the waste that results in animals dying after consuming plastic objects. Since plastics are persistent, inert, non-biodegradable, and contribute significantly to environmental contamination, they are regarded as one of the main environmental issues of our day (Schwarz et al., 2019). As plastic breaks down into small pieces in both marine and terrestrial environment, so it could be consumed by various species which causes bioaccumulation that might cause serious health consequences (Cole et al., 2011). Since soil is thought of as a sink for plastic waste, it is more susceptible to plastic pollution, which can have a harmful effect on the ecology of the soil by altering the physical and chemical properties of the soil, as well as the flora and fauna, and nutrients of soil (Ya et al., 2021). Plastic pollution on land increases the likelihood that it may contaminate the food chain, which could have detrimental consequences on human health in addition to the more obvious harm it does to the environment (Dissanayake et al., 2022). Plastics are produced using a variety of conventional materials and waste products, such as formaldehyde, phthalates, and other plasticizers (found in PVC and other materials), benzene (found in polystyrene), vinyl chloride (found in PVC), dioxins (found in PVC), and bisphenol-A, or BPA (found in polycarbonate). Many of these materials are neurotoxic, hormone-disrupting, and carcinogenic, and they eventually contaminate our air, land, and water, entering our ecosystem through contamination (Proshad et al., 2018). However, efforts to create sustainable alternatives and enhance recycling procedures are being driven by the growing concern over the environmental effects of plastic waste, particularly single-use plastics throughout the world.

11.2 Status of Plastic Pollution in the World, Southeast Asia, and Pakistan

The world's expanding population is one of the main factors contributing to rising environmental degradation. The amount of waste generation increases as the number of people increases. People easily disposed of items like water bottles and soda cans to live better lifestyle. We have become a disposable state and nation, although it is a harsh fact that hoarding these things has increased the amount of plastic pollution in the world. Currently, according to the United Nations, 80% of ocean pollution originates from land consisting of 8 million tons of plastic waste. Over 320 million tons of plastic are produced annually worldwide, of which 40% is used for single-use packaging (Wright & Kelly, 2017). Approximately 9% of the plastic waste produced in a year is recycled, 12% is burned, and 79% is disposed of in open spaces or landfills (Suzuki et al., 2022). Consequently, an astounding 31.9 million metric tons of plastic trash are dumped into the natural environment annually (Bucci et al., 2020).

Southeast Asia, despite being a highly biodiverse region, is thought to be responsible for one-third of the world's total plastic pollution (Omeyer et al., 2023). More than half of the total plastic waste is produced worldwide by the Philippines, Malaysia, Indonesia, Myanmar, Vietnam, and Thailand, with the Philippines producing a third of the total 3,56,371 metric tons annually. Nearly 60% of the world's population lives in Asia, which also produces more than 80% of the world's ocean plastic debris (Arumdani et al., 2021). South Asian nations such as India, Bangladesh, Sri Lanka, Pakistan, Nepal, Bhutan, Maldives, and Afghanistan are considered developing nations based on their economic capacities. Plastic pollution has grown significantly during the past few decades. For example, in 2018 the production of plastics in South Asian countries alone amounted to 17–20 million tons (MT). In 2016, these countries disposed of roughly 26.72 MT of plastic waste (Amrutha et al., 2021). This amounts from 1.7 to 4.6% of the total plastic waste generated by the participating countries. Between 4.8 and 12.7 million metric tons of plastic waste ended up in the ocean out of an estimated 99.5 million metric tons of plastic waste generated in coastal regions of Southeast Asian countries (Jambeck et al., 2015; Ng et al., 2023).

11.2.1 India

The amount of plastic used worldwide is growing exponentially every year. A significant amount of plastic waste is being transferred from terrestrial ecosystems to the seas and oceans via connected water channels, according to a research on plastic contamination in Mumbai's South Juhu Creek (Amrutha et al., 2021). The size range of the plastic debris at Silver Beach, Southern India, is 0.5–1 mm, with an average of 204 pieces per kilogram (Jeyasanta et al., 2020). The Arabian Sea borders, the Indian Ocean on the West, the Bay of Bengal on the East, and India on the South. Thirteen States are touched by its about 7516.66 km (including islands) of coastline. The coastal zone is considered a major hotspot for microplastics since almost half of the world's population lives near beaches. Plastic debris, including fibres and films, has heavily polluted the Ganga River that affect human and environmental health according to the experts (Amrutha et al., 2021). Researchers have been aware of plastic pollution as a result of recent observations of microplastic in a variety of human commodities, seafood, salt, drinking water, and tap water in India (Chinglenthoiba et al., 2023; Vaid et al., 2021).

11.2.2 Bangladesh

Bangladesh is among the rapidly developing nations whose use of plastic products has skyrocketed, seriously threatening biodiversity and the environment. In Bangladesh, microplastic pollution is a relatively new occurrence, and neither customers nor producers are aware of the dangers it poses. In 2002, the Bangladeshi government outlawed the manufacture, distribution, and use of polythene packs thinner than 55 microns on account of the potential for plastic contamination. In 2010, the government enacted new laws mandating the use of jute fiber instead of polythene bags for bundling commodities, in support of the 2002 prohibition (Niloy, 2021). Nonetheless, polythene was continuously produced, traded, and utilized throughout the country at the same time because of lax enforcement of the legislation and a dearth of reasonably priced, environmentally suitable substitutes. Data show that the percentage of plastic waste in Dhaka's whole landfill rose from 1.74% in 1992 to 6.5% in 2014. Plastic garbage increased more than 3.5 times in Dhaka city alone, from 178 tons per day in 2005 to 646 tons per day in 2020 (Rifa & Hossain, 2022). Findings from a second investigation corroborated the identification of many polymer forms (PET, PE, and PP) and microplastic types (fibers, microbeads, bits, etc.) in beach sand samples from Kuakata, North Bay of Bengal, Bangladesh (Banik et al., 2022).

11.2.3 Maldives

The Maldives are known for having breathtaking blue waters, which have contributed to their status as a veritable beach paradise. However, microplastics, which are prevalent below the surface, pose a threat that is essentially invisible to the unaided eye. On nearby islands, the quantity of trash created per person has surged by 58% in the last ten years. Transporting waste to Thilafushi, an artificial island constructed as a municipal landfill, where 500 tons of rubbish are put every day, is now the greatest alternative for waste management (Stevens & Froman, 2019). However, plastic pollution is one of the major stresses on this coral reef system, contributing to disease outbreaks (Montano et al., 2015), coral bleaching events (Perry & Morgan, 2017), and the extinction of corallivorous predators (Montalbetti et al., 2019). In particular, Naifaru has a concentration of microplastics ranging from 55 to 1127.5 microplastics per kilogram, which is significantly higher than the rate found in Tamil Nadu, India, which is 3–611 microplastics/kg.

11.2.4 Sri Lanka

Sri Lanka's southern coastline is home to several well-known resorts and beaches, yet this formerly unspoiled area hasn't been spared from the worldwide plastic rubbish crisis. Researchers found that fishing ports and more rural beaches had high levels of microplastic pollution and plastic debris, even when leisure beaches had considerable microplastic levels (Rifa & Hossain, 2022).In terms of weight (157 microplastic bits per square meter) and count (5.98 g per square meter), the busy public beach at Weligama was the dirtiest. Although tourism in the southern part of Sri Lanka has increased, appropriate waste management strategies have not been put in place. Research commissioned by the National Aquatic Resources Agency (NARA) and sponsored by the Norwegian Institute of Marine Research (IMR) found that rivers and canals were the primary source of roughly four-fifths of the tiny particles of plastic waste found in Sri Lanka's territorial waters (Seethawaka, 2023).

11.2.5 Pakistan

Nearly 0.2 million tons of plastic debris have been thrown into the Arabian Sea by Pakistani citizens living along the coast and along the Indus River. In Pakistan, 6,000 manufacturers are producing 0.6 million tons of plastic of which 55 billion plastic bags are produced annually. Most of these single-use, non-biodegradable plastic items end up in open landfills, rubbish dumps, or municipal sewers. Regretfully, plastics account for 65% of all garbage in Pakistan, and this percentage is predicted to rise by 15% yearly (Rifa & Hossain, 2022). According to an estimation, 9% of the

30 million tons of solid trash produced annually in Pakistan is made up of plastics from which 8 million tons of plastic waste is purposefully discarded into the world's oceans. In South Asia, Pakistan has the highest rate of mismanagement of plastic waste and improperly disposed of plastic (Mukheed & Alisha, 2020).

11.3 Treatment Technologies for Combating Plastic Pollution in the World

In view of the growing concern about the negative effects of plastics on environmental and human health, several governments are increasingly responding to this problem at the local, national, and international levels (Adam et al., 2020). At least 28 international policies were established from the year 2000 till now in order to deal with increasing plastic pollution, three of which are legally binding on member states: the Antarctic Treaty, amendments to the London Convention and Protocol, and Annex V of the International Convention for the Prevention of Pollution from Ships (MARPOL) (Karasik et al., 2020). In addition to these international initiatives, regional groups of governments have begun to address the problem of plastic pollution (in different geographic locations) these efforts can be seen in the form of 39 regional policy documents that have been adopted, mostly in Europe (Karasik et al., 2020). Apart from the aforementioned global and regional initiatives, there has been a surge in national and subnational government reactions in the past ten years, with a primary focus on prohibiting plastic bags and to a lesser degree, single-use plastics in general for example through reduce and reuse campaigns or by putting taxes or fees, etc. Many countries have banned products that include plastic particles, such as cosmetic beads, seed coatings, and agricultural capsules. The primary goal of these strategies is to reduce the 42,400 tons of plastic that are released into the environment annually. These responses have been observed in many countries such as the United States, Canada, the United Kingdom, New Zealand, Finland, France, Iceland, Ireland, Italy, Luxemburg, Norway, and Sweden (Lam et al., 2018). Numerous local, national, and regional solutions have been proposed to solve this issue because there are numerous affordable ways and effective technologies to control plastic pollution in various social settings and geographical areas (Lau et al., 2020). Such technologies addressing these problems aim to either (1) collect current plastic waste or (2) directly prevent plastic from leaking into waterbodies (Mitrano & Wohlleben, 2020). These innovative techniques work for lowering the amount of plastic pollution in the world by focusing on the various stages of the plastic life cycle, such as manufacture, consumption, and waste management (Nielsen et al., 2020).

11.3.1 Conventional Technology for Plastic Waste Management

Recycling, disposing of plastic waste in landfills, and incinerating waste are the conventional technologies for handling plastic waste (Fig. 11.1).

11.3.1.1 Recycling Methods

The recycling of plastics needs to be done in a way that reduces pollution levels, increases process efficiency and preserves energy. Following steps involved in the recycling process are

i. **Selection:** The recyclers need to select plastic waste which are suitable for recycling.
ii. **Segregation:** The selected plastic waste is separated according to its chemical and physical properties.
iii. **Processing:** Direct recycling of the pre-consumer waste is required following selection and segregation. Used plastic trash that has been post-consumer must be cleaned, shredded, agglomerated, extruded, and granulated.

Primary, secondary, tertiary, and quaternary are the four broad categories into which plastic recycling methods have been separated (Hopewell et al., 2009).

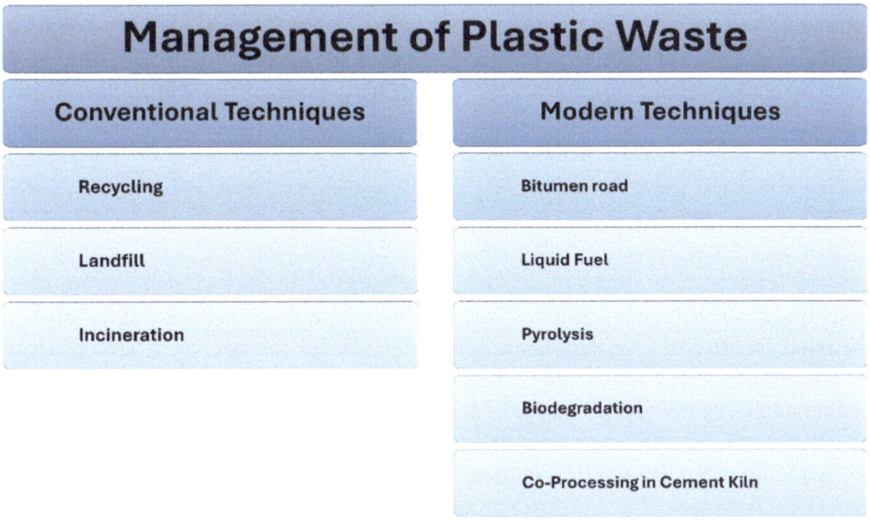

Fig. 11.1 Overview of the handling of plastic trash (Modified from Siddiqui & Pandey, 2013)

- **Primary** recycling includes mechanical processing of plastic waste into a product with features similar to the original product. For instance, a consumer plastic bottle is mechanically washed, sterilized, refilled and recycled in a market.
- **Secondary** recycling involves mechanical processing of plastic waste into products that have characteristics dissimilar from those of original plastic products. For instance, a consumer plastic bottle is mechanically converted into a new plastic product, but its quality is inferior in comparison to original plastic product.
- **Tertiary** recycling is producing basic chemicals and fuels from plastic scrap as part of the municipal waste stream or as a segregated waste. This method applies when plastic waste cannot be modified/recycled through primary and secondary recycling. A waste is converted into basic monomers through depolymerization process.
- **Quaternary** recycling reclaims the energy content of the scrap plastics by burning/incineration/pyrolysis. However, the product has little value along with environmental consequences in the form of greenhouse gas emissions and aesthetic value.

11.3.1.2 Landfilling

This is a conventional method of managing plastic waste, however in certain nations, there isn't enough room to build landfills. A properly maintained landfill site limits immediate environmental damage beyond the effects of collection and transportation. However, there are long-term risks of soil and groundwater contamination by certain additives and degradation by plastic products, which can become persistent organic pollutants (Hopewell et al., 2009). The fact that no material resource utilized in the manufacturing of plastic is recovered, the material flow is linear rather than cyclic which is a major disadvantage of landfills from a sustainability perspective (Oehlmann et al., 2009).

11.3.1.3 Incineration

Here are less needs to dump plastic waste in landfills, thanks to incineration technique, that has significant benefit of reducing various types of trash by about 80% to 90% but there are concerns that potentially dangerous substances might be discharged into the atmosphere. For instance, plastic waste frequently contains PVC and halogenated additives which increases the risk of dioxins, furans, and other polychlorinated biphenyls being released into the environment when burnt (Gilpin et al., 2003). Two techniques for eliminating plastic waste are energy recovery and incineration (Fig. 11.2), both are subject to strict regulatory regulations because if not adequately managed and monitored, they might gravely affect the environment and human health (Ágnes & Kuti, 2016). On the other hand, expense of treating the gasses exceeds the amount of energy recovered. With the help of contemporary incineration technology, every incinerator issue may be resolved without endangering the environment.

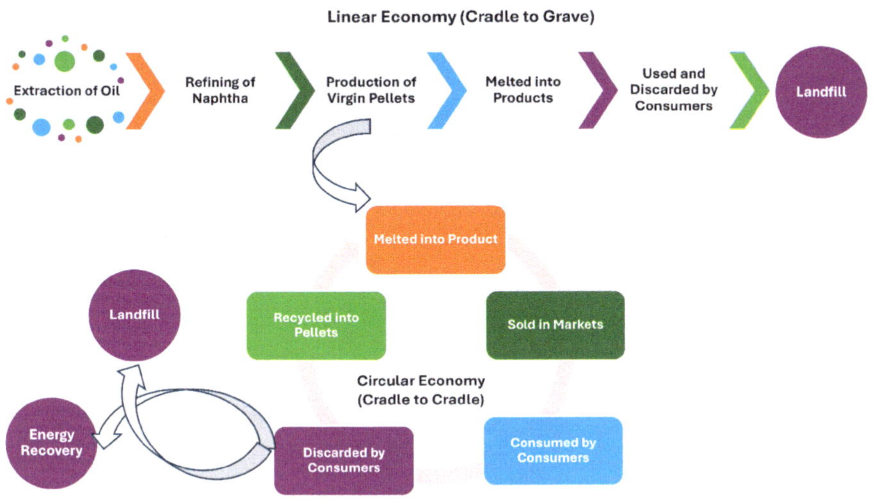

Fig. 11.2 Plastic life cycle in two different economies: circular economy (cradle-to-cradle), and linear economy (cradle-to-grave) (modified from Prata et al., 2019)

11.3.2 Recent Technologies for Plastic Waste Management

The recent technologies for plastic waste management are listed here.

11.3.2.1 Polymer Blended Bitumen Road

The technology for constructing flexible roadways using waste plastics has been developed and is currently being successfully applied in several locations across the world including India and Pakistan (Fig. 11.3).

11.3.2.2 Co-processing of Plastic Waste in Cement Kiln

Recycling plastic waste for energy recovery is a highly effective method, particularly in cement kilns. This method not only creates new products and conserves scarce crude oil resources. The high temperature used in the cement kilns gives a scope for use of even some type of plastic waste contaminated with toxic chemicals like pesticides and some other harmful materials without creating any increased emissions in air and water. No segregation or cleaning is required for disposal in this method. For example: at 10% replacement rate, 170 cement kilns in India could dispose of the entire plastic waste generated in the country today, reducing the use of fossil fuels like coal.

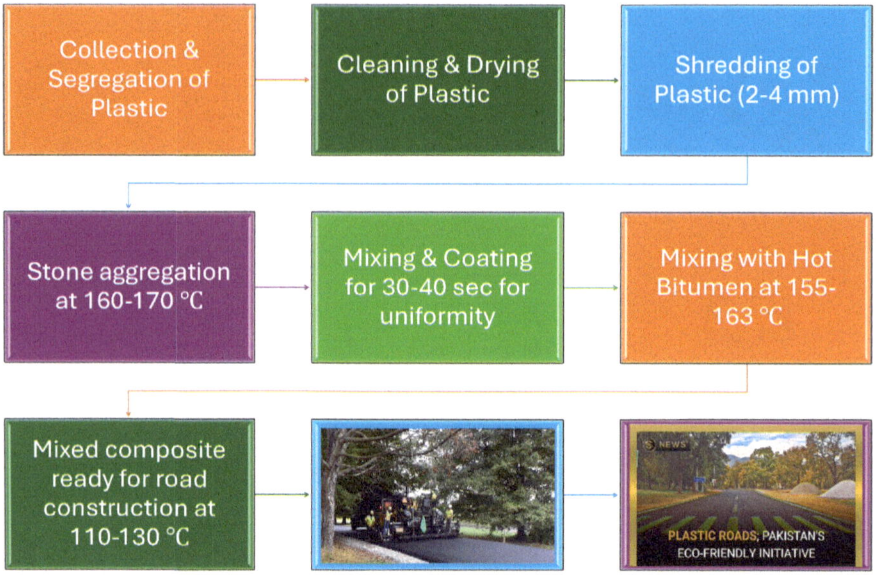

Fig. 11.3 Schematic flow diagram of plastic coated bitumen road construction (modified from Hikmath & Arachchige, 2020)

11.3.2.3 Plasma Pyrolysis Technology (PPT)

Plasma pyrolysis is a technology that combines plasma's thermo-chemical properties with pyrolysis. It involves feeding plastic waste into a primary chamber at 850 °C, which dissociates into carbon monoxide, hydrogen, methane, and other hydrocarbons. The waste is then drained into a secondary chamber, where the gases are combusted in excess air to prevent toxic gases. The secondary chamber maintains a temperature of 10,500 °C, converting hydrocarbons and CO into water and safe carbon dioxide. The extreme and versatile heat generation ability of plasma pyrolysis technology enables it to dispose of all types of plastic waste.

11.3.2.4 Conversion of Plastics Waste into Liquid Fuel

The technology converts plastic waste (unwashed and hard-to-recycle plastics) into liquid fuel. The waste is weighed into a hopper and materials like car bumpers, fuel tanks and packaging products are loaded. Natural gas is burned to generate heat, and a catalyst breaks plastic hydrocarbons into shorter chains. Off-gases are used to produce heat. Fuel oil and diesel are condensed into liquid form that are collected and placed in temporary fuel tanks. The process is controlled by an automated system, ensuring efficient and environmentally friendly waste management.

11.3.2.5 Biodegradation

Landfills, recycling, and incineration are common methods of removing plastic waste, but they proved insufficient for eliminating plastic waste entirely. As a result, there is growing interest in using effective microorganisms to remove plastic polymers entirely through a process called biodegradation (Kale et al., 2015). The term "biodegradation" refers to the biological breakdown of materials by microbes (*Aspergillus, Pseudomonas, Streptomyces, Corynebacterium, Arthrobacter, Bacillus* and *Rhodococcus,* etc.) which involves converting the materials into carbon dioxide, methane, and other salts as a source of energy for growth (Qin et al., 2021). Normally, aerobic microorganisms undergo plastic degradation on land and in water, turning large plastic polymers into monomers, oligomers, metabolites, water, and carbon dioxide, etc. But sometimes anaerobic microorganisms in landfills or deep sediments undergo plastic degradation in anoxic conditions to produce methane, water, and salts as an end product (Plakunov et al., 2020).

11.4 Treatment Technologies for Combating Plastic Pollution in Pakistan

Our lives are now totally dependent on plastics. When compared to solid wastes in Western nations, the qualities of plastic waste debris in Pakistan differ greatly in terms of both toxicity and structure. Pakistan has one of the highest rates of improperly handled plastic. Pakistan wastes more than 3.3 million tons of plastic annually, most of it ends up in uncontrolled dumps, landfills, or scattered throughout the nation's land and waterways, endangering both the environment and public health. The combined height of these plastic dumps, when dumped all at once, is 16,500 m equal to the height of two K2 mountains, the second-highest peak in the world. While the programs for plastic waste management have been tested and proved to work well in developed countries while Pakistan still has challenges with financing sustainability and feasibility.

11.4.1 Strategies in Pakistan to Combat Plastic Pollution

- In 2017, Pakistan implemented Oxo-biodegradable technology as a mean of addressing the problem of plastic pollution. After their useful lives, oxo-biodegradable polymers fully decompose in the environment with no negative environmental effects. The legislation was created after all available possibilities were carefully considered and supported by scientific data (Umar, 2020).
- In 2019, the Islamabad Capital Territory introduced an alternative model that prohibited plastic bags and encouraged the use of biodegradable bags. The Prime

Minister's aim for a Clean and Green Pakistan inspired the Ministry of Climate Change (MoCC) to adopt this effort. MoCC is striving to guarantee that the prohibition is fully implemented by enforcing stiff penalties and fines on anyone who violate it (Umar, 2020).

Thus, to address the problem of plastic pollution, federal authorities must come up with a just plan of action. If a ban is decided upon, it should apply to all varieties of single-use plastics, such as WPP, BOPP, and PP non-woven bags, and scientifically shown biodegradable alternatives should be encouraged (Mukheed & Alisha, 2020; Umar, 2020).

Treatment methods that are used to manage plastic waste generate in Pakistan are given below:

11.4.2 Chemical Treatment Methods

These techniques work by using chemicals to disrupt the polymeric bonds that bind plastics together and change the polymers into non-hazardous forms. Large-scale enterprises in Pakistan that employ these techniques, generate an enormous amount of chemical waste due to the use of chemical to degrade plastic. There are more drawbacks than advantages to Pakistan's traditional methods for handling and disposing of plastic garbage. Burning of plastic waste with chemicals releases dangerous fumes that can lead to a variety of respiratory conditions. The primary issue in Pakistan originates from the dumping of plastic waste in municipal waste disposal systems, which releases toxic leachate into the land.

11.4.3 Physical Treatment Methods

These procedures include the physical processes like compressing, crushing, and incinerating plastic waste in order to reduce its volume. Among the physical waste treatment techniques that are often employed in Pakistan at smaller level is compressing and crushing of plastic waste while at larger level incineration and UV or photooxidant degradation is being used.

11.5 Thermal Treatment Techniques

In Pakistan, the most popular methods for thermally treating solid waste are pyrolysis and incineration. Thermal waste treatment, such as incinerating garbage, can be utilized as a feedstock for raw or unprocessed waste. About 850 °C is the air temperature during incineration procedures, which convert disposal wastewater into carbon dioxide, water, and solid waste fuel. Nevertheless, burning plastic garbage creates several dangerous chemicals that we should avoid breathing in since they are the primary source of Pakistan's growing air pollution. Another popular thermo-chemical conversion technique is pyrolysis, which burns solid fuel without the need for an oxidizing agent. Facilities for pyrolysis use a lot of energy. They also release sulphur dioxide, nitrogen oxides, and carbon monoxide, which pollutes the air and water. Plant owners in Pakistan who need to purchase appropriate pollution control equipment and conduct routine emission monitoring would bear a heavy financial burden because of these concerns.

11.5.1 Biological Treatment Methods

The plastic polymers break down by micro-organisms without creating any hazardous byproducts. The process of bio-methanation involves using microbes to transform organic polymers or solid wastes into manure and methane. Anaerobic digestion is the technique used to carry out this process in the absence of oxygen. However, because there has been little research related to this field, it is uncommon in Pakistan for plastic trash to be broken down by bacteria that consume plastic.

11.6 Management Strategies in Pakistan

Among South Asian countries, Pakistan has one of the highest rates of badly managed plastic (UNDP, 2021). Rethinking, Pakistan's relationship with plastics, UNDP, 2021 report states that more than 3 million tons of plastic are wasted annually in Pakistan. The plastic debris is 16,500 m high, or twice the height of a K2, world's second highest mountain, if it is disposed of all at once! The majority of it winds up dispersed throughout the nation's land and waterways, in landfills, or in poorly maintained dumps. Plastic does not decompose naturally like other materials do. Even after decomposing for up to a millennium, it never really vanishes.

In addition to choking our land, oceans, and wildlife, the tidal wave of plastic also endangers our personal health. They result from the open burning of plastic garbage, which releases a range of hazardous chemicals. Under certain circumstances, compounds like monomers, antimony, 2-aminobenzamide, and non-intentionally added substances (NIAS) may diffuse from PET bottles into water (Muhamedbegović et al., 2023).In addition, a study published in the medical journal Annals of Global Health discovered that, among other illnesses, low birth weight, asthma, childhood leukemia, and lung cancer are more common in areas close to plastic manufacturing facilities and trash disposal sites (The Guardian, 2023). In humans, plastic pollution and its by-products induce negative sequels in different organs, including the lungs, skin, liver, brain and gastrointestinal system, which triggers an immune response, cell toxicity, cancer and neurological disorders, along with other health issues (Yusuf et al., 2022). Concerns regarding microplastics, which are tiny plastic particles as small as 5 mm in diameter that get into everything we eat, drink, and even shed by our clothing, are also growing. They have been discovered in a number of human organs, including the placenta of a baby (UNEP, 2021).

11.6.1 Recycling is the Best Management

The global plastic recycling rate remains quite low. According to a report published by the OECD in 2022, only about 9% of plastic waste is effectively recycled worldwide (OECD, 2022). This underscores the ongoing challenges in plastic waste management, with most plastic waste ending up in landfills, incinerated, or leaking into the environment. Additionally, Visual Capitalist's analysis from 2023 provides further insight into the movement of global plastic waste, highlighting the disparity between plastic waste production and recycling efforts (Visual Capitalist, 2023). Plants for Recycling Plastic: The Pakistani plastic recycling plants directory lists businesses that transform plastic trash into new commodities in Pakistan (Table 11.1).

Table 11.1 Plastic recycling plants in Pakistan

Company name	Area	Materials accepted	Recycled products
Al-Quraish Group of Industries	Punjab	PET	Flakes
International Industrial Services	Sindh	ABS, HDPE, HIPS, LDPE, PE, PET, PP, PVC	Granules/Pellets, Flakes
Marhaba Plastic	Punjab	PET	Flakes
Noman International	Sindh	PET	Flakes
Purely Green Industry Co, Ltd	Sindh	PET	Flakes
Sanak Pakistan	Baluchistan	PET	Flakes
SF Globally	Punjab	PET	Flakes
Shazil Pakistan Private Limited	Sindh	PET	Flakes
Spinzers	Sindh	PET	Flakes
Star Flakes	Sindh	PET	Flakes
TechnoPET	Punjab	PET	Granules/Pellets, Flakes
United Exporters	Sindh	ABS, HIPS, PC, PET	Flakes
Universal Exporters	Sindh	PET, PC, POM	Flakes

Source Plastic Recycling Plants in Pakistan, Directory of Plastic Recycling Companies ENF (https://www.enfplastic.com/directory/plant/Pakistan?plastic_products=Flakes)

11.6.2 Plastic Bag Ban

Pakistan joined the 128 nations that have outlawed single-use plastic (polythene) bags as of 2019. The annual manufacturing of 600,000 kg was stopped by this ban. The success of the ban hinged on several factors, including public awareness raised by Ministry of Climate Change-led education campaigns for manufacturers and citizens, free alternative bag distribution, media coverage of the Pakistan Environmental Protection Agency's (EPA) strict enforcement measures, and easy access to affordable alternatives made possible by the identification of reusable bags made in Pakistan. Crucially, though, the EPA started to regulate offenders at the manufacturer and consumer levels, which altered behavior. Plastic bag manufacturing and sales were subject to fines, which ranged from 500 to 500,000 rupees.

11.6.3 Clean Green Pakistan

One example is the Clean Green Pakistan Index (CGPI). The CGPI, which was introduced in November 2019, aims to encourage healthy competition between cities and their residents in order to enhance sanitation services throughout Pakistan. Cities are ranked annually based on the Index, which includes well over 35 indicators of performance on five components, including management of solid waste. Together with the municipalities, the indicators were developed, and the latter defined the aims, which were eventually focused on better governance and boosting the reliability of the data gathered. Using an ICT-based champions initiative, citizens may record their voluntary efforts to maintain clean streets, enhance public spaces, and foster a sense of ownership over their cities and environments. This allows citizens to compete in the competition. The prime minister gives rewards to residents and cities that perform well in order to maintain positive momentum and ongoing progress. In order to guarantee that public sector capabilities are developed to attain a clean and green Pakistan, the competition was extended to all provinces and regions (93 cities) following the pilot's successful completion.

11.6.4 Extended Producer Responsibility

A policy tool called extended producer responsibility (EPR) attempts to reduce the risks connected to waste management. Producers are financially and legally obligated under EPR to reduce the environmental effects of their goods at every stage of their lifespan. EPR can assist in reducing the negative effects that plastic goods have on society, the environment, health, and safety (Eriksen et al., 2018). This may aid in reducing and preventing plastic pollution. Enforcing EPR could incentivize producers to create and produce environmentally friendly plastics. For instance, the California-based Natural Resources Defence Council created policy ideas in 2013 to hold some product manufacturers accountable for both litter prevention and mitigation in addition to recycling. They had to cut the overall volume of their goods in the environment by 75% in 6 years and 95% in 11 years as a result of the new policy (Eriksen et al., 2018).

11.6.5 Waste-To-Energy

The process of turning waste into fuel, steam, heat, or electricity is known as waste-to-energy. Plastic waste-to-energy technologies enable energy recovery and get over recycling barriers including polymer separation requirements and competition from virgin plastics. On the other hand, it keeps the economy linear. Energy is required during the reformulation of plastics into feed stocks in order to recover components that can be utilized to create gasoline-like fuels (waste-to-energy) or to create chemicals, lubricants, and carbon black (Liu et al., 2018; Poulikakos et al., 2017).

11.6.6 Microbial Degradation

Since micro-organisms biodegrade a variety of organic and inorganic compounds, including lignin, starch, cellulose, and hemicelluloses, many researchers have expressed a significant interest in the microbial breakdown of plastic. A few soil-dwelling microorganisms use polyester polyurethane as a single source of carbon and nitrogen. Similar to this, recent studies have shown that a variety of bacteria found in mangrove soil cultures are capable of biodegrading plastic effectively, albeit slowly (Auta et al., 2022). Similar to bacteria, several fungi can also break down plastic by infiltrating the polymer solids. Through fungus colonization, plasticized polyvinyl plastic has been shown to biodegrade both in-situ and ex-situ. Numerous enzymes released by microorganisms into soil water initiate the degradation of polymers. Microbes release enzymes known as intracellular and extracellular polymerases, which break down complex polymers into smaller, more easily absorbed monomers that can be used as a source of carbon and energy. This process is known as depolymerization. Moreover, this process is referred to as mineralization when it produces products like carbon dioxide, water, or methane.

11.6.7 Other Efforts

- Take a bag with you, if you have any, and always reuse plastic bags.
- Throw away bottled water. Individuals ought to consume copious amounts of water daily, and plastic water bottles have emerged as an excellent means of

staying hydrated all day. Nevertheless, the majority of these are only meant to be used once, so each time a bottle is empty, it ends up in the garbage.

- These plastic bottles take more than 400 years for natural decomposition because they are usually composed of polyethylene terephthalate, or PET. Reusable water bottles are becoming more widely available from businesses, which helps to cut down on plastic waste and exposure to leaky bottles. The ideal solution is to always have a reusable metal bottle in your luggage.
- Food containers are manufactured and packaged using plastic. The coffee cafe uses plastic for its drink cups. As an insulation measure, it's probably lined with plastic (try pouring some coffee on some cardboard and seeing what happens). Reusable containers are an easy way to replace plastic lids and utensils, and they will drastically reduce the amount of waste bottles from even a single meal that ends up in your bag.
- Look for products with less plastic packaging.
- Steer clear of makeup and personal hygiene items that contain microbeads—those tiny specks in toothpaste—and face scrubs, which are essentially made of microplastic. According to recent studies, microbeads are causing more harm to human health and marine life.
- Inform companies: Talk to nearby eateries and companies about other choices for product packaging, storage, and bagging. A lot of businesses are beginning to provide great, affordable alternatives, like bamboo cutlery in place of plastic ones.
- Engage in dialogue with legislators and participate in any aspect of governance, and you will observe how numerous special interest organizations have contributed to our unnecessary reliance on plastic. Promote the creation of things and, when appropriate, offer substitutes.
- Avoid pieces of clothing with artificial microfibers. After being scrubbed, these frequently leak microfibers into the water which enters ocean and eventually consumed by fish and other marine life.

11.6.8 Efforts by Karandaaz

KARANDAAZ PAKISTAN is a non-profit special purpose organization that was established in August 2014 under section 42. The UK's Foreign, Commonwealth & Development Office (FCDO) has designated Karandaaz as its implementation collaborator for the Sustainable Energy and Economic Development (SEED) and Enterprise and Asset Growth (EAGR) programmes. Karandaaz's Green Challenge Fund is a ray of hope in the global arena of environmental sustainability and innovation, supporting ground-breaking approaches to address the urgent problem of managing plastic waste. Before the GreenFin Innovations Programme was formally established, Karandaaz played a significant role in sponsoring four innovative green technology firms and SMEs over the course of the previous two years, especially in plastic waste management.

One of these amazing endeavors is **Suftech Innovations**, an organization that is leading the way in recycling waste plastics into premium recycled granules that may efficiently replace virgin polymer materials. Suftech revitalizes different plastic waste streams by using cutting-edge chemical treatments, turning them into translucent sheets with properties comparable to those of virgin polymers. Suftech, which has an amazing yearly capacity to recycle 864 metric tons of plastic waste, is a prime illustration of how Karandaaz's support can significantly affect sustainable change (Khalid, 2024).

Another company that benefited from Karandaaz's investment is **Concept Loop**, which creates construction materials and lifestyle goods out of recycled plastic waste, a prime example of creativity in plastic waste recycling. Interestingly, Concept Loop uses plastic trash to create concrete pavers, which are useful for building infrastructure like parking lots and walkways as well as building construction. Concept Loop highlights the observable outcomes of Karandaaz's dedication to promoting sustainable practices. It has the capacity to recycle about 1,404 tons of plastic waste annually and produce 1.4 million square feet of concrete products.

With the help of Karandaaz, **Davaam Life** has developed dispensing machines for household goods as a novel solution to the problem of single-use plastic. Davaam Life lowers carbon emissions and the expenses related to plastic packaging and shipping by encouraging the reuse of plastic using practical dispensing methods. Davaam Life, which has already installed 32 dispensing devices, is a prime example of how sustainable solutions can be incorporated into daily life.

11.6.9 Case Studies in Pakistan

11.6.9.1 Rawal Lake Study

Pollution from microplastics is becoming a bigger environmental concern. Information about the presence of microplastics in freshwater, particularly in Pakistan, is scarce. The goal of the study was to find out how common and concentrated microplastics were in the sediments and surface water of Pakistan's capital city, Rawal Lake (Irfan et al., 2020). In water, the average microplastic abundance was 0.142 items/0.1 L, whereas in sediments, it was 1.04 items/0.01 kg. The report revealed that the most common forms of microplastics were fibres and pieces. Blue, crimson, black, and transparent were the main hues. Due to the appearance of these polymers' distinctive peaks, FTIR analysis of visible microplastic particles revealed a higher degree of similarity with polyesters, polyethylene, polypropylene, polyethylene terephthalate, and polyvinyl chloride. Additionally, the study found that the sediments had higher concentrations of microplastics than the Rawal Lake's water. The lake's microplastic contamination may be mostly caused by inappropriate waste disposal, tourism, recreational activities, and a high population density surrounding the lake.

11.6.9.2 WWF's Efforts in Lahore

The Coca-Cola Foundation and the Pakistan chapter of World Wildlife Fund (WWF) for Nature have initiated a month-long trial project to collect and recycle plastic. Packages Mall, Fortress Square, and Emporium Mall have been designated for the construction of plastic recovery facilities. The recovery facility will accept used PET bottles from the public, which will subsequently be collected and transferred to Amal, a WWF partner waste collector, for recycling. According to a 2019 study from the University of Newcastle, humans may be consuming five grams of plastic each week on average—the weight of a credit card—through their food, drink, and particularly the air they breathe.

It implies that approximately 2,000 microplastic particles are consumed by humans each week. That's a little over 250 g annually, or around 21 g per month. Pakistan's approach to addressing plastic pollution in 2017, Pakistan implemented oxo-biodegradable technology as a means of addressing the problem of plastic pollution, after their useful lives, oxo-biodegradable polymers fully decompose in the environment with no negative environmental effects. The regulation was created after all available possibilities were carefully considered and supported by scientific data.

In 2019, the Islamabad Capital Territory introduced an alternative model that prohibited polyethylene bags and encouraged the use of biodegradable bags. The Prime Minister's aim for a Clean and Green Pakistan inspired the Ministry of Climate Change to adopt this effort. MoCC is striving to guarantee that the prohibition is fully implemented by enforcing stiff penalties and fines on those who violate it. Bags made of polypropylene are not biodegradable because, as two of the most widely used polymers, microorganisms cannot break down any of these materials. Currently, the majority of vendors are taking advantage of the way non-woven PP bags seem. To address the problem of plastic pollution, federal authorities ought to come up with a just plan of action. If it is decided to outlaw plastics altogether, then all single-use plastics, including WPP, BOPP, and PP non-woven bags, should be included. Additionally, choices that have been scientifically shown to be biodegradable ought to be encouraged.

11.6.9.3 Plastic Roads in Pakistan

Pakistan is ranked third in Asia and sixth globally in terms of the amount of plastic waste it produces. Every year, it generates about 49.6 million tons of solid trash. Plastics make up almost 9% of it, which includes a variety of single-use bags, packaging materials, and silverware (Naeem et al., 2024). The recycled plastic has successfully been exploited to make plastic roads in Pakistan (Fig. 11.4).

Fig. 11.4 Construction of metal road with recycled plastic in Pakistan

Islamabad

Recycled plastic debris was used to carpet Pakistan's "first plastic road," which was officially opened in Islamabad on Monday, according to a news statement issued by three organizations (Capital Development Authority, TeamUp and Coca Cola Company) involved in the project with an estimated cost of Rs. 21 million. In South Asia, Pakistan has the highest rate of improperly disposed of plastic. These plastics utilized in the construction of road is the most effective approach toward management of plastic waste. In addition, the process of producing plastic roads is less expensive and takes less time than that of regular asphalt roads. They also require less maintenance, which makes them a durable and sensible choice for infrastructure.

Lahore

The plastic road's ability to withstand precipitation is one of its main benefits, since it guarantees a minimum of ten years of use. These creatively designed roadways are impervious to harm from rain, in contrast to traditional highways. Furthermore, the cost of building a plastic road has been drastically lowered; this project will only cost Rs. 20 million, compared to the usual Rs. 60 million for a traditional road. The fact that the plastic utilized in these roads is made from recycled materials, so recycling plastic trash that would otherwise contribute to environmental contamination, is a significant environmental benefit. This environmentally friendly method of building roads reduces expenses while providing a long-term fix. Mohsin Naqvi, the acting chief minister of Punjab, opened the 900-m Zafar Ali plastic road, which runs from Gymkhana Mall Road to Jail Road. These roads would not be constructed without the use of shopping bags and other plastic materials. The Chief Minister emphasized the accomplishments of this innovative project and disclosed intentions to broaden its reach. "Further expansions, including the conversion of Mall Road into a plastic road, are on the horizon following an extensive one-month review of the current plastic road," he continued.

Karachi

An attempt has been made to lessen pollution in Karachi by completing the construction of a road composed of recycled plastic. The 730 feet long and 60 feet broad road was framed using more than 6,000 plastic recycled pieces, totaling about 2.5 tons. According to project manager Zunair Bin Hasan, all recycled plastic bottles were used in the construction of the environmentally friendly road. According to experts, plastic roads have a lifespan that is almost three times longer than that of conventional roads, he continued. In a similar vein, the cost of building plastic roads is lower than that of conventional roads since polymers are less expensive to obtain than the materials needed for conventional roads. A commuter using the recently constructed road applauded the initiative, stating that it feels flexible and clean.

11.7 Policy Statement

The first stage in reducing plastic pollution is plastic management, which calls for robust legislative support Lau et al., 2020.

11.7.1 Single Use Plastic Prohibition Regulations, 2023 (EPA, 2023)

The regulation said that these rules were to take effect throughout Pakistan's capital territory.

- As per deadlines, it is forbidden in ICT to manufacture, import, distribute, sell, supply, stock, buy, exchange, or consume single-use plastic items listed in Schedule I.
- In accordance with Schedule II, beverage manufacturers and producers of beverage containers are prohibited from manufacturing, importing, and using single-use plastic beverage containers manufactured wholly of virgin plastic in ICT.
- No one may distribute, import, manufacture, buy, sell, store, provide, trade, or use single-use plastics once the deadlines have passed, and any permit or permission granted prior to the start of these restrictions will no longer be valid.

These bans were not applied on following purposes: industrial packaging, medical purposes, scientific purposes, polythene bags for hospital waste, polythene bags for municipal waste, and polythene bags for hazardous waste. The party seeking permission to import or produce goods would apply along with a recycling strategy. After that, the Federal Agency decides on the application and notifies the applicant in writing.

11.7.2 Recycling Plan

(1) To encourage the collection and recycling of single-use plastic, each producer and importer creates a recycling plan.
(2) The manufacturer or importer shall carry out the authorized recycling strategy following the Federal Agency's approval of it.

Each manufacturer or importer of single-use plastics is required to maintain a log of the amount of materials produced, imported, or exported, and to report to the Federal Agency.

11.7.3 Enforcement

Under the guidelines provided by the Act, any authorized officer tasked with enforcing these regulations may carry out the following activities, subject to Act provisions:

(a) Searching any land, building, premises, vehicle, vessel, or other location where there are reasonable grounds to believe that an offence under the Act and these regulations has been or is being committed.
(b) Making any examination, investigation, or inspection that may be required; and
(c) Searching any of these locations.

11.7.4 Penalty

Penalties under these regulations will be assessed in accordance with the Compounding of Offences and Payment of Administrative Penalty Rules 2015 (Table 11.2).

Table 11.2 Penalties for violation of polythene bag regulations

Sr. No.	Person/Entity violating the regulations	1st time violation	2nd time violation	Violation of Regulations more than twice
1	Manufacturer/importer/wholesaler	Rs. 100,000/-	Rs. 200,000/-	Rs. 500,000/-
2	Shop keeper/hawker/ stallholder	Rs. 10,000/-	Rs. 20,000/-	Rs. 50,000/-
3	Any other person involved in offense under these regulations	Rs. 5000/-		Rs. 5000/-

11.8 Conclusion and Future Recommendations

Pakistan faces a severe challenge in managing plastic waste. The country's strategies to combat plastic pollution include a variety of measures: recycling initiatives, plastic bag bans, construction of plastic roads from plastic waste, the Clean Green Pakistan Index, Extended Producer Responsibility (EPR) programs, waste-to-energy projects, and microbial degradation research. Efforts by organizations like Karandaaz, along with specific projects like plastic roads, demonstrate innovative approaches for recycling and repurposing plastic waste. Furthermore, regulatory frameworks such as the Single Use Plastic Prohibition Regulations, 2023, and penalties for violations underscore the Government's commitment to reducing plastic pollution in Pakistan. Pakistan has successfully constructed plastic roads in Lahore, Karachi and Islamabad with the help of private organizations like TeamUp, BRR Enterprise, the Coca Cola Company and the Shell Pakistan.

However, it is crucial to invest in advanced recycling technologies to improve the quality and scope of recycled products, making recycling more economically viable. Public awareness and participation can be bolstered through continuous educational campaigns and community involvement initiatives that reward active participation. Stricter regulations and enforcement, along with the full implementation of Extended Producer Responsibility schemes, will hold producers accountable for their products' lifecycle. Innovation in alternative materials, supported by R&D investments and funding for startups, will provide eco-friendly solutions. Upgrading waste management systems to integrate recycling, waste-to-energy, and advanced sorting technologies will further reduce plastic waste. Policy and economic incentives, such as subsidies, tax benefits, and green procurement policies will encourage sustainable practices. Lastly, international collaboration and regional cooperation in South Asia will allow Pakistan to leverage global best practices and resources to tackle plastic pollution more effectively.

References

Adam, I., Walker, T. R., Bezerra, J. C., & Clayton, A. (2020). Policies to reduce single-use plastic marine pollution in West Africa. *Marine Policy, 116*, 103928. https://doi.org/10.1016/j.marpol.2020.103928.

Ágnes, N., & Kuti, R. (2016). The environmental impact of plastic waste incineration. *Academic and Applied Research in Military and Public Management Science. Budapest, 15*(3), 231–237. https://doi.org/10.32565/aarms.2016.3.3.

Amrutha, K., Unnikrishnan, V., Shajikumar, S., & Warrier, A.K. (2021). Current state of microplastics research in SAARC Countries—A review. In: S. S. Muthu (Ed.), *Microplastic Pollution. Sustainable Textiles: Production, Processing, Manufacturing & Chemistry* (pp. 27–63). Springer. https://doi.org/10.1007/978-981-16-0297-9_2.

Andrady, A. L., & Neal, M. A. (2009). Applications and societal benefits of plastics. *Philosphical Transactions Royal Society of London B Biological Sciences, 364*(1526), 1977–1984. https://doi.org/10.1098/rstb.2008.0304.

Arumdani, I., Puspita, A., & Budihardjo, M. (2021). MSW handling of top 5 leading waste-producing countries in Southeast Asia. In *Paper Presented at the IOP Conference Series: Earth and Environmental Science.*

Auta, H. S., Abioye, O. P., Aransiola, S. A., Bala, J. D., Chukwuemeka, V. I., Hassan, A., & Fauziah, S. H. (2022). Enhanced microbial degradation of PET and PS microplastics under natural conditions in mangrove environment. *Journal of Environmental Management, 304*, 114273.

Banik, P., Hossain, M. B., Nur, A.-A.U., Choudhury, T. R., Liba, S. I., Yu, J., Noman, M. A., & Sun, J. J. (2022). Microplastics in sediment of Kuakata Beach, Bangladesh: Occurrence, spatial distribution, and risk assessment. *Frontiers in Marine Science, 9*, 860989. https://doi.org/10.3389/fmars.2022.860989.

Bucci, K., Tulio, M., & Rochman, C. M. (2020). What is known and unknown about the effects of plastic pollution: A meta-analysis and systematic review. *Ecological Applications, 30*(2), e02044.

Chinglenthoiba, C., Pukhrambam, B., Chanu, K. T., Devi, K. S., Meitei, N. J., Devika, Y., & Valiyaveettil, S. (2023). A review on microplastic pollution research in India. *Regional Studies in Marine Science, 58*, 102777.

Cole, M., Lindeque, P., Halsband, C., & Galloway, T. S. (2011). Microplastics as contaminants in the marine environment: A review. *Marine Pollution Bulletin, 62*(12), 2588–2597. https://doi.org/10.1016/j.marpolbul.2011.09.025.

Eriksen, M., Thiel, M., Prindiville, M., & Kiessling, T. (2018). Microplastic: What are the solutions? In: M. Wagner, & S. Lambert, (Eds.), *Freshwater Microplastics. The Handbook of Environmental Chemistry* (Vol. 58). Springer. https://doi.org/10.1007/978-3-319-61615-5_13.

Dissanayake, P. D., Kim, S., Sarkar, B., Oleszczuk, P., Sang, M. K., Haque, M. N., Ahn, J. H., Bank, M. S., & Ok, Y. S. (2022). Effects of microplastics on the terrestrial environment: A critical review. *Environmental Research, 209*, 112734. https://doi.org/10.1016/j.envres.2022.112734.

EPA. (2023). The Gazette of Pakistan, Single use Prohibition regulation 2023.

Gilpin, R. K., Wagel, D. J., & Solch, J. G. (2003). Production, distribution, and fate of polychlorinated dibenzo-p-dioxins, dibenzofurans, and related organohalogens in the environment. In: A. Schecter, & T.A., Gasiewicz (Eds.), *Dioxin and Health* (pp. 55–87). https://doi.org/10.1002/0471722014.ch2.

Hikmath, J., & Arachchige, U. S. P. R. (2020). Future strategies in plastic waste management in Sri Lanka. *Journal of Research Technology and Engineering, 1*, 62–71.

Hopewell, J., Dvorak, R., & Kosior, E. (2009). Plastics recycling: Challenges and opportunities. *Philosphical Transactions Royal Society of London B Biological Sciences, 364*(1526), 2115–2126. https://doi.org/10.1098/rstb.2008.0311.

Irfan, T., Khalid, S., Taneez, M., & Hashmi, M. Z. (2020). Plastic driven pollution in Pakistan: The first evidence of environmental exposure to microplastic in sediments and water of Rawal Lake. *Environmental Science and Pollution Research, 27*, 15083–15092.

Iroegbu, A. O. C., Ray, S. S., Mbarane, V., Bordado, J. C., & Sardinha, J. P. (2021). Plastic pollution: A perspective on matters arising: challenges and opportunities. *ACS Omega, 6*(30), 19343–19355. https://doi.org/10.1021/acsomega.1c02760.

Jambeck, J. R., Geyer, R., Wilcox, C., Siegler, T. R., Perryman, M., Andrady, A., Narayan, R., & Law, K. L. (2015). Marine pollution. Plastic waste inputs from land into the ocean. *Science, 347*(6223), 768–671. https://doi.org/10.1126/science.1260352.

Jeyasanta, K. I., Sathish, N., Patterson, J., & Edward, J. K. P. (2020). Macro-, meso-and microplastic debris in the beaches of Tuticorin district, Southeast coast of India. *Marine Pollution Bulletin, 154*, 111055. https://doi.org/10.1016/j.marpolbul.2020.111055.

Joel, F. R. (1995). *Polymer Science and Technology* (pp. 1–509). Prentice Hall. USA.

Kale, S. K., Deshmukh, A. G., Dudhare, M. S., & Patil, V. B. (2015). Microbial degradation of plastic: a review. *Journal of Biochemical Technology, 6*(2), 952–961.

Karasik, R., Vegh, T., Diana, Z., Bering, J., Caldas, J., Pickle, A.,Rittschof, D., & Virdin, J. (2020). *20 years of government responses to the global plastic pollution problem: The plastics policy inventory* (p. 309). Nicholas Institute for Environmental Policy Solutions, Duke University. https://hdl.handle.net/10161/21347.

Khalid, J. (2024). Resolving plastic waste crisis in Pakistan. *Business Recorder*. https://www.brecorder.com/news/40297564/resolving-plastic-waste-crisis-in-pakistan.

Lam, C. S., Ramanathan, S., Carbery, M., Gray, K., Vanka, K. S., Maurin, C., Bush, R., & Palanisami, T. (2018). A comprehensive analysis of plastics and microplastic legislation worldwide. *Water Air Soil Pollution, 229*, 345. https://doi.org/10.1007/s11270-018-4002-z.

Lau, W. W. Y., Shiran, Y., Bailey, R. M., Cook, E., Stuchtey, M. R., Koskella, J., Velis, C. A., Godfrey, L., Boucher, J., Murphy, M. B., Thompson, R. C., Jankowska, E., Castillo, A. C., Pilditch, T. D., Dixon, B., Koerselman, L., Kosior, E., Favoino, E., Gutberlet, J., et al. (2020). Evaluating scenarios toward zero plastic pollution. *Science, 369*(6509), 1455–1461.

Liu, Z., Adams, M., & Walker, T. R. (2018). Are exports of recyclables from developed to developing countries waste pollution transfer or part of the global circular economy? *Resources, Conservation and Recycling, 136*, 22–23.

MacLeod, M., Arp, H. P. H., Tekman, M. B., & Jahnke, A. (2021). The global threat from plastic pollution. *Science, 373*(6550), 61–65. https://doi.org/10.1126/science.abg5433.

Mitrano, D. M., & Wohlleben, W. J. (2020). Microplastic regulation should be more precise to incentivize both innovation and environmental safety. *Nature Communications, 11*(1), 5324.

Montalbetti, E., Saponari, L., Montano, S., Maggioni, D., Dehnert, I., Galli, P., & Seveso, D. (2019). New insights into the ecology and corallivory of Culcita sp. (Echinodermata: Asteroidea) in the Republic of Maldives. *Hydrobiologia, 827*, 353–365.

Montano, S., Strona, G., Seveso, D., Maggioni, D., & Galli, P. (2015). Widespread occurrence of coral diseases in the central Maldives. *Marine and Freshwater Research, 67*(8), 1253–1262.

Mukheed, M., & Alisha, K. (2020). Plastic pollution in Pakistan: Environmental and health Implications. *Journal of Pollution Effects & Control, 4*, 251–258.

Muhamedbegović, B., Aldžić, A., Ibrahimpašić, J., & Alihodžić, D. (2023). *Chemical migrations from polyethylene terephthalate bottles into drinking water*.

Naeem, Z. (2024, May 12). *Pakistan's plastic predicament*. https://www.thenews.com.pk/tns/detail/1187968-pakistans-plastic-predicament.

Ng, C. H., Mistoh, M. A., Teo, S. H., Galassi, A., Ibrahim, A., Sipaut, C. S., Foo, J., Seay, J., Taufiq-Yap, Y. H., & Janaun, J. (2023). Plastic waste and microplastic issues in Southeast Asia. *Frontiers in Environmental Science, 11*, 1142071. https://doi.org/10.3389/fenvs.2023.1142071.

Nielsen, T. D., Hasselbalch, J., Holmberg, K., & Stripple, J. (2020). Politics and the plastic crisis: A review throughout the plastic life cycle. *Energy and Environment, 9*(1), e360. https://doi.org/10.1002/wene.360.

Niloy, A. C. (2021). Jute: Solution to global challenges and opportunities of Bangladesh. *Science Business Review, 1*(2), 59–75. https://doi.org/10.33215/sbr.v1i2.633.

OECD. (2022). Plastic pollution is growing relentlessly as waste management and recycling fall short. https://www.oecd.org/en/about/news/press-releases/2022/02/plastic-pollution-is-growing-relentlessly-as-waste-management-and-recycling-fall-short.html.

Oehlmann, J., Schulte-Oehlmann, U., Kloas, W., Jagnytsch, O., Lutz, I., Kusk, K. O., Wollenberger, L., Santos, E. M., Paull, G. C., Van Look, K. J., & Tyler, C. R. (2009). A critical analysis of the biological impacts of plasticizers on wildlife. *Philosphical Transactions Royal Society of London B Biological Sciences, 364*(1526), 2047–2062.

Omeyer, L. C., Duncan, E. M., Abreo, N. A. S., Acebes, J. M. V., AngSinco-Jimenez, L. A., Anuar, S. T., et al. (2023). Interactions between marine megafauna and plastic pollution in Southeast Asia. *Science of the Total Environment, 874*, 162502. https://doi.org/10.1016/j.scitotenv.2023.162502.

Perry, C. T., & Morgan, K. (2017). Post-bleaching coral community change on southern Maldivian reefs: Is there potential for rapid recovery? *Coral Reefs, 36*(4), 1189–1194. https://doi.org/10.1007/s00338-017-1610-9.

Plakunov, V., Gannesen, A., Mart'Yanov, S., & Zhurina, M. V. (2020). Biocorrosion of synthetic plastics: Degradation mechanisms and methods of protection. *Microbiology, 89*, 647–659. https://doi.org/10.1134/S0026261720060144.

Prata, J. C., Silva, A. L. P., Da Costa, J. P., Mouneyrac, C., Walker, T. R., Duarte, A. C., & Rocha-Santos, T. (2019). Solutions and integrated strategies for the control and mitigation of plastic and microplastic pollution. *International Journal of Environmental Researchand Public Health, 16*(13), 2411. https://doi.org/10.3390/ijerph16132411.

Poulikakos, L. D., Papadaskalopoulou, C., Hofko, B., Gschösser, F., Falchetto, A. C., Bueno, M., & Partl, M. N. (2017). Harvesting the unexplored potential of European waste materials for road construction. *Resources, Conservation and Recycling, 116*, 32–44.

Proshad, R., Kormoker, T., Islam, M. S., Haque, M. A., Rahman, M. M., & Mithu, M. M. R. (2018). Toxic effects of plastic on human health and environment: A consequences of health risk assessment in Bangladesh. *International Journal of Health, 6*(1), 1–5. https://doi.org/10.14419/ijh.v6i1.8655.

Qin, Z. H., Mou, J. H., Chao, C. Y. H., Chopra, S. S., Daoud, W., Leu, S. Y., Ning, Z., Tso, C. Y., Chan, C. K., Tang, S., Hathi, Z. J., Haque, M. A., Wang, X., & Lin, C. S. K. (2021). Biotechnology of plastic waste degradation, recycling, and valorization: Current advances and future perspectives. *Chemsuschem, 14*(19), 3981. https://doi.org/10.1002/cssc.202101825.

Rifa, T., & Hossain, M. B. (2022). Micro plastic pollution in South Asia: The impact of plastic pollution over the unsustainable development goals. *Lex Publica, 9*(2), 1–28. https://doi.org/10.58829/lp.9.2.2022.01-28.

Schwarz, A. E., Ligthart, T. N., Boukris, E., & Van Harmelen, T. (2019). Sources, transport, and accumulation of different types of plastic litter in aquatic environments: A review study. *Marine Pollution Bulletin, 143*, 92–100. https://doi.org/10.1016/j.marpolbul.2019.04.029.

Seethawaka, D. (2023). Restoring the ecosystems affected by climate change: An analysis of the legislative framework on the protection of ecosystems in Sri Lanka. *Sri Lanka Journal of International Law, 29*, 81.

Seymour, R. B. (1989). Polymer science before and after 1899: Notable developments during the lifetime of Maurits Dekker. *Journal of Macromolecular Science: Part A - Chemistry, 26*(8), 1023–1032.

Siddiqui, J., & Pandey, G. (2013). A review of plastic waste management strategies. *International Research Journal of Environment Sciences, 2*(12), 84.

Stevens, G. M., & Froman, N. (2019). *The Maldives Archipelago. World Seas: An Environmental Evaluation* (Vol. II, 2nd ed., pp. 211–236). The Indian Ocean to the Pacific. https://doi.org/10.1016/B978-0-08-100853-9.00010-5.

Suzuki, G., Uchida, N., Tanaka, K., Matsukami, H., Kunisue, T., Takahashi, S., Viet, P. H., Kuramochi, H., & Osako, M. (2022). Mechanical recycling of plastic waste as a point source of microplastic pollution. *Environmental Pollution, 303*, 119114. https://doi.org/10.1016/j.envpol.2022.119114.

The Guardian. (2023). Plastics cause wide-ranging health issues from cancer to birth defects, landmark study finds. https://www.theguardian.com/environment/2023/mar/29/plastics-cause-wide-ranging-health-issues-from-cancer-to-birth-defects-landmark-study-finds.

UNDP. (2021, October). Rethinking Pakistan's relationship with plastics. https://pakngos.com.pk/rethinking-pakistans-relationship-with-plastics/.

Urbanek, A. K., Rymowicz, W., & Mirończuk, A. M. (2018). Degradation of plastics and plastic-degrading bacteria in cold marine habitats. *Applied Microbiology Biotechnology, 102*, 7669–7678.

Umar, I. (2020, January 30). Blog: plastic pollution—What is the real solution? https://dailytimes.com.pk/549058/plastic-pollution-what-is-the-real-solution.

UNEP. (2021, October). From pollution to solution: A global assessment of marine litter and plastic pollution. https://malaysia.un.org/en/171922-pollution-solution-global-assessment-marine-litter-and-plastic-pollution.

Vaid, M., Mehra, K., & Gupta, A. (2021). Microplastics as contaminants in Indian environment: A review. *Environmental Science and Pollution Research, 28*, 68025–68052.

Vijayakumar, A., & Sebastian, J. (2018). Pyrolysis process to produce fuel from different types of plastic–a review. In *Paper Presented at the IOP Conference Series: Materials Science and Engineering*.

Visual Capitalist. (2023). Charting the movement of global plastic waste. https://www.visualcapitalist.com/cp/charting-the-movement-of-global-plastic-waste/.

Windsor, F. M., Durance, I., Horton, A. A., Thompson, R. C., Tyler, C. R., & Ormerod, S. J. (2019). A catchment-scale perspective of plastic pollution. *Global Change Biology, 25*(4), 1207–1221.

Wright, S. L., & Kelly, F. J. (2017). Plastic and human health: A micro issue? *Environmental Science and Technology, 51*(12), 6634–6647.

Ya, H., Jiang, B., Xing, Y., Zhang, T., Lv, M., & Wang, X. (2021). Recent advances on ecological effects of microplastics on soil environment. *Science of the Total Environment, 798*, 149338. https://doi.org/10.1016/j.scitotenv.2021.149338.

Yusuf, A. A., Ampah, J. D., Soudagar, M. E. M., Veza, I., Kingsley, U., Afrane, S., Jin, C., Liu, H., Elfasakhany, A., & Buyondo, K. A. (2022). Effects of hybrid nanoparticle additives in n-butanol/waste plastic oil/diesel blends on combustion, particulate and gaseous emissions from diesel engine evaluated with entropy-weighted PROMETHEE II and TOPSIS: Environmental and health risks of plastic waste. *Energy Conversion and Management., 264*, 115758.

Part III
Technologies and Products

Chapter 12
Value Addition of Plastic Waste Through Plastic Oil Production for Future Energy Security by Hybrid Pyrolysis

K. T. Vadiraj, Alfin K. Tom, and Raghu Ram Achar

Abstract Generation of plastic wastes has increased steeply over the past few years. Solutions like landfill deposits and recycling waste are the only initiative that the government is concerned about. Due to the enormous deposition of plastic wastes, landfill deposit turns out a problem and thus is not feasible. Many researchers have undertaken alternate solution to convert plastic waste into a useful fuel source via the pyrolysis technique. This paper focuses on converting low-density polyethylene, high-density polyethylene, polypropylene, and a mixture of all into oil via pyrolysis in a self-designed reactor without any catalyst. The generated oil's physical characteristics such as gross calorific value, specific gravity, and viscosity were later compared with the conventional fuels. The oil was further subjected to the GC-MS analysis to understand the chemical composition present. When compared with conventional fuel, plastic waste oil obtained from thermal pyrolysis showed promising results to be used as an alternative fuel, by changing waste into valuable resources and management of plastics waste for a sustainable future.

Keywords Waste plastic · Plastic pyrolysis · Pyrolyzed oil · Calorific value · Waste management

12.1 Introduction

Plastic production started in bulk during the second world war and ever since then plastic has become a habituated part of our lives. From micro to macro levels, they are been manufactured all across the world (Fok et al., 2019). Plastic is commonly used

K. T. Vadiraj (✉) · A. K. Tom
Department of Environmental Sciences, JSS Academy of Higher Education and Research, Mysuru, India
e-mail: vadirajkt@jssuni.edu.in

R. R. Achar
Division of Biochemistry, School of Life Sciences-Mysuru, JSS Academy of Higher Education and Research, Mysuru, India

S. Suriyanarayanan et al. (eds.), *Combating Plastic Pollution in Terrestrial Environment*,
https://doi.org/10.1007/978-981-96-2343-3_12

in packaging, automotive, construction, etc., in bulk qualities. Due to the increase in human population, the global manufacture of plastics has touched 368 million metric tons (9,200 tonnes per day). Over the past 4 decades, manufacturing has quadrupled (Anuar Sharuddin et al., 2016). Due to their non-biodegradable property, plastics are one of the major contributors to environmental pollution (Shen et al., 2020). Plastic waste contributes much to industrial and municipal making it the third-largest contributor, after the food and paper industry at primary and secondary levels respectively. According to studies between 1915 and 2015, 9% of plastic wastes were recycled globally, out of which only 10% was recycled more than once, 79% was dumped into water bodies and landfills, and only 12% was incinerated. In developing countries, the practice of landfills has become a common management of plastic wastes. Recent studies have also stated that plastic waste in the oceans is creating the pollution in many cities (Chen et al., 2021). Bulk plastic waste continues to grow and increase making it one of the biggest global issues (Ncube et al., 2020). Proper management for effective dumping of plastic wastes should be concern of the hour.

According to research, polyethylene has the highest mandate among other plastics in India (33%), subsequently followed by polypropylene (32%) (Sivadas et al., 2022). A very general form of management of plastics is the RRR practice, i.e., reduce, reuse, and recycle. Though many researchers do not agree that recycling plastic waste is the only solution, because after the fourth or fifth recycling process they become unfit for use, thus resulting in landfills. Therefore, the energy recovery process is the developing solution for the complete disposal solution of plastic waste (Gopi Kiran et al., 2022). Amid various recycling methods, chemical-based recycling is one the most productive methods that result in the yield of products like liquid fuel and gas through pyrolysis for economical use (Varsha et al., 2022). In India though, many companies use technologies for the disposal of plastics, for example in the construction of roads, co-processing in cement kilns, etc. (Tejaswini et al., 2022).

Also, petroleum is a non-renewable resource, and the decrease in demands for petrochemicals is increasing at a very high rate for their application in industry and energy production. Research study shows that a scarce amount of petroleum used for the fabrication of household materials could be filched out through the thermal decomposition method of plastic wastes (Kasar et al., 2020).

The process of thermal decomposition of plastics carried out in absence of oxygen which converts plastic wastes into useful products is termed as pyrolysis. Useful products can be liquid oil, ono = condensable gas, and solid residues rich in hydrocarbons (Adeniyi et al., 2018). The resultant yield varies among each other depending upon reaction rates and the type of feedstock utilized. Plastic pyrolysis is considered a green technology because the gaseous by-product produced has a significant calorific value which can be used to stabilize the energy requirement for pyrolysis and also it does not contribute to water contamination (Sipra et al., 2018). The production, transportation, and storage cost of the obtained liquid are quite low, with high energy density (Zhang et al., 2018).

The present work focuses on designing and developing a pyrolytic reactor for the recovery of liquid oil from plastic wastes and the physio-chemical characterization of derived oil from designed pyrolytic reactor.

12.2 Materials and Method

12.2.1 Material Used

Polypropylene (PP), Low-Density polyethylene (LDPE) and High-Density polyethylene (HDPE) polymer granules of 1000 g and their sizes of less than 0.5 mm each were bought commercially.

12.2.2 Reactor Design

Reactor setup is shown in Fig. 12.1. Five liters of a self-made reactor which is constructed using stainless steel. A home made T-shaped pressure valve was introduced on the top of the reactor. One end of the valve was connected with 2 feet long condenser circulating the water. Temperature of the reactor was measured using thermocouple device from the top, till it reaches material.

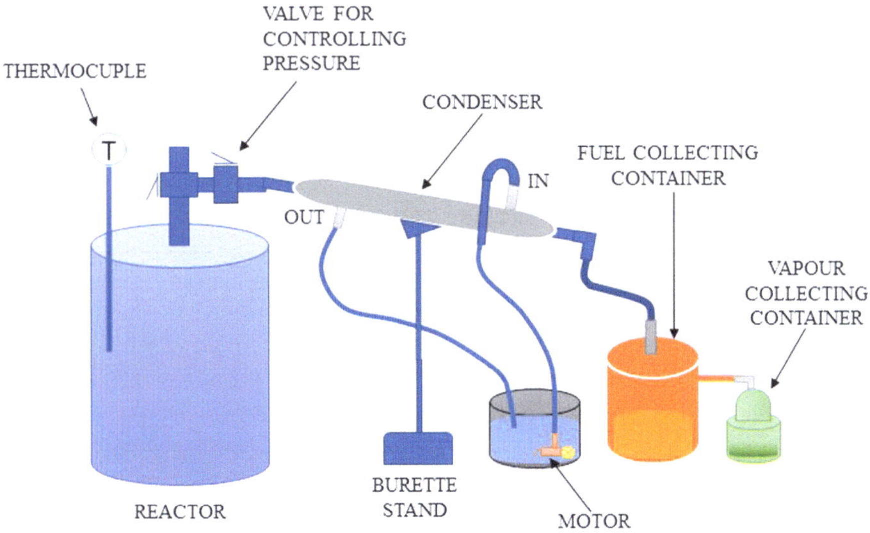

Fig. 12.1 Pyrolysis experimental setup

12.2.3 Analytic Technique

12.2.3.1 Pyrolysis

In the open space, the experimentation setup was carried out. The experimentation was done in different ratios using the plastic pellets of PP, HDPE, and LDPE being weighed individually at 1 kg and different ratios ways 1 kg each and the granules of plastic, HDPE, PP, and LDPE. They were weighted in different ratios like increasing the ratio of one certain plastic granule and changing the ratios of granules at a different set of experiments and also taking the individual plastics as 1 kg. These granules of plastics were removed to a self-made reactor after that using high-temperature silicon, the reactor was sealed and it was left to dry for at least 15 h. Using the vacuum pump, the other gases and air molecules that exist inside the reactor were removed. Then the reactor was kept on the electric coil stove and the condenser was coupled by the silicon tube, the next end was used for collecting the yield. Figure 12.1 shows the experimental setup.

Around 4–8 h the experimentation was conducted and in the initial stage, till the temperature reached 100° the pressure valves were kept closed and the formation of vapors started at 150°, at around 200° gasses flow started through condenser. At the collection end of the condenser, water was placed so that the gas vapor dissolves in it. The liquid formed at 230 °C and the water in the collection end of the condenser was detached and a conical flask was put for the oil collection and when the temperature began to increase the liquid flow also increases. The condenser and temperature were maintained for nearly an hour depending upon the plastic granules type and maximum yield was gained. The electrical coil cooktop was turned off and the apparatus slowly cools down, the formation of the liquid at the condenser also stopped.

12.2.3.2 Separation of Pyrolyzed Oil Mixture

Distillation is a conventional procedure for purifying samples that are used practically everywhere in labs and industry. Because waste plastics are typically a combination of diverse sorts, mixed plastics were given more focus in the experiment. The oil obtained from the pyrolysis process was distilled with the help of heating system source, condenser, round bottom flask in which the sample placed, thermometer, and distillate bottle. It was decided to distill 300 ml of pyrolyzed oil, which was heated in a round bottom flask. To identify the physical and chemical features, around 6 purified samples were obtained at various temperature ranges.

12.2.4 Characterization Method

The samples were characterized using the MassHunterGC-MS instrument, and their chemical and physical characteristics were observed.

12.3 Results and Discussion

12.3.1 Amount of Yield from Plastics

High-density polyethylene (HDPE): At 200 °C the vapor started to form and between 260 and 280 °C, the liquid from the condenser began to flow. At 280 °C, the highest yield was obtained. For 1000 g of plastics (HDPE), 710 ml of liquid, and the remaining residue (wax) were collected.

Low-density polyethylene (LDPE): At 200 °C the vapor started to take place and between 220 and 260 °C the liquid from the condenser began to flow. At 260 °C, the highest yield was obtained. For 1000 g of plastics (LDPE), 220 ml of liquid and left some residue (wax) behind and collected.

Polypropylene (PP): At 200 °C, vapor formation began. Between 230 and 290 °C, from the condenser the liquid started flowing, and at 290 °C, the highest yield was attained. For 1000 g of plastics (PP), 390 ml of oil was obtained, and some wax was left in reactor.

Different mixed ratios of plastics (HDPE, LDPE, PP): At 200 °C, the vapor started to form, and between 240 and 290 °C, inside the condenser the flow of the oil started and at 290 °C, the highest yield was obtained. For 1000 g of plastics which was mixed in the ratio (HDPE 250 g, LDPE 500 g, PP 250 g). Nearly 250 oil was collected and left some residue behind (wax) in the reactor. At 200 °C, from the reactor the vapor started to form, between 250 and 290 °C, the formation of oil took place inside the condenser and began to move and at 290 °C, the highest yield was obtained. For 1000 g of plastics which was mixed in the ratio (HDPE 500 g, LDPE 250 g, PP 250 g). Nearly 310 ml of oil was collected and left some residues behind (wax) in the reactor.

Municipal waste plastics: In this experiment, the waste plastics were collected, especially the HDPE, LDPE and PP. At 200 °C, from the reactor, the vapors formation began, and between 240 and 290 °C, inside the condenser the formation of liquid happened and began to flow. The highest yield of oil is obtained at 290 °C. Thousand grams of plastics in ratio which are LDPE, HDPE, and PP. Nearly 350 ml of oil is extracted and collected, leaving some residue (wax) behind in the reactor.

12.3.2 Characteristics of Pyrolyzed Oil

The Physiochemical properties of mixed plastics and municipal waste plastics which were gained at 290 °C are shown in the table below:

12.3.2.1 Gross Calorific Value °C

Gross calorific value or GCV represents the heat amount given out by the burning of plastic waste for a volume unit. This property is a very important parameter in fuel analysis. The calorific value for the pyrolyzed mixed pellet plastic was found to be 10,027 kcal/Kg and for municipal waste pyrolyzed plastic it came to 9397 kcal/Kg. Figure 12.2 showcases the comparison of gross calorific value for the different kinds of fuel. Tables 12.1 and 12.2 carve out the GCV of the pyrolyzed fuel analyzed.

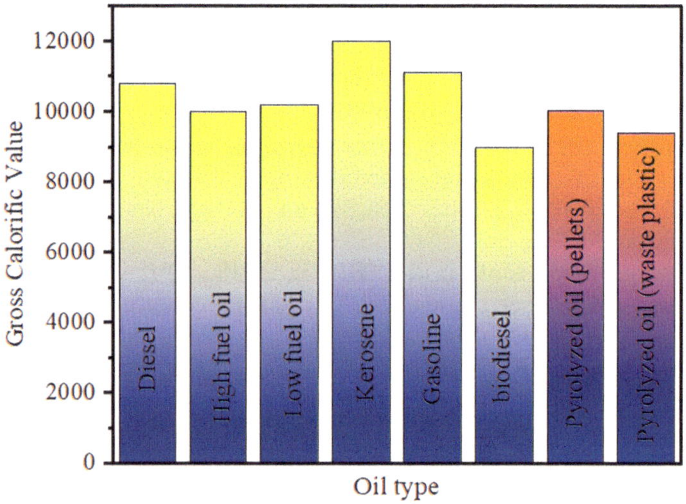

Fig. 12.2 Gross calorific value of different oil types

Table 12.1 Physiochemical properties of mixed pellet plastics

Sl. No	Test	Units	Results	Instruments used
1	Gross calorific value	Cal/g	10,027	Bomb calorimeter
2	Specific gravity at 29 °C	–	0.7276	Hydrometer
3	Viscosity at 40 °C	CST	0.52	Viscometer

Table 12.2 Physiochemical properties of municipal waste plastics

Sl. No	Test	Units	Results	Instruments used
1	Gross calorific value	Cal/g	9397	Bomb calorimeter
2	Specific gravity at 29 °C	–	0.7230	Hydrometer
3	Viscosity at 40 °C	CST	0.47	Viscometer

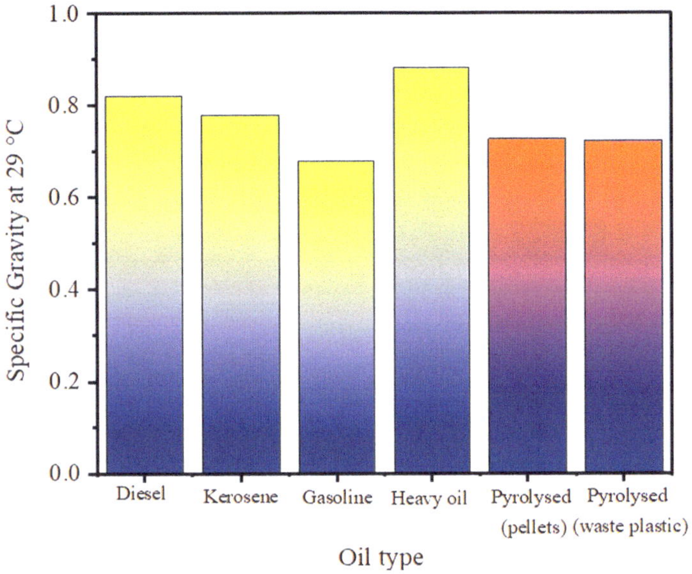

Fig. 12.3 Specific gravity of different oil types

12.3.2.2 Specific Gravity

Specific gravity is defined to be a specific analyzed value of a material under normal pressure and temperature. This is also an important parameter for identification of a fuel. The specific gravity of the two pyrolyzed fuel sample was analyzed by hydrometer at 29 °C. The specific gravity of pyrolyzed fuel for mixed pellet plastic came to be 0.7276 and for municipal waste pyrolyzed fuel was around 0.7230. A comparison graph is created for comparing the specific gravity value showcased in Fig. 12..3.

12.3.2.3 Viscosity

Values of viscosity are different according to temperature, pyrolysis condition, type of feedstock utilized, etc. Higher the range of viscosity, more will be the consumption of fuel, more will be the engine load, and rise in engine temperature. Thin density oil can flow smoothly at lower temperatures and in comparison, to thicker density oils. Thus,

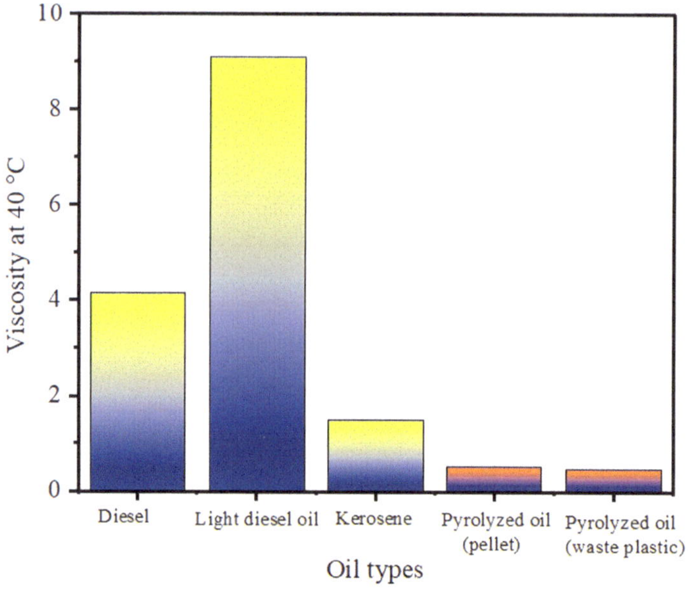

Fig. 12.4 Density of different oil types

the viscosity of oil should not be either exceedingly high or exceedingly low which sometimes very friction less or extra friction correspondingly (Khan et al., 2016). The viscosity analysis showed that pellet pyrolyzed plastic was 0.52 and pyrolyzed municipal waste plastics to be 0.47. Figure 12.4 showcases viscosity comparison of different types of pyrolyzed oil.

12.3.3 GC-MS Analysis

GC-MS results were measured using Agilent MassHunter done at 5 °C with heat rate of 30°C/min. This analysis is a powerful tool to chemically identify the polymers through which degraded products. The GC-MS analysis of both pyrolyzed mixed pellets and waste plastic is described in these sections.

12.3.3.1 GC-MS Analysis of Mixed Pellet Plastic

The pyrolyzed oil samples for mixed pellets were subjected to GC-MS method. The resulted data is ascribed in Table 12.3. The GC-MS graph is showcased in Fig. 12.5. The graph showcases specific peak, and each peak represents a classified compound characterized by the pyrolyzed oil. The compounds identified are listed in the table with its retention time period. The presence of Methane and Ethane is due to the

Table 12.3 Showcasing GC–MS analysis for pyrolyzed plastic pellets

Peak no	Compound	Retention time	Group	Formula
1	Methane	2.046	Alkane	CH_4
2	Ethane	4.688	Alkane	C_2H_6
3	n-heptanol	5.73	Alcohol	$C_7H_{16}O$
4	Octanol	6.90	Alcohol	$C_8H_{18}O$
5	n-octyl acetate	6.94	Amine	$C_{10}H_{20}O_2$
6	Cyclopropane	9.107	Alkane	C_3H_6
7	Benzaldehyde	9.27	Aldehyde	C_7H_6O
8	Propene	9.28	Alkene	C_3H_6
9	Propane	9.776	Alkane	C_3H_8

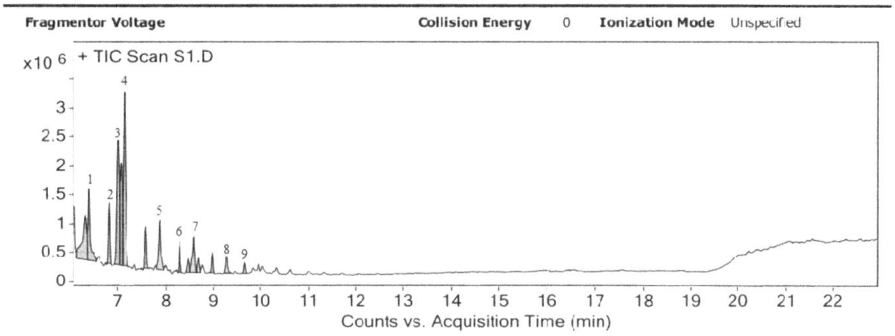

Fig. 12.5 GC-MS analysis of pyrolyzed pellet plastic oil

increase in temperature in the pyrolysis step, which has induced an increase in the gaseous hydrocarbons during the yield of fuel (Lee et al., 2021). The results were matched with similar GC-MS analysis of diesel fuel (Liang et al., 2005).

12.3.3.2 GC-MS Analysis of Municipal Waste Plastic Oil

The samples of pyrolyzed oil for municipal waste plastic were subjected to GC-MS analysis. The resulting data is ascribed in Table 12.4. The GC-MS graph is showcased in Fig. 12.6. The graph showcases a specific peak, and each peak represents a classified compound characterized by the pyrolyzed oil. The compounds identified are listed in the table with its retention time period. The results were matched with similar GC-MS analysis of diesel fuel (Liang et al., 2005).

Table 12.4 GC-MS for municipal waste plastic oil

Peak no	Compound	Retention time	Group	Formula
1	Cumene	6.03	Aromatic hydrocarbon	C_9H_{12}
2	p-xylene	6.29	Aromatic hydrocarbon	C_8H_{10}
3	m-xylene	6.36	Aromatic hydrocarbon	C_8H_{10}
4	o-xylene	6.58	Aromatic hydrocarbon	C_8H_{10}
5	n-octanone	6.90	Ketone	$C_8H_{16}O$
6	n-octyl acetate	6.94	Amine	$C_{10}H_{20}O_2$
7	Thiourea	7.12	Phenyl	CH_4N_2S
8	Trichloroethylene	7.33	Halocarbon	C_2HC_{13}
9	Benzene	7.99	Aromatic Hydrocarbon	C_6H_6

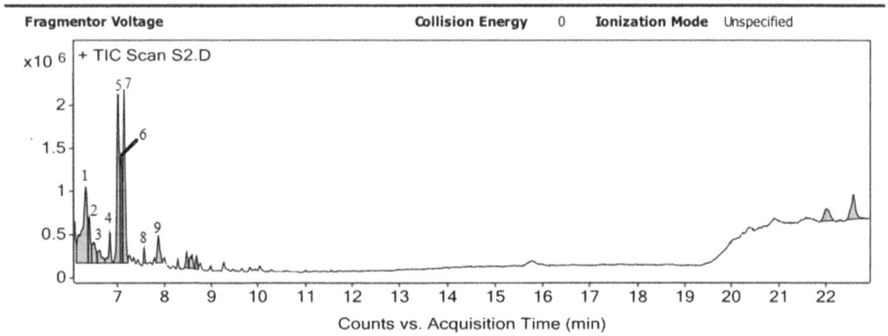

Fig. 12.6 GC-MS analysis of pyrolyzed municipal plastic oil

12.4 Conclusion

This study demonstrates that energy production via plastics is appealing since it addresses management of waste difficulties, lessens the problem of plastic waste disposal, and provides an alternative source of energy. Since the late 1960s, increased use of plastic as a result of urbanization has led to pollution of plastic. Despite the fact that reducing, reusing, recycling, incinerating, blast furnacing plastics, and other techniques processes are used, they are still undervalued internationally owing to their high cost and harmful gas emissions. We may infer that pyrolysis is most effective method that reduces the waste plastic pollution, as a result, producing alternative fuels on a small and big scale.

It was deduced from the aforementioned pyrolysis procedures used by the researchers that:

- Pyrolysis assisted in the extraction of over 75% of the fuel from unwanted plastics.
- Using of catalyst that results in a higher production of fuel.
- Both thermal and catalytic processes might be carried out.

- It contributed to a decrease in the number of plastics entering landfills, resulting in less pollution.
- It helped to minimize the quantity of unwanted plastics that ended in landfills, resulting in reduced pollution.
- The fuel possesses features comparable to fossil fuels, which aid in the generation of more energy.

Fuel oil was produced from the pyrolysis of mixed polymers, which is a treasured energy resource recapture. Testing was conducted without usage of catalyst because catalysts are expensive and its regeneration is a time-consuming process. There was no char in the output from mixed plastics. All of these polymers aided in the gradual development of aliphatic C-C and aromatic and groups of C-H. When the temperature was kept steady for at least 1–2 h, the liquid production rose. At 2 h, the highest yield of was about 85%. When the aforesaid physicochemical features of pyrolyzed oil are compared to those of other commercial fuels, it is clear that the resulting fuel may be used to produce extremely effective fuel at a lower rate. However, more research and experiments are needed before the pyrolyzed oil may be used as a fuel.

Acknowledgements CFTIR Mysuru has helped in the characterization of GC-MS analysis. Physical and chemical properties of sample were done by Ganesh Laboratories, Mysuru.

Authors Contribution All authors critically gave their feedback and supported the work. A.K.T.: data interpretation and article preparation, K.T.V.: designing, conceptualization, and direction of the entire work.

Funding No funding.

Compliance with Ethical Standards

Availability of Data and Materials All data generated or analyzed in the study are included in the published article.

There is no potential conflict of interest.

Competing Interests There is no conflict of interest among authors.

References

Adeniyi, A. G., Eletta, O. A. A., & Ighalo, J. O. (2018). Computer aided modelling of low density polyethylene pyrolysis to produce synthetic fuels. *Nigerian Journal of Technology, 37*, 945–949. https://doi.org/10.4314/njt.v37i4.12.

Anuar Sharuddin, S. D., Abnisa, F., Wan Daud, W. M. A., & Aroua, M. K. (2016). A review on pyrolysis of plastic wastes. *Energy Conversion and Management, 115*, 308–326.

Chen, Y., Awasthi, A. K., Wei, F., et al. (2021). Single-use plastics: Production, usage, disposal, and adverse impacts. *Science of the Total Environment, 752*, 141772. https://doi.org/10.1016/J.SCITOTENV.2020.141772.

Fok, L., Cheng, I. N. Y., & Yeung, Y. Y. (2019). Mismanaged plastic waste: Far side of the moon 57–71. https://doi.org/10.1007/978-981-13-9173-6_5.

Gopi Kiran, K., Dixit, S., & Vadiraj, K. T. (2022). Valorization of plastic waste to generate hydrocarbons by hybrid pyrolysis. *Environmental Engineering Science, 39*, 662–670. https://doi.org/10.1089/ees.2021.0437.

Kasar, P., Sharma, D. K., & Ahmaruzzaman, M. (2020). Thermal and catalytic decomposition of waste plastics and its co-processing with petroleum residue through pyrolysis process. *Journal of Cleaner Production, 265*, 121639. https://doi.org/10.1016/J.JCLEPRO.2020.121639.

Khan, M. Z. H., Sultana, M., Al-Mamun, M. R., & Hasan, M. R. (2016). Pyrolytic waste plastic oil and its diesel blend: fuel characterization. *Journal of Environmental and Public Health, 2016*, 7869080. https://doi.org/10.1155/2016/7869080.

Lee, N., Joo, J., & Lin, K. Y. A. (2021). Lee J (2021) Waste-to-fuels: Pyrolysis of low-density polyethylene waste in the presence of H-ZSM-11. *Polymers, 13*, 1198. https://doi.org/10.3390/POLYM13081198.

Liang, F., Lu, M., Keener, T. C., et al. (2005). The organic composition of diesel particulate matter, diesel fuel and engine oil of a non-road diesel generator. *Journal of Environmental Monitoring, 7*, 983–988. https://doi.org/10.1039/b504728e.

Ncube, L. K., Ude, A. U., Ogunmuyiwa, E. N., et al. (2020). (2020) Environmental impact of food packaging materials: A review of contemporary development from conventional plastics to polylactic acid based materials. *Materials, 13*, 4994. https://doi.org/10.3390/MA13214994.

Shen, M., Song, B., Zeng, G., et al. (2020). Are biodegradable plastics a promising solution to solve the global plastic pollution? *Environmental Pollution, 263*, 114469. https://doi.org/10.1016/J.ENVPOL.2020.114469.

Sipra, A. T., Gao, N., & Sarwar, H. (2018). Municipal solid waste (MSW) pyrolysis for bio-fuel production: A review of effects of MSW components and catalysts. *Fuel Processing Technology, 175*, 131–147. https://doi.org/10.1016/J.FUPROC.2018.02.012.

Sivadas, S. K., Mishra, P., Kaviarasan, T., et al. (2022). Litter and plastic monitoring in the Indian marine environment: A review of current research, policies, waste management, and a roadmap for multidisciplinary action. *Marine Pollution Bulletin, 176*, 113424. https://doi.org/10.1016/J.MARPOLBUL.2022.113424.

Tejaswini, M. S. S. R., Pathak, P., Ramkrishna, S., & Ganesh, P. S. (2022). A comprehensive review on integrative approach for sustainable management of plastic waste and its associated externalities. *Science of the Total Environment, 825*, 153973. https://doi.org/10.1016/J.SCITOTENV.2022.153973.

Varsha, M., Senthil Kumar, P., & Senthil Rathi, B. (2022). A review on recent trends in the removal of emerging contaminants from aquatic environment using low-cost adsorbents. *Chemosphere, 287*, 132270. https://doi.org/10.1016/J.CHEMOSPHERE.2021.132270.

Zhang, X., Pan, L., Wang, L., & Zou, J. J. (2018). Review on synthesis and properties of high-energy-density liquid fuels: Hydrocarbons, nanofluids and energetic ionic liquids. *Chemical Engineering Science, 180*, 95–125. https://doi.org/10.1016/J.CES.2017.11.044.

Chapter 13
From Trash to Treasure: The Rise of Plastic Construction

Pushpa Tuppad, Sinchana S. Gowda, N. Srivatsa, H. D. Rakshitha, M Apoorva, and N. Y. Rakshitha

Abstract The utilization of plastics as construction raw material represents an innovative approach to reusing waste plastic in an environmentally friendly and cost-effective manner. Plastic waste, in various forms such as fibers, granules, flakes, or melted, can potentially be blended with concrete or bitumen for structures and roadways. This plastic-blended concrete offers characteristics like lightweight, corrosion resistance, chemical resistance, lower production costs, and increased durability. However, investigations suggest that while increased plastic content can enhance concrete's compressive strength, it may reduce tensile and flexural strength. Additionally, the density of plastic-blended products tends to be lower compared to conventional methods. HDPE (High-Density Polyethylene) is commonly used as fibers in concrete, enhancing properties such as decreased shrinkage and improved ductility, flexural toughness, and tensile fatigue strength. LDPE (Low-Density Polyethylene) is utilized in LDPE bonded sand blocks and asphalt concrete, enhancing stiffness and rutting resistance. Waste PET (Poly Ethylene Terephthalate) bottles can be recycled into fibers and added to concrete mixes, improving tensile strength and crack resistance. PP (Polypropylene) fibers added to concrete improve split tensile and flexural strength. Overall, these findings highlight the diverse applications and benefits of integrating plastics into construction materials, including enhanced durability and performance, reduced environmental impact, and cost savings. Addressing challenges in plastic waste management requires a multifaceted approach involving the development of alternative materials, reduced plastic production and consumption, and promoting responsible consumer behavior.

Keywords Plastic · HDPE · LDPE · PET · PP · Concrete · Pavement · Strength · Construction material

P. Tuppad (✉) · S. S. Gowda · N. Srivatsa · H. D. Rakshitha · M. Apoorva · N. Y. Rakshitha
JSS Science and Technology University (JSSSTU), Mysuru, India
e-mail: ptuppad@jssstuniv.in

S. Suriyanarayanan et al. (eds.), *Combating Plastic Pollution in Terrestrial Environment*,
https://doi.org/10.1007/978-981-96-2343-3_13

13.1 Introduction

It is almost impossible to imagine the world without the usage of plastic. Due to its versatility, durability, flexibility, lightweight, resistance to corrosion and cost effectiveness, plastic has a wide range of applications across various sectors: packaging—containers, bags, wraps for food, beverages and other consumer goods, etc.; construction—pipes, fittings, insulation materials, roofing, etc.; automotive industry—interior parts, exterior panels, bumpers, etc.; Agriculture—irrigation pipes, mulch films, packaging for fertilizers and pesticides; consumer goods—toys, kitchenware, furniture and appliances; textiles—polyester and nylon for clothing, upholstery and carpets; sports equipment—protective gear, shoes, etc. However, the advantages of plastic utilization are overshadowed by its detrimental effects on the environment, raising widespread concerns leading to efforts to find more sustainable alternatives and improve recycling methods.

Globally, more than 400 Mt (Metric tonnes) of plastic waste is generated annually (Hossain et al., 2019). Disposal of plastic waste presents several challenges: non-biodegradability, formation of microplastics via photodegradation and mechanical erosion, inappropriate disposal practices leading to pollution of atmosphere and waterbodies, landfill overflows and leaching. The situation has been exacerbated by a lack of awareness and education, restricted recycling caused by contamination, mixed or multilayered materials, and insufficient infrastructure. In India, approximately 95% of stray bovine animals in urban areas endure various illnesses primarily due to the ingestion of plastic bags, which often become lodged in their intestines. The immature eating behavior and lack of essential minerals render cattle vulnerable to the inadvertent ingestion of these harmful foreign substances (Murray, 2009). Disposal options of landfilling, incineration, pyrolysis, and waste-to-energy conversion plants are not devoid of further pollution. Instead, the favored choice involves recycling, whereby new raw materials are acquired through mechanical means (often resulting in re-granulation) or chemical processes (typically resulting in monomer building blocks). Each ton of plastic waste recycled leads to a preservation of roughly 3.8 barrels of petroleum (TERI, 2018).

India ranks within the top 12 nations responsible for 52% of the global mismanaged plastic waste (Pandey, 2023). Quantum of specific plastic type is presented in Fig. 13.1.

Plastic is the predominant material utilized in almost 50% of all packaged goods, as indicated by various studies (Banerjee et al., 2014; NPWMTF, 1997; SGCCI, 2000). Packaging represents the most substantial sector of plastic usage in India, accounting for 42–52% of the overall plastics consumption. This consumption trend mirrors those of other countries including USA and UK, where packaging constitutes the largest portion of overall plastic usage (NPWMTF, 1997). Besides packaging, plastics find extensive applications in consumer products, building and construction, and industrial goods (Fig. 13.2).

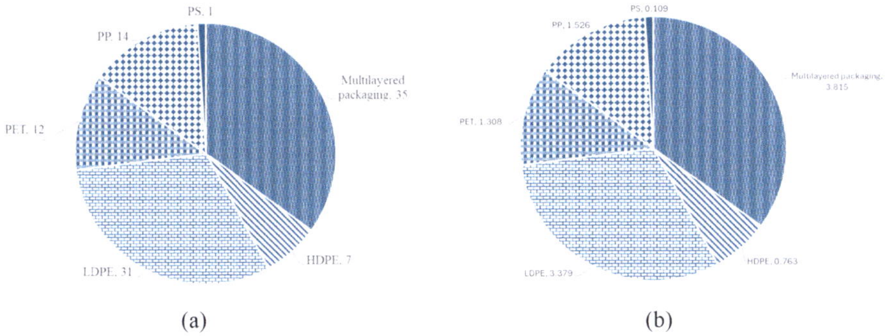

Fig. 13.1 Plastic waste generation in India in 2018–19. **a** Quantum of specific plastic type **b** quantity of plastic waste (million tonne per year). *Source* Singh & Biswas, 2022

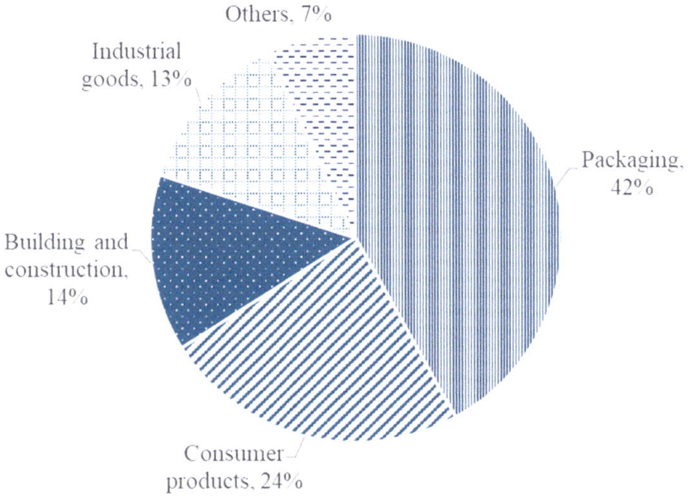

Fig. 13.2 Plastics used in different sectors (Gupta, 2021)

Per capita generation of plastic waste in India has increased from 1.25 kg/year to 3.07 kg/year (CPCB, 2021). Plastic waste generated by States and Union Territories in India is being recycled and/or used for various purposes including road construction, waste to energy/oil and in cement plants for co-processing. State of Odissa tops in percentage utilization of plastic waste followed by Tripura and Tamil Nadu. While rag pickers manage to retrieve a significant proportion of this waste, yet much of it remains either contaminated with organic matter or deemed unsuitable for further processing. Plastic is classified into seven types: HDPE (High density polyethylene), LDPE (Low density polyethylene), PET (Polyethylene terephthalate), PS (Polystyrene), PP (Polypropylene), PVC (Polyvinyl chloride), and Others (CSE, 2019). In the Indian context, Polyethylene (PE), PP, and PVC dominate the market primarily due to their affordability, chemical composition, physical properties, and durability. Polyolefins comprise 60% of India's total plastic consumption (Mutha et al., 2006).

This paper seeks to delve into the feasibility of integrating plastic into construction materials and evaluate how this incorporation influences various properties of the resulting products. By scrutinizing the potential applications of plastic in construction, the study aims to explore both the advantages and limitations of utilizing plastic in this context. Through a comprehensive analysis, it endeavors to highlight the implications of incorporating plastic on the durability, strength, flexibility, sustainability, and other pertinent characteristics of the construction materials. This examination is crucial for understanding the broader implications of utilizing plastic in construction and for informing future decisions regarding material selection and construction practices.

13.2 Important Properties of Plastics

Properties such as tensile strength and elasticity are crucial considerations in assessing the suitability of plastics for structural applications, while thermal stability, chemical resistance, water absorption and density significantly influence their performance in various environments. Important properties of different forms of plastics are summarized in Table 13.1.

13.3 Plastics Recycling

Refuse, reuse, recycle, resource (waste-to-energy) recovery and ultimate disposal (landfill) is the preferred plastic management hierarchy. Major challenges to recycle plastic is a post-consumer collection, cross-contamination and availability of clean materials, versatility in form and characteristics. Although, the use of waste plastic in the construction industry is a major application, alternate strategies of plastic waste management are summarized in Table 13.2. Owing to the fact that some plastics such as polypropylene, polystyrene, polycarbonates, nylon, etc. cannot be recycled, alternate option could include processing to produce energy. Plastics can be converted to fuel in the form of hydrogen, crude oil, diesel, and sulphur (Kabeyi and Olanrewaju, 2023).

Table 13.1 Important properties of different plastic forms

	LDPE	HDPE	PET	PP	PS
Elongation (%)	13–17.5	15	30–70	15–600	1–70
Tensile strength (683106Nm22)	10–15	25–30	45–70	15–45	5.2–52.4
Tensile modulus (GPa)	0.13–0.3	0.5–1.1	2–2.9	1.1–1.6	3.2
Impact strength (ASTM D256, kJ/m^2)	23–147	158–505	4.6	3–30	2.09–4.8
Chemical resistance	• Shows good resistance to alcohols, dilute alkalis and acids • Exhibits moderate resistance to mineral oils, oxidizing agents aliphatic and aromatic and halogenated hydrocarbons (HCs)	• Exceptional resistance to most solvents • Very good resistance to dilute acids, alcohols, and alkalis • Moderate resistance to greases and oils • Poor resistance to aliphatic, aromatic, and halogenated HCs	• Excellent resistance to alcohols, aliphatic HCs, oils, grease and diluted acids • Moderate resistance to diluted alkalis, aromatic and halogenated HCs	• Excellent resistance to dilute and concentrated acids, alcohols, and bases • Good resistance to aldehydes, esters, aliphatic HCs, ketones • Limited resistance to oxidizing agents and aromatic and halogenated HCs	• Not resistant to aromatic, aliphatic, and chlorinated organic solvents, cyclic ethers, ketones, acids, and bases • Moderately resistant to higher molecular weight aliphatic alcohols, dilute aqueous acids and bases, and bleach
Transparency	100% in thin film form	80%	70–90%	0–100%	88%
Density, g/cm^3	0.917–0.940	0.94 -0.97	1.3–1.4	0.904–0.908	1.0–1.2
Water absorption 24 h, %	0.005–0.015	0.005–0.01	0.1–0.2	0.01–0.1	0.03–0.05
Melting temperature	105–115 °C	120–140 °C	240–270 °C	160–165 °C	177–280 °C

Table 13.2 Application of waste plastic in non-construction sectors

	Application	Advantage	Disadvantage
Recycled plastic products	Reprocess and produce new products such as Water bottles, cosmetic bottles, detergent bottles, sewer pipes, buckets, toys, car interiors, tubes, agricultural mulch films, heavy duty sacks, cling films, milk crates, textile products such as bags, T-shirts and other garments, cloth hangers, seeding containers, picture frames, furniture, benches, shoe soles, etc.	Waste to usable product is follows zero waste concept	Requirement of clean materials which otherwise make this option expensive as its requires a lot of cleaning
Co-processing of plastic waste as alternate fuel and raw material	Co-generation in large combustion plants	Substitution for primary fuel leading to savings in fossil fuel; zero waste practice	Ensuring consistency in waste composition; characteristics such as high moisture content and heavy metals; high capital investment
Plastic to fuel: hydrogen, crude oil, diesel	Pyrolytic conversion technology—high temperature (300–900 °C) thermal depolymerization in the absence of oxygen	Alternate source of energy	Variation in yield depending on the quality of plastic feedstock; high sulphur content in pyrolysis oil; high capital and operation and maintenance cost
Plastic waste incineration	Combustion at temperature between 800 and 1200 °C	Reduce volume of waste and generate energy	Release of air pollutants like dioxins, furans, polychlorinated biphenyls, mercury, halogenated compounds, etc., and generation of toxic ash

13.4 Plastic Waste Utilized in Construction

Various forms of plastics have been utilized across different sectors within the construction industry, facilitating the recycling of plastic waste while maintaining material properties conducive to construction purposes (Table 13.3).

Table 13.3 Plastic waste utilized in construction

Plastic Type	Common uses	Advantages of Using Plastic Waste	Disadvantages of Using Plastic Waste	Plastic form	Application	Performance	References
HDPE	Pipes, lids, toys	Durable, Weatherproof, good chemical resistance	Difficult to work with, not aesthetically pleasing	Fibers	Used in structural concrete	Improved flexural toughness and durability	Pešić et al. (2016), Horta et al. (2017)
				Granules	Used as substitute to natural sand in concrete	Sustainable alternative to sand	Abdelhak et al., (2018)
LDPE	Bags, liners	Flexible, waterproof	Lower strength, tear easily	Water sachets (molten form)	Blended with sand to make LDPE-bonded sand blocks	High strength, reduced rate of heat transfer	Jnr et al., 2018; Zulkernain et al., (2021)
				Flakes	Used in Asphalt concrete mixtures	Enhanced asphalt properties	Arminda et al., (2020)
				Mix with demolition waste	Used for road construction	Improved road strength and durability	Arulrajah et al. (2019)
PP	Containers, textiles	Chemically resistant, good heat tolerance	Can be brittle at low temperatures	Mix with crumbs rubber (molten form)	Added to base bitumen in road pavements	Increased resistance to cracking	Khan et al., (2016), Ahmed et al. (2020); Hazha et al. (2020)
				Aggregates	Construction or renovation of conventional timber beam flooring systems	Improved structural properties	Rubio-de Hita et al. (2018)

(continued)

Table 13.3 (continued)

Plastic Type	Common uses	Advantages of Using Plastic Waste	Disadvantages of Using Plastic Waste	Plastic form	Application	Performance	References
PET	Bottles, containers, fabrics	Lightweight, Good insulator, readily available	May leach chemicals, lower fire resistance	Fibers	Added to concrete	Reinforced concrete, reduces shrinkage	Ghernouti et al. (2015), Zulkernain et al. (2021)
				Flakes	Added to light weight concrete	Improved insulation properties	Hameed and Hamza (2019)
				Aggregates	Applied to road pavements	Enhanced road durability	Nazmi et al. (2013)
				Molten Form	Mixed with clayed sand to form bricks	Enhanced brick strength and durability	Akinwumi et al. (2019)
Composite				Particles mixed with waste crumb tire rubber	Sustainable concrete	Improved durability	Saman et al., (2016)
				Polymer powders	Stabilization of aggregates in local roads	Stabilization of degraded aggregates	Cameron et al. (2016)
				Polycarboxylic ether-polymer (molten form)	Used in concrete mixes	Better resistance to abrasion	Sofi (2018)

(continued)

Table 13.3 (continued)

Plastic Type	Common uses	Advantages of Using Plastic Waste	Disadvantages of Using Plastic Waste	Plastic form	Application	Performance	References
				Polyester resins (molten form)	Polymer mortars and cement replacement	Enhanced mortar properties	Martínez-Barrera et al. (2019)
				Polycarbonates	Polyester/polymer mortars	Improved compressive strength	Paihtea et al. (2019)

13.4.1 HDPE

High-density polyethylene (HDPE) is a thermoplastic polymer derived from the ethylene monomer, renowned for its remarkable strength-to-density ratio. It finds extensive application in manufacturing plastic bottles, plastic lumber, geomembranes, and corrosion-resistant piping. HDPE is widely recyclable and is identified by the resin code '2' (ACC, 2020).

HDPE is employed as fibers in concrete to create Fibre Reinforced Concrete (FRC), which exhibits favorable characteristics such as reduced shrinkage and enhanced ductility, flexural toughness, and tensile fatigue strength. FRC finds application across a spectrum of structures ranging from industrial floors, foundation slabs, and roadways to tunnels and bridges. While the inclusion of steel fibers in concrete enhances its flexural and shear strength, the benefits conferred by plastic fibers primarily pertain to enhancing the serviceability of concrete. The incorporation of plastic fibers does not adversely affect the elastic modulus or compressive strength. Both the quantity and diameter of the fibers used have no significant impact on the compressive strength, akin to plain concrete mixtures. The addition of up to 1.25% of plastic fibers reduces water permeability, forming the basis for the assertion that HDPE fibers significantly enhance concrete durability. By diminishing water intake, concrete reinforced with HDPE fibers demonstrates enhanced resistance to deterioration processes associated with water movement through voids, such as freeze/thaw action, salt ingress, and carbonation. Although the tensile strength and modulus of elasticity of HDPE fibers derived from recycled materials are inferior to those of engineering-grade HDPE, they still improve several performance properties of concrete (Pešic et al., 2016).

HDPE serves as an alternative to natural sand in concrete blocks. As the HDPE content increases, there is a reduction in the apparent density compared to regular concrete. The apparent density of HDPE-sand mortars is lower than that of typical concrete, with a notable decrease observed particularly when the apparent density of HDPE-sand reaches 0.362 g/cm^3. Consequently, the grain size distribution between the medium-sized particles of HDPE-sand and natural sand contributes to an increase in pore volume, leading to a decline in bending strength (Abdelhak et al., 2018).

In road construction materials, HDPE is blended with recycled asphalt concrete (RAC) and crushed bricks (CB). However, the addition of high quantities of plastic granules diminishes the bearing capacity, stiffness, and resilient modulus of plastics-RAC and plastics-CB. This reduction is attributed to the smooth surface of HDPE particles in CB and RAC, which contrasts with the typically high surface roughness of these materials (Arulrajah et al., 2017). The optimal proportion of plastics is determined to be 3–5%.

HDPE was utilized as a partial substitute for cement across varying ratios: 0, 2.5, 5, 7.5, 10, and 20%. The incorporation of controlled quantities of recycled plastic waste demonstrated potential in producing lightweight, porous composites exhibiting enhanced thermal properties. Notably, the partial replacement of coarse aggregates with HDPE yielded promising outcomes, leading to decreased thermal conductivity and improved thermal resistance. These findings suggest the potential for enhancing the thermal performance of buildings and reducing energy consumption (El-Metwally et al., 2023).

Ali et al. (2017) illustrated that incorporating 3.0% HDPE into bricks yielded greater compressive strength compared to the control, consequently reducing sand utilization and environmental pollution. Future investigations might entail altering water-cement ratios to optimize strength, examining the use of HDPE in clay bricks, and evaluating smaller HDPE bead sizes to enhance compressive strength.

Ahmad (2017) found that the optimal proportion of waste HDPE in concrete mixture is 35%, beyond which the compressive strength decreases. This decrease in strength is ascribed to the decrease in adhesive strength between the surface of waste plastic and cement. The density of the mixture varied based on the percentage of waste HDPE in the concrete mix design, with a maximum density of 1.972 gm/cm^3, lower than that of normal cement concrete. The resulting material exhibits characteristics of a semi-crystalline polymer, rendering it flexible rather than brittle like cement concrete. Additionally, mixtures containing 25–30% waste plastic demonstrate good workability, allowing for the creation of holes without difficulty. However, deviations from this range weaken the workability of the mixture (Ahmad, 2017).

13.4.2 LDPE

Low-density polyethylene (LDPE) is a thermoplastic derived from the ethylene monomer. LDPE is widely recyclable and is identified by the resin code "4" (ACC, 2020). Despite facing competition from newer polymers, LDPE remains a significant plastic grade. Unlike HDPE, LDPE exhibits greater branching (occurring in roughly 2% of the carbon atoms), resulting in weaker intermolecular forces, lower tensile strength and higher resilience. Additionally, its molecules are less densely packed and less crystalline due to these side branches, resulting in a lower density.

Melted LDPE residues were fused with sand particles to fabricate LDPE-bonded sand blocks. LDPE typically exhibits a density of 0.91–0.94 g/cm^3. Compressive strength decreased whereas porosity increased by the utilization of larger sand particle sizes. Particularly, samples incorporating sand particles with diameters less than 500 mm attained the highest compressive strengths. Moreover, addition of sand in higher proportions elevated the density, which ranged between 1.46 and 1.91 g/cm^3. Incrementing the sand content augmented the compressive strength (27.3 MPa) for additions up to 75% by weight. The experiments elucidate that the mechanical attributes of LDPE-bonded sand depend on the size of sand particles and the ratio of sand to plastic. Sand incorporations surpassing 75 wt% lead to a decline in compressive strength because the volume of LDPE binder becomes insufficient to adequately coat and thus bind the sand grains together (Jnr et al., 2018).

The rutting resistance and stiffness of asphalt concrete (AC) increase with the addition of recycled waste LDPE in the form of flakes. With and without plastic, aged

AC mixtures presented higher stiffness and better rutting resistance. Considering the aging effect, the AC with plastic performed better for moderate and low strain levels. Incorporating LDPE as a binder modifier partially compensated the adverse effect of aging, i.e., fatigue cracking resistance (Arminda et al., 2020).

Dhawan et al. (2019) created a novel composite binder by blending LDPE with a pure binder. Increasing proportions of LDPE generally exert a substantial influence on the elastic properties of the modified binder. Introducing 10% LDPE into the binder yields an optimal value for rutting resistance across all temperatures. Additionally, incorporating 10% LDPE enhances the performance grade of neat bitumen at elevated temperatures, raising it from 64 to 70 °C. By enhancing its elastic properties, inclusion of LDPE in the pristine binder significantly prolongs the service life of pavements by reducing susceptibility to rutting and cracking (Khan et al., 2016).

The LDPE wastes incorporated into tiles reduced flammability and improved tensile strength. Structures built with these tiles would be lightweight, resistant to corrosion, chemically resistant, low cost, and have increased service life (Dhawan et al., 2019).

Binding properties of LDPE are made use of in paving blocks in walkways, footpaths, and building premises which generally are low weight bearing surfaces. LDPE blend is particularly advantageous in highly moist environments because of their minimal water absorption capacity. Besides increase in the compressive strength, significant time savings in the manufacturing process are realized, as these blocks achieve over 80% of their final strength within 24 h. Thus, their swift curing and low water absorption properties are promising in waterlogged areas and other situations requiring rapid construction, rendering them less susceptible to mechanical damage and chemical attack (Agyeman et al., 2019).

13.4.3 PET

Polyethylene Terephthalate (PET), a member of the polyester family, is a thermoplastic polymer resin characterized by its non-biodegradability and semi-crystalline structure. Approximately 60% of PET production is allocated to synthetic fibers, with PET bottle manufacturing accounting for 30% of total production. Despite a notable portion of PET being recycled into new products, it remains a significant contributor to the materials disposed of in landfills.

There exists vast potential for utilizing PET as composites in the construction industry (Emin and Kar, 2018).In construction, PET bottles can be recycled and transformed into fibers, commonly known as PET fibers. These fibers are added to concrete mixes as a reinforcement material, enhancing the tensile strength and crack resistance of concrete. Studies have shown that incorporating PET fibers in concrete can improve its flexural and impact strength properties, making it a viable option for applications in construction such as pavements, slabs, and precast elements (Siddique et al., 2008).

Shredded PET waste is blended with clayey-sand soil in different ratios to produce compressed earth bricks (CEB). Among these, CEB incorporating 1% waste plastic particles smaller than 6.3 mm demonstrated the highest compressive strength, showing an increase of 244.4% compared to CEB without shredded waste plastic (Akinwumi et al., 2019).

Upon comparing plastic bottles filled with crushed recycled aggregates and an appropriate amount of water with conventional clay bricks in terms of compressive strength, Paihtea et al. (2019) reported that bottles filled with aggregates of size < 425 μm demonstrated superior compressive strength compared to those filled with larger aggregates. A water content of 5% proved to be the most optimal and the study observed minimal impact on compressive strength due to exposure to saline conditions (Paihtea et al., 2019).

Modified asphalt mixtures were created using recycled PET pallets, ranging from 5 to 25% of the weight of the asphalt mixture, with sieve sizes ranging from 2.36 to 1.18 mm, and 5% of the weight of bitumen content. The results of Axial Test and Indirect Tensile Stiffness Modulus Test indicated that PET has the ability to enhance the permanent deformation properties of asphalt mixtures, making it suitable for use in road pavements (Nazmi et al., 2013).

There exists potential for employing granules of shredded waste PET as aggregate in structural concrete (Akçaözoğlu & Ulu, 2014).

Akçaözoğlu et al. (2010) investigated two sets of mortar samples: (1) comprising solely of PET aggregates, and (2) adding sand with PET aggregates. The PET granules used in the mortar mixtures were upto 4.0 mm in size. Additionally, blast-furnace slag was introduced as a replacement for cement on a mass basis, at a ratio of 50%, aiming to decrease cement usage and offer cost savings. The findings revealed that mortars containing solely PET aggregates or a blend of PET and sand aggregate or those with slag as a cement replacement could all be grouped as structurally lightweight concrete based on their unit weight and strength properties. Blending PET with Portland cement also yields lightweight concrete. Incorporating PET with Portland cement (at a rate of 1% by the weight of cement) led to a notable increase in compressive strength by 58%. This combination also resulted in optimal splitting tensile strength (Hameed and Ahmed, 2019).

Incorporation of PET and PU in interlocking bricks showed excellent impact strength. The PET/PU ratio of 60/40 achieved the highest impact strength at 23.3 J/m, which makes these bricks ideal for applications such as highway and column crash barriers (Alaloul et al., 2020).

13.4.4 PP

Das et al. (2018) investigated the effects of adding 0.5, 0.75, and 1% PP fibers to natural aggregate concrete (NAC) as well as recycled aggregate concrete (RAC). As the amount of PP fibers increased, the density of RAC decreased further compared to NAC due to its inherently lower density. While fiber content had negligible effects on compressive strength, it was a significant factor for split tensile and flexural strength properties of both NAC and RAC.

However, Hossain et al. (2019) observed an initial increase in compressive strength with a 10% replacement level of Recycled Concrete Aggregate (RCA), 5% Crumb Rubber (CR), and 2% PP fiber. However, this increase diminished at higher replacement levels (30% RCA, 10% CR, and 2% fiber). Additionally, compressive strength increased with higher fiber content. Overall, rubberized concrete with these mix ratios exhibited superior strength compared to normal concrete in terms of compressive strength, split tensile strength, and flexural strength tests.

Rubio-de Hita et al. (2018) experimented on manufacturing beam-filling pieces for traditional timber-beam floor structures wherein cement mortar was blended with aggregates comprising mixed PP waste from urban waste collection plants. Increased utilization of recycled plastic (15%) enhanced the eco-efficiency of the proposed piece, albeit impairing workability in its fresh state unless additives were incorporated. Like many other studies, it was observed that incorporating more plastic waste reduced the density, flexural, and compressive strength of the resulting composites (Rubio-de Hita et al., 2018).

13.4.5 Miscellaneous Plastic

Mixing hydrophobic dry powder polymers (DPP) with lime offers a cost-effective and rapid method for enhancing road bases. DPP plays a significant role in stabilizing degraded aggregates. Upon testing for hydraulic conductivity, repeated loading triaxial, and large column capillary rise, it was found that DPP decreases the maximum dry density while increasing the optimum moisture content (Cameron et al., 2016).

Sofi (2018) examined the incorporation of rubber tyre waste into concrete. As the proportion of crumb rubber in concrete increased, both compressive and flexural values decreased. However, compared to the control mixture, the rubberized concrete displayed superior abrasion resistance. Moreover, the water penetration of rubberized concrete exceeded that of the control mix.

Utilizing waste tyre rubber, up to 12.5% by weight, as a partial substitute for fine aggregate in high-strength concrete is feasible. This approach may find applicability in structures prone to brittle failure. Furthermore, high-strength concrete incorporating crumb rubber exhibits enhanced resistance to abrasion relative to the control mix, making it suitable for use in floors, pavements, concrete highways, hydraulic structures including tunnels and dam spillways, or other surfaces subject to abrasive forces from moving objects (Sofi, 2018).

Using Gamma rays (at doses of 100 and 200 kGy) to polyester mortars during post-curing process found that the ionizing radiation improved the mechanical properties of irradiated mortars with PET particles and mortar with polycarbonate particles by 46 and 37%, respectively, compared with control mortar (Martínez-Barrera et al., 2019).

Specific variations in the properties of blended plastics are summarized in Tables 13.4a–c.

Plastic mixed bituminous roads are laid in several cities in developing countries. This is gaining popularity because of enhanced road quality in terms of higher strength and water resistance. NHAI (2022) has issued policy guidelines (Policy

Table 13.4 **a** Variation in the property of the materials with addition of waste plastics as fine aggregates. **b** Variation in the property of the materials with addition of waste plastics as coarse aggregates. **c** Variation in the property of the materials with addition of waste plastics as fibers

Sl. No	Properties	Percentage of plastic waste as fine aggregate					References
		0%	5%	10%	15%	20%	
1	Compressive strength, N/mm2- HDPE	16.5	18.88	19.33	17.77	–	Lumina et al. (2022)
2	Compressive strength, N/mm2- PP	36.6	34.9	32.8	–	–	Nabajyoti and de Brito (2012)
3	Water absorbed, %—HDPE	23.08	14.22	11.43	9.56	–	Lumina et al. (2022)
4	Water absorbed, %—PP	0.85	1.09	1.35	–	–	Nabajyoti and de Brito (2012)
5	Flexural Strength, Mpa—PET	4.4	6.1	4.9	4.8	4.3	Azhdarpour et al. (2016)
6	Tensile Strength, MPa—PET	2.5	3.1	3.3	2.9	2.8	Azhdarpour et al. (2016)
Sl. No	**Properties**	**Percentage of Plastic Waste as Coarse Aggregate**					**References**
		0%	5%	10%	15%	0%	
1	Compressive strength, N/mm2—PET	35	51	38.1	31	29	Azhdarpour et al. (2016)
2	Compressive strength, N/mm2- E plastic	47.18	–	44.07	–	24.69	Manjunath (2016)
3	Compressive strength, N/mm2– miscellaneous plastic	25.93	–	28.15	30.3	–	Bhandari and Tajne (2016)
4	Split Tensile strength, N/mm2—E plastic	4.9	–	4.8	–	5.4	Manjunath (2016)
5	Flexural Strength, N/mm2-PET	4.35	–	4.4	–	4.3	Manjunath (2016)
6	Flexural Strength, N/mm2—miscellaneous plastic	3.82	–	3.68	3.94	–	Bhandari and Tajne (2016)
Properties		**Percentage of Plastic Waste as Coarse Aggregate**					**References**
		0%	0.30%	0.60%	0.90%	1.20%	
	Compressive strength, N/mm2—miscellaneous plastic	20.15	22.21	24.06	25.7	23.92	Ananthi et al. (2017)

Circular No. 13.24/2022 dated 2nd June 2022) which mandate the use of waste plastic for making 5% of the length of the road in any highway project. Indian Road Congress has published guidelines for the use of waste plastic in hot bituminous mixes in wearing coats (IRC SP:98—2013). Significant road lengths are constructed using waste plastic in urban and rural India (Biswas et al., 2020; Heriawan, 2020; Conlon, 2022). India based company, Rhino Machines has launched the Silica Plastic Blocks—a sustainable building brick that is made from recycling foundry dust/sand waste (80%) and mixed plastic waste (20%).

13.5 Conclusions

Utilization of plastics as construction raw material is one of the newer approaches for re-utilization of waste plastic material in an environmentally friendly and economical way. Plastic waste as fibers, granules, flakes and in melted form can potentially be used with concrete or bitumen for structures/roadways. Concrete blended with plastics have characteristics such as light weight, corrosion resistant, chemically resistant, less cost of production, increased durability, etc. Investigations have indicated that increased plastic content increased the compressive strength of the concrete but decreased properties such as tensile strength, and flexural strength. Density of the plastic blended product is lesser than the product from conventional methods. Reuse of plastic waste as a construction material is promising. Further studies are necessary to investigate the short-term/long-term leaching characteristics of plastic blended construction materials to understand their impacts on environmental attributes.

HDPE:

- HDPE is commonly used as fibers in concrete to create Fiber Reinforced Concrete (FRC), which enhances properties such as reduced shrinkage and improved flexural toughness, ductility, and tensile fatigue strength.
- Inclusion of HDPE fibers in concrete does not significantly affect its compressive strength but reduces water permeability, thus enhancing durability against deterioration processes associated with water transfer.
- HDPE can serve as a partial substitute for cement in concrete blocks, although higher HDPE content reduces bending strength due to increased pore volume.
- HDPE blended with recycled asphalt concrete and crushed bricks for road construction materials diminishes bearing capacity, stiffness, and resilient modulus with high plastic granule content.

LDPE:

- LDPE, characterized by weaker intermolecular forces and higher resilience compared to HDPE, is used in the production of LDPE-bonded sand blocks, exhibiting reduced compressive strength with larger sand particles.
- Incorporation of recycled LDPE plastic waste flakes in asphalt concrete enhances stiffness and rutting resistance, particularly beneficial for aged mixtures.

- LDPE serves as a modifier for asphalt binders, improving rutting resistance and extending pavement service life.
- LDPE can be utilized in the production of lightweight concrete and tiles, offering reduced flammability and improved tensile strength.

PET:

- PET bottles can be recycled into fibers and added to concrete mixes, enhancing tensile strength and crack resistance.
- Shredded PET waste can be incorporated into compressed earth bricks and plastic-filled bottles to improve compressive strength, making them suitable for construction applications.
- Modified asphalt mixtures containing recycled PET pallets exhibit enhanced permanent deformation properties, suitable for road pavements.
- PET waste can serve as aggregate in structural lightweight concrete, showing promise for various construction applications.

PP:

- PP fibers added to concrete improve split tensile and flexural strength, although their impact on compressive strength varies with replacement levels of recycled concrete aggregate and crumb rubber.
- Utilizing mixed PP waste in cement mortars reduces density but impairs workability unless additives are used.

From a socioeconomic standpoint, blending plastic waste into building materials provides affordable housing solutions in developing countries, reducing construction expenses and enhancing accessibility for low-income communities. This approach also generates job opportunities, particularly in recycling and construction sectors, fostering local entrepreneurship and economic growth. Additionally, initiatives to recycle plastic waste encourage small businesses and cooperatives, promoting community involvement. Utilizing plastic waste improves infrastructure, leading to better living conditions with the construction of roads and public amenities. On the environmental front, repurposing plastic waste into construction materials offers an effective waste management solution, reducing the strain on landfills and minimizing plastic pollution. It also contributes to energy efficiency through improved thermal insulation, conserves natural resources by lessening reliance on traditional materials like timber, and mitigates environmental pollution risks. These plastic-based building materials are durable, long-lasting, and eco-friendly, presenting a sustainable solution that addresses both socioeconomic and environmental challenges in developing regions (Bai and Givens, 2021).

Overall, these findings underscore the diverse applications and benefits of incorporating plastics in construction materials, ranging from enhanced durability and performance to reduced environmental impact and cost savings. Although increased recycling efforts help in effective plastic waste management, addressing the challenges needs a multi-faceted approach involving developing alternative materials, reduced plastic production and consumption, creating awareness and promotion of responsible consumer behavior among others. It is high time, that structured recycling approach following cradle-to-cradle model in plastic management aims to create a sustainable and circular economy where plastics are continuously reused, recycled, and regenerated, contributing to the reduction of waste, conservation of resources, and preservation of the environment for future generations.

References

Abdelhak, B., Benosman, A. S., Senhadji, Y., & Mouli, M. (2018). Thermo-physical and mechanical characteristics of sand-based lightweight composite mortars with recycled high-density polyethylene (HDPE). *Construction and Building Materials, 163*, 40–52.

Agyeman, S., Obeng-Ahenkorac, N. K., Assiamahd, S., & Twumasid, G. (2019). Exploiting recycled plastic waste as an alternative binder for paving blocks production. *Case Studies in Construction Materials, 11*, 112–118.

Ahmad, K. J. (2017). Recycling of polyethylene waste to produce plastic cement. *Procedia Manufacturing, 8*, 635–642.

Ahmed, T. W., Ali, A. A. M., & Zidan, R. S. (2020). Properties of high strength polypropylene fiber concrete containing recycled aggregate. *Construction and Building Materials, 241*, 118010. https://doi.org/10.1016/j.conbuildmat.2020.118010.

Akçaözoğlu, S., & Ulu, C. (2014). Recycling of waste PET granules as aggregate in alkali-activated blast furnace slag/metakaolin blends. *Construction and Building Materials, 58*, 31–37.

Akçaözoğlu, S., Atiş, C. D., & Akçaözoğlu, K. (2010). An investigation on the use of shredded waste PET bottles as aggregate in lightweight concrete. *Waste Management, 30*(2), 285–290.

Akinwumi, I. I., Domo-Spiff, A. H., & Salami, A. (2019). Marine plastic pollution and affordable housing challenge: Shredded waste plastic stabilized soil for producing compressed earth bricks. *Case Studies in Construction Materials, 11*, e00241.

Alaloul, W. S., Syed, A. A., Al-Attar, H. S., & Mohammed, T. T. (2020). Mechanical and thermal properties of interlocking bricks incorporating PET and polyurethane binder using response surface methodology. *International Journal of Concrete Structures and Materials, 14*(1), 24–36.

Ali, N., Din, N., Khalid, F. S., Shahidan, S., Abdullah, S. R., Abdul Samad, A. A., & Mohamad, N. (2017). Compressive strength and initial water absorption rate for cement brick containing high-density polyethylene (HDPE) as a substitutional material for sand. *IOP Conference Series: Materials Science and Engineering, 271*, 012083.

American Chemistry Council. (2020). Plastic Resin Codes PDF. https://plastics.americanchemistry.com.

Ananthi, A., Eniyan, A. J. T., & Venkatesh, S. (2017). Utilization of Waste Plastics as a Fiber in Concrete. *International Journal of Concrete Technology, 3*(1), 1–5.

Arminda, A., Capitão, S., Bandeira, R., Fonseca, M., & Picado-Santos, L. (2020). Performance of AC mixtures containing flakes of LDPE plastic film collected from urban waste considering ageing. *Construction and Building Materials, 232*, 112–118.

Arulrajah, A., Yaghoubi, E., Wong, Y. C., & Horpibulsuk, S. (2017). Recycled plastic granules and demolition wastes as construction materials: Resilient moduli and strength characteristics. *Construction and Building Materials, 147*, 639–647.

Arulrajah, A., Mohammadinia, A., Maghool, F., & Horpibulsuk, S. (2019). Tire-derived aggregates as a supplementary material with recycled demolition concrete for pavement applications. *Journal of Cleaner Production, 230*, 129–136.

Azhdarpour, A. M., Nikoudel, M. R., & Taheri, M. (2016). The effect of using polyethylene terephthalate particles on physical and strength-related properties of concrete; a laboratory evaluation. *Construction and Building Materials, 109*, 55–62.

Bai, Y., & Givens, J. E. (2021). Ecologically unequal exchange and global trade in plastic waste: A world-systems analysis. *Journal of World-Systems Research, 27*(1), 253–287. https://doi.org/10.5195/JWSR.2021.1026.

Banerjee, T., Srivastava, R. K., & Hung, Y.-T. (2014). Plastics waste management in India: An integrated solid waste management approach. In *Handbook of Environmental and Waste Management* (Vol. 2).

Bhandari, P., & Tajne, D. K. (2016). Use of foundry sand in conventional concrete. *International Journal of Latest Trends in Engineering and Technology, 3*, 249–254.

Biswas, A., Goel, A., & Potnis, S. (2020). Performance comparison of waste plastic modified versus conventional bituminous roads in Pune city: A case study. *Case Studies in Construction Materials, 13*, e00411. https://doi.org/10.1016/j.cscm.2020.e00411.

Cameron, D., Hopkins, C., & Rahman, M. (2016). Hydrophobic polymer additive for stabilization of aggregates in local government roads. *Procedia Engineering, 143*, 26–33.

Conlon, K. (2022). Plastic roads: Not all they're paved up to be. *International Journal of Sustainable Development & World Ecology, 29*(1), 80–83. https://doi.org/10.1080/13504509.2021.1915406.

CSE. (2019). Fact sheet.

Das, C. S., Dey, T., Dandapat, R., BhusanMukharjee, B., & Kumar, J. (2018). Performance evaluation of polypropylene fibre reinforced recycled aggregate concrete. *Construction and Building Materials, 189*, 649–659.

Dhawan, R., Bisht, B. M. S., Kumar, R., Kumari, S., & Dhawan, S. K. (2019). Recycling of plastic waste into tiles with reduced flammability and improved tensile strength. *Process Safety and Environment Protection, 124*, 299–307.

El-Metwally, Y., Dewidar, K., Ismail, M., & El-Mahallawi, I. (2023). Optimization of plastic waste integration in cement bricks. *Journal of Engineering and Applied Science, 70*, 55.

Emin, Ç. M., & Kar, F. (2018). Characterization of composite produced from waste PET and marble dust. *Construction and Building Materials, 163*, 734–741.

Ghernouti, Y., Rabehi, B., Bouziani, T., Ghezraoui, H., & Makhloufi, A. (2015). Fresh and hardened properties of self-compacting concrete containing plastic bag waste fibers (WFSCC). *Construction and Building Materials, 82*, 89–100.

Gupta, P. (2021). Plastic waste management, a concern for community. *The Holistic Approach to Environment, 11*(2), 49–66.

Hameed, A. M., & Ahmed, B.-F. (2019). Employment of plastic waste to produce lightweight concrete. *Energy Procedia, 57*, 30–38.

Hameed, A. M., & Hamza, M. T. (2019). Characteristics of polymer concrete produced from wasted construction materials. *Energy Procedia, 157*, 43–50.

Hazha, M., Sadique, M., Shaw, A., & Bras, A. (2020). The influence of incorporating plastic within concrete and the potential use of microwave curing, A review. *Journal of Building Engineering, 32*, 101824. https://doi.org/10.1016/j.jobe.2020.101824.

Heriawan, A. (2020). Upcycling plastic waste for rural road construction in India an alternative solution to technical challenges. In *Asian Development Bank South Asia Working Paper Series*. https://doi.org/10.22617/WPS200097-2.

Horta, J. F., Simões, F. J., & Mateus, A. (2017). Study of wood-plastic composites with reused high density polyethylene and wood sawdust. *Procedia Manufacturing, 12*, 221–229.

Hossain, F. M. Z., Shahjalal, M., Islam, K., Tiznobaik, M., & Shahria, A. M. (2019). Mechanical properties of recycled aggregate concrete containing crumb rubber and polypropylene fiber. *Construction and Building Materials, 225*, 983–996.

Jnr, A.-L., Yunana, D., Kamsouloum, P., Webster, M., Wilson, D. C., & Cheeseman, C. (2018). Recycling waste plastics in developing countries: Use of low-density polyethylene water sachets to form plastic bonded sand blocks. *Waste Management, 80*, 112–118.

Kabeyi, M. J. B., & Olanrewaju, O. A. (2023). Review and design overview of plastic waste-to-pyrolysis oil conversion with implications on the energy transition. *Journal of Energy* Article ID 1821129. https://doi.org/10.1155/2023/1821129.

Khan, I. M., Kabir, S., Alhussain, M. A., & Almansoor, F. F. (2016). Asphalt design using recycled plastic and crumb-rubber waste for sustainable pavement construction. *Procedia Engineering, 145*, 1557–1564.

Lumina, P., Pallavi, G. A., Bhavya, C. H., Vanishree, S., & Supriya, C. B. (2022). Use of waste plastic as fine aggregate substitute in concrete. *IOP Conference Series: Materials Science and Engineering, 1255*, 012010.

Manjunath, B. T. A. (2016). Partial replacement of E-plastic waste as coarse-aggregate in concrete. *Procedia Environmental Sciences, 35*, 731–739. https://doi.org/10.1016/j.proenv.2016.07.079.

Martínez-Barrera, G., Martínez-López, M., del Coz-Díaz, J. J., López-Gayarre, F., & Varela-Guerrero, V. (2019). Waste polymers and gamma radiation on the mechanical improvement of polymer mortars: Experimental and calculated results. *Case Studies in Construction Materials, 11*, e00273.

Murray, G. R. (2009). Environmental implications of plastic debris in marine settings entanglement, ingestion, smothering, hangers-on, hitch-hiking and alien invasions. *Philosophical Transactions of the Royal Society, 364*, 2013–2025.

Mutha, N. H., Patel, M., & Premnath, V. (2006). Plastics materials flow analysis for India. *Resources, Conservation and Recycling, 47*(3), 222–244.

Nabajyoti, S., & de Brito, J. (2012). Mechanical properties and abrasion behaviour of concrete containing shredded PET waste bottle aggregate as a partial substitution of natural aggregate. *Construction and Building Materials, 34*(2012), 385–401.

Nazmi, W. M., Rahman, W. A., & Wahab, A. F. A. (2013). Green pavement using recycled Polyethylene Terephthalate (PET) as partial fine aggregate replacement in modified asphalt. *Procedia Engineering, 53*, 124–128.

NPWMTF. (1997). National Plastics Waste Management Task Force (NPWMTF) 16 Report. Ministry of Environment and Forest, Government of India, New Delhi.

Paihtea, L. P., Lalngaihawmaa, A. C., & Saini, G. (2019). Recycled aggregate filled waste plastic bottles as a replacement of bricks. *Materials Today: Proceedings, 15*, 663–668.

Pandey, K. (2023). Down to Earth. Retrieved March 21, 2024, from https://www.downtoearth.org.in/news/waste/india-among-the-12-countries-responsible-for-52-of-the-world-s-mismanaged-plastic-waste-report-90927.

Pešić, N., Živanović, S., Garcia, R., & Papastergiou, P. (2016). Mechanical properties of concrete reinforced with recycled HDPE plastic fibres. *Construction and Building Materials, 115*, 362–370.

Rubio-de Hita, P., Pérez-Gálvez, F., Morales-Conde, M. J., & Pedreño-Rojas, M. A. (2018). Reuse of plastic waste of mixed polypropylene as aggregate in mortars for the manufacture of pieces for restoring jack arch floors with timber beams. *Journal of Cleaner Production, 198*, 1515–1525.

Saman, R. R., Iman, N. M., Allahyari, H., & Samira, H. T. (2016). Sustainable approach for recycling waste tire rubber and polyethylene Terephthalate (PET) to produce green concrete with resistance against sulfuric acid attack. *Journal of Cleaner Production, 126*, 166–177.

SGCCI. (2000). The Indian Plastic Industry. PlastIndiaFoundation, Southern Gujarat Chamber of Commerce and Industry (SGCCI), India.

Siddique, R., Khatib, J., & Kaur, I. (2008). Use of recycled plastic in concrete: A review. *Waste Management, 28*(10), 1835–1852.

Singh, S. G., & Biswas, A. (2022). The plastic life-cycle. Centre for Science and Environment, New Delhi.

Sofi, A. (2018). Effect of waste tyre rubber on mechanical and durability properties of concrete—A review. *Ain Shams Engineering Journal, 9*(4), 2691–2700.

TERI. (2018). Fact Sheet on plastic waste in India. Retrieved March 21, 2024, from https://www.teriin.org/sites/default/files/files/factsheet.pdf.

Zulkernain, N. H., Gani, P., Chuan, N. C., & Uvarajan, T. (2021). Utilisation of plastic waste as aggregate in construction materials: A review. *Construction and Building Materials, 296*(16), 123669. https://doi.org/10.1016/j.conbuildmat.2021.123669.

Chapter 14
Heterogeneous Microbial Biofilms: A Promising Solution for Combating Terrestrial Microplastic Pollution

Sharmila Jayasena, Madushika Perera, Dilrukshi Wijayarathna, Gamini Seneviratne, Sulochana Wijesundera, and S. D. M. Chinthaka

Abstract Environmental pollution by primary and secondary microplastics is currently recognized as a major threat to planetary health. Transferring among organisms through food webs, they accumulate in humans and other organisms resulting in detrimental effects across the biosphere. Bacteria and fungi are known to degrade petroleum hydrocarbons. Heterogeneous communities of microbial biofilms benefit from an extended genetic repertoire and consequent metabolic and survival capacity. Our previous work demonstrated high efficiency degradation of hexadecane and crude oil by naturally occurring fungal-bacterial biofilms. These studies provide a platform for investigating the use of biofilms for combating microplastic pollution in terrestrial environments. Microorganisms were isolated from a municipal landfill in Sri Lanka. Fungal–bacterial communities that appeared during screening formed a biofilm during static culture in hexadecane and crude oil at 1% as sole carbon source. Biofilm formation was confirmed by scanning electron microscopy. Degradation of alkanes in cultures was quantitatively estimated using GC-MS. In silico investigation of putative *Aspergillus flavus* alkane monooxygenases was carried out using

S. Jayasena (✉) · M. Perera · D. Wijayarathna · S. Wijesundera
University of Colombo, Colombo, Sri Lanka
e-mail: sharmila@bmb.cmb.ac.lk

M. Perera
e-mail: madushika@bmb.cmb.ac.lk

D. Wijayarathna
e-mail: dilruksh@chem.cmb.ac.lk

S. Wijesundera
e-mail: sulochana@bmb.cmb.ac.lk

G. Seneviratne
National Institute of Fundamental Studies, Kandy, Sri Lanka
e-mail: gamini.se@nifs.ac.lk

S. D. M. Chinthaka
University of Sri Jayawardenepura, Nugegoda, Sri Lanka
e-mail: sdmchin@sjp.ac.lk

S. Suriyanarayanan et al. (eds.), *Combating Plastic Pollution in Terrestrial Environment*,
https://doi.org/10.1007/978-981-96-2343-3_14

the NCBI protein database and computational tools such as PSI-BLAST, SWISS MODEL and Autodock Vina. A biofilm-forming consortium comprising *Bacillus cereus* and *A. flavus* demonstrated highly efficient synergistic degradation of the alkanes in hexadecane and crude oil in liquid culture (~99% in 7 days). Several alkane monooxygenases were identified in *A. flavus* as distal homologs of bacterial long chain alkane monooxygenases LadA and AlmA and the short to medium alkane monooxygenase Cyp52. Our results suggest that heterogeneous microbial biofilms are promising candidates for the application of 'green solutions' to combating terrestrial microplastic pollution. Future directions for consideration are improvement of the bioremediation potential of (polluted) soils through inclusion of plastic degraders in the soil microbiota. Development of stable bioformulations of biofilm forming microbes is the next step. Exploitation of the identified extracellular fungal alkane monooxygenases as biocatalysts in bioremediation formulations needs to be additionally considered.

Keywords Biofilm · Heterogeneous · Microplastic · Bioremediation · Alkane degradation

14.1 Introduction

Plastics are a global pollutant that have now become a threat to planetary health. Their low production cost, durability and convenience have led to their extensive use and disposal in the domestic sector. In particular, single use plastics such as grocery bags, food containers, cutlery, water bottles and straws constitute a significant proportion of urban waste. Even though some plastics may be recyclable, most of the plastic waste (~95%) is not recycled (Zhou et al., 2022a) due to a variety of reasons including inadequate infrastructure for waste management, as well as cost of collection, sorting and separation and cleaning (Vogt et al., 2021). As a result this waste becomes accumulated in garbage dumps or enters waterways and oceans.

Plastics are manufactured from petrochemicals originating from crude oil. Their chemically inert hydrocarbon structure makes them recalcitrant compounds, taking 100–1000s of years to degrade through natural weathering processes (Zhou et al., 2022a). Through these physical, chemical and biological processes, plastics disintegrate into smaller fragments. Although macroplastic contamination is easily visible, a large proportion of contamination comprises microplastics which are defined as fragments of plastic <5 mm in diameter. Plastic fragments <1 μm in diameter are further defined as nanoplastics. As such, beyond the easily visible problem of plastic pollution, lies an even larger, 'hidden' threat to living organisms (Wang et al., 2022), in the barely visible-invisible category. Microplastic contamination is thus becoming an emerging global concern.

Microplastics found in the environment may be of primary or secondary origin. Primary microplastics are those which are synthesized directly for use and subsequently discarded (eg: beads found in certain personal care products). Microplastic beads also known as nurdles, which form the seed plastic for manufacturing industries may also pollute the environment through accidental spills. Secondary microplastics are those which are derived through degradation of larger plastics through combined mechanical breakage and natural weathering processes (Akdogan & Guven, 2019). Microplastic fibres from synthetic textiles are released into the environment during washing as well as recurrent abrasion during wear (Napper & Thompson, 2016). While more dense plastics entering the terrestrial environment may become buried and thereby remain in the soil, lighter plastics and microplastics may be carried across land by the wind and eventually end up in water bodies.

Microplastics in the environment may be inhaled or ingested directly by humans and animals (Lwanga et al., 2016; Wu et al., 2022). Living organisms have been reported to accumulate microplastics in various cells and tissues leading to a potential health hazard. These microplastics may be transferred among organisms through food webs and become accumulated in humans, resulting in detrimental effects on health. Inadvertent ingestion may also occur through the use of plastic food and drink packaging.

14.2 Use of Microorganisms for Bioremediation

Microorganisms, both bacteria and fungi are known to colonize and degrade petroleum hydrocarbons. They have been investigated extensively as a potential bioremediation strategy for polluted environments. An advantage of microbial bioremediation is its potential as an 'in situ' strategy with either or combined use of bioaugmentation and biostimulation.

Microbial biodegradation of microplastics in soil has been investigated through microcosm studies, particularly in the past few years (Sarker et al., 2021; Sun et al., 2022). However, it is noted that individual organisms are generally limited in their ability to utilize the entire spectrum of alkanes and additional components present in petroleum products (Jayasena & Perera, 2021). Mixed communities of microbes, on the other hand, benefit from an extended genetic repertoire and consequent metabolic capacity. Therefore, the development of heterogeneous microbial communities for degradation of hydrocarbons has been a popular area of investigation (Malik and Ahmed, 2012; Hamzah et al., 2013; Poddar et al., 2019).

However, development of microbial consortia for bioremediation can be challenging due to possible antagonistic behavior among the participating organisms which might in turn affect the rate of biodegradation. This is clearly demonstrated in the experiments of Patowary et al. (2016) who described the development and investigation of 14 different consortia from five selected bacterial isolates obtained from three different hydrocarbon-contaminated sites. The organisms included both biosurfactant producers as well as non-producers. While all five isolates used in the study were selected based on efficient growth on hydrocarbons, biodegradation of total petroleum hydrocarbon varied among the consortia, while just one combination demonstrated relatively higher rates of degradation (~68%) after three weeks, compared to the others which varied between ~40 and 51%. These experimental results reflect the impact of the interaction between the microorganisms in the community. More recently however, it has been shown that hydrocarbon-degrading bacteria may benefit from the use of non-degrading fungi in a mixed community (Chen et al., 2024). In this study, the fungus was shown to improve accessibility to the alkanes by adsorbing it onto the mycelium.

Heterogeneous communities of microbes that form naturally in the environment, particularly those that form biofilms may provide a promising strategy for terrestrial bioremediation. The formation of multispecies biofilms is a widespread phenomenon in natural environments where organisms co-exist and partake in mutually beneficial interactions. Such indigenous communities are likely to have developed optimal adaptation behaviors for survival, especially under harsh conditions. Microbial populations form biofilms on both biotic and abiotic solid surfaces as well as at oil–water interfaces.

Biofilms are three dimensional, dynamic microbial communities comprising single or multiple species attached to a biotic or abiotic surface and embedded in a matrix of extracellular polymeric substances (EPS). The EPS often comprises polysaccharides, proteins, glycoproteins, glycolipids and sometimes extracellular DNA as well. Biofilm formation has been shown to involve the expression of distinct genes indicating an adaptation to a particular mode of life (Mounier et al., 2014). The biofilm structure enables the intercellular exchange of metabolites, signal molecules, as well as genetic material among the organisms in the biofilm. The organisms embedded in the polymeric matrix are further protected from predators and antimicrobial agents, making them upto thousand times more resistant to antibiotics than planktonic bacteria.

Due to the ability of microorganisms to form biofilms on particulate substrates, biofilm formation has been considered as a strategy for combatting pollution (Sivadon et al., 2019). Thus, exploiting such naturally occurring consortia is likely to provide more promising results over synthetically developed consortia.

Microplastics found in the environment comprise polyethylene (PE), polypropylene (PP), polystyrene (PS), polyethylene terephthalate (PET), or polyvinyl chloride (PVC), or mixtures of these along with various additives such as stabilizers, colorants, plasticizer and flame retardants that are often included into products. Their chemical properties, particularly their chemical inertness have impacted the environment, affecting the health of living organisms across the biosphere. The slow rate of degradation is the main challenge that currently limits microbial bioremediation technology.

The chemical structure of polyethylene is among the simplest among the synthetic plastics (Fig. 14.1). They are non-polar, high molecular weight saturated hydrocarbons composed of polymers of two carbon units (ethylene) having the chemical structure $(C_2H_4)n$. Their hydrophobicity and lack of functional groups in its chemical structure make them highly inert and therefore recalcitrant in the environment.

Degradation of alkanes requires the initial activation of a terminal or subterminal carbon through oxidation. It is known that both biotic as well as abiotic factors such as enzymes and UV irradiation have the ability to initiate the degradation of alkanes

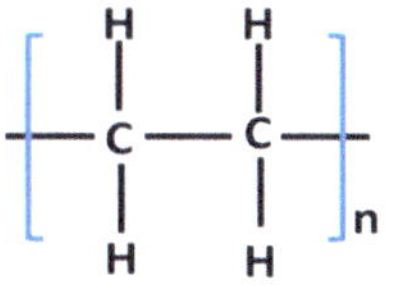

Fig. 14.1 Basic chemical structure of polyethylene

such as those found in polyethylene. However, the polyethylene fragments need to be reduced to ~10–50 carbons for enzyme action (Restrepo-Flórez et al., 2014). Thus, the photo-oxidative fragmentation of plastics is an important initial step in the degradation process. The UV-mediated photo-oxidation also leads to the increase of carbonyl residues which have been shown to be utilized by *Rhodococcus ruber* during degradation of polyethylene (Orr et al., 2004). The biodeterioration and biodegradation of polyethylene and methods of investigation have been comprehensively reviewed by Restrepo-Flórez et al. (2014).

14.3 Use of Heterogeneous Microbial Biofilms for Degradation of Liquid Alkanes

The need for high efficiency biodegradation of plastics continues to drive the exploration of novel microbial species and strains as well as the investigation of genetically modified microorganisms and mixed microbial communities with a high capacity for degradation of hydrocarbons. The use of genetically modified organisms is aimed at overcoming the challenges of acclimatization of microorganisms in new environments, which may sometimes occur (Rebello et al., 2021).

Our laboratory investigations using liquid petroleum products, namely hexadecane and crude oil, demonstrated the ability of three naturally occurring fungal–bacterial communities that formed biofilms (*Aspergillus—Bacillus* Consortia, designated ABC1, 2 and 3) to metabolize alkanes with remarkable efficiency, when provided as the sole source of carbon in minimal media (Perera et al., 2019, 2021).

Microorganisms capable of degrading hydrocarbons were isolated from samples of waste polyethylene with attached soil that were collected from a municipal landfill in Colombo district in Sri Lanka. Organisms capable of utilizing hydrocarbons as a sole carbon source were purified by enrichment culture (1% hexadecane in Bushnell and Haas minimal medium) at 40 °C followed by plating of serially diluted cultures to obtain isolated colonies.

This initial screening process led to the isolation of three naturally occurring fungal–bacterial communities that appeared during culture and they were further investigated for hydrocarbon degradation potential. All three communities were observed to form a biofilm at the oil–water interface, when grown as static cultures. The individual organisms in the communities were identified through phenotype, biochemical characteristics as well as molecular level analysis (16S ribosomal gene in bacteria or internal transcribed spacer (TS) region in fungi). The bacteria in the three

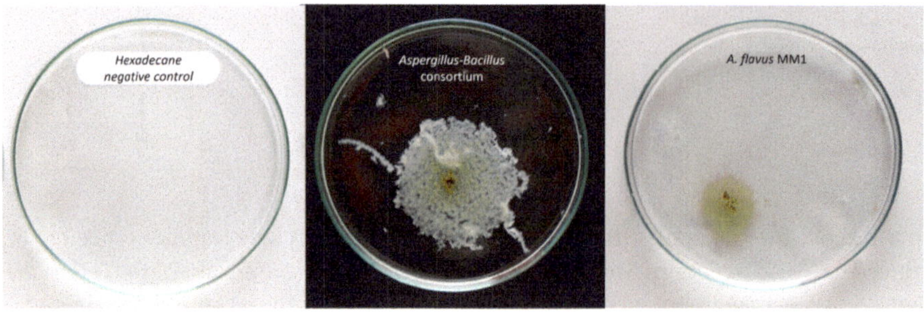

Fig. 14.2 Growth of ABC1 and *A. flavus* MM1 monoculture in 1% *n*-hexadecane in minimal medium (liquid medium) after 14 days (*Note* cultures were grown in conical flasks and poured into petri dishes for better visualization after the incubation period)

communities were *Bacillus* spp and *Lysinibacillus* spp while the fungi comprised *Aspergillus* spp. The two organisms in ABC1 (*Aspergillus-bacillus* consortium 1) were further characterized as *Bacillus cereus* and *Aspergillus flavus* MM1. ABC1 consortium was selected for further investigations based on its efficient degradation of crude oil and the simplicity of the consortium (Fig. 14.2).

The fungus in ABC1 consortium, *A. flavus* MM1 was observed to form the mycelial mat at the oil–water interface. The aqueous phase which became turbid upon initial growth of the bacilli was then observed to clear, concomitant with the development of the fungal mat. Alongside this, a reduction of the layer of hexadecane or crude oil was observed, indicating its likely utilization by the biofilm. Filamentous fungi, such as the isolated *A. flavus* are known to produce hydrophobic proteins facilitating the attachment to hydrophobic surfaces (Kershaw & Tablbot, 1998). This, together with the fact that the bacillus counterpart in the biofilm was an active, motile bacteria, provides explanation to the observations. The bacteria would be capable of overcoming the nutrient limitation in the aqueous minimal medium, by traveling toward the fungus and attaching onto the mycelia, forming a biofilm. Thus, obtaining access to the carbon source.

Biofilm formation in the ABC1 community was confirmed through scanning electron microscopy (Fig. 14.3) in which bacteria were observed on the fungal hyphae embedded in the EPS (Perera et al., 2019). In fungal-bacterial biofilms, the fungal hyphae serve as a physical and metabolic habitat for the interacting bacteria, allowing the bacterial cells to colonize the surface of the hyphae, obtain nutrients that may be directly or indirectly secreted by them, and 'travel' along with the growing hyphae (Guennoc et al., 2018).

Utilization of hexadecane and the alkanes in crude oil by the biofilm as well as the individual organisms when grown in monoculture was quantitatively estimated in ABC1 using gas chromatography-mass spectrometry (GC-MS). Residual alkanes in 7- or 14-day cultures grown at 30 °C without shaking, were extracted into hexane or dichloromethane, quantified by GC-MS and percentage utilization was calculated. Uninoculated culture media was used as control to estimate abiotic loss of alkanes (e.g. through vaporization).

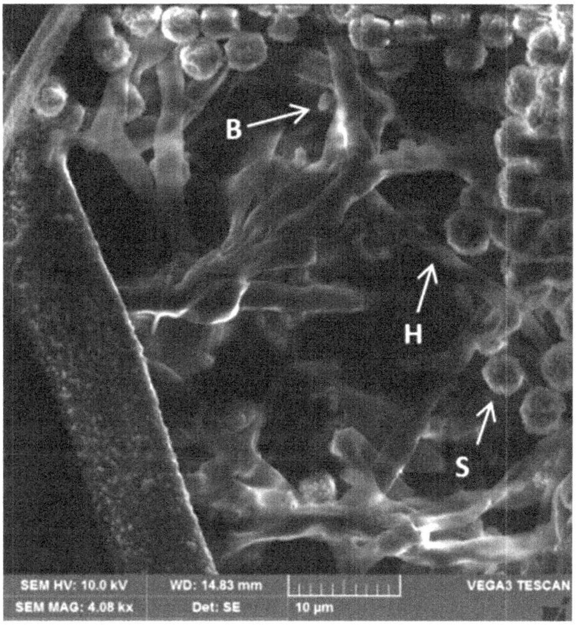

Fig. 14.3 SEM micrograph of *Aspergillus—Bacillus* biofilm. Magnification 4080x. B; *Bacteria*, H; fungal hyphae, S; fungal spores

The ABC1 *Aspergillus-bacillus* biofilm demonstrated highly efficient, synergistic degradation of hexadecane as well as a spectrum of alkanes in crude oil, when provided as the sole source of carbon. Biodegradation of hexadecane by the ABC1 biofilm and the individual organisms in monoculture were estimated in 14-day static cultures by GC-MS analysis. Briefly, residual hexadecane in the culture medium at day 14 was estimated by analysis of the hexane extract and the percentage utilization was calculated.

The ABC1 biofilm demonstrated ~99% utilization of hexadecane (1% in 20 ml at 40 °C) in 14 days. When grown in monoculture however, the fungus utilized ~52% and the bacterium utilized only ~9% hexadecane in 14 days (Perera et al., 2019). Therefore, it is clear that in the biofilm mode, the organisms demonstrate synergistic behavior, where together, the degradation was greater than the sum of their individual degradation.

The ABC1 biofilm, when cultured in 1% crude petroleum oil (8.25 g/L)(CEYPETCO oil refinery, Sri Lanka) in 20 ml minimal medium degraded ~99% of the crude oil within 7 days. The crude oil sample obtained was found to have a range of alkanes from ~C9 to C30 or more (Perera et al., 2021). The fungus demonstrated efficient degradation of the alkanes in crude oil when grown in monoculture as well, reaching degradation levels similar to that observed when in biofilm mode, when estimated at day 7, while the degradation of crude oil by the bacterium was very poor. This begged to question the role of the bacterial counterpart in the degradation of crude oil. In order to obtain answers to this question, the degradation of crude oil was estimated at two additional time points prior to day 7.

Table 14.1 Degradation of straight chain alkanes (C12–C30) in crude oil by ABC community organisms

Organisms	Degradation of alkanes in crude oil		
	Day 3	Day 5	Day 7
ABC1 biofilm	24 ± 1.4%	66 ± 7%	99 ± 0.2%
A. flavus MM1	4.1 ± 0.4%	14 ± 6%	98 ± 2%
B. cereus MM1	–	–	20 ± 4%

These experiments demonstrated that utilization of crude oil by the biofilm was ~ 66% in 5 days compared to 14% by the fungus in monoculture. This demonstrated that although *A. flavus* could degrade the crude oil in monoculture to ~98% by day 7, there was a delay in commencement of the degradation, which was overcome by the presence of the bacillus in biofilm mode (Perera et al., 2021). This observation has implications, particularly if bioremediation is used to mitigate a sudden high exposure (e.g.: due to an accidental oil spill or seepage), where early removal of the contaminant is vital to minimize ecological damage (Table 14.1).

Percentage degradation was calculated by GC-MS analysis of the residual alkanes at the indicated time of incubation (Results of triplicate experiments).

The observation of this rapid degradation of alkanes in the isolated fungus suggested that it processed a repertoire of alkane degrading enzymes, in particular enzymes capable of oxidizing large alkanes (~C30) which are rare in nature.

14.4 Identification of Alkane Oxidizing Enzymes in a Filamentous Fungus

Several aerobic enzyme systems that carry out alkane oxidation have been identified in microorganisms. The methane monooxygenases (Wang et al., 2017) and the related butane monooxygenases oxidize short chain alkanes. The AlkB family of alkane hydroxylases which are found in bacteria, as well as the cytochrome P450 system which is reported in both bacteria and fungi, are known to oxidize medium chain alkanes (~C7–C17). Long chain alkane degrading enzymes, LadA and AlmA systems had been previously reported only in some extremophilic bacteria, such as the thermophile *Geobacillus* and the halophile *Amycolicicoccus* (Feng et al., 2007; Boonmak et al., 2014, Bowman & Deming, 2014). However, investigations in our laboratory and previous research have shown that filamentous fungi are able to degrade long-chain alkanes.

Our research therefore aimed at seeking out the fungal long chain alkane monooxygenases. An in silico investigation of putative fungal enzymes responsible for alkane oxidation was carried out using the *A. flavus* protein sequences in the NCBI non-redundant protein database using a variety of NCBI search tools such as PSI-BLAST (homolog searches) (Altschul et al., 1997), SWISS MODEL (Waterhouse et al., 2018), I-TASSER (3D modeling) (Zhou et al., 2022b), ORCA (geometric optimization) (Neese, 2012) and Autodock Vina (enzyme-ligand docking).

The publicly available databases at the National Center for Biotechnology Information (NCBI) were screened. Probing of the NCBI protein data bank using Position-Specific Iterative Basic Alignment Search Tool (PSI-BLAST) to search distant homologs, with the query sequences led to the identification of homologs of the flavin-binding long chain alkane monooxygenases LadA and AlmA in *A. flavus* NRL3357.

The LadA homologs formed a family of five enzymes designated Af1-5. Their 3D structures were modeled and all except Af2 were found to accommodate the coenzyme flavin mononucleotide (FMN) in the active site. All four were shown to bind alkanes upto C30 in the active site (Perera et al., 2022) and further confirmation is achieved through the conservation of the active pocket residues with the *Geobacillus thermodenitrificans* LadA (3B9O_A) (Li et al., 2008). This data represents the first demonstration of the presence of distant homologs of the bacterial LadA monooxygenases in a eukaryotic organism.

A similar experimental procedure led to the discovery of AlmA homologs in *A. flavus*. Similar to the LadA family, the AlmA also comprised a multi-enzyme family and in silico analysis of one homolog showed putative binding of alkanes upto C40 (published in abstract form). Our investigations further led to the isolation of the medium chain alkane-oxidizing monooxygenase of the cytochrome P450 family, Cyp52 in *A. flavus* MM1 and alkane binding to the active site was demonstrated in silico (unpublished data).

Together these results clearly show that the fungus *A. flavus* MM1 in the isolated biofilm-forming heterogenous consortium, produces the enzymes required for initiating the oxidation of alkanes. Thus, they have the potential to be exploited for biodegradation of polyethylene.

14.5 Discussion and Future Directions

Our investigations in *Aspergillus flavus* demonstrate that these filamentous fungi have the potential to carry highly active long chain alkane degrading monooxygenases. Unlike bacteria where generally only one enzyme is encoded, the fungal genomes appear to encode multi-enzyme families. Together these enzymes provide efficient alkane-degrading capacity to the organism.

Microbial biofilms are highly stable, dynamic entities often with enhanced characteristics obtained through combined characteristics of the resident organisms as well as horizontal gene transfer. Our results further demonstrated that naturally occurring biofilm forming bacterial-fungal communities with a capacity for highly efficient degradation of liquid petroleum hydrocarbons are present in contaminated soils. It is highly likely that this capacity also extends to degradation of microplastics.

A previous study on biofilm formation by *Marinobacter* spp. demonstrated that it formed a biofilm on the oil–water surface only if it was able to metabolize the oil (Mounier et al., 2014; Klein et al., 2008). Microplastics provide a solid surface for biofilm formation. Indeed, *Rhodococcus ruber* C208 has been shown to form a biofilm on a polyethylene film while utilizing the polyolefin as a sole carbon source (Orr et al., 2004).

Our results (Perera et al., 2019) clearly demonstrated the importance of biofilm formation for rapid degradation of the hydrocarbons. The investigations of oil degradation by *Marinobactor* spp. also demonstrated that when biofilm formation was perturbed by agitation, hexadecane degradation was decreased (Klein et al., 2008; Mounier et al., 2014).

These investigations provide a platform for the investigation of this and similar naturally occurring fungal–bacterial biofilms in combating microplastic pollution in terrestrial environments. Microbial biofilms have the advantage of providing a 'green' solution for purifying contaminated terrestrial environments. It is envisaged that natural microbial biofilms may potentially offer a high efficiency biological system for microplastic degradation. Bioaugmentation of the soil microbiome with such organisms will improve the bioremediation potential of the soil and may provide a path to minimize the accumulation of microplastics in the soil. Exploitation of the identified extracellular fungal alkane monooxygenases for use in bioremediation formulations needs to be additionally considered. This includes the development and optimization of a stable bioformulation for improvement of the bioremediation potential of polluted soils.

Future directions for research and development:

- Investigation of the capacity for microplastic degradation by previously isolated consortia as well as exploration for novel biofilm forming heterogeneous microbial consortia.
- Preparation of a stable bioformulation comprising live organisms for bioaugmentation purposes.
- Investigation of extracellular monooxygenases for potential use as biocatalysts in bioformulations.

Acknowledgements This work was funded by the research grant AP/3/2/2014/RG/12, University of Colombo.

References

Akdogan, Z., & Guven, B. (2019). Microplastics in the environment: A critical review of current understanding and identification of future research needs. *Environmental Pollution, 254*, 113011. https://doi.org/10.1016/j.envpol.2019.113011.

Altschul, S. F., et al. (1997). Gapped BLAST and PSI-BLAST: A new generation of protein database search programs. *Nucleic Acids Research 25*(17), 3389–3402. https://doi.org/10.1093/nar/25.17.3389

Boonmak, C., Takahashi, Y., & Morikawa, M. (2014). Cloning and expression of three ladA-type alkane monooxygenase genes from an extremely thermophilic alkane-degrading bacterium *Geobacillus thermoleovorans* B23. *Extremophiles, 18*(3), 515–523. https://doi.org/10.1007/s00792-014-0636-y.

Bowman, J. S., & Deming, J. W. (2014). Alkane hydroxylase genes in psychrophile genomes and the potential for cold active catalysis. *BMC Genomics, 15*(1), 1120. https://doi.org/10.1186/1471-2164-15-1120.

Chen, A., Wang, C., Cheng, Z., Kennes, C., Qiu, S., & Chen, J. (2024). Enhancing bacterial biodegradation of n-hexane by utilizing the adsorption capacity of non-degrading fungi. *Chemosphere, 363*, 142900. https://doi.org/10.1016/j.chemosphere.142900. Epub 2024 Jul 17. PMID: 39029712.

Feng, L., et al. (2007). Genome and proteome of long-chain alkane degrading *Geobacillus thermodenitrificans* NG80-2 isolated from a deep-subsurface oil reservoir. *Proceedings of the National Academy of Sciences of the United States of America, 104*(13), 5602–5607. https://doi.org/10.1073/pnas.0609650104.

Guennoc, C. M., Rose, C., Labbe, J., & Deveau, A. (2018). Bacterial biofilm formation on the hyphae of ectomycorrhizal fungi: a widespread ability under controls. *FEMS Microbiology Ecology, 94*, fiy093. https://doi.org/10.1093/femsec/y093.

Hamzah, A., Phan, C. W., Abu Bakar, N. F., & Wong, K. K. (2013). Biodegradation of crude oil by constructed bacterial consortia and the constituent single bacteria isolated from Malaysia. *Bioremediation Journal, 17*(1), 1–10. https://doi.org/10.1080/10889868.2012.731447.

Jayasena, S., & Perera, M. (2021). Microbial bioremediation of petroleum hydrocarbons. In D. G. Panpatte & Y. K. Jhala (Eds.), *Microbial Rejuvenation of Polluted Environment* (Vol. 2, Chap. 11, pp. 263–291). Springer.https://doi.org/10.1007/978-981-15-7447-4_11.

Kershaw, M. J., & Talbot, N. J. (1998). Hydrophobins and repellents: Proteins with fundamental roles in fungal morphogenesis. *Fungal Genetics and Biology, 23*(1), 18–33. https://doi.org/10.1006/fgbi.1997.1022.

Klein, B., Grossi, V., Bouriat, P., Goulas, P., & Grimaud, R. (2008). Cytoplasmic wax ester accumulation during biofilm-driven substrate assimilation at the alkane–water interface by Marinobacter hydrocarbonoclasticus SP17. *Research in Microbiology, 159*(2), 137–144. https://doi.org/10.1016/j.resmic.2007.11.013.

Li, Y. Q., et al. (2008). Diesel pollution biodegradation: Synergetic effect of Mycobacterium and filamentous fungi. *Biomedical and Environmental Sciences, 21*(3), 181–187. https://doi.org/10.1016/S0895-3988(08)60026-4.

Lwanga, E. H., Gertsen, H., Gooren, H., et al. (2016). Microplastics in the terrestrial ecosystem: implications for *Lumbricus terrestris* (Oligochaeta, Lumbricidae). *Environmental Science and Technology, 50*, 2685e2691. https://doi.org/10.1021/acs.est.5b05478.

Malik, Z., & Ahmed, S. (2012). Degradation of petroleum hydrocarbons by oil field isolated bacterial consortium. *African Journal of Biotechnology, 11*(3), 650–658. https://doi.org/10.5897/AJB11.036.

Mounier, J., Camus, A., Mitteau, I., et al. (2014). The marine bacterium *Marinobacter hydrocarbonoclasticus* SP17 degrades a wide range of lipids and hydrocarbons through the formation of oleolytic biofilms with distinct gene expression profiles. *FEMS Microbiology Ecology, 90*(3), 816–831. https://doi.org/10.1111/1574-6941.12439.

Napper, I. E., & Thompson, R. C. (2016). Release of synthetic microplastic plastic fibres from domestic washing machines: effects of fabric type and washing conditions. *Marine Pollution Bulletin, 112*, 39e45. https://doi.org/10.1016/j.marpolbul.2016.09.

Neese, F. (2012). The ORCA program system. *Abstract WIREs Computational Molecular Science, 2*(1), 73–78. https://doi.org/10.1002/wcms.v2.110.1002/wcms.81

Orr, I. G., Hadar, Y., & Sivan, A. (2004). Colonization, biofilm formation and biodegradation of polyethylene by a strain of *Rhodococcus ruber*. *Applied Microbiology and Biotechnology, 65*, 97–104. https://doi.org/10.1007/s00253-004-1584-8.

Patowary, K., Patowary, R., Kalita, M. C., & Deka, S. (2016). Development of an efficient bacterial consortium for the potential remediation of hydrocarbons from contaminated sites. *Frontiers in Microbiology, 7*, 1092. https://doi.org/10.3389/fmicb.2016.01092.

Perera, M., Wijayarathna, D., Wijesundera, S., et al. (2019). Biofilm mediated synergistic degradation of hexadecane by a naturally formed community comprising *Aspergillus flavus* complex and *Bacillus cereus* group. *BMC Microbiology, 19*, 84. https://doi.org/10.1186/s12866-019-1460-4.

Perera, M., Chinthaka, S. D. M., Wijesundera, W. S. S., et al. (2021). Reduction of lag in crude oil degradation by *Aspergillus* when it is in synergy with *Bacillus* in biofilm mode. *Bioprocess and Biosystems Engineering, 44*(7), 1501–1510. https://doi.org/10.1007/s00449-021-02534-6.

Perera, M., Wijesundera, S., Wijayarathna, C. D., Seneviratne, G., & Jayasena, S. (2022). Identification of long-chain alkane-degrading (LadA) monooxygenases in *Aspergillus flavus* via *in silico* analysis. *Frontiers in Microbiology, 13*, 898456. https://doi.org/10.3389/fmicb.2022.898456.

Poddar, K., Sarkar, D., & Sarkar, A. (2019). Construction of potential bacterial consortia for efficient hydrocarbon degradation. *International Biodeterioration and Biodegradation, 144*, 104770. https://doi.org/10.1016/j.ibiod.2019.104770.

Rebello, S., Nathan, V. K., Sindhu, R., Binod, P., Awasthi, M. K., & Pandey, A. (2021). Bioengineered microbes for soil health restoration: Present status and future. *Bioengineered, 12*(2), 12839–12853. https://doi.org/10.1080/21655979.2021.2004645.

Restrepo-Flórez, J.-M., Bassi, A., & Thompson, M. R. (2014). Microbial degradation and deterioration of polyethylene—A review. *International Biodeterioration and Biodegradation, 88*, 83–90. https://doi.org/10.1016/j.ibiod.2013.12.014.

Sarker, R. K., Chakraborty, P., Sarkar, S., Ghosh, M. M., & Tribedi, P. (2021). Bioaugmentation of Enterobacter cloacae AKS7 causes an enhanced degradation of low-density polyethylene (LDPE) in soil: A promising approach for the sustainable management of LDPE waste. *Archives of Microbiology, 204*(1), 74. https://doi.org/10.1007/s00203-021-02645-4. PMID: 34951695.

Sivadon, P., Barnier, C., Urios, L., & Grimaud, R. (2019). Biofilm formation as a microbial strategy to assimilate particulate or non-aqueous droplet substrates. *Environmental Microbiology Reports, 11*, 749–764.

Sun, Y., Duan, C., Cao, N., Ding, C., Huang, Y., & Wang, J. (2022). Biodegradable and conventional microplastics exhibit distinct microbiome, functionality, and metabolome changes in soil. *Journal of Hazardous Materials, 424*(Pt A), 127282. https://doi.org/10.1016/j.jhazmat.2021.127282.

Vogt, B. D., Stokes, K. K., & Kumar, S. K. (2021). Why recycling of postconsumer plastics is so challenging. *ACS Applied Polymer Materials., 3*(9), 4325–4346. https://doi.org/10.1021/acsapm.1c00648.

Wang, C., Tang, J., Yu, H., et al. (2022). Microplastic pollution in the soil environment: characteristics, influencing factors, and risks. *Sustainability, 14*(20), 13405. https://doi.org/10.3390/su142013405.

Wang, Y., et al. (2017). Functional characterization of two alkane hydroxylases in a versatile Pseudomonas aeruginosa strain NY3. *Annals of Microbiology, 67*(7), 459–468. https://doi.org/10.1007/s13213-017-1271-5.

Waterhouse, A. et al. (2018). SWISS-MODEL: Homology modelling of protein structures and complexes. *Nucleic Acids Research, 46*(W1), W296–W303. https://doi.org/10.1093/nar/gky427

Wu, P., Lin, S., Cao, G., et al. (2022). Absorption, distribution, metabolism, excretion and toxicity of microplastics in the human body and health implications. *Journal of Hazardous Materials, 437*, 129361. https://doi.org/10.1016/j.jhazmat.2022.129361.

Zhou, Y., Kumar, M., Sarsaiya, S., et al. (2022a). Challenges and opportunities in bioremediation of micro-nano plastics: A review. *Science of the Total Environment., 802*, 149823. https://doi.org/10.1016/j.scitotenv.2021.149823.

Zhou, X., Zheng, W., & Li, Y., et al. (2022b). I-TASSER-MTD: A deep-learning-based platform for multi-domain protein structure and function prediction. *Nature Protocols, 17*, 2326–2353. https://doi.org/10.1038/s41596-022-00728-0

Chapter 15
Preparation and Characterization of Wood Plastic Composite from Plastic Wastes

Thida Aung, Thin Thin Hlaing, and Phyu Phyu Win

Abstract In today's world, the waste plastic materials used by humans are having a serious impact on the environment. We are experiencing air pollution, water pollution and soil damage due to waste plastic materials. Therefore, efforts are being made to reduce the use of waste plastic materials by recycling them and using them repeatedly. Now, waste plastic materials are mixed with wood powder to produce composite materials. Wood Plastic Composites are used as floors and as roof panels and it is also used as furniture. Collecting teak sawdust to use as wood powder and drying and grinding of waste plastics. After measuring the solubility of plastic powder, it was found that waste plastic (R-PE) is 120 °C and (P-PE) is 115 °C. The melting point of MAPP used as a coupling agent is 75 °C. 500 μm was used to select the size of plastic powder and sawdust for making composite board. Boards were tested by mixing 500 μm fine plastic powder and teak sawdust in 1:1, 2:1, 3:1 ratios and using different temperatures (140, 160, 180 °C). At 160 °C, composite boards were obtained, and then two types of Recycled PE and Pure PE were tested. According to the quality measurements of the obtained composite boards, it was found that there was little difference. Composite boards made of Recycled PE and Pure PE were found to have the best ratio of plastic powder and teak sawdust (3:1) in bending strength (psi), impact strength (KJ/m^2) and density (g/cm^3). According to the tests, the use of MAPP (Coupling Agent) based on sawdust was found to be better. We continued testing using 3, 6 and 9% MAPP, based on the weight of sawdust. It was found that the tested composite board with 3% MAPP had better physical properties and mechanical properties.

Keywords Recycled polyethylene (R-PE)) · Pure polyethylene (P-PE) · Sawdust · Wood plastic composite (WPC) · Maleic anhydride-grafted Polypropylene (MAPP)

T. Aung (✉) · T. T. Hlaing · P. P. Win
Polymer Research Department, Department of Research and Innovation, Chemical Technology Research Center, Ministry of Science and Technology, Yangon, Myanmar
e-mail: aungthida915@gmail.com

S. Suriyanarayanan et al. (eds.), *Combating Plastic Pollution in Terrestrial Environment*,
https://doi.org/10.1007/978-981-96-2343-3_15

15.1 Introduction

We are experiencing air pollution, water pollution and soil damage due to waste plastic materials. Therefore, efforts are being made to reduce the use of waste plastic materials by recycling them and using them repeatedly now, waste plastic materials are mixed with wood powder to produce composite materials (Bouaf et al., 2009; Carus et al., 2012; Caulfield et al., 2005). The weather enduring low water absorption composite materials are available. Resistant to sunlight and fungus, low water absorption, long-lasting harder compost materials are available. Wood plastic composites are used as floors, as roof panels and it is also used as seat and table (Hull, 1996; Opportunities for natural fibers in plastic composites, 2000). Wood plastic composite requires very little care as it has a strong resistance against weather conditions. Moreover, due to the wood/plastic mixture, it does not attract bugs. Wood plastic composites (WPCs) are composite materials made of wood fiber (Klyosov, 2007) and thermoplastic(s) such as polyethylene (PE), polypropylene (PP), polyvinyl (RPE), recycled polyethylene polyvinyl chloride (PVC), or polylactic acid (PLA). In addition to wood fiber and plastic, WPCs can also contain other ligno-cellulosic (Felix & Gatenholm, 1991) and/or inorganic filler materials. Wood plastic composite is panel or lumber product made from recycled plastic and small wood particles or fibers. At physical test, due to its water restraints, WPC has low moisture content, water absorption and swelling. The blend of wood powder and hard-setting plastics creates a durable material that can last 25–30 years or more with regular maintenance. Plastic wood is a lot more durable than real wood. Plastic wood does not swell or shrink due to exposure to moisture like real wood does. One advantage of WPC is that WPC flooring is 100% waterproof.

This means that it can be also installed in places of typically in spaces such as commercial bathrooms, commercial kitchens, and areas below ground level. These WPC boards consist of stabilizing coupling agents, foaming agents and such as contain mixing ratio (Lu et al., 2000; Maldas & Kokta, 1994). Due to the perfect blend of high-quality materials, WPC boards become high-quality materials. WPC can be molded into different shapes and sizes. The first and foremost reason why WPC is better is due to the safe ingredients used in making WPC. Unlike plywood that uses harmful synthetic resins like urea formaldehyde and phenol formaldehyde mixed in adhesives, WPC is also a comparatively safer option. WPC is a lot more durable and weather resistant than wood, which needs constant care and maintenance to keep it looking good and protected (Kazayawoko et al., 1999). Termite and fungi resistant WPC is the perfect option for replacing standard plywood, and particle boards in the kitchen. Termite and fungi resistant WPC is the perfect option for replacing standard plywood, and particle boards in the kitchen.

It is perfect flooring for any room in the house. Surface hardness is one of the qualities of WPC Boards. A good surface hardness ensures higher durability for furniture applications and ensures the furniture last longer in spite of rough and tough use. WPC is the most durable and long-lasting material. It is water-resistant, fire-resistant, termite and borer-resistant, fungus-resistant, and resistant to almost all weather conditions (Kuan et al., 2003). WPC also doesn't split when exposed to heat and moisture. WPC floors are stronger than PVC floors panels due to the presence of wood fibers also WPC floor panels are more durable than PVC floor panels, as they are less prone to cracking or breaking. Whereas, PVC floor panels are less strong & durable than WPC floor panels, as they are more prone to cracking and warping. Wood components are generally used in sawdust form or small fibers, of the WPC. Oak, bamboo, peanut hulls and pine are common types of sawdust, and tend to be relatively low cost. Wood plastic composite is used more than any other material; it is a perfect substitute for plywood and water proof ply. It is a low maintenance product, does not change its original shape over time, is moisture resistant, durable and strong, environmentally friendly and fully recyclable (Raj et al., 1989).

15.2 Materials and Methods

15.2.1 Materials

15.2.1.1 Raw Materials that Are Necessary

(a) Maleic anhydride-grafted Polypropylene (MAPP)
(b) Sawdust
(c) Pure Polyethylene (P-PE)
(d) Recycled Polyethylene (R-PE)

15.2.1.2 Mechanical Equipment to Be Used

(a) Hydraulic Hot Press
(b) Refiner
(c) Oven
(d) Mixer
(e) Sieve Shaker
(f) Impact Tester
(g) Hardness Tester
(h) Density meter
(i) Bending Strength Tester

15.2.2 Methods

Production of wood plastic composite from plastic wastes is the step of the research work is as following.

(a) (b)

Fig. 15.1 **a** Pure polyethylene and **b** recycled polyethylene

15.2.2.1 Collection of the Wood Powders

Wood sawdust powder is collected from hkayat saann yain Furniture Factory, industrial zone (1), South Dagon Township. The obtained sawdust powder is put into the oven, placed at a temperature of 60 °C, and dried the moisture content is until it runs out.

15.2.2.2 Drying and Grinding of Polyethylene and Recycled Polyethylene

The research uses pure polyethylene as a new plastic, and recycled polyethylene as a glucose containers used in hospitals as waste plastic. The raw materials are washed and cleaned, cut to a size that can be milled in the refiner, and then dried. Pure polyethylene and a piece of recycle plastics cut and dried grind with a grinder. The milled plastics were dried at room temperature (Adhikary et al., 2008) and moisture was removed in an oven at 60 °C for 2 h. It is shown in Fig. 15.1.

15.2.2.3 Sieving the Raw Materials to Obtain the Required Fiber Size

The dried and ground plastic powders and sawdust powders were filtered and sized using a Sieve Shaker into above and below 500 μm (Migneault et al., 2009). Size-separated plastic powder and sawdust are shown in Fig. 15.2.

15.2.2.4 Measuring and Checking the Quality of Raw Materials

Melting point and Moisture content of the raw materials used were measured. The result is shown in Table 15.1.

(a) (b)

Fig. 15.2 500 μm size separation of **a** plastic powder and **b** sawdust

Table 15.1 The qualities of the teak sawdust and plastic powder

Name	Melting point (°C)	Moisture content	Density (g/cm^3)
Teak sawdust	–	5.6395%	–
PE (Pure)	115	–	0.913
PE (Recycled)	120	–	0.850
MAPP	75	–	–

15.2.2.5 Making Wood Plastic Composite Board

Moisture removed and size-separated below 500 μm teak sawdust and plastic powders were tested in three ratios of (1:1, 2:1, 3:1). In the mixer stir for about 5 min. The mixed raw materials are placed in a cold molding cup. After cold forging, heat for 15 min at 160 °C in hydraulic hot press with a pressure of 2240 psi. If you take it out immediately from machine after applying heat, it may break; the composite board is taken out after the temperature of the mold is about 60 °C (Fig. 15.3).

15.2.2.6 Making Wood Plastic Composite Board by Using Maleic Anhydride-Grafted Polypropylene (MAPP)

We tested using MAPP to improve the quality of wood plastic composite board. MAPP is used as a coupling agent for making wood plastic composite board and provides to improve mechanical strength. Melting point of MAPP is 75 °C, there were two types of testing: using it as a solid and grinding as powder. MAPP was tested as a proportion test on the total weight of teak sawdust and plastic powder and as a proportion test on the weight of teak sawdust powder. Then, taking the ratio based on the weight of better teak sawdust, 3, 6 and 9% were tested.

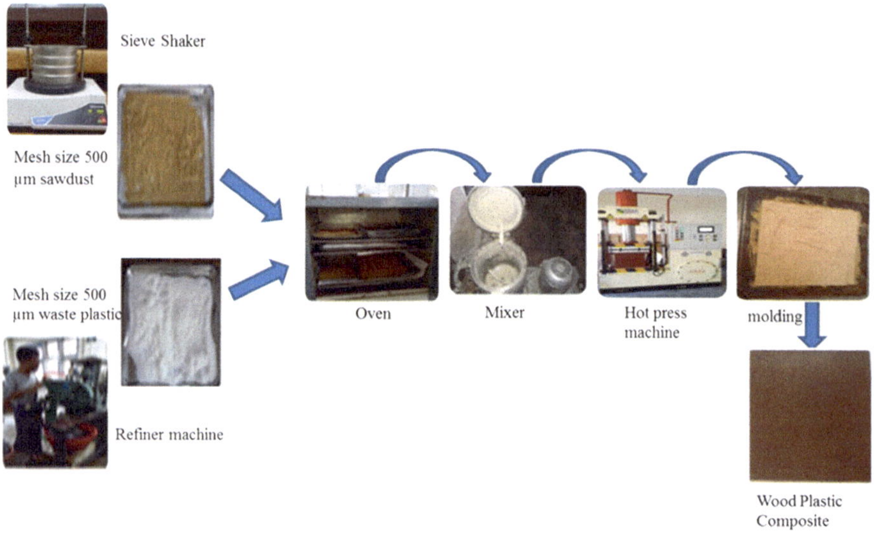

Fig. 15.3 Step by step of making composite board

15.2.2.7 Measuring the Quality of Resulting Composite Boards

Physical properties and mechanical properties were measured in order to compare the quality of the obtained Composite Board sheets (Jayaraman & Bhattacharyya, 2004). Physical properties were measured the water solubility and oil solubility of the composite board. Mechanical properties were measured bending strength, impact strength, hardness, thickness and density.

15.3 Results and Discussion

After measuring the solubility of plastic powder, it was found that waste plastic (R-PE) is 120 °C and (P-PE) is 115 °C. The melting point of MAPP used as coupling agent is 75 °C. 500 µm was used to select the size of plastic powder and sawdust for making composite board. Boards were tested by mixing 500 µm fine plastic powder and teak sawdust in 1:1, 2:1, 3:1 ratios and using different temperatures (140, 160, 180 °C). At 140 °C, it was found that the plastic did not melt because the temperature was low, so it did not come out as a composite board. Also, it was found that the sawdust burned because the heat was too high at 180 °C. At 160 °C, composite boards were obtained, and then two types of Recycled PE and Virgin PE were tested (Yeh et al., 2009). The ratio of plastic powder and sawdust used, temperature, time and pressure are shown in Tables 15.2 and 15.3 (Figs. 15.4 and 15.5).

According to the quality measurements of the obtained composite boards, it was found that there was little difference. Composite boards made of Recycled PE and

Table 15.2 Pure PE and sawdust made with 3 ratios

No	Name	Pure PE: sawdust	Temperature (°C)	Pressure (Psi)	Time (minutes)
1	VP-1	1:1	160	2240	15
2	VP-2	2:1	160	2240	15
3	VP-3	3:1	160	2240	15

Table 15.3 Recycled PE and sawdust made with 3 ratios

No	Name	Recycled PE: sawdust	Temperature (°C)	Pressure (Psi)	Time (minutes)
1	RP-1	1:1	160	2240	15
2	RP-2	2:1	160	2240	15
3	RP-3	3:1	160	2240	15

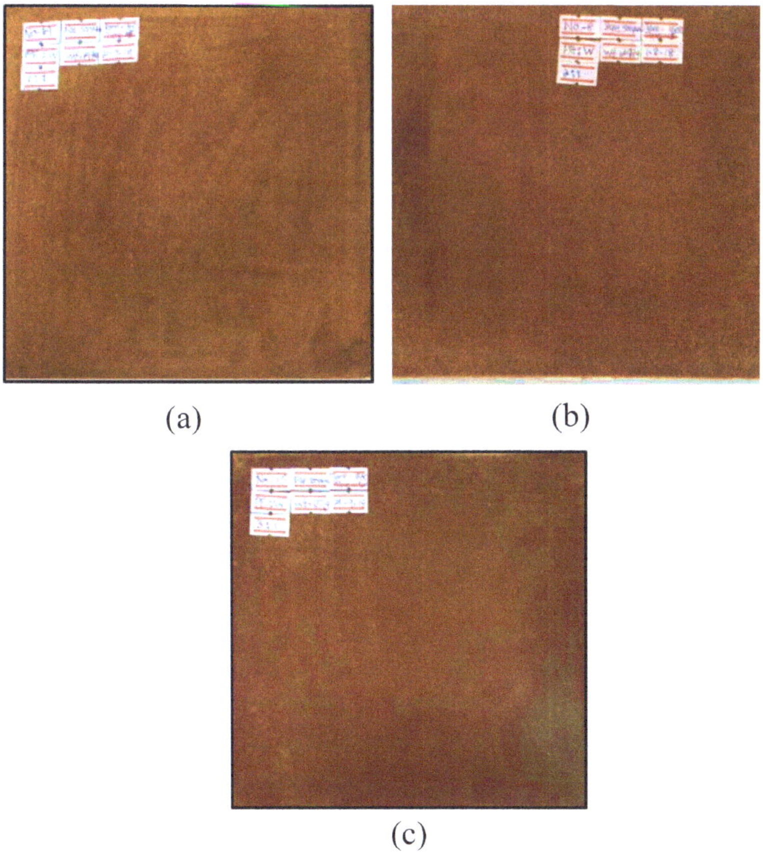

Fig. 15.4 Composite board made using a ratio of Pure PE and sawdust **a** 1:1 **b** 2:1 **c** 3:1

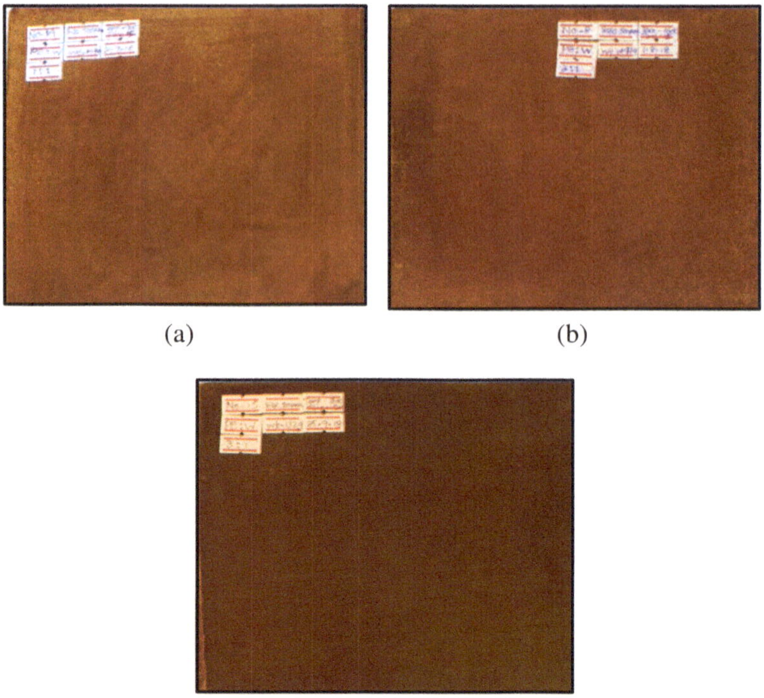

(a) (b) (c)

Fig. 15.5 Composite board made using a ratio of Recycled PE and sawdust **a** 1:1 **b** 2:1 **c** 3:1

Pure PE were found to have the best ratio of plastic powder and teak sawdust (3:1) in bending strength (psi), impact strength (KJ/m^2) and density (g/cm^3). The findings are presented in Table 15.4. It was found that there was little difference in the hardness (Shore-D) for all three ratios of both recycled PE and Pure PE composite board. In terms of oil absorption (%) and water absorption (%), the ratio (3:1) is the least (Figs. 15.6 and 15.7).

Therefore, in order to reduce environmental pollution, we tested using recycled PE, which is slightly different in quality. MPPE was used as the coupling agent. MAPP was tested in two ways: using it as a solid and using it after grinding it into powder with refiner machine. MAPP solid was used for no difference. Two tests were conducted: 3% MAPP (RPM-1) as a proportion test on the total weight of teak sawdust and plastic powder and 3% MAPP (RPM-2) as a proportion test on the weight of teak sawdust powder (Janigova et al., 2001). It is shown in Table 15.5 that the composite board was tested by taking the weight of the sawdust and was found to have better quality (Fig. 15.8).

According to the tests, the use of MAPP (Coupling Agent) based on sawdust was found to be better. We continued testing using 3, 6 and 9% MAPP, based on the weight of sawdust. It was found that the tested composite board with 3% MAPP had

Table 15.4 According to the ratio of polyethylene plastic (pure, recycled) and sawdust, composite boards were measured the properties of physical and mechanical

No	Measured physical properties and mechanical properties	The ratio of polyethylene plastic powder and sawdust					
		1:1		2:1		3:1	
		PPE-1	RPE-1	PPE-2	RPE-2	PPE-3	RPE-3
1	Bending strength (psi)	4303	5182	5487	6264	NB	7363
2	Impact strength (KJ/m^2)	38	147	50	155	62	150
3	Hardness (Shore-D)	96	95	97	95	95	96
4	Density (g/cm^3)	0.96	0.80	0.73	0.85	0.99	0.96
5	Oil absorption (%)	11.53	4.93	5.78	3.25	1.14	0.43
6	Water absorption (%)	9.48	12.4	5.82	3.6	1.6	0.36
7	Thickness (cm)	0.43	0.46	0.49	0.45	0.40	0.43

* NB = Not Breakable

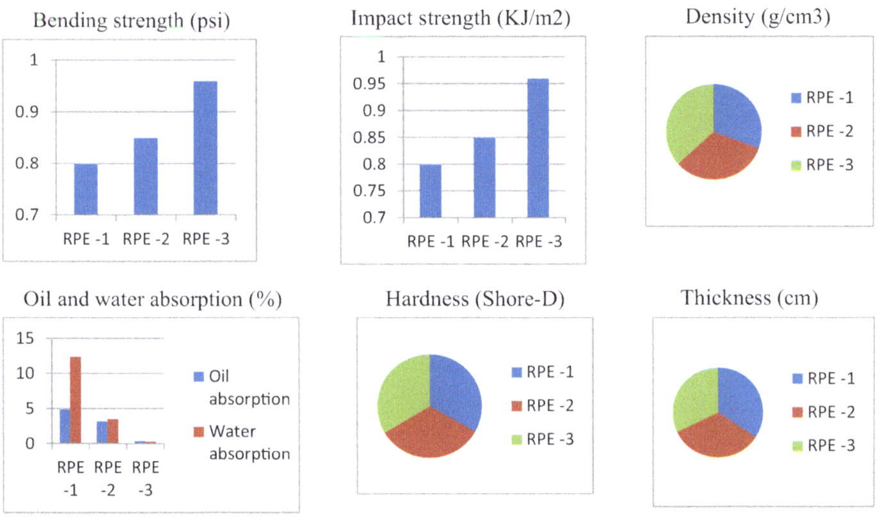

Fig. 15.6 Measured the physical and mechanical properties of (RPE: sawdust/1:1, 2:1 and 3:1) wood plastic composite

better physical properties and mechanical properties. The measured results are listed in Table 15.6 (Fig. 15.9).

The greatest advantage of WPC is its environmentally friendly approach of using waste wood and recycled plastic material. The addition of sawdust to the composite can improve its strength and reduce its weight, making it a popular choice for applications such as flooring, root panels, seat and table. Wood plastics composite have low maintenance cost as compared to that of solid wood. Plastic wood is extremely durable, rot-resistant and will withstand the outdoor elements (Buehlmann et al., 2001). It can be observed that the addition of sawdust caused a reduction in the

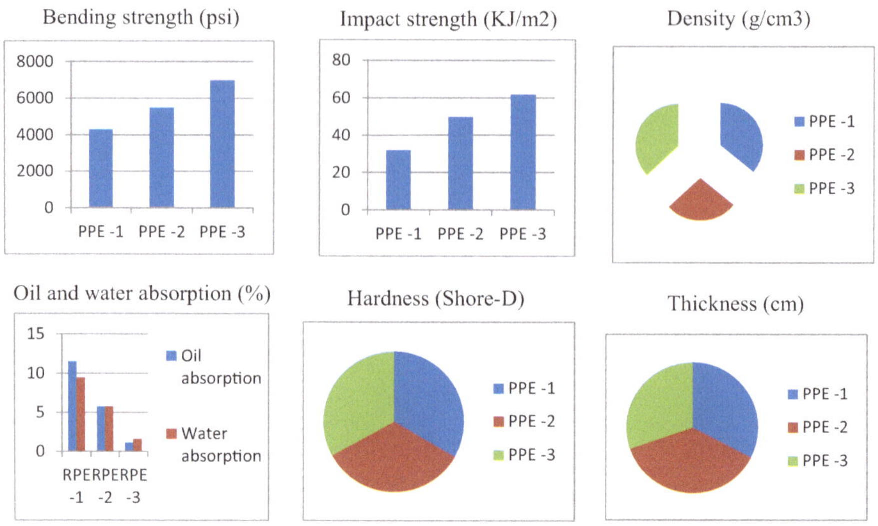

Fig. 15.7 Measured the physical and mechanical properties of (PPE: sawdust/1:1, 2:1 and 3:1) wood plastic composite

Table 15.5 Comparisons of the physical and mechanical properties of composite boards were made using a coupling agent (MAPP) at 3% by weight of both RPE and sawdust, using a coupling agent (MAPP) at 3% by weight of sawdust

No	Bending strength (psi)	Impact strength (KJ/m^2)	Hardness (Shore-D)	Density (g/cm^3)	Oil absorption (%)	Water absorption (%)	Thickness (cm)
RPM-1	5045	153	98	0.81	2.07	2.65	0.45
RPM-2	NB	155	97.5	0.97	0.34	0.44	0.40

* NB = Not Breakable

(a) (b)

Fig. 15.8 MAPP has used **a** weight of both RPE and sawdust and **b** comparatively, composite board were made using MAPP at 3% by weight of the only sawdust

Table 15.6 Comparisons of the physical and mechanical properties of composite boards were made using a coupling agent (MAPP) at 3, 6 and 9% by weight of sawdust

MAPP (%)	Bending strength (psi)	Impact strength (KJ/m^2)	Hardness (Shore-D)	Density (g/cm^3)	Oil absorption (%)	Water absorption (%)	Thickness (cm)
3	NB	155	97.5	0.97	0.34	0.44	0.40
6	NB	55.05	97.5	0.98	0.37	0.45	0.38
9	8107	54.20	98	0.99	0.42	0.73	0.41

* NB = Not Breakable

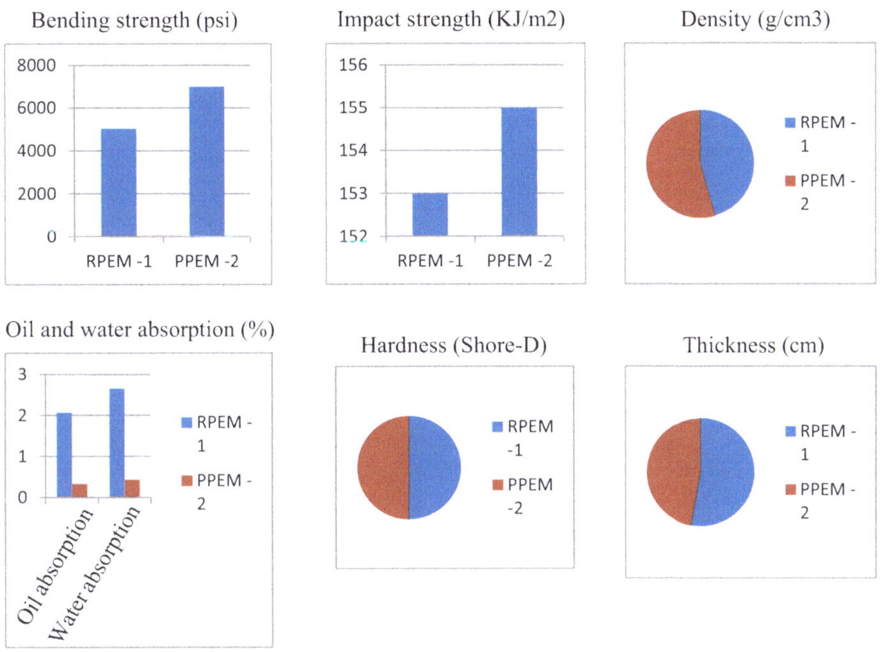

Fig. 15.9 Measured the physical and mechanical properties of (RPE + MAPP) wood plastic composite

impact strength and an increase in the values for the hardness of the composites. Better mechanical properties were obtained when coupling agent was used, for all sawdust contents. The formulations with MAPP were found to have higher strength and better water resistance. Addition of the MAPP into the WPC significantly improved the water resistance. The coupling agents improve the quality of adhesion between plastics and fibers. Coupling agents improve the adhesion of natural fibers and plastic, which promotes greater resistance and reduces water absorption and improves the mechanical properties of products (Rude, 2007). Recycling plastics is the best way to reduce environmental water and food pollution. Plastic pollution is causing irreparable damages to the planet and is getting worse every day. WPC products use recycled plastic and can produce graceful, highly resistant materials for the building industry and other sectors. By using recycled plastic from landfills, rivers and oceans, plastic pollution can be reduced.

15.4 Conclusions

According to the results, composite boards made of Recycled PE and Pure PE were found to have the best ratio of plastic powder and teak sawdust (3:1) in bending strength (psi), impact strength (KJ/m^2) and density (g/cm^3). It was found that there was little difference in the hardness (Shore-D) for all three ratios of both recycled PE and Pure PE composite board. In terms of oil absorption (%) and water absorption (%), the ratio (3:1) is the least. The use of MAPP (Coupling Agent) was tested using 3% depending on the weight of the sawdust in the ratio (3:1) of waste plastic and sawdust of the composite board with the best properties. According to the tests, it was found that the tested composite board using 3% MAPP (Coupling Agent) had better physical properties and mechanical properties. The use of MAPP as a coupling agent has been found to be more resistant in bending strength (psi). In terms of oil absorption (%) and water absorption (%), it was found that the use of MAPP was more resistant than the composite sheet without MAPP. In summary, there is pollution in the environment due to waste plastic, so if we can produce composite board using this waste plastic, not only can we reduce environmental pollution, but we can turn it into a high-value product, and it can be economically profitable. By mixing waste plastic materials with sawdust to produce composite materials, it is able to withstand the effects of weather, sunlight and fungi, and has less water absorption; harder, composite materials are available. Composite materials can be used as flooring and furniture, and can be widely used locally and commercially produced.

Acknowledgements The author is grateful to the laboratory and technical staff of the Polymer Research Department, for the support rendered in carrying out the work. The author would like to thank deeply the Director General of Chemical Technology Research Center for giving technical advices. The author also would like to express thanks to Director General Dr. Phyu Phyu Win Department of Research and Innovation for laboratory facility and financial support. Finally, the author deeply appreciates the NAM (S&T) Centre.

References

Adhikary, K. B., Pang, S., & Staiger, M. P. (2008). Dimensional stability and mechanical behaviour of wood–plastic composites based on recycled and virgin high- density polyethylene (HDPE). *Composites: Part B, 39*, 807–815.

Bouaf, H., Koubaa, A., Perre, P., & Cloutier, A. (2009). Effects offiber characteristics on the physical and mechanical properties of wood plastic composites. *Composites Part A Apply Science and Manufacturing, 40*, 1975–1981.

Buehlmann, U., Saloni, D., & Lemaster, R. L. (2001). *Wood fiber-plastic composite: Machining and surface quality*. North Carolina State University.

Carus, M., Eder, A., Dammer, L., Korte, H., Scholz, L., Essel, R., Breitmayer, E., & Barth, M. (2015). "Wood-plastic composites (WPC) and natural fiber composites (NFC)": European and global markets 2012 and future trends in automotive and construction.

Caulfield, D. F., Clemons, C., Jacobson, R. E., & Rowell, R. M. (2005). Wood thermoplastic composites. In R. M. Rowell (Ed.), *Handbook of wood chemistry and wood composites* (pp. 365–378). CRC Press.

Felix, J., & Gatenholm, P. (1991). The nature of adhesion in composites of modified cellulose fibers and polypropylene. *Journal of Apply Polymer Science, 42*(3), 609–620.

Hull, D., & Clyne, T. W. (1996). *An introduction to composite materials* (2nd ed.). Cambridge University Press.

Janigova, I., Lednicky, F., Nogellova, Z., Kokta, B. V., & Chodak, I. (2001). The effect of crosslinking on properties of low-density polyethylene filled with organic filler. In *Macromolecular Symposia* (pp. 149–158).

Jayaraman, K., & Bhattacharyya, D. (2004). Mechanical performance of woodfibre– waste plastic composite materials. *Resources, Conservation and Recycling Journal, 41*(4), 307–319.

Kazayawoko, M., Balatinecz, J. J., & Matuana, L. M. (1999). Surface modification and adhesion mechanisms in wood fiber-polypropylene composites. *Journal of Material Science, 34*(24), 6189–6199.

Kuan, H. C., Huang, J. M., Ma, C. C. M., & Wang, F. Y. (2003). Process ability, morphology and mechanical properties of wood reinforced high density polyethylene composites. *Plastics, Rubber and Composites, 2*(3), 122–126.

Lu, J. Z., Wu, Q., & McNabb, H. S. (2000). Chemical coupling in wood fiber and polymer composites: a review of coupling agents and treatments Jr.

Maldas, D., & Kokta, B. V. (1994). Role of coupling agents on the performance of wood flour filled polypropylene composites. *International Journal of Polymeric Materials, 27*(1–2), 77–88.

Migneault, S., Koubaa, A., Erchiqui, F., & Chaala, A. (2009). Effects of processing method and fiber size on the structure and properties. *Composites: Part A, 40*, 80–85.

Eckert, C. (2000). Opportunities for natural fibers in plastic composites. In: Progress in wood fiber– plastic composites. University of Toronto.

Raj, R. G., Kokta, B. V., Maldas, D., Daneault, C. (1989). Use of wood fibers in thermoplastics. VII. The effect of coupling agents in polyethylene-wood fiber composites. *Journal of Apply Polymer Science, 7*(4), 1089–1103.

Rude, E. F. (2007). Evaluation of coupling mechanism in wood plastic composites. Washington State University.

Klyosov, A. (2007). *Wood-plastic composites*. Wiley.

Yeh, S.-K., Agarwal, S., & Gupta, R. K. (2009). Wood–plastic composites formulated with virgin and recycled ABS. *Composites Science and Technology, 69*, 2225–2230.

Chapter 16
Kitchen and Plastic Waste Used in Emergency Biogas Plant Construction During 2022 Energy Crisis in Sri Lanka

K. J. Sirikumara and S. Khanam

Abstract During Sri Lanka's 2022 energy crisis, an innovative solution was developed to combat both plastic pollution and LPG shortages. This approach involved using discarded plastic containers to construct biogas digesters, collectors, and bubblers. Household kitchen waste, which was often improperly disposed of, and plastic waste, including chemical storage containers, were repurposed to produce biogas. This biogas served as a substitute for expensive imported LPG, thereby conserving foreign currency and meeting the country's cooking energy needs. The study targeted 396 randomly selected households that had adapted plastic container biogas digesters. Data were collected using four-stage cluster sampling to obtain cross-sectional information. The first step involved the random selection of 11 districts out of the 25 districts in Sri Lanka. In the second stage, six towns were randomly chosen from each selected district. The third stage involved the random selection of three villages from each town, resulting in a total of 198 villages. Because households in Sri Lanka quickly installed emergency biogas digesters made from plastic containers during the energy crisis of 2022, there wasn't enough time to establish a comprehensive database of digester owners. As a result, the snowball sampling method was used to randomly select two adopters from each village, resulting in a sample size of 396. Through a structured questionnaire, 396 biogas digester adaptor households were identified and interviewed, with the heads of each selected household participating. Moreover, the biogas systems produced biofertilizers for home gardening and farming, providing a triple benefit of energy production, waste management, and agricultural support. The study revealed that a significant number of used plastic containers could be effectively recycled into biogas technologies. These systems provided a renewable cooking fuel, reducing reliance on expensive imported LPG and alleviating economic strain. Moreover, biogas production generated biofertilizers, enhancing the value of these systems for home gardening

K. J. Sirikumara (✉)
Sri Lanka Standards Institution, Colombo, Sri Lanka
e-mail: jayantha@slsi.lk

S. Khanam
Bangladesh University of Professional, Dhaka, Bangladesh

S. Suriyanarayanan et al. (eds.), *Combating Plastic Pollution in Terrestrial Environment*,
https://doi.org/10.1007/978-981-96-2343-3_16

and farming. The adoption of biogas technology was influenced by household demographics, such as age, education level, occupation, and income. Higher acceptance rates were observed among households with better education and economic stability, highlighting the importance of these factors in promoting sustainable practices. In summary, the research highlights the potential of biogas technology to mitigate energy crises and combat plastic pollution simultaneously, offering a sustainable alternative to conventional cooking fuels. It showcases the adaptability and resilience of local communities in addressing environmental and economic challenges through innovative solutions.

Keywords Domestic biogas plants · Plastic containerized biogas digester · Plastic pollution

16.1 Introduction

16.1.1 Background and Context

The study highlights the evolution and increasing adoption of biogas technology in Sri Lanka, particularly during recent economic crisis in Sri Lanka in 2022 (DeVotta, 2022; Fernando, 2022; Gandhi, 2022; Jiang, 2022; OCHA, 2022). It emphasizes innovative solutions involving plastic barrels and containers for biogas production. The research aims to assess the effectiveness of utilizing these waste plastics in biogas generation and their role in addressing plastic pollution on land. In Sri Lanka, the improper disposal of used plastic containers is identified as a critical environmental concern exacerbated by limited reuse options and inadequate disposal methods. Various types of plastic items, such as bottles, tanks, and chemical containers, significantly contribute to terrestrial pollution.

16.1.2 Plastic Waste and Wildlife

Plastic waste in terrestrial environments has far-reaching effects on wildlife. Sri Lankan elephants, for instance, are often attracted to areas with dumped plastic containers as they search for food as shown in Fig. 16.1. People encroaching on elephants' territory exacerbate this issue (Bandara et al., 2023; Jayasinghe, 2012).

The ingestion of plastic by elephants can lead to severe health issues, including digestive tract blockages, internal injuries, and even death. This situation highlights the urgent need to manage plastic waste more effectively and explore sustainable alternatives such as biogas technology, which can repurpose plastic waste into valuable resources. Implementing biogas technology using plastic containers not only addresses energy needs but also reduces the amount of plastic waste that can

Fig. 16.1 Sri Lankan elephants visit to forage for food and inadvertently consume food contained in plastic containers

harm wildlife. By transforming plastic barrels into biogas digesters, collectors, and bubblers, the risk of wildlife consuming harmful plastic is mitigated.

The situation depicted in Fig. 16.1 illustrates the issue of open dumping in Sri Lanka, where plastic bottles, barrels, and containers are improperly disposed of. This irresponsible waste disposal not only pollutes the environment but also attracts wildlife like elephants, who may visit in search of food due to habitat encroachment by human activities. Elephants, as they roam for food, are increasingly forced to navigate through areas where human development has infringed upon their natural habitats. This encroachment often leads them into contact with human settlements or areas where waste is improperly discarded, such as open dump sites. The presence of easily accessible food waste in these areas can draw elephants closer to human habitation, leading to potential conflicts and risks for both elephants and humans. Addressing this issue requires comprehensive waste management practices to reduce open dumping and better manage waste disposal sites. Additionally, measures to mitigate human-wildlife conflicts, such as creating buffer zones or corridors for wildlife, are crucial for preserving the natural habitats of animals like elephants while ensuring human safety and reducing environmental degradation (Bekchanov et al., 2019; Musafer et al., 2020).

Figure 16.2 illustrates a positive aspect of recycling and reuse, where used plastic containers have been cleaned and prepared for sale in shops. These containers are intended for various applications, particularly in constructing biogas plants and for other necessary uses.

Fig. 16.2 Used plastic containers ready for sale in Shop (BothalPaththarakade) (*Source* Field Survey, 2022/2023)

- **Recycling Benefits**: By cleaning and preparing used plastic containers for resale, the figure promotes recycling efforts. This helps reduce the amount of plastic waste that would otherwise end up in landfills or open dumping sites, mitigating environmental pollution.
- **Application in Biogas Plants**: Biogas plants utilize organic waste materials to produce methane gas, which can be used as a renewable energy source. Plastic containers may be used in the construction or operation of biogas plants, highlighting their versatility and contribution to sustainable energy solutions.
- **Other Applications**: Besides biogas plants, cleaned plastic containers can serve various purposes in different industries or household applications. They might be used for storage, transportation, or even as components in other construction projects.
- **Economic Opportunity**: Selling cleaned and reusable plastic containers creates economic opportunities for businesses and individuals involved in recycling and waste management. It promotes a circular economy where materials are reused, reducing the need for new plastic production.

Overall, Fig. 16.2 demonstrates a proactive approach to managing plastic waste by emphasizing recycling and reuse, thereby contributing to environmental sustainability and economic efficiency.

16.1.3 Seizing Opportunity During Economic Crisis

The energy crisis of 2022 in Sri Lanka highlighted the vulnerabilities of relying heavily on imported liquefied petroleum gas (LPG). As shortages and high costs persisted, biogas emerged as a viable alternative for household cooking needs. Biogas

Fig. 16.3 People queue to buy liquefied petroleum gas on June 28, 2022 in Sri Lanka (*Source* www.Newsfirst.lk)

utilizes organic waste and repurposed plastic barrels to produce sustainable energy, addressing both energy needs and plastic pollution (*Sri Lanka' s Economic Crisis*, 2022; Fernando, 2022; Gandhi, 2022).

At the same time, gas stations were beginning to restrict fuel supply and offer drivers up to 150 waiting tickets in one go (Fig. 16.3). According to the (*Sri Lanka' s Economic Crisis*, 2022; Fernando, 2022; Gandhi, 2022; Jiang, 2022; Sharma, 2022), there are even longer gas lines at some gas stations. According to the outlet, one of the drivers said he thought he was the 300th person on the line.

Due to lack of LP gas people suffered a lot during the crisis. As there was lots of power shedding during the crisis using electric cookers was a dream only (Agency report, 2022; Fernando, 2022). Therefore, there was no cooking fuel except fuel wood. So people in the urban area started constructing biogas plant using waste plastic barrel. Lots of innovative ideas emerge and most of them were put into practice during the crisis related to biogas technologies as well as and related to other cooking energy technologies. Some innovative people use WhatsApp groups, facebook groups, telegram groups, etc., to share their innovation during the crisis. So this LP gas shortage become an opportunity for new innovation to happen in biogas industry and brought a solution to combat plastic pollution in terrestrial environment for years.

16.1.4 Innovations in Biogas Technology Using Plastic Barrels as a Biogas Digester, Collector and Bubbler

Amidst the crisis in 2022, innovative uses of waste plastic barrels, containers, and pet bottles were explored for biogas production. This study investigates the efficiency of various types and sizes of plastic barrels as biogas digesters, collectors, and bubblers, aiming to assess their effectiveness in biogas production and their role in reducing

Fig. 16.4 180 L and 200 L two used plastic containers are used in constructing a biogas plant (*Source* Field Survey, 2022/2023)

plastic waste (*Sri Lanka' s Economic Crisis*, 2022; de Alwis, 2002; De Alwis, 2012; Katuwal, 2022; Lall et al., 2023).

16.1.5 Types of Used Plastic Barrels and Their Applications

This study examines the suitability of small, medium, and large plastic barrels as biogas digesters and collectors. Examples of these applications in biogas systems during the crisis are illustrated in Fig. 16.3.

16.1.5.1 Small Plastic Barrel Used as Biogas Collector and Digester

Large plastic barrels or containers can serve as the primary chamber for anaerobic digestion, in which organic waste such as kitchen scraps, agricultural waste, or animal manure is fermented to generate biogas. These barrels should be airtight to create the anaerobic conditions required for biogas generation. Biogas digesters are typically 200–3000 L or larger, depending on the size of the operation. 200 L capacity waste plastic barrel is used for biogas digester while 180 L upside down container and 200 L water filled used plastic container are used as collector as shown in Fig. 16.4.

16.1.5.2 Medium Size Plastic Barrel Used as Biogas Digester and Small Size Containers Used as Collector

Utilizing a 500-L capacity old plastic water tank as a biogas digester, alongside an 180-L upside-down container and a 200-L water-filled used plastic container as

Fig. 16.5 180 L, 200 L and 500 L used plastic containers are used in constructing a biogas plant (*Source* Field Survey, 2022/2023)

collectors, exemplifies an innovative and practical application of biogas technology. These components not only facilitate efficient biogas production and collection but also contribute to sustainability efforts through the recycling and repurposing of materials. Figure 16.5 visually captures the integration and functioning of these components within the biogas system, highlighting their roles in sustainable energy generation and combatting plastic waste management.

Using a 500-L capacity old plastic water tank as a biogas digester, along with an 180-L upside-down container and a 200-L water-filled used plastic container as collectors, as depicted in Fig. 16.5, represents a practical and innovative approach to biogas technology.

16.1.5.3 Large Size IBC Plastic Tanks Used as Biogas Digester

Converting 2000-L capacity old plastic Intermediate Bulk Containers (IBCs) into biogas digesters, as depicted in Fig. 16.6, represents a significant adaptation of large-scale plastic containers for sustainable biogas production. The conversion of 2000-L capacity old plastic IBC tanks into biogas digesters, as shown in Fig. 16.6, exemplifies a robust approach to sustainable energy generation and combating plastic waste management. These tanks are repurposed effectively to facilitate anaerobic digestion, converting organic waste into biogas while promoting environmental sustainability through recycling initiatives. Figure 16.6 visually conveys the layout and functionality of these adapted IBC tanks within the biogas system, highlighting their role in supporting large-scale biogas production efforts.

Fig. 16.6 Used IBC plastic tanks are converted to Biogas digester (*Source* Field Survey, 2022/2023)

16.1.5.4 Small Plastic Bottles Used as Bubbler

This clause discusses the role and placement considerations of bubblers in biogas systems, highlighting several benefits and optimal configurations. Bubblers are strategically placed in biogas units to serve multiple purposes:

- **Measurement of Biogas Production Rate**: Bubblers act as gauges to measure the rate at which biogas is produced, providing valuable data on the efficiency and performance of the biogas unit.
- **Biogas Purification**: By dissolving $CO2$ in biogas, bubblers function as purifiers, enhancing the quality of biogas by removing impurities.
- **Safety Valve Against Backfire**: Positioned correctly, bubblers prevent the risk of backfire, safeguarding the biogas unit and gas collection system.

The effectiveness of bubblers depends on their location within the biogas circuit. Placing them between the digester and collector optimizes their function by allowing sufficient time for gas interaction, thereby maximizing the above first two benefits. Conversely, positioning a bubbler between the collector and stove may compromise these benefits due to rapid gas flow, necessitating additional measures to regulate pressure.

In conclusion, while bubblers primarily enhance safety and efficiency in biogas systems, their optimal placement between the digester and collector ensures comprehensive functionality, serving as critical indicators of biogas production and system behavior.

16.1.6 Different Layout Diagram for the Use of Used Plastic Barrels in Construction of a Biogas Plant

There are several varieties of digesters, with variations in their configuration, building materials, and setting up needs, among other features. For domestic use, underground digesters are usually no bigger than 12 m^3. Other than the digester, a biogas plant also has inlet tanks, an outlet tank, valves, collector, bubbler, Fe filter, mixture and weight to pressurize a collector, etc.

The three tank separate digester and collector system is depicted in Fig. 16.8 as being the most popular amid Sri Lanka's energy crisis in 2022.

16.1.7 Fundamental of Anaerobic Digestion

Anaerobic digestion (AD) is a biological process where microorganisms break down organic matter in the absence of oxygen. This process produces biogas, primarily composed of methane (CH4) and carbon dioxide (CO2), along with digestate (a nutrient-rich fertilizer). The key stages of anaerobic digestion include hydrolysis, acidogenesis, acetogenesis, and methanogenesis (Gadirli et al., 2024; Thenabadu et al., 2015).

16.1.8 Objectives

The overall objective of the study is to explore the opportunity created during crisis in Sri Lanka for combating plastic pollution in Terrestrial Environment.

The specific objectives are;

(i) To understand the socioeconomic, environment and technological characteristic of the household in the sampled area
(ii) To analyze how much plastic barrels, containers and pet bottles can be reused using reason development of biogas technology during crisis for combating plastic pollution
(iii) To analyze the motivating factor for the adoption of the biogas technology to reuse the plastic barrels, containers and pet bottles

16.2 Data Collection and Methodology

The study on biogas technology adoption in Sri Lanka focuses on evaluating its efficiency and impact, particularly amid economic challenges. The research employs a carefully selected sample comprising 11 out of the country's 25 districts, encompassing 396 households. This approach aims to ensure a representative cross-section of Sri Lanka's diverse geographic, demographic, and socio-economic landscape, essential for comprehensive and nuanced analysis (Seem et al., 1988; Smajic et al., 2022).

16.2.1 Representativeness

The choice of 11 districts facilitates a thorough examination of variations across different province, enhancing the study's ability to generalize findings to the broader population. This selection not only supports the statistical robustness of the research but also aligns with its overarching goal of assessing biogas technology's adoption and its implications during the energy crisis of 2022 (Civelek, 2018; Rahman et al., 2014; Sayer, 1984).

16.2.2 Statistical Power

Methodologically, the study employs a four-stage cluster sampling technique. Initially, districts were randomly selected, followed by the random selection of towns within these districts and villages within those towns. Due to the emergent nature of biogas digester installations during the crisis, snowball sampling was employed to identify households using these technologies. This methodological approach ensures that the sample is representative and captures the diversity of households adopting biogas digesters across different local contexts. The sample size of 396 households is statistically robust, offering sufficient power to detect meaningful relationships and differences within and between districts. This enhances the precision and reliability of the study's findings, enabling accurate estimation of population parameters such as adoption rates, socio-economic impacts, and environmental considerations related to biogas technology (Civelek, 2018; Devkota et al., 2014).

16.2.3 Cost and Feasibility

Moreover, the practical considerations of cost and feasibility were crucial in determining the sample size and district selection. Conducting research across all 25 districts would have been resource-intensive and impractical, highlighting the importance of balancing comprehensive data collection with available resources (Albuam & Oppenheim, 1993; Keiser & Tortora, 2022).

In conclusion, the study's methodology, including its sample size and district selection, is grounded in methodological rigor aimed at producing reliable and applicable findings. By focusing on 11 districts and 396 households, the research effectively addresses its objectives of evaluating biogas technology adoption amidst an energy crisis, providing insights that are valuable for policymaking, sustainable development initiatives, and environmental management in Sri Lanka and potentially beyond.

16.2.4 Questionnaire Design

The Household Biogas Survey conducted in Sri Lanka in 2022 was designed as a robust tool to gather extensive data on the adoption and impacts of biogas technology in households across different districts (Albuam & Oppenheim, 1993; Smajic et al., 2022). Here's a detailed summary of the questionnaire's design, question types, implementation process, and the insights gathered:

Questionnaire design:

- **Purpose**: The questionnaire aimed to assess multiple dimensions related to household biogas plants, including socio-economic impacts, environmental benefits, technological aspects, and user satisfaction.
- **Sections**: It was structured into several sections:
  - **Household Information**: Demographics of biogas users, socio-economic status, and basic household characteristics.
  - **Social Impact**: Questions on changes in daily routines, health impacts, and social dynamics due to biogas adoption.
  - **Economic Impact**: Financial aspects such as income generated or saved, expenditure related to biogas use, and economic benefits perceived.
  - **Environmental Impact**: Queries regarding environmental benefits like reduction in deforestation, improvements in soil quality, and impacts on local ecosystems.
  - **Technological Details**: Specific questions about biogas plant specifications, operational details, maintenance practices, and technical challenges faced.
  - **Dependent-Satisfaction**: Perceptions and satisfaction levels of biogas plant owners regarding various aspects of biogas technology.
- **Question Types**: Included both closed-ended (e.g., multiple-choice, Likert scale) and open-ended questions to capture nuanced responses and additional comments from respondents.
- **Language**: The questionnaire was primarily in Sinhala, ensuring clarity and comprehension among the respondents.

Implementation process:

- **Pilot Testing:** Before full deployment, the questionnaire underwent pilot testing to identify and address any issues such as unclear questions or logistical challenges in data collection.
- **Revision:** Based on pilot test feedback, revisions were made to improve clarity, relevance, and effectiveness in capturing intended data.
- **Training:** Enumerators and field staff were trained to administer the questionnaire consistently and accurately, ensuring standardized data collection procedures.
- **Data Collection:** Conducted across various districts in Sri Lanka, the survey employed both electronic and paper-based methods as per respondent accessibility and preferences.

Evaluation of information gathered:

- The questionnaire's structured approach facilitated systematic data collection on diverse variables related to biogas adoption and its impacts.
- Comprehensive insights were gained into socio-economic conditions, energy usage patterns, biogas plant specifications, feedstock availability, and the gender dynamics of household energy management.
- The use of closed-ended questions facilitated easier data processing, analysis, and comparison across different households and regions.
- Insights gathered from the survey provided valuable evidence on the benefits, challenges, and policy implications associated with biogas technology adoption in Sri Lankan households.

In summary, the Household Biogas Survey in Sri Lanka exemplifies a well-designed instrument tailored to capture comprehensive data on biogas technology adoption and its multifaceted impacts. Through rigorous implementation and careful questionnaire design, the survey effectively contributed to understanding the complex dynamics surrounding biogas adoption in the region. The findings from this survey are instrumental for informing future research endeavors, policy formulation, and sustainable energy initiatives in Sri Lanka and beyond.

16.2.5 Description of Study Variables

These variables are crucial for understanding the socio-economic context and potential impacts of biogas technology adoption within households in Sri Lanka. Each variable's measurement and expected effects provide a framework for analyzing how demographic, economic, and behavioral factors influence the adoption and outcomes associated with biogas technology in the study (Jöreskog et al., 2006).

Table 16.1 provides a description of the study variables used in the investigation:

Table 16.1 Description of the study variables

Variable name	Description	Measurement	Expected sign/ effect
Sex	Sex of household head	Binary (1 = male, 2 = female)	±
Age	Age of household head	Categorical (1 = less than 30 years, 2 = 30–39, 3 = 40–49, 4 = 50–59, 5 = above 59)	±
Family size	Size of the family	Categorical (1 = less than or equal to 3, 2 = 4, 3 = 5, 4 = 6, 5 = equal and above 7)	±
Educ	Education level of household head	Categorical (1 = less than grade 5, 2 = Primary, 3 = Secondary, 4 = OL passed, 5 = AL passed, 6 = Graduation, 7 = Post-gr.)	±
Occupation	Occupational status of household head	Categorical (1 = Agriculture, 2 = Business, 3 = monthly pay jobs, 4 = Remittance)	±
Income	Annual average income of HH	Continuous	±
Source fuel	Source of fuel for cooking	Categorical (1 = fuelwood, 2 = LPG, 3 = Biogas,)	±

- **Sex**: This variable denotes the sex of the household head, categorized as binary (1 = male, 2 = female). The expected sign/effect (±) indicates that the variable may have different effects depending on the context being tested.
- **Age**: Represents the age of the household head, categorized into five groups: 1 = less than 30 years, 2 = 30–39 years, 3 = 40–49 years, 4 = 50–59 years, 5 = above 59 years. The expected sign/effect (±) suggests that age may influence outcomes positively or negatively, depending on the study's focus.
- **Family size**: Refers to the size of the household family, categorized as: 1 = less than or equal to 3 members, 2 = 4 members, 3 = 5 members, 4 = 6 members, 5 = 7 members or more. The expected sign/effect (±) implies that family size could impact outcomes in various ways, possibly affecting resource allocation or consumption patterns.
- **Educ. (Education)**: Indicates the education level of the household head, categorized as: 1 = less than grade 5, 2 = Primary education, 3 = Secondary education, 4 = Ordinary Level (OL) passed, 5 = Advanced Level (AL) passed, 6 = Graduation, 7 = Post-graduation. The expected sign/effect (+) suggests a positive relationship between higher education levels and certain outcomes, reflecting potential knowledge or skill advantages.

- **Occupation**: Describes the occupational status of the household head, categorized as: 1 = Agriculture, 2 = Business, 3 = Monthly pay jobs, 4 = Remittance. The expected sign/effect (+) indicates that different occupations may lead to varied impacts on outcomes, such as income generation or lifestyle choices.
- **Income**: Represents the annual average income of the household. This variable is continuous, reflecting the expected sign/effect (+), indicating a positive relationship between income levels and various outcomes, such as access to resources or standard of living.
- **Source fuel**: Denotes the primary source of fuel used for cooking in the household, categorized as: 1 = Fuelwood, 2 = Liquefied Petroleum Gas (LPG), 3 = Biogas. The variable description does not specify an expected sign/effect, but it suggests the relevance of energy source choice in influencing environmental, economic, and health outcomes.

16.3 Results and Discussion

The following sections provide objective explanations of the research findings. The sample size and responses to each question serve as the foundation for the statistics in this section. The researcher claims that the findings are reliable enough to draw broad conclusions and make recommendations. A summary of the empirical findings from this investigation is provided.

16.3.1 Background Characteristic of Respondent

16.3.1.1 Age Distribution

Regarding the age of the household head who embraced emergency biogas plant technology during 2022 energy crises in Sri Lanka, the survey's results are summed up in Table 16.2.

Table 16.2 Population pattern of sampled biogas plant

Age group	Number	Percent (%)
<30	15	3.8
30–39	132	33.3
40–49	131	33.1
50–59	96	24.2
>59	22	5.6
Total	**396**	**100.0**

(*Source* Field Survey, 2022/2023)

Sri Lanka was going through a severe economic crisis in 2022 as a result of the country's declining foreign exchange reserves. Sri Lanka does not have LP gas resources. This issue led to a reduction in LP gas imports before they were completely stopped in the middle of 2022. Young people in the nation therefore conducted experiments to discover LP gas substitutes. In towns and semi-urban areas where it is difficult to get fuel wood for cooking, emergency biogas plants have been established using plastic water tanks, used plastic barrels, bus tubes, etc., to address specially the cooking fuel issue. Table 16.1 explains why more than 50% of respondents in Sri Lanka are between the ages of 30–39 and 40–49. The data above are consistent with (Chekol et al., 2022; Hafeez et al., n.d.; Kibria, 2015; Mitra, 2017; Nevzorova & Kutcherov, 2019; Tembo et al., 2023) findings, which demonstrated a connection between household age and biogas technology adoption. The current study differs from that of (Katuwal, 2022; Lall et al., 2023; Riek et al., 2012; Shimahata et al., 2020; Wawa & Mwakalila, 2021), who discovered that younger farmers were more likely than older farmers to adopt structured land conservation methods.

16.3.1.2 Family Composition

Family composition, as detailed in Table 16.3, plays a significant role in influencing the adoption of emergency biogas plants, particularly due to the varying costs associated with cooking fuel across different family sizes.

Family Size Distribution

- The majority of households (44.7%) in the sampled area consist of 4 members, reflecting a common family size according to local norms.
- Smaller families, with 3 members or fewer, constitute 21.5% of the sample.
- Larger families of 5 members or more are less represented, with 22.2% having 5 members, 9.3% having 6 members, and only 2.3% having 7 members or more.

Table 16.3 Family size distribution in sampled households

Family size	Number	Percent (%)
Less than or equal to 3	85	21.5
4	177	44.7
5	88	22.2
6	37	9.3
Equal and above 7	9	2.3
Total	**396**	**100.0**

(*Source* Field Survey, 2022/2023)

Implications for Biogas Adoption

Cost considerations:

- **Cost Efficiency**: Larger families typically incur higher costs for cooking fuel due to increased energy demands. Biogas, derived from organic waste, offers a cost-effective alternative by reducing reliance on conventional cooking fuels like LPG or fuelwood.
- **Adoption Stimulus**: The prevalence of medium-sized families (4 members) suggests a substantial potential market for biogas technology adoption, as they represent nearly half of the surveyed households.
- **Challenges with Larger Families**: The lower representation of larger families may indicate challenges in scaling biogas adoption to meet the energy needs of households with higher consumption requirements.

In summary, the distribution of family sizes in the surveyed households highlights the relevance of biogas technology in addressing varying energy needs and costs across different family compositions. Understanding these dynamics is crucial for promoting sustainable energy solutions and optimizing biogas adoption strategies tailored to diverse household contexts (Hasan et al., 2020; Masud et al., 2020; Wang et al., 2019).

16.3.1.3 Literacy Profile

The survey results are summarized in Table 16.4 with regard to the educational level of the household head who constructed the emergency biogas plant during the energy crisis in the sampled biogas adopter community. Education is the most important catalyst when it comes to the acceptability and dissemination of emergency biogas plant.

When examining Table 16.4, it is found that the number of years of learning of the head of the household is expected to have upward trend in Sri Lanka with adoption of new technology except for postgraduate in this case. When household head's educational level increases naturally he/she is exposed to source of information and also aware of environmental disaster that use plastic has brought to the

Table 16.4 Biogas literacy profile

Education	Number	Percent (%)
O/A passed	56	14.1
Certificate/Diploma	102	25.8
Graduation or equivalent	173	43.7
Post-graduation equivalent	65	16.4
Total	**396**	**100.0**

(*Source* Field Survey, 2022/2023)

Table 16.5 Occupation in sampled households

Occupation	Number	Percent (%)
Government sector	113	28.5
Private sector	270	68.2
Business sector	5	1.3
Others	8	2.0
Total	**396**	**100.0**

(*Source* Field Survey, 2022/2023)

society and negative fossil fuel impact (depleting and non-renewable energy sources) usage on the environment and human. So they consider renewable energy (biogas), non-depleting energy and clean energy resources for their cooking needs. As a result, higher educated household heads are more likely than their less educated counterparts to adopt biogas (Atinkut et al., 2020; Bedi et al., 2018; M. Jayasinghe et al., 2021; Nevzorova & Kutcherov, 2019; Yasmin & Grundmann, 2019). The education and training on the proper use of biogas digesters, as well as how to get rid of all residues from them, is often lacking in households. It has been found that most biogas users only use biogas for cooking, while the use of residues from digesters is low (Ariyasingha, 2024; Cheng et al., 2011; Wakjira et al., 2024).

16.3.1.4 Occupation of the Respondent

In this study, occupation was classified into government sector, private sector, business sector and others.

Occupation is also an important factor for adaption of emergency biogas digester. It is seen from Table 16.5 that the highest proportion (about 68%) of biogas digester adaptors' occupation was private sector. Major reason for this is the gravity of the risk understood by the private sector employees and also taking high risk in difficult situations. About 29% of the biogas digester adaptors' occupation was government sector. Further, the production of plastic biogas digesters is also considered to have a positive impact on the economy, creating new green jobs and increasing household income (Anderman et al., 2015; Bharadwaj et al., 2023; Feiz et al., 2022; Shaibur et al., 2021).

16.3.1.5 Family Income

Family income is also one of the most important factors for influencing people to adapt biogas digester. Table 16.6 shows how household income influences the use of technology. Middle-income households are more likely than low and higher-income households to deploy this emergency plastic containerized biogas technology. As a

Table 16.6 Monthly households' income distribution in USD

Per month family income (USD)	Total HH	Percent (%) of HH
0–200	13	3.3
201–400	166	41.9
401–600	85	21.5
601–800	59	14.9
801–1000	64	16.2
1001–1400	9	2.3
Total	**396**	**100.0**

(*Source* Field Survey, 2022/2023)

result, the decision to build a biogas digester at home is projected to be adversely connected with household income in this specific situation.

The above finding contradicts the hypothesis that higher-income families are more likely to utilize biogas technology than low-income households. Installing a biogas digester at home is projected to be favorably connected with household income (Buysman & Mol, 2013; Cheng et al., 2011; Salam et al., 2020a, 2020b; Sarker et al., 2020; van Groenendaal & Gehua, 2010).

16.3.1.6 Used Plastic Barrels, Containers, and Pet Bottles for Combatting Plastic Pollution

Used Plastic Barrels, and Containers Are Employed for Constructing Biogas Digester

The amount of waste generated each day and the household's daily gas consumption are used to calculate the digester's capacity.

Sri Lanka has 11 varieties of plant sizes ranging from 200 to 3000 L among the households surveyed as per Table 16.7. Even though they have gained knowledge about biogas technology from social media sites like Facebook, WhatsApp groups, and YouTube, some households lack faith in biogas plants and are hesitant to spend money on expensive models if they fail. As can be seen from Table 16.7, they first constructed a small size test plant with digesters that were 200 and 220 L in size. Later, they raised the size by adding additional plants, such as 220, 500, and 1000 L. Total number of used plastic containers from 200 to 3000 L capacity range as shown in Table 16.7 employed by sampled HH is 396. Total number of household in Sri Lanka is approximately 5.5 million. If 10% of the total Sri Lankan household built biogas plant, 550,000 used plastic container from 200 to 3000 L capacity range can be reused without exposing to environment as plastic waste for 10–20 years in Sri Lanka (Bandara et al., 2023; Kajavatan Eng & Nitharsan, 2016).

Table 16.7 Plant size and household in Sri Lanka

Plant size (in Litres)	Number	Percent (%)
200	42	10.6
220	30	7.6
400	9	2.3
440	8	2.0
500	72	18.2
660	11	2.8
720	40	10.1
1000	146	36.9
1220	12	3.0
2000	11	2.8
3000	15	3.8
Total	**396**	**100**

(*Source* Field Survey, 2022/2023)

To modify plastic containers for use in biogas systems, especially as digesters in anaerobic digestion processes, several technical adjustments are typically made. Here's a breakdown of the modifications and their functions;

Function: Digesters are the primary vessels where anaerobic digestion takes place, converting organic materials into biogas and digestate.

Design Modification: Plastic containers can serve as anaerobic digesters by optimizing their shape, volume, and inlet/outlet configurations. The container should be airtight to maintain anaerobic conditions. Inlets allow for organic substrate to be added to the liquid phase, while outlets are placed at the highest point of the digester to take biogas out from the gas phase. The design may include insulation to maintain optimal temperature conditions for microbial activity. Additionally, the container may have an integrated mixing mechanism (mixture) to ensure homogenous mixing of the substrate and microorganisms as well as breaking the scum formation between liquid and gas surface (Kabeyi & Olanrewaju, 2022; Nye, 2023).

In summary, modifying plastic containers for use as anaerobic digesters in biogas systems involves optimizing their design to ensure functionality, durability, and efficiency. These adaptations enable effective anaerobic digestion processes, converting organic materials into valuable biogas and digestate while promoting environmental sustainability through plastic reuse. Proper design and implementation are critical to achieving optimal performance and longevity in biogas technology applications.

16.3.1.7 Used Plastic Barrels, and Containers Are Employed for Constructing Biogas Collector

Smaller plastic containers or barrels can be repurposed as biogas collectors. To collect the biogas produced, these containers shall be connected to an outlet of a digestion device. They should also be airtight to prevent biogas leakage. The size of the collector containers varies according to the amount of biogas produced and the frequency of collection. Various collector types were found in biogas digesters built during the 2022 energy crisis. There are two major categories. In biogas digesters, one old bus and automobile tube are employed as a collector, while used plastic barrels are used as collectors.

Collector Are Made Using Used Plastic Barrels

Using old plastic barrels of various sizes as biogas collectors amid an economic crisis in Sri Lanka in 2022 is a creative and practical approach to meet energy demands while efficiently reusing used plastic containers without exposing the environment to plastic garbage. A sense of ownership and cooperation in times of crisis may be promoted through the involvement of local communities in collecting organic waste materials and used plastic barrels for biogas production.

Maintaining a fish pond in a collector tank to avoid mosquito breeding in stagnant water is a brilliant method shown during economic crisis innovation in the biogas business. The layout and design of the collector tank with the fish pond in it is likely to be shown in Fig. 16.9. In order to meet the needs of fish, such as sufficient oxygenation and space, it would be necessary to ensure that appropriate design considerations are taken into account.

Additionally, constant monitoring and upkeep of the fish pond would be required to maintain its efficiency in decreasing mosquito populations. Overall, incorporating fish ponds into biogas collector tanks provides a long-term and environmentally benign solution to mosquito reproduction, aiding in the overall management of vector-borne diseases and promoting ecological balance.

Total number of used plastic containers from 200 to 3000 L capacity range as shown in Table 16.7 employed by sampled HH is 792. Total number of household in Sri Lanka is approximately 5.5 million. If 10% of the total Sri Lankan household built biogas plant, 1,100,000 used plastic container from 200 to 3000 L capacity range can be reused without exposing to environment as plastic waste, for 10–20 years in Sri Lanka.

To modify plastic containers for use in biogas systems, especially as collectors in anaerobic digestion processes, several technical adjustments are typically made. Here's a breakdown of the modifications and their functions;

Function: The primary function of a collector is to gather biogas generated through anaerobic digestion in the digester. Once collected, the biogas is stored temporarily in the collector. This storage is essential to accommodate fluctuations in biogas production rates, ensuring a continuous and reliable supply of gas for use. Collectors also help regulate the pressure of the biogas. Biogas production can vary based on factors like temperature and feedstock availability. The collector allows for the adjustment of pressure to ensure efficient utilization of biogas downstream in the system.

Design Modification: Plastic containers can be adapted to act as biogas collectors by incorporating gas outlets at the top of the container. These outlets are connected to a piping system that directs biogas to biogas stove. The container must be sealed effectively to prevent gas leakage.

Collectors are integral components of biogas systems, facilitating the efficient gathering, storage, and regulated release of biogas produced in digesters. Their role in maintaining a steady biogas supply and ensuring system safety underscores their importance in sustainable energy production from organic waste.

Collector Are Made Using Used Bus and Car Tubes

Using bus and automobile tubes to build biogas collectors amid Sri Lanka's 2022 energy crisis is a realistic and resourceful solution. During an energy crisis, finding alternate sources of energy is critical, particularly in locations where biomass fuel sources are rare or expensive. Using recycled materials lowers the cost of building biogas collectors, making them more affordable to communities experiencing financial constraints during an energy crisis.

Due to the energy crisis, there was a shortage of conventional building materials such as steel and cement for the construction of biogas plants. An alternative source of energy that is readily available is the recycling of used bus and car tubes, which helps to overcome the need for cooking energy. Using locally available materials such as recycled tubes enables communities to take ownership of their energy solutions and promotes independence and resilience in the face of crises.

16.3.1.8 Used Plastic Pet Bottles Are Used as Bubbler

In a unit where the biogas unit and the gas collection system are located separately, there are several benefits that the biogas unit can get from the bubbler.

1. Bubbler can be used as a gauge to measure the biogas production rate
2. As CO2 in biogas dissolve in bubbler, it can be used as biogas purifier
3. Bubbler can be used as a safety valve to prevent backfire

Depending on the location of the bubbler, this advantage can be used in a variety of ways. Using bubbler between the stove and the gas collection point can somehow prevent the fire from the stove to the gas collection point. But when using bubbler between the stove and the gas collection point, gas flows so fast between the points it doesn't pass through the water long enough to pick up water vapor or carbon dioxide molecules, and it is difficult to measure its velocity, it misses benefits No.1 and No.2, I mentioned earlier. In addition, the bubbler gives some pressure to the gas collector, so an additional force (weight) has to be applied to the gas collector to avoid that pressure.

When using the bubbler between the collector and the digester as shown in Fig. 16.10, the speed of gas production and the time the gas travels through the bubbler is higher, and because there is no need to apply pressure, the main function of the bubbler is to measure the production and behavior of biogas units. Ideally bubbler should be placed between the biogas unit and the gas collection point as shown in Fig. 16.7.

According to Fig. 16.11, 75.1% of the survey household had connected a bubbler to the biogas circuit while 24.8% of the survey household had no bubbler installed in Sri Lanka.

Fig. 16.7 Bubbler between digester and collector (*Source* Field Survey, 2021/2022)

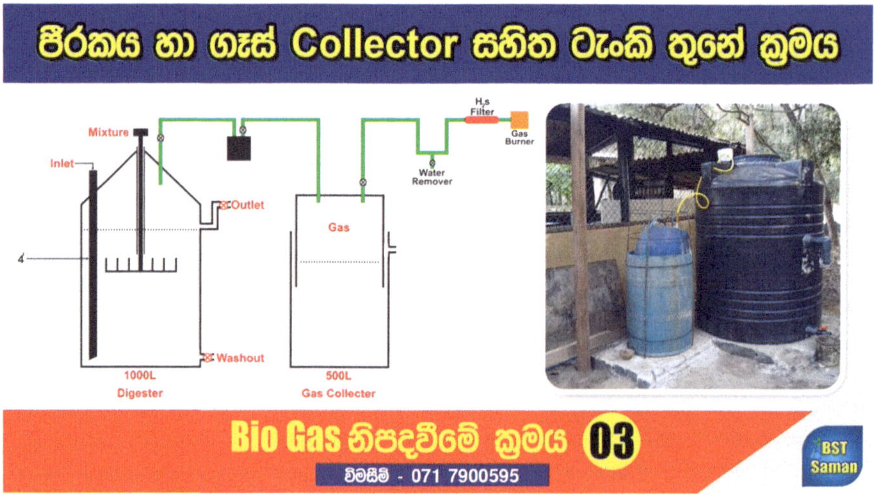

Fig. 16.8 Sri Lankan three tanks separate digester and collector biogas system

Fig. 16.9 Fish pond is in collector tank to avoid mosquito breeding (*Source* Field Survey, 2022/2023)

Fig. 16.10 Used bus tubes as collectors of biogas (*Source* Field Survey, 2022/2023)

A bubbler in a biogas circuit is made from a variety of used plastic bottles with capacities ranging from 500 ml to 2 L. The main types of bottles used for the construction of the bubble in the biogas plant circuit are used plastic beverage bottles. According to Fig. 16.9, 75.1% of the survey households had connected a bubbler to the biogas circuit while 24.8% of the survey household had no bubbler installed in Sri Lanka. Therefore, as per the above finding, 297 HH had installed the different capacities of pet bottles as bubbler in their biogas plant. Total number of household in Sri Lanka is approximately 5.5 million. If 10% of the total Sri Lankan household built biogas plant, 412,500 used pet bottles with different capacity range can be reused without exposing to environment as plastic waste for 10–20 years in Sri Lanka.

To modify plastic containers for use in biogas systems, especially as bubblers in anaerobic digestion processes, several technical adjustments are typically made. Here's a breakdown of the modifications and their functions;

Bubblers:

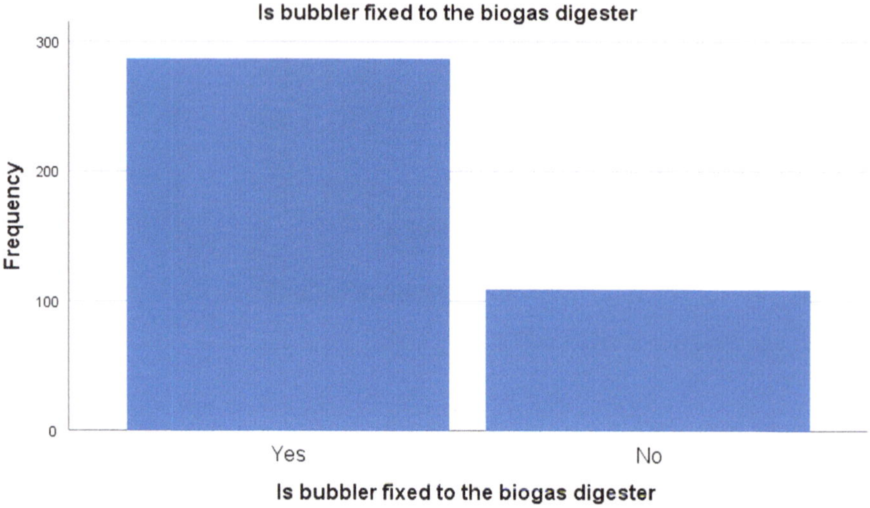

Fig. 16.11 Availability of bubbler set up in the biogas plant in the study area in Sri Lanka (*Source* Field Survey, 2022/2023)

Function: Bubblers in biogas systems are used primarily for monitoring and safety purposes. They allow the visualization of biogas production rates and act as a safety device to prevent backflow of flames into the digester.

Design Modification: Plastic bottle can be modified by adding inlet and outlet ports at appropriate heights to facilitate gas circulation (Fig. 16.11). This bottle should be sealed except for inlet and outlet ports to ensure gas flows through the water column. Marking or calibrating the container to indicate biogas production levels based on bubble frequency or height.

16.3.2 Tipple Benefits Derived from Used Plastic Containerized Biogas Plant

The biogas technology can help Sri Lanka in a variety of ways, as this technology gives triple benefits. These benefits include long-term environmental protection, energy generation, and agricultural and farming support.

16.3.2.1 Biogas Generated Using Plastic Barrels and Containers Used as Cooking Energy

Biogas is a renewable energy source that consists mostly of methane and carbon dioxide. It may be used to cook food, generate energy, and even fuel vehicles. Biogas

Table 16.8 Saving of Conventional Fuel after the Installation of Biogas Plant

Particulars	Quantity used before the installation of biogas plant (Kg/month)	Quantity used after the installation of biogas plant (Kg/month)	Saving/deficit (kg/month)
LP gas			
Total quantity used (Assuming 1 cylinder = 13 kg)	3563	1282	2280
Average quantity used per HH	8.996	3.23	5.76
Firewood			
Total quantity used	17,210	8579	8631
Average quantity used per HH	43.46	21.66	21.80

(*Source* Field Survey, 2022/2023)

production, which captures methane from organic waste, reduces greenhouse gas emissions and dependence on fossil fuels. The 396 sampled households in Sri Lanka directly and economically benefit from the biogas plant's ability to save fuel used for cooking and/or lighting.

According to Table 16.8, the field finding from 396 sampled household from Sri Lanka revealed that 187.48 cum of biogas is produced daily. This saves about 299 kg of fuel wood per day. As a result, Sri Lanka saved 578 kg/year/hh of firewood on average in the sampled HH.

Yearly monetary saving due to use of biogas plants in the sampled household can also be calculated based on the quantity of saving of conventional fuel resources in Table 16.8.

16.3.2.2 Biofertilizer Generated Using Plastic Barrels and Containers Used in Sustainable Agriculture

Biogas production not only creates energy, but it also produces nutrient-rich digestate, which can be utilized as fertilizer. This closed-loop waste management system recovers valuable resources from organic waste while also encouraging sustainable agriculture practices and soil health.

According to Table 16.9, 28% of households (110) produced 3–4 L of liquid fertilizer each day. According to statistics, the average amount of liquid fertilizer produced by one family is 7.1 L per day, while the total amount produced by all 396 households per day is 2813 L. The majority of this biofertilizer comes from 100 L digester capacity (Fig. 16.12).

Table 16.9 Litres of liquid fertilizer taken from the Biogas Plants per day

Liters of liquid fertilizer	Number of households	% of total households
1–2 L	51	12.8
3–4 L	110	27.8
5–6 L	79	19.9
7–8 L	36	9.0
10–2 L	81	20.4
13–20 L	24	6.0
21–30 L	15	3.7
Total	396	100

(*Source* Field Survey, 2022/2023)

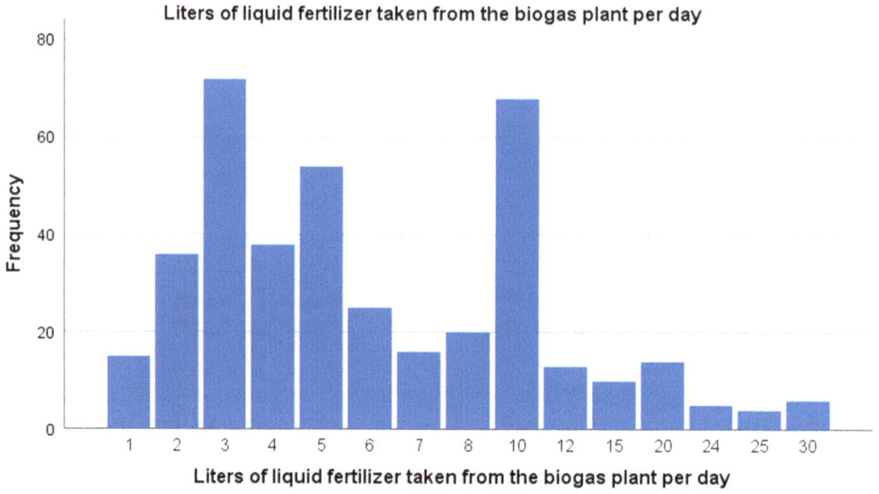

Fig. 16.12 Liters of liquid fertilizer taken from the biogas plant per day

Due to their lack of technical training, some household biogas users are finding it difficult to integrate biogas technology with eco-agriculture technologies (Gosens et al., 2013).

16.3.2.3 Biogas Generated Using Plastic Barrels and Containers Used as Environment Management Tool

Using plastic barrels and containers to generate biogas can be a valuable environmental management strategy. Biogas production from organic waste can assist address a variety of environmental challenges while also supplying the previously mentioned renewable energy source and biofertilizer. For the management of organic waste from the kitchen, biogas production provides a sustainable solution. Materials such as food scraps, agricultural residues, sewage sludge, and other organic matter can be converted into biogas instead of ending up in open dumping near poor people's home where they would decompose anaerobically, releasing methane, a potent greenhouse gas. Plastic barrels and containers can be used as inexpensive and easily accessible digesters for small-scale biogas generation. These containers can be used to generate anaerobic environments in which organic matter decomposes and produces biogas. Plastic barrels are ideal for residential biogas building since they are simple and easy to use.

16.4 Conclusion

In 2022, Sri Lanka faced a severe economic crisis marked by fuel shortages and a disrupted LPG supply network. This situation led to the widespread adoption of innovative biogas technology using repurposed plastic containers. Traditional underground biogas plants, which had been in use since the mid-20th century, could not meet the urgent demand, prompting a shift to more accessible and adaptable solutions. This summary explores the context, methodology, findings, and implications of this innovative approach to biogas production in Sri Lanka.

16.4.1 Context and Necessity

Sri Lanka's lack of fossil fuel reserves made it highly dependent on imported Liquefied Petroleum Gas (LPG), where LPG supply halted during the crisis. Firewood, another alternative, was not readily available in urban homes. In response, biogas production using waste plastic tank emerged as a crucial solution to meet cooking energy needs and reduce the economic strain caused by fuel shortages. The crisis also highlighted an environmental challenge: the increasing nuisance of kitchen waste, often dumped by roadsides or near poor households. The need to address both energy and waste management issues created a fertile environment for biogas technology innovations.

16.4.2 Innovative Approaches

Biogas technology utilizing repurposed plastic containers gained popularity as a practical and sustainable solution. Various sizes of plastic containers, ranging from 200 to 3000 L, were transformed into biogas digesters and collectors. This approach not only helped combating plastic pollution in terrestrial environment but also contributed to sustainable energy production. Different innovative configurations of biogas systems were developed, using different types of plastic containers. Smaller containers were repurposed as collectors, larger barrels as primary digestion chambers, and 500 ml to 5 L plastic bottles as bubblers. These adaptations demonstrated ingenuity in using everyday items for sustainable energy solutions.

16.4.3 Methodology

To assess the adoption and effectiveness of these biogas systems, a national survey was conducted with 396 randomly selected households. A four-stage cluster sampling technique was employed to gather comprehensive data from 2022 to 2023. Structured questionnaires captured socio-economic, environment and technological characteristics of the adopters. The data were analyzed using statistical techniques, including descriptive statistics and frequency tables, with SPSS Version 26.

16.4.4 Key Findings

The study revealed that a significant number of used plastic containers could be effectively recycled into biogas technologies. These systems provided a renewable cooking fuel, reducing reliance on expensive imported LPG and alleviating economic strain. Moreover, biogas production generated biofertilizers, enhancing the value of these systems for home gardening and farming. The adoption of biogas technology was influenced by household demographics, such as age, education level, occupation, and income. Higher acceptance rates were observed among households with better education and economic stability, highlighting the importance of these factors in promoting sustainable practices.

16.4.5 Triple Benefits and Holistic Impact

The biogas systems offered triple benefits: energy generation, waste management, and agricultural support. They conserved LP gas and biomass, combating plastic pollution in terrestrial environment, and provided valuable biofertilizers. This holistic impact underscored the significant contribution of biogas technology to environmental and economic resilience during the crisis. By converting plastic waste into biogas infrastructure, households were able to secure a renewable energy source, reduce environmental pollution, and support local agriculture. This innovative approach exemplifies a practical response to both environmental and economic challenges, showcasing the potential of locally adaptable solutions in addressing global sustainability issues.

16.5 Summary

In summary, Sri Lanka's response to the 2022 energy crisis highlights how crises can drive innovative and sustainable practices. The integration of used plastic containers into biogas technology provided a practical solution to immediate energy needs while addressing long-term environmental sustainability. This study underscores the potential of innovative, locally adaptable solutions to address global sustainability challenges, demonstrating how crises can stimulate beneficial responses that support both communities and the environment.

References

Agency report. (2022). Sri Lanka economic crisis: Five other countries facing a similar catastrophe. *Outlook Web Desk*.

Albuam, G., & Oppenheim, A. N. (1993). Questionnaire Design, Interviewing and Attitude Measurement. *Journal of Marketing Research, 30*(3). https://doi.org/10.2307/3172892.

Anderman, T. L., DeFries, R. S., Wood, S. A., Remans, R., Ahuja, R., & Ulla, S. E. (2015). Biogas cook stoves for healthy and sustainable diets? A case study in Southern India. *Frontiers in Nutrition, 2*, 28. https://doi.org/10.3389/fnut.2015.00028.

Ariyasingha U. D. H. P. (2024). Public opinion survey on the status of waste management in Sri Lanka. In *Proceedings of International Forestry and Environment Symposium* (Vol. 27). https://doi.org/10.31357/fesympo.v27.6570.

Atinkut, H. B., Yan, T., Zhang, F., Qin, S., Gai, H., & Liu, Q. (2020). Cognition of agriculture waste and payments for a circular agriculture model in Central China. *Scientific Reports, 10*(1). https://doi.org/10.1038/s41598-020-67358-y.

Bandara, K. G. M. B., Gowsiga, M., Asmone, A. S., & Dilogini, R. A. A. (2023). Applicability of recycling and resource recovery for solid waste of Sri Lankan supermarkets. In *World Construction Symposium* (Vol. 1, pp. 211–223). https://doi.org/10.31705/WCS.2023.18.

Bedi, A. S., Sparrow, R., & Tasciotti, L. (2018). Indonesia's domestic biogas programme—Household panel survey data. *Data in Brief, 17*, 1388–1390. https://doi.org/10.1016/j.dib.2018.02.083.

Bekchanov, M., Mondal, M. A. H., de Alwis, A., & Mirzabaev, A. (2019). Why adoption is slow despite promising potential of biogas technology for improving energy security and mitigating climate change in Sri Lanka? *Renewable and Sustainable Energy Reviews, 105*, 378–390. https://doi.org/10.1016/j.rser.2019.02.010.

Bharadwaj, B., Kambo, A., Arratia-Solar, A., & Ashworth, P. (2023). Awareness of biogas increases its social acceptance. *Journal of Cleaner Production, 421*, 138432. https://doi.org/10.1016/j.jclepro.2023.138432.

Buysman, E., & Mol, A. P. J. (2013). Market-based biogas sector development in least developed countries —The case of Cambodia. *Energy Policy, 63*, 44–51. https://doi.org/10.1016/j.enpol.2013.05.071.

Chekol, F., Tsegaye, A., Mazengia, T., & Hiruy, M. (2022). Small scale biogas technology adoption behaviours of rural households and its effect on major crop yields in East Gojjam Zone of Ethiopia: propensity score matching approach. *Cogent Engineering, 9*(1). https://doi.org/10.1080/23311916.2022.2133343.

Cheng, S., Li, Z., Shih, J., Du, X., & Xing, J. (2011). A field study on acceptability of 4-in-1 biogas systems in Liaoning Province, China. *Energy Procedia, 5*, 1382–1387. https://doi.org/10.1016/j.egypro.2011.03.239.

Civelek, M. E. (2018). Essentials of Structural Equation Modeling. *Zea Books*. https://doi.org/10.13014/k2sj1hr5.

de Alwis, A. (2002). Biogas—A review of Sri Lanka's performance with a renewable energy technology. *Energy for Sustainable Development*, *6*(1). https://doi.org/10.1016/S0973-0826(08)60296-3.

De Alwis, A. (2012). A tool for sustainability: A case for biogas in Sri Lanka. *Journal of Tropical Forestry and Environment*, *2*(1). https://doi.org/10.31357/jtfe.v2i1.564.

Devkota, J. U., Singh, S., Prajapati, C., & Hada, B. (2014). Principal components analysis of the socioeconomic conditions of biogas users—With example from Nepal. *International Journal of Renewable Energy Research*, *4*(3). https://doi.org/10.20508/ijrer.23129.

DeVotta, N. (2022). Sri Lanka's agony. *Journal of Democracy, 33*(3), 92–99.

Feiz, R., Metson, G. S., Wretman, J., & Ammenberg, J. (2022). Key factors for site-selection of biogas plants in Sweden. *Journal of Cleaner Production, 354*, 131671.

Fernando, J. (2022). The citizen's uprising in Sri Lanka: A watershed moment against racial capitalism. *Human Geography(United Kingdom)*. https://doi.org/10.1177/19427786221123004.

Gadirli, G., Pilarska, A. A., Dach, J., Pilarski, K., Kolasa-Więcek, A., & Borowiak, K. (2024). Fundamentals, operation and global prospects for the development of biogas plants—A review. *Energies, 17*(3). Multidisciplinary Digital Publishing Institute (MDPI). https://doi.org/10.3390/en17030568.

Gandhi, K. (2022). Issue of Srilankans—An overview. *Trends in Banking, Accounting and Business, 1*(1). https://doi.org/10.46632/tbab/1/1/9.

Gosens, J., Lu, Y., He, G., Bluemling, B., & Beckers, T. A. M. (2013). Sustainability effects of household-scale biogas in rural China. *Energy Policy, 54*, 273–287.

Hafeez, A. S. M. G., Roy, D. R., Majumder, S., & Mitra, S. (n.d.). *Adoption of Biogas for Household Energy and Factors Affecting Livelihood of the Users in Rural Bangladesh*. https://doi.org/10.22004/ag.econ.284894.

Hasan, A. S. M. M., Kabir, M. A., Hoq, M. T., Johansson, M. T., & Thollander, P. (2020). Drivers and barriers to the implementation of biogas technologies in Bangladesh. *Biofuels, 13*, 643–655. https://doi.org/10.1080/17597269.2020.1841362.

Jayasinghe, M., Selvanathan, E. A., & Selvanathan, S. (2021). Energy poverty in Sri Lanka. *Energy Economics, 101*, 105450. https://doi.org/10.1016/j.eneco.2021.105450.

Jayasinghe, R. P. C. K. (2012). Solid waste management in Sri Lanka: A case study of maharagama urban council. In *Proceedings of International Conference on Business Management*, *8*(2005), 667–678. http://journals.sjp.ac.lk/index.php/icbm/article/view/322.

Jiang, Q. (2022). Debt crisis and national bankruptcy: Evidence from Sri Lanka. *BCP Business & Management*, *29*. https://doi.org/10.54691/bcpbm.v29i.2288.

Jöreskog, K. G., Sörbom, D., & Yang-Wallentin, F. (2006). Latent variable scores and observational residuals. *Lisrel, 1996*.

Kabeyi, M. J. B., & Olanrewaju, O. A. (2022). Technologies for biogas to electricity conversion. *Energy Reports, 8*, 774–786. https://doi.org/10.1016/j.egyr.2022.11.007.

Kajavatan Eng, A., & Nitharsan, U. (2016). Containerizing Biogas: Design and development of portable low cost Biogas bottling system. *International Journal of Scientific and Research Publications*, *6*(4).

Katuwal, H. (2022). Biogas adoption in Nepal: Empirical evidence from a nationwide survey. *Heliyon, 8*(8), e10106. https://doi.org/10.1016/j.heliyon.2022.e10106.

Keiser, S., & Tortora, P. G. (2022). Qualitative data. *The Fairchild Books Dictionary of Fashion.* . https://doi.org/10.5040/9781501365287.2181.

Kibria, G. (2015). Sustainable energy for rural development in Bangladesh—Economic, social and environmental benefits of renewable energy. *Sydneybashi-Bangla; Australia Science & Technology Article*, *37*.

Lall, O., Bowatte, G., Dharmaratne, S., Lowe, A. J., Vakalopoulos, A., Ambrose, I., Jayasinghe, P., Yasaratne, D., Heyworth, J., & Dharmage, S. C. (2023). Household use of biomass fuel, especially traditional stove is associated with childhood wheeze and eczema: a cross sectional

study of rural communities in Kandy, Sri Lanka. *Journal of Asthma*, *60*(2). https://doi.org/10.1080/02770903.2022.2043360.
Masud, M. H., Nuruzzaman, M., Ahamed, R., Ananno, A. A., & Tomal, A. N. M. A. (2020). Renewable energy in Bangladesh: current situation and future prospect. *International Journal of Sustainable Energy*, *39*(2). https://doi.org/10.1080/14786451.2019.1659270.
Mitra, S. (2017). A review of potential renewable energy preference in rural area of Bangladesh. *Journal of Energy and Natural Resources*, *6*(5). https://doi.org/10.11648/j.jenr.20170605.12.
Musafer, N., Samaratunga, N., & Kumara, P. G. A. (2020). The applications of appropriate renewable energy technologies by the refugees and displaced persons under humanitarian assistance programmes. *International Journal of Research and Innovation in Social Science*, *04*(12). https://doi.org/10.47772/ijriss.2020.41203.
Nevzorova, T., & Kutcherov, V. (2019). Barriers to the wider implementation of biogas as a source of energy: A state-of-the-art review. *Energy Strategy Reviews*, *26*. Elsevier Ltd. https://doi.org/10.1016/j.esr.2019.100414.
Nye, C. D. (2023). Reviewer resources: Confirmatory factor analysis. *Organizational Research Methods*, *26*(4). https://doi.org/10.1177/10944281221120541.
OCHA. (2022). Sri Lanka multi-dimensional crisis. Situation Report No. 1I2 Are, 1.
Rahman, K. M., Woodard, R., Manzanares, E., & Harder, M. K. (2014). An assessment of anaerobic digestion capacity in Bangladesh. *Renewable and Sustainable Energy Reviews*, *32*. https://doi.org/10.1016/j.rser.2014.01.026.
Riek, I., Rücker, A., Schall, T., & Uhlig, M. (2012). Renewable energy generation from biomass-biogas in India. Center for Applied International Finance and Development (CAIFD), University of Applied Sciences Nuremberg, Seminar Paper.
Salam, S., Parvin, R., Salam, M. A., & Azad, S. M. N. (2020a). Feasibility study for biogas generation from household digesters in Bangladesh: Evidence from a household level survey. *International Journal of Energy Economics and Policy*, *10*(4), 23–30. https://doi.org/10.32479/ijeep.9206.
Salam, S., Parvin, R., Salam, M. A., & Nasim Azad, S. M. (2020b). Feasibility study for biogas generation from household digesters in bangladesh: Evidence from a household level survey. *International Journal of Energy Economics and Policy*, *10*(4). https://doi.org/10.32479/ijeep.9206.
Sarker, S. A., Wang, S., Adnan, K. M. M., & Sattar, M. N. (2020). Economic feasibility and determinants of biogas technology adoption: Evidence from Bangladesh. *Renewable and Sustainable Energy Reviews*, *123*, 109766. https://doi.org/10.1016/j.rser.2020.109766.
Sayer, A. (1984). *Method in Social Science: A Realist Approach*. https://doi.org/10.2307/2070081.
Seem, J., Nachmias, D., & Nachmias, C. (1988). Research methods in the social sciences. *Teaching Sociology*, *16*(2). https://doi.org/10.2307/1317432.
Shaibur, M. R., Husain, H., & Arpon, S. H. (2021). Utilization of cow dung residues of biogas plant for sustainable development of a rural community. *Current Research in Environmental Sustainability*, *3*, 100026. https://doi.org/10.1016/j.crsust.2021.100026.
Sharma, R. (2022, May). Sri Lanka crisis 2022: The biggest economic debacle & political turmoil since 1948. Research Institute for European and American Studies.
Shimahata, A., Farghali, M., & Fujii, M. (2020). Factors influencing the willingness of dairy farmers to adopt biogas plants: A case study in Hokkaido, Japan. *Sustainability (Switzerland)*, *12*(18). https://doi.org/10.3390/SU12187809.
Smajic, E., Avdic, D., Pasic, A., Prcic, A., & Stancic, M. (2022). Mixed methodology of scientific research in healthcare. *Acta Informatica Medica*, *30*(1). https://doi.org/10.5455/aim.2022.30.57-60.
Sri Lanka's economic crisis, Takshashila case study 2022–01 (2022)
Tembo, A., Rahman, M. M., & Jerin, T. (2023). Barriers to development and adoption of biogas in Mokambo peri-urban of Mufulira, Zambia: How does local government fail to provide renewable energy? *Biofuels*, *14*(6). https://doi.org/10.1080/17597269.2022.2156055.

Thenabadu, M., Abeyweera, R., Jayasuriya, J., & Senanayake, N. S. (2015). Anaerobic digestion of food and market waste; waste characterisation and biomethane potential: A Case study in Sri Lanka. *SLEMA Journal*, *18*(2). https://doi.org/10.4038/slemaj.v18i2.18.

van Groenendaal, W., & Gehua, W. (2010). Microanalysis of the benefits of China's family-size bio-digesters. *Energy, 35*(11), 4457–4466. https://doi.org/10.1016/j.energy.2009.05.028.

Wakjira, L., Tolera, M., & Shifaraw, G. (2024). Contribution of biogas technology in forest conservation and carbon emission reduction, Gimbi District, Western Ethiopia. *Science Development*. https://doi.org/10.11648/j.scidev.20240501.11.

Wang, Q., Dogot, T., Wu, G., Huang, X., & Yin, C. (2019). Residents'willingness for centralized biogas production in Hebei and Shandong provinces. *Sustainability (Switzerland), 11*(24), 7175. https://doi.org/10.3390/su11247175.

Wawa, A., & Mwakalila, S. (2021). Factors affecting the adoption and non adoption of biogas technology in semi-arid areas of Tanzania. *Journal of the Geographical Association of Tanzania*, *37*(1). https://doi.org/10.56279/jgat.v37i1.145.

Yasmin, N., & Grundmann, P. (2019). Adoption and diffusion of renewable energy—The case of biogas as alternative fuel for cooking in Pakistan. *Renewable and Sustainable Energy Reviews*, *101*, 255–264. https://doi.org/10.1016/j.rser.2018.10.011.

Part IV
Impact of Plastic Pollution: Case Studies

Chapter 17
A Comprehensive Review of Microplastic Contamination in Sri Lanka: Sources, Sinks, and Pathways

R. R. M. K. P. Ranatunga and D. S. Wijetunge

Abstract Microplastics (MPs) have become one of the evilest environmental pollutants worldwide. Due to its minute size and omnipresence, its potential to impact environmental systems is immeasurable. Sri Lanka reported the first evidence of MP contamination in 2017 from beach sand, and thenceforth numerous studies have been looked at MP contamination, from sources to potential impacts. This review is derived from information collected from a systematic literature survey covering more than fifty peer-reviewed authentic publications. The majority of MP studies have focused on marine environments in particular beach sand, surface water, sediment, and residing organisms including finfish and shellfish. Other studies have been conducted in rivers, estuaries, lagoons, and slatterns as well as cosmetics, personal care products, bottled water, open-air, and lichens have provided invaluable insight into MP contamination in the Sri Lankan environment. In 2021, the worst maritime disaster in Sri Lankan history encountered in containership MV *X-Press Pearl* significantly accelerated MP contamination in Sri Lankan coastal waters due to the spillage of tons of virgin plastic pellets together with a myriad of plastic products. Due to irresponsible littering behavior of people, open-dumping of unsorted waste, and poor waste-collecting and recycling mechanisms of local authorities a larger fraction of waste find its way to coastal waters. Improving waste-collecting facilities, discouraging use and manufacture of plastics and products made of non-degradable materials specifically single-use plastics, and encouraging the 4R (Refuse-Reduce-Reuse-Recycle) concept can be best practices to curb the harmful impacts of MP pollution in Sri Lanka. This work will discuss an overview of the present level of understanding of MP contamination in Sri Lanka, impacts of point sources such as the MV *X-Press Pearl* disaster, institutional involvement towards plastic-waste mitigation, regulation, challenges, and way forward.

R. R. M. K. P. Ranatunga (✉) · D. S. Wijetunge
Centre for Marine Science and Technology, Department of Zoology, University of Sri Jayewardenepura, Nugegoda, Sri Lanka
e-mail: ranatunga@sci.sjp.ac.lk

D. S. Wijetunge
e-mail: dilhara.sw@sci.sjp.ac.lk

S. Suriyanarayanan et al. (eds.), *Combating Plastic Pollution in Terrestrial Environment*,
https://doi.org/10.1007/978-981-96-2343-3_17

Keywords Microplastics · MV X-press pearl · Plastic pollution · Sri Lanka

17.1 Introduction

Plastics are synthetic polymers with high molecular weight and specific features such as durability, water resistance, versatility, bio-inertness, low production cost, etc. With the skyrocketing consumer demand, in 2022, the world plastic production exceeded 400 million tons where 32% is manufactured by China, the biggest plastic producer in the world (Plasticsthefastfacts, 2023).

Microplastics top the world's worst environmental pollutants list contributing 60–80% of marine debris (Derraik, 2002). According to Hidalgo-Ruz et al. (2012), the term "microplastics" was first used in 2004 referring to very small fragments of plastic (~50 μm) in the water column and sediments. Approximately 4–12 million metric tons of plastic that enter the ocean turn into MPs eventually (Geyer et al., 2017), and they spatially and temporally distribute due to their intrinsic properties and oceanic phenomenon. Wind patterns, oceanic surface currents, and surface drifting events strongly influence the horizontal distribution of MPs in marine waters. Verticle distribution is influenced by densities of MPs, movements of inhabitants, biofouling, turbulence, etc.

According to the origin, these are categorized as primary, secondary, and tertiary MPs where manufactured sphere-shaped, micro-sized plastic particles (i.e. nurdles, pellets, beads) are primary MPs, breakdown particles (due to weathering and fragmentation) of macro-plastics are secondary MPs, and tertiary MPs are preproduction pellets used to mold plastic goods (Carbery et al., 2018; GESAMP, 2015). According to shapes or types, different categorizations are reported including films, fragments, fibres, nurdles/beads, sheets, foam, etc. MPs are found in vibrant colors with different chemical compositions. The most common polymer types identified include polyethylene (PE), polyethylene terephthalate (PET), polypropylene (PP), polyamide (PA), polystyrene (PS), polybutylene adipate-terephthalate (PBAT), poly-lactic acid (PLA), etc.

Major pathways of MPs reaching the oceans are surface runoff, river influxes, atmospheric wind actions, and rainfall in freshwater and terrestrial systems. Approximately 80% of marine debris contains river contributions. Compared to a vast number in marine environments relatively less number of research focused on freshwater and terrestrial ecosystems (Wu et al., 2018). MPs were first recorded in freshwater lakes in 2013 (Rochman, 2018). Since then, MPs have been reported on freshwater beaches, lakes and rivers around the world as well as in freshwater organisms including insects, worms, clams, fish, and birds. Although these systems are less dilute than oceans, MP concentrations reported from freshwater systems are similar to those of marine environments. However, contamination tends to be greater with larger populations, but MPs have been reported in remote locations with lesser populations too. Experts describe this situation as a result of atmospheric deposition (Eerkes-Medrano et al., 2015; Rochman, 2018).

MPs pose devastating impacts on wildlife, environment, and human health due to their resistance to degradation, ubiquity, bioavailability, minute size, buoyancy, dispersion, persistence, etc. The presence of MPs reduces aesthetic value and causes ecological and economic loss in environmental systems. MPs can act as transport vectors for contaminants such as Persistent Organic Compounds (POCs) and heavy metals. Chemical substances and microorganisms on MPs act like pathogens, and reduce water quality making it unsuitable for drinking, bathing, recreational purposes, etc. (Van der Meulen et al., 2014). MPs have been recorded in different biological samples such as the gastrointestinal tract, gut content, gills, muscles, and skins of organisms, and cause difficulties in feeding, breathing, reproduction, and behavioral activities upon ingestion. MPs cross gastrointestinal walls and retain in the body tissues of organisms like nematodes, annelids, molluscs, and arthropods. When the majority of the organisms in the ecosystem are affected, there is a potential for the natural cycles as well as food web mechanics to be affected and eventually bioaccumulate. The uptake of microplastic particles by humans can occur through the consumption of terrestrial and aquatic food products, drinking water, salt consumption, atmospheric deposition, and inhalation. The direct impact of MPs on the health of organisms has not yet been studied in Sri Lanka.

17.2 Estimating Microplastic Contamination

MP assessments are time-consuming and costly. Apart from analyzing, sample collection and preparation should also be carried out avoiding contamination via unintended MP sources such as cleaning agents, chemicals, atmospheric deposition, etc. Modified versions of the standard MP analyzing protocols [including Wet Peroxide Oxidation (WPO), density separation, vacuum filtration, and microscopic identification] introduced by Gago et al. (2018) and Masura et al. (2015) were followed in MP assessments for surface waters, beach sand, sediment, and biological samples.

Most studies used 0.05 M Fe(II), H_2O_2 (30%) (Fenton's reagent), and/or NaOH/KOH (10%) under 60–75 °C for WPO, followed by density separation using NaBr/NaCl (1–1.2 M). Vacuum filtration was done to extract MPs from the sample. Stereomicroscopy, binocular light microscopy, or dissecting microscopy are employed in MP identification in terms of shape, color, or type. A hot needle test or Nile red addition is performed for MP confirmation. Attenuated Total Reflectance-Fourier Transform Infrared (ATR-FTIR) spectroscopy, Micro FTIR (m-FTIR), and/or RAMAN spectroscopy, coupled with Thermo Gravimetric Analysis (TGA) have been conducted for polymer type identification (Athapaththu et al., 2020; Kapukotuwa et al., 2022; Koongolla et al., 2018; Lawan et al., 2024; Ranatunga et al., 2021).

17.3 The Present Status of MP Pollution in Sri Lanka

17.3.1 Scientific Assessments of MP Contamination in Sri Lanka

The first discovery of MPs contamination was recorded in 2017 from local beaches of Negombo, Sri Lanka (Ranatunga & Karunarathna, 2018; Ranatunga et al., 2021). Another empirical study conducted by Jang et al. (2018) analyzed marine debris stranded in Sri Lankan beaches and reported 93.3% of large debris (>25 mm) and 98.7% of small debris (5–25 mm) were plastics by number. In terms of the country of origin, only 0.9% of large marine debris was of foreign origin leaving approximately 99% of large debris being locally produced. Since then, investigations have been carried out to find the occurrence of MPs in different environmental compartments reporting a significantly higher number of experiments in 2023.

The majority of MP assessments have been conducted focusing on western, southwestern, and southern coasts while north, northwestern, northeastern, and eastern coasts contain comparatively less number of records. A few assessments are reported from southeastern coastal areas. The reason might be inaccessibility to local beaches due to land coverage from wildlife national parks in the area (Fig. 17.1). Compared to other cities in the country, Negombo and adjacent coastal areas have gained a lot of attention from the scientific community. Negombo is the major economic and ecological hub of the west coast harboring the lagoon, artisanal and industrial fisheries, as well as recreational and tourism activities.

Surface waters, beach sand, coastal sediment, and marine organisms adjacent to local beaches and coastal waters are mostly subjected to scientific experiments compared to freshwater bodies, and other components in Sri Lanka mainly due to the vast discussions regarding the contribution of marine debris to marine plastic and MP pollution. Table 17.1 contains MP abundance values reported in the surface waters of Sri Lanka. Though differences in units make it difficult to compare abundance levels, the highest MP contamination levels in surface waters were reported from highly urbanized locations on the west coast of Sri Lanka.

Table 17.2 contains MP abundance in sediment, beach sand, and other dry samples collected from different locations in Sri Lanka. Despite of abundances presented in different units, raw salt from the Hambantota saltern reported extremely high amounts (3345.7 ± 311.4 items/kg) of MP contamination followed by beach sand and sediment samples collected from Dondra harbor (738 ± 195 items/m^2), Maha Oya estuary (656 items/m^2), and Colombo (329 particles/m^3). Perera et al. (2022) reported MP contamination in air samples for the first time in Sri Lanka. Indoor and outdoor air from selected locations (i.e. urban, rural, coastal, inland, industrial, national park/wetland) was passed through a 1 μm stainless-steel mesh screen using an air sampler, and mesh screens were directly observed under a stereomicroscope for MP identification.

The level of MP contamination in organisms is a function of habitat preference and feeding behavior. MP abundance values reported in edible and non-edible biota

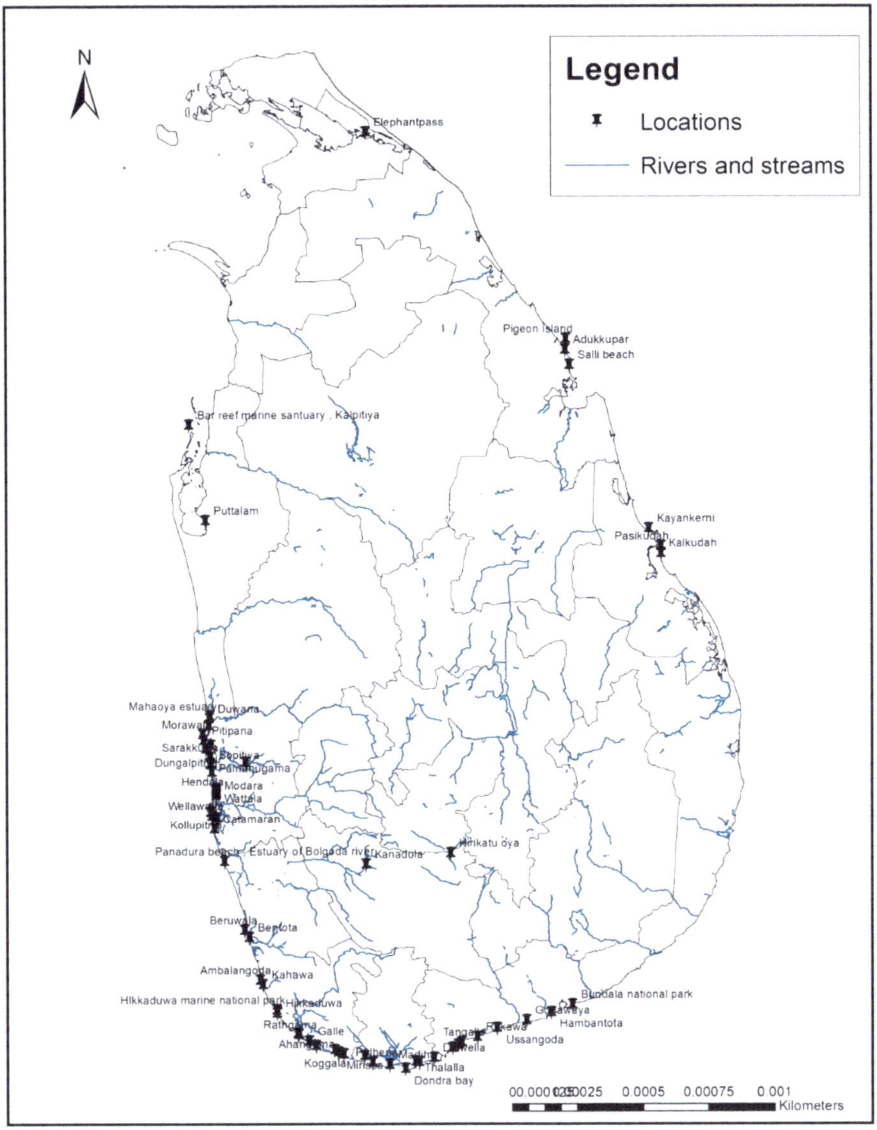

Fig. 17.1 Map of Sri Lanka showing the locations of MP studies conducted in Sri Lanka by the end of 2023

alarm the potential of indirect MP contamination in humans during consumption. Table 17.3 contains finfish, shellfish, and other types of living organisms tested for MP contamination in Sri Lanka. The digestive tract, gut content, gills, muscles, and skins of respective organisms have been tested for MP contamination. Out of all edible species, big-eye trevally from the Dondra lagoon reported the highest MP abundance

Table 17.1 MP abundance values in surface waters of Sri Lanka

Site	Abundance of MP	Reference(s)
Wellawatta	229.25 ± 46.39 items/m^3*	Athawuda et al. (2020)
Modera	201.13 ± 62.13 items/m^3*	
Hendala	161 items/L*	Weerasekara et al. (2022)
Kollupitiya	143.79 ± 52.11 items/m^3*	Athawuda et al. (2020)
Kerawalapitiya	139.75 ± 15.10 items/m^3*	
Bambalapitiya	113.85 ± 28.67 items/m^3*	
Dikkowita	76.86 ± 12.78 items/m^3	
Uswetakeiyawa	76.52 ± 9.22 items/m^3	
Hirikatu oya	69 ± 22 items/100 m^3	Lahiru et al. (2023)
Dondra Harbor	29 items/m^3	Koongolla et al. (2018)
Hambantota	23.46 ± 3.63 items/m^3	Athapaththu et al. (2020)
Tangalle	22.39 ± 5.00 items/m^3	
Mirissa	20.67 ± 3.75 items/m^3	
Pigeon Island Marine Sanctuary	20.47 ± 15.8 items/m^3 (2022)	Hansani et al. (2023)
Ambalangoda	15 items/m^3	Koongolla et al. (2018)
Hikkaduwa	15 items/m^3	
Rekawa	10.52 ± 3.10 items/m^3	Athapaththu et al. (2020)
Ahangama	10.47 ± 0.82 items/m^3	Hansani et al. (2023)
Dondra lagoon	7.11 ± 1.53–9.83 ± 2.56 items/L	Bandara et al. (2023)
Kelani river estuary	7 ± 0.28–2.5 ± 1.323 items/m^3	Thakshila and Ranatunga, (2019)
Weligama	5 items/m^3	Koongolla et al. (2018)
Negombo lagoon	2.46 ± 1.13 items/m^3	Athukorala et al. (2023b)
Dondra reef	2.38 ± 1.6 items/m^3	Hansani et al. (2023)
Browns Beach	2 items/L	Weerasekara et al. (2022)
Kepungoda	2 items/L	
Attanagalu River	1.12–17.45 items/m^3	Karunarathna and Ranatunga, (2021)
Koggala	1 items/m^3	Koongolla et al. (2018)
Bar Reef Marine Sanctuary	0.60–30.81 p/m^2 (2019)	Weerakoon and Wimalasiri, (2020)
Bar Reef Marine Sanctuary	0.54–30.43 p/m^2 (2018)	
Hikkaduwa marine national park	0.515 ± 0.054 MPs/m^3	Dharmadasa et al. (2021)
Northwest coast	0.85 particles/m^3	Weerakoon et al. (2019)
Bundala national park	0.276 ± 0.077 MPs/m^3	Dharmadasa et al. (2021)
Central-east coast	0.27 particles/m^3	Weerakoon et al. (2019)

(continued)

Table 17.1 (continued)

Site	Abundance of MP	Reference(s)
Pigeon Island Marine Sanctuary	0.26–17.09 p/m^2 (2019)	Weerakoon and Wimalasiri, (2020)
Southeast coast	0.25 particles/m^3	Weerakoon et al. (2019)
Pigeon Island Marine Sanctuary	0.24–16.83 p/m^2 (2018)	Weerakoon and Wimalasiri, (2020)
Northeast coast	0.18 particles/m^3	Weerakoon et al. (2019)
South coast	0.12 particles/m^3	
Maha-oya estuary	11.84 ± 2.62 p/dm^3	Meegahakotuwa et al. (2020)
Kahawa	0 items/m^3	Koongolla et al. (2018)
Rathgama	0 items/m^3	
Galle fort	0 items/m^3	
Unawatuna	0 items/m^3	
Thalalla	0 items/m^3	

* Top six MP abundance values in surface waters of Sri Lanka

value (531 ± 155 MP/kg). Scleractinian corals from eastern and southern coastal reef areas also reported higher MP abundances suggesting corals act as MP sinks in the sea (Hansani et al., 2023). Though numbers are revealed, the inconsistency of final data reporting makes it difficult to come to conclusions or use individual data in decision-making.

In most studies, gastrointestinal tract and gut contents contained the highest amounts of MPs (Kandeyaya et al., 2023). The larger the organism higher the potential to get contaminated via trophic transfer and/or accidental ingestion through the gills or digestive system. According to McIlwraith et al. (2021), in larger stomachs food tends to retain longer times permitting minute MP particles to accumulate and translocate into muscle tissue or the bloodstream. However, Rochman, (2018) suggests an equal possibility for minute plastic particles to be accidentally ingested as well as to intestinally excrete. Therefore it is important to focus on MP contamination in edible parts of the organisms rather than expressing as an amount for the whole fish since non-edible parts like skin, gut content, and gills most of the time are removed while preparing for human consumption.

Furthermore, the highest MP concentrations reported in Sri Lanka were from salterns (Table 17.2). However, commercial salt reported comparatively lesser contamination (11–193 MP/kg) due to purification during commercialization (Kapukotuwa et al., 2022). Therefore, the MP concentration in table salt is what matters when it comes to human consumption. Lenz et al. (2016) emphasized the importance of involving animals in MP studies under real concentrations to yield more realistic estimates of sub-lethal effects since higher concentrations experimented in laboratories are different than that of real environments.

Table 17.2 MP abundance in beach sand, sediment, and other dry samples collected from Sri Lanka

Site	Sample type	Abundance of MP	Reference(s)
Hambantota saltern	Raw salt	3345.7 ± 311.4 items/kg*	Kapukotuwa et al. (2022)
Dondra Harbor	Beach sand (10 m)	738 ± 195 items/m^2*	Koongolla et al. (2018)
Maha-oya estuary	Sediment	656 items/m^2*	Weerasekara et al. (2022)
Colombo	Beach sand	329 particles/m^3*	Weerakoon et al. (2018)
Puttalam saltern	Raw salt	272.3 ± 10.6 items/kg	Kapukotuwa et al. (2022)
Beruwala	Beach sand	183 particles/m^3	Weerakoon et al. (2018)
Weligama	Beach sand (20 m)	157 ± 94 items/m^2	Koongolla et al. (2018)
Hikkaduwa	Beach sand	135 particles/m^3	Weerakoon et al. (2018)
Rathgama	Beach sand (20 m)	133 ± 63 items/m^2	Koongolla et al. (2018)
	Beach sand (10 m)	125 ± 50 items/m^2	
Hikkaduwa Marine National Park	Sediment	111 ± 29 MPs/m^2	Dharmadasa et al. (2021)
Ussangoda	Sediment	103.33 ± 30.5 items/kg	Hansani et al. (2023)
Bundala National Park	Sediment	102 ± 16 MPs/m^2	Dharmadasa et al. (2021)
Hikkaduwa	Beach sand (10 m)	97 ± 93 items/m^2	Koongolla et al. (2018)
Negombo lagoon	Sediment	62.33 ± 45.16 items/kg	Athukorala et al. (2023b)
Dondra lagoon	Dry sediment	56.25 ± 12.6–116.67 ± 44.9 items/kg	Bandara et al. (2023)
Galle fort	Beach sand (20 m)	54 ± 23 items/m^2	Koongolla et al. (2018)
Ambalangoda	Beach sand (20 m)	48 ± 6 items/m^2	
Adukkuparu	Sediment	46.3 ± 29.7 items/kg	Hansani et al. (2023)
Elephant Pass saltern	Raw salt	36.3 ± 4.5 items/kg	Kapukotuwa et al. (2022)
Kahawa	Beach sand (20 m)	21 ± 21 items/m^2	Koongolla et al. (2018)
Salli beach	Sediment	20 ± 13.2 items/kg	Hansani et al. (2023)
Hirikatu oya	Sediment	top: 15 ± 5.92 items/kg	Lahiru et al. (2023)
Attanagalla River	Sediment	11.38–129.47 items/kg	Karunarathna and Ranatunga, (2021)

(continued)

Table 17.2 (continued)

Site	Sample type	Abundance of MP	Reference(s)
Hikkaduwa	Beach sand (20 m)	8 ± 8 items/m^2	Koongolla et al. (2018)
Pitipana	Beach sand	7.2 ± 7.66 particles/m^3	Ranatunga and Karunarathna, (2018), Ranatunga et al. (2021)
Hirikatu oya	Sediment	bottom: 7 ± 3.28 items/kg	Lahiru et al. (2023)
Ambalangoda	Beach sand (10 m)	6 ± 6 items/m^2	Koongolla et al. (2018)
Catamaran	Beach sand	5.2 ± 3.77 particles/m^3	Ranatunga and Karunarathna, (2018), Ranatunga et al. (2021)
Kelani river estuary	Sediment	4 ± 0.5–1 ± 0.43 item/kg	Thakshila and Ranatunga, (2019)
Dungalpitiya	Sediment	3 items/m^2	Weerasekara et al. (2022)
Dūwana	Beach sand	2.8 ± 1.30 particles/m^3	Ranatunga and Karunarathna, (2018), Ranatunga et al. (2021)
Indoor sampling	Air	0.13–0.93 n/m^3	Perera et al. (2022)
Outdoor sampling	Air	0.01–0.23 n/m^3	
Kahawa	Beach sand (10 m)	0 ± 0 items/m^2	Weerasekara et al. (2022)
Galle fort	Beach sand (10 m)	0 ± 0 items/m^2	Koongolla et al. (2018)
Unawatuna	Beach sand (20 m)	0 ± 0 items/m^2	
	Beach sand (10 m)	0 ± 0 items/m^2	
Weligama	Beach sand (10 m)	0 ± 0 items/m^2	
Dondra Harbor	Beach sand (20 m)	0 ± 0 items/m^2	
Thalalla	Beach sand (20 m)	0 ± 0 items/m^2	
	Beach sand (10 m)	0 ± 0 items/m^2	
Koggala	Beach sand (20 m)	0 ± 0 items/m^2	
	Beach sand (10 m)	0 ± 0 items/m^2	

* Top four MP abundance values in beach sand, sediment, and other dry samples collected from Sri Lanka

Table 17.3 MP abundance reported in edible and non-edible organisms in Sri Lanka

	Test organism	Location	MP abundance	Reference(s)
Finfish	Spotted herring (*Herklotsichthys quadrimaculatus*)	Weilgama	9.40 ± 3.05 MP/ind	Gamage et al. (2023)
		Tangalle	2.6 ± 0.894 MP/ind	
		Kottegoda	5.80 ± 4.60 MP/ind	
	Big-eye trevally (*Caranx sexfasciatu*)	Dondra lagoon	531 ± 155 MP/kg*	Bandara et al. (2023)
	Stinging catfish (*Heteropneustes fossilis*)	Dondra lagoon	11.07 ± 5.66 MP/kg	Bandara et al. (2023)
		Bellanwila-Attidiya urban wet land	3.0 ± 2.1 items/ind	Weerasuriya et al. (2023)
	Flathead grey mullet (*Mugil cephalus*)	Negombo lagoon	2.68 ± 0.23 MP/ind	Athukorala et al. (2023a)
	Skipjack tuna (*Katsuwonus pelamis*)	Negombo lagoon	1.42 ± 0.29 MP/g	Kandeyaya et al. (2023)
	Jumping halfbeak (*Hemiramphus archipelagicus*)	Negombo lagoon	1.17 ± 0.14 MP/g	
	Commerson's anchovy (*Stolephorus commersonnii*)	Negombo lagoon	0.83 ± 0.10 MP/g	
	Goldstripe sardine (*Sardinella gibbosa*)	Negombo lagoon	0.82 ± 0.30 MP/g	
	Nile tilapia (*Oreochromis niloticus*)	Bellanwila-Attidiya urban wet land	19.0 ± 3.0 items/ind	Weerasooriya et al. (2023)
Shell fish	Rock oyster (*Saccostrea cucullata*)	Godawaya, Rekawa, Tangalle, Dickwella, Polhena, Mirissa, Weligama, Unawatuna, Galle fort, Dodanduwa, Hikkaduwa	7.2–2.8 MP/g	Wijethunga et al. (2019)
	Indian backwater oyster (*Magallana bilineata)*	Panadura estuary	5.133 ± 4.37 MP/ind	Nawarathna et al. (2023)
	Periwinkle (*Littorina* sp.)	Godawaya, Rekawa, Tangalle, Dickwella, Polhena, Mirissa, Weligama, Unawatuna, Galle fort, Dodanduwa, Hikkaduwa	7.2–2.8 MP/g	Wijethunga et al. (2019)

(continued)

Table 17.3 (continued)

	Test organism	Location	MP abundance	Reference(s)
	Limpets (*Patella* sp.)	Godawaya, Rekawa, Tangalle, Dickwella, Polhena, Mirissa, Weligama, Unawatuna, Galle fort, Dodanduwa, Hikkaduwa	7.2–2.8 MP/g	
	Brown mussel (*Perna perna*)	Pallikudawa, Seeni modara, Galle harbor, Unawatuna	1.75 ± 0.5 MP/g	Dissanayaka and Rathnasuriya, (2022)
		Negombo lagoon	1.4 ± 0.06 MP/g	Kandeyaya et al. (2023)
	Mud crab (*Scylla serrata*)	Negombo lagoon	1.8 ± 0.21 MP/g	Kandeyaya et al. (2023)
		Negombo lagoon	Gills: 0.70 ± 0.52 items/g Digestive tracts: 0.71 ± 0.52 items/g	Dias et al. (2023)
	Giant tiger prawn (*Penaeus monodon*)	Negombo lagoon	1.7 ± 0.29 MP/g	Kandeyaya et al. (2023)
			8.29 ± 4.63 items/g in (GI and GL)	Lawan et al. (2024)
	Indian prawn (*Penaeus indicus*)	Negombo lagoon	5.52 ± 3.78 items/g of (GI and GL)	Lawan et al. (2024)
	Cuttlefish (*Sepia* sp.)	Negombo lagoon	0.04 ± 0.02 MP/g	Kandeyaya et al. (2023)
Other	Scleractinian Coral (*Acropora* sp.)	Eastern and southern coasts	675.52 ± 149 items/kg*	Hansani et al. (2023)
		Ahangama reef	866.57 ± 525.60 items/kg*	
	Sclerectinian Coral (*Montipora* sp.)	Ahangama reef	1728.94 ± 395.33 items/kg*	
	Scleractinian Coral	Pasikudah reef	546.7 ± 170.3 items/kg*	
	Scleractinian Coral (*Acropora* sp.)	Salli Beach reef	361.03 ± 323.98 items/kg*	
	Needle seagrass (*Halodule pinifolia*)	Negombo lagoon	Leaf surfaces: 8 ± 3 items/g	Rathnayake et al. (2023)
	Macro algae (*Chaetomorpha* sp.)	Negombo lagoon	Leaf surfaces: 6 ± 1 items/g	

* Top six MP contamination values in edible and non-edible organisms in Sri Lanka

When considering prominent MP types, the majority (80%) of the literature (Ranatunga & Karunarathna, 2018; Koongolla et al., 2018; Viraj et al., 2019; Thakshila & Ranatunga, 2019; Wijethunga et al., 2019; Weerakoon et al., 2019; Athapaththu et al., 2020; Meegahakotuwa et al., 2020; Kapukotuwa et al., 2020; Dharmadasa et al., 2021; Ranatunga et al., 2021; Karunarathna & Ranatunga, 2021; Kapukotuwa et al., 2022; Weerasekara et al., 2022; Dissanayake and Rathnasuriya, 2022; Perera et al., 2022; de Vos et al., 2022; Sewwandi et al. 2022a, 2022b; Athukorala et al., 2023a; Athukorala et al., 2023b; Dias et al., 2023; Bandara et al., 2023; Rathnayake et al., 2023; Weerasuriya et al., 2023; Kandeyaya et al., 2023; Hansani et al., 2023; Nawarathna et al., 2023; Gamage et al., 2023; Lawan et al., 2024) recorded the presence of fragments and fibers in Sri Lanka (Fig. 17.2) however aftermath the MV *X-Press Pearl* disaster, more nurdle/pellet sightings were reported specifically from the west and south-west coastal areas.

Pellets/nurdles/microbeads are spherical-shaped, industrially produced primary MPs. Fragments, fibres, films, foam, and rod-like MP are secondary MPs. Fragments and rod-like MP include small broken-out particles from larger items. Monofilament and nylon fishing gear discarded in fishing grounds may create thousands of MP fibres and also artificial clothing fibres brought via surface runoff. The worm-like shape of fibers/filaments increases the ability to be misidentified as food prey. Foam is primarily made out of PS and disperses horizontally in the water column due to lower density. Mostly foam MPs were generated from single-use Styrofoam food packaging, fishing buoys, and other packaging material. Plate 1 contains stereomicroscopic images of major MP types recorded from Sri Lanka.

The color of MPs determines the potential to get ingested by foraging inhabitant organisms. In experimental set-ups the color makes it easy to visually identify the particles and also to identify potential sources. The most dominant color of MP reported from Sri Lanka was blue (85.71%) followed by transparent (52.38%), white, and red (47.62%) (Fig. 17.3) (Ranatunga & Karunarathna, 2018; Thakshila & Ranatunga, 2019; Athapaththu et al., 2020; Meegahakotuwa et al., 2020; Ranatunga et al., 2021; Dharmadasa et al., 2021; Dissanayaka and Rathnasuriya, 2022; Weerasekara et al., 2022; Perera et al., 2022; Gamage et al., 2023; Bandara et al., 2023; Athukorala et al., 2023a; Athukorala et al., 2023b; Weerasuriya et al., 2023; Dias et al., 2023; Rathnayake et al., 2023; Kandeyaya et al., 2023; Hansani et al., 2023; Lahiru et al., 2023; Nawarathna et al., 2023; Lawan et al., 2024).

Most of the literature (78.26%) reported polypropylene (PP) as the dominant polymer type recorded from Sri Lanka followed by polystyrene (PS) (47.83%), and polyethylene (PE) (43.48%) (Fig. 17.4) (Ranatunga & Karunarathna, 2018; Koongolla et al., 2018; Thakshila & Ranatunga, 2019; Athapaththu et al., 2020; Meegahakotuwa et al., 2020; Kapukotuwa et al., 2020; Kapukotuwa et al., 2022; Dharmadasa et al., 2021; Ranatunga et al., 2021; Karunarathna & Ranatunga, 2021; Weerasekara et al., 2022; Perera et al., 2022; de Vos et al., 2022; Sewwandi et al., 2022a, 2022b; Athukorala et al., 2023a; Athukorala et al., 2023b; Weerasuriya et al., 2023; Kandeyaya et al., 2023; Hansani et al., 2023; Lahiru et al., 2023; Nawarathna et al., 2023; Lawan et al., 2024). According to Napper and Thompson (2016), domestic and textile-washing effluents contain acrylic, PES, and PA MPs.

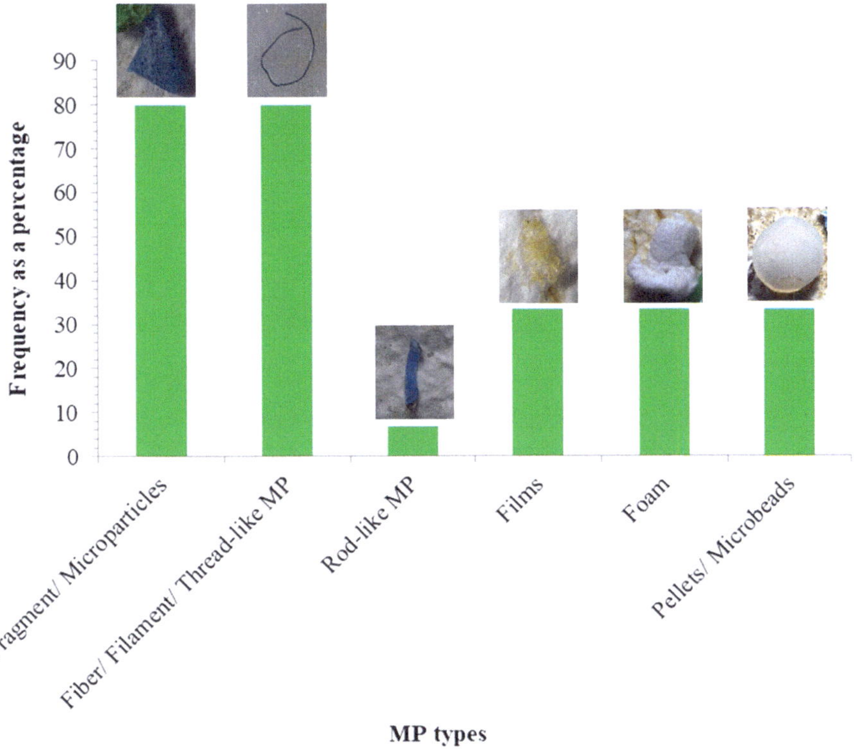

Fig. 17.2 Analysis of major types of MP shapes reported in Sri Lanka with reference to Ranatunga & Karunarathna, 2018, Koongolla et al., 2018, Viraj et al., 2019, Thakshila & Ranatunga, 2019, Wijethunga et al., 2019, Weerakoon et al., 2019, Athapaththu et al., 2020, Meegahakotuwa et al., 2020, Kapukotuwa et al., 2020, Dharmadasa et al., 2021, Ranatunga et al., 2021, Karunarathna & Ranatunga, 2021, Kapukotuwa et al., 2022, Weerasekara et al., 2022, Dissanayaka and Rathnasuriya, 2022, Perera et al., 2022, de Vos et al., 2022, Sewwandi et al. 2022a, 2022b, Athukorala et al., 2023a, Athukorala et al., 2023b, Dias et al., 2023, Bandara et al., 2023, Rathnayake et al., 2023, Weerasuriya et al., 2023, Kandeyaya et al., 2023, Hansani et al., 2023, Nawarathna et al., 2023, Gamage et al., 2023, and Lawan et al., 2024

The polymer type of the MP particle determines at which strata of the water body it tends to accumulate. The density of the particle changes with chemical composition as well as reactions onset of other chemicals. Therefore, foam which is mostly made of PS is less likely to be expected in sediment or bottom strata of a water body or inside bottom dwellers.

Plate 1 Stereomicroscopic images of common MP types recorded in Sri Lanka (2 × 30 × 1) (*a-j*: fragments/microparticles, *k-n*: rod-like MP, *o-r*: fibers/filaments/thread-like MP, *s*: nurdle/pellet/bead, *t-v*: foam, *w,x*: film)

17.3.2 Impact of X-Press Pearl Maritime Disaster

The West coast is valued for artisanal and industrial fishing in Sri Lanka. On 20th May 2021, the Singapore-flagged MV *X-Press Pearl* containership carrying approximately 1680 tons of virgin plastic pellets, with a nitric acid leak caught fire in Sri Lankan waters 18 km off the west coast. Scientists reported this as the worst maritime disaster in Sri Lankan history. Over nine months, this was the second maritime disaster Sri Lanka experienced after the MT *New Diamond* crude oil carrier

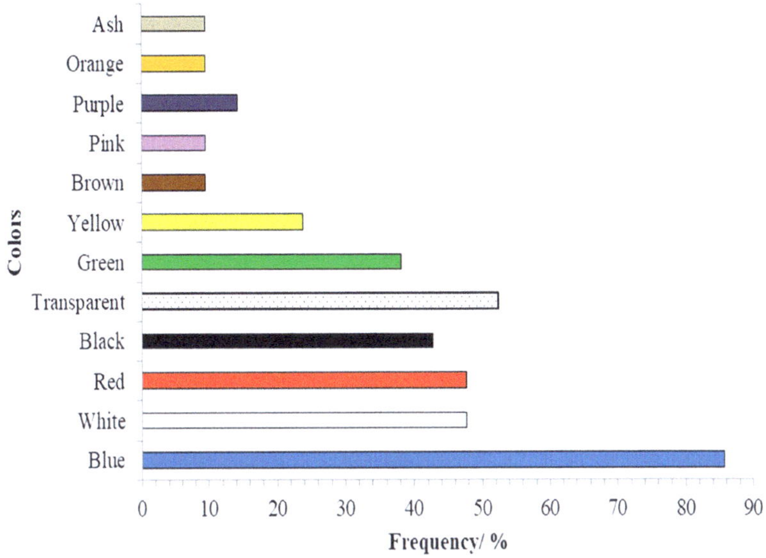

Fig. 17.3 Colors of MP reported in Sri Lanka

in September 2020 which leaked approximately 1700 million tons of fuel. And less than 2 months after the MV *X-Press Pearl* disaster, another containership MSC *Messina*, caught fire 480 miles outside the Sri Lankan coast (Rubesinghe et al., 2022; de Vos et al., 2022; UN Environment Programme and UN Office for the Coordination of Humanitarian Affairs, 2021).

The immediate impact of the accident was waves of industrial pellets hoarding ashore approximately a 100 km stretch between Negombo and Kalutara. The fire and explosions caused cargo to fall aboard including containers of nurdles. Sarakkuwa in Gampaha district was the first and the most impacted area by the mass washing up of nurdles. Sri Lankan navy, coast guard, air force troops, conservationists, environmentalists, and the general public initiated massive cleanup operations at the beginning to control obvious plastic pollution by nurdles under the supervision of MEPA (Marine Environmental Protection Authority). According to Sewwandi et al. (2022b), a mechanized beach cleaning unit called the blue treatment facility was effectively used for the physical separation of MP from sand. Initially, manual sieving and separation by sea water were used which was highly labor consuming. Near the inlets of Negombo Lagoon, booms were deployed along entrance channels to protect the fishing ground from plastic pollution. Since then to date, even after more than 2.5 years, although in small quantities, nurdles probably from the incident are found on beaches and the coast of Sri Lanka.

Among the major environmental impacts of the disaster are (Fig. 17.5),

1. Chemical pollution due to hazardous and potentially toxic chemicals.
2. Oil pollution due to bunker fuel oil spill.

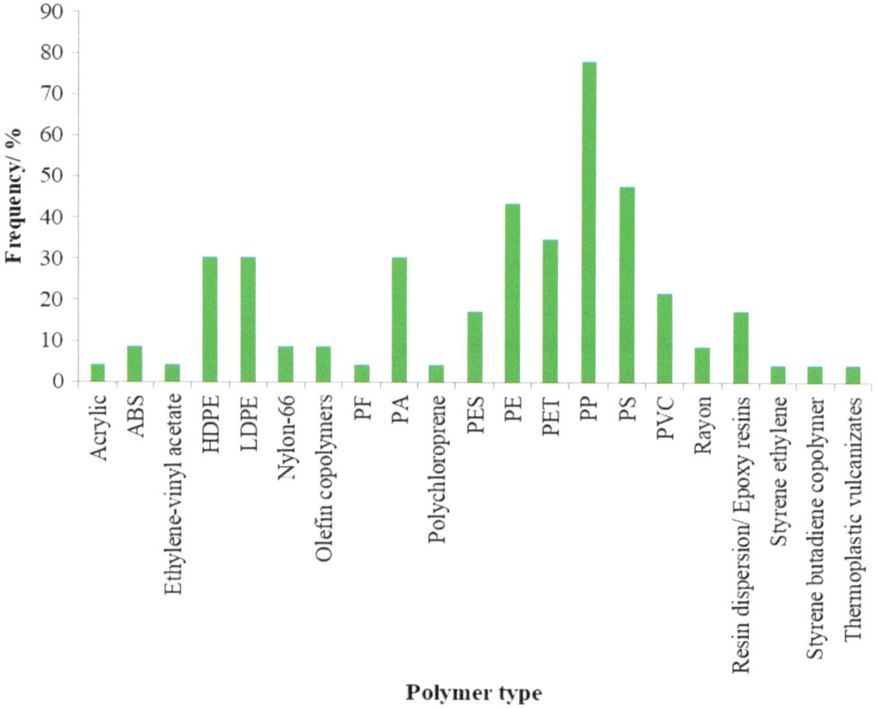

Fig. 17.4 Polymer types of MPs reported in Sri Lanka (ABS: Acrylonitrile Butadiene Styrene; HDPE: High-density polyethylene; LDPE: Low-density polyethylene; PF: Phenol formaldehyde; PA: Poly Amide; PES: Polyester; PE: Polyethylene; PET: Polyethylene terephthalate; PP: Polypropylene; PS: Polystyrene; PVC: Polyvinyl chloride)

3. Plastic pollution due to microplastics, macro-plastics, and burnt plastics.
4. Fire residue.
5. Bulk debris including shipwreck, cargo, and damaged containers.
6. Air pollution.
7. Impacts on wildlife and sensitive environments.
8. Impacts on fishery resources and fishing communities
9. Legislative issues due to shipwreck and lost containers.

17.3.2.1 Chemical Pollution Due to Hazardous and Potentially Toxic Chemicals

The hazardous and toxic chemicals contained in the vessel included nitric acid (>25 tons), caustic soda (1040 tons), methanol (210 tons), sodium methoxide, sulfur, powder urea, sodium methylate, vinyl acetate, epoxy resins (9700 tons), etc., along with heavy metals such as Cu, Pb, Li, Mo and Cr (Sewwandi et al., 2022a; UN Environment Programme and UN Office for the Coordination of Humanitarian Affairs,

Fig. 17.5 Graphical representation of the summary of impacts of the MV *X-Press Pearl* disaster

2021). The damaged or leaked chemical containers caused immediate impacts while lost, sunken or displaced containers pose a risk of future contamination.

Nitric acid and caustic soda have higher densities than seawater and also dissolve in water. The chemical reaction might have created a highly corrosive plume of nitric acid and caustic soda on the seabed. The corrosive nature of the sinking chemical plume might be the reason for observations like damaged fishing gear and burnt or bleached carapace of some dead sea turtles washed up ashore. Their shells and skins might have disintegrated by contact with the chemical plume. Once spilled, methanol floats on the subsurface forming a plume of contaminated seawater. This eventually might turn into a toxic vapor cloud above the water surface due to evaporation processes and might adversely impact surface-dwelling marine organisms. Epoxy resins are harmful depending on the type of formulation, either liquid, solid, paste, or gel. The liquid form is the most dangerous to the environment. If it was in liquid form, it would sink due to higher density and create a plume near the sea floor. The toxic plume will float with currents and interact with suspended material in the surrounding water column creating long-lasting impacts on residing marine fauna. If ingested these chemicals can cause severe skin and eye burns.

17.3.2.2 Oil Pollution Due to Bunker Fuel Oil Spill

An approximate 348 tons of bunker fuel oil was spilled creating a possible oil slick of an approximate area of 0.51 km^2 and a length of 3.23 km. The slick expanded around 4.3 km in 6 days and continuously leaked for nearly a month. Sentinel-1-SAR satellite imagery showed a long plume heading northeast toward the coast in late June-early July times. The oil spill might adversely impact mangroves since their aerial roots capture floating debris and removal is difficult.

17.3.2.3 Plastic Pollution Due to Microplastics, Macro-Plastics (5–50 mm), and Burnt Plastics

The containership carried 87 containers ~ 1680 tons of plastic pellets. According to de Vos et al. (2022), plastics could be observed in three major forms namely un-burnt nurdles (white, opaque, homogenous, spherical, smooth), burnt plastic pieces (completely black/dark green, similar to pyro-plastics, rough, brittle), and degraded plastic pellets (yellowed, darkened, speckled with black inclusions). According to ATR-FTIR spectroscopy plastics washed ashore were confirmed as PE. Distorted plastics after exposure to heat, chemicals, and combustion were agglomerated, fragmented, charred, and chemically modified, and accumulated in adjacent local beaches. Though the cargo details didn't report any kind of different colored pellets other than white, the presence of yellow and black colored nurdles may be due to chemical reactions and combustion effect between sulphur.

According to Sewwandi et al. (2022b) in August 2020, before the MV *X-Press Pearl* incident, only 5 pellets were recorded within $0.25 \times 0.25m^2$ quadrat showing moderate pollution. In May 2021, onsite of the incident, 711 pellets were reported within a quadrat exhibiting heavy pollution. In June 2021, 1 month after the incident, 329 pellets were reported within a quadrat exhibiting a considerable reduction of pollution due to manual and mechanical cleaning processes, though still in the very high pollution range. Figure 17.6 exhibits the drastic changes in calculated Pellet Pollution Index (PPI) values by Sewwandi et al. (2022b).

By October 2021 the Sarakkuwa beach, Negombo was contaminated with 2–5 mm partially pyrolyzed LDPE fragments and nurdles with a peak abundance of 13.30 g/kg and 1 mm–500 μm sized LDPE fragments up to 2 m depth in the sand (Sewwandi et al., 2022a). Nurdle pollution is predicted to become worse with the onset of the southwest monsoon and might be a future threat to turtle nesting behavior. MPs impact the temperature of beach sand which is critical in the sex determination of turtles. In the presence of UV radiation plastics degrade further and generate more and more micro and nano particles and the threat will keep on existing leading to biomagnification and bioaccumulation upon ingestion.

Nurdles undergo natural weathering and achieve a higher tendency to carry toxic contaminants. Similarly, biofilm colonization of MPs leads to higher exposure to various potential pathogens Sewwandi et al. (2022a).

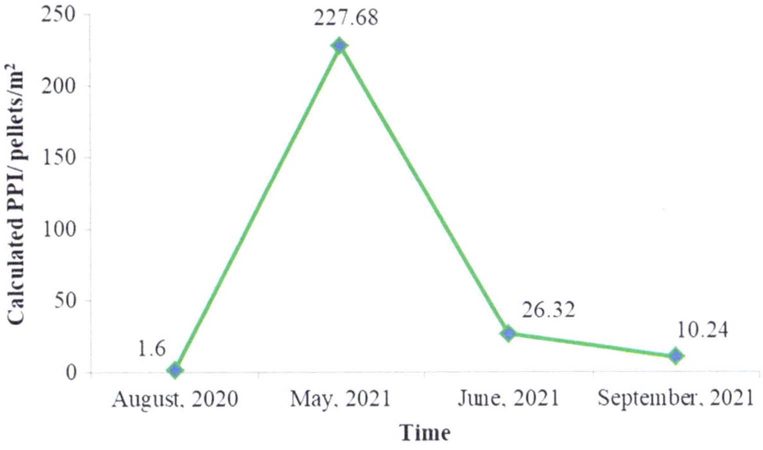

Fig. 17.6 Difference of calculated PPI index values before and after the incident

17.3.2.4 Fire Residue

The containership burnt for 13 days, and the fire was controlled by dropping foam, a dry chemical powder, and water by air (Rubesinghe et al., 2022). The impact of fire residue has not been evaluated.

17.3.2.5 Bulk Debris Including Shipwreck, Cargo, and Damaged Containers

Following the outbreak of fire and explosions, the ship generated assorted bulk debris including cargo, damaged and non-damaged containers, and the shipwreck itself. The shipwreck created a maritime hazard as well as port traffic in the Colombo harbor, which is one of the top 20 busiest ports in the world. Lost containers create navigational risk at the sea since they might drift below the sea surface and collide with merchant ships. The presence of the wreck in a known popular, fishing ground impacts the livelihoods of the local fishing community due to disturbances as well as the immediate fishing ban implemented along an 80 km littoral stretch adjacent to the incident pausing their day-to-day livelihood.

17.3.2.6 Air Pollution

The massive fire and explosions caused air pollution that lasted for 10 days, and the smoke plume was highly visible spreading from the incident site. An estimated 8,000–13,000 tons of toxic air pollutants including nitrogen oxide, carbon monoxide, sulphur dioxide, soot, particulate matter, furans, heavy metals, etc., are being released.

17.3.2.7 Impacts on Wildlife and Sensitive Environments

Potential threats to turtles, sea birds, crows, dogs, and other residing animals by actively or passively interacting and ingesting floating plastics were reported. Though it was not certain if the major spike in turtle and marine mammal (dolphin) deaths was due to a tragic incident or a coincident or higher reporting due to an increase of concern, Rubesinghe et al. (2022) reported deaths of 258 turtles, 43 dolphins, and 6 whales by the end of July 2021. Dead animal carcasses washed ashore and started getting collected daily. Turtle mortality in different stages was observed.

1. Turtle carapaces are being bleached or sheared off due to potential chemical burns.
2. Juvenile turtles with no sign of damage to the carapace, specifically green turtle (*Chelonia mydas*) and olive ridley (*Lepidochelys olivacea*) were in apparently good health with guts full of food.

Potential fish death was reported due to the observation of pellets lodging around their gills obstructing breathing functions. By September 2021 more than 800 marine animal strandings were reported. According to modeling studies, the spilled pellets may reach Indonesia, Malaysia, to Somalia crossing the Indian Ocean coastlines (UN Environment Programme and UN Office for the Coordination of Humanitarian Affairs, 2021).

The entire coastal area is critically economically important for fisheries, aquaculture, and tourism. The area was known for nursery grounds for fish, spiny lobsters, shrimps, and crabs. Seagrass meadows and mangroves in the Negombo lagoon, turtle nesting beaches, and marine aquarium fish spawning areas were heavily affected. Long-term deposition of MPs might affect microbiological and physiochemical properties of the soil influencing soil fertility, and availability of nutrients (Sewwandi et al., 2022a).

17.3.2.8 Impacts on Fishery Resources and Fishing Communities

The impacted area being a known fishing ground made the biggest loss. Among the victims of the tragic disaster were boat owners, beach-seine and traditional *theppam* boat owners and operators, fish sorting individuals, multiday-day boat operators, lagoon fishermen, small-scale and large-scale fish sellers, tour guides, ornamental fish sellers, etc. Immediate banning of fishing activities economically impacted local coastal fishing communities. At the same time, the reduction of consumer demand for "contaminated fish" due to health risks after pollution, and damaged and dissolved nets and other fishing gear caused reduced fish supply. An estimated 20,000 fishing families were impacted by the incident. Other than that, complaints regarding decreased catch, changes in the sea, and allergic reactions to contact with seawater were lodged (Rubesinghe et al., 2022).

17.3.2.9 Legislative Issues Due to Shipwreck and Lost Containers

Following the accident, Sri Lanka called on the International Maritime Organization (IMO) to classify plastic pellets as hazardous substances after considering the impact caused in the coastal waters of Sri Lanka.

The Center for Environmental Justice, a public interest environmental organization in Sri Lanka filed a lawsuit with a plea toward the authorities involved in mitigation responses under case no. CA/WRT/383/21. And also, the case SC FR 168/21 was filed representing the fishing communities.

As of 4th of August 2021, 899 tons of unburnt nurdles, burnt plastics, and other debris were dispatched from Sri Lankan beaches (Rubesinghe et al., 2022). Continuous reporting of potential MV *X-Press Pearl* debris is essential to calculate the magnitude of the disaster. Therefore educating, and empowering citizen scientists to continue safe reporting of nurdle sightings is important. Revising existing response efforts, protocols, rules, and regulations must be done in order to urgently act in future similar scenarios. Scientific research must be encouraged to continue to find out potential long-term impacts on the environment, organisms, and human health.

17.4 Discussion

Sri Lanka annually imports 160,000 metric tons of plastic raw materials, and another 100,000 metric tons of finished and intermediate plastic products (Premarathna et al., 2023). With no proper mechanism to recycle plastic waste in the country, most of the plastic waste eventually ends up in landfills causing devastating environmental hazards. Being a developing country under a critical economic crisis, allocating funds for waste management itself is a challenge.

17.4.1 Institutional Mediation Toward Plastic Waste Mitigation

Among the major contributing factors to MP pollution in Sri Lanka are, poor littering behavior of citizens, rapid urbanization of cities with populated residencies, unregulated industrial and domestic discharges, unregulated river and water channel openings to the coast, unsustainable tourism activities in populated tourist and recreational destinations, lack of sound and systematic waste collection mechanisms, rules and regulations, unsustainable fishing practices in local fishery harbors and adjacent waters, unsustainable development projects like port construction, harbor operational activities, hydrodynamics and vertical and horizontal ocean current circulation, intermonsoon and monsoonal rainfall patterns, wind conditions directions, etc. (Athapaththu et al., 2020; Kandeyaya et al., 2023).

One of the main responsibilities of any government is to maintain a proper waste management system within the country that directly involves waste generation, waste disposal as well as waste recycling. According to local authorities, less than 59% of the waste generated within the country is properly collected and discarded. Nijamdeen et al. (2020) reported municipal waste generation in Sri Lanka as 6500–7000 metric tons per day, which varies from 0.4 to 1 kg/day/capita based on living standards. However, the municipal solid waste collection rate is 3500 metric tons per day, which is almost 50% of the waste generated. This implies that 50% of the waste generated at the municipal level becomes mismanaged. Kariyawasam et al. (2019), reported that 0.75 kg/ppd of solid waste is generated by approximately 2.7 million of urban population in coastal cities of Sri Lanka.

Government authorities, non-government organizations, public and private waste recycling companies, academics, and citizen scientists including the public are active stakeholders in the plastic-free movement of the country. In Sri Lanka, several regulatory institutions are involved in solid waste management such as the National Solid Waste Management Support Centre (NSWMSC), Central Environmental Authority (CEA), Urban Development Authority (UDA), Marine Environment Protection Authority (MEPA), Ministry of Mahaweli Development and Environment, Ministry of Megapolis and Western Development, Ministry of Environment and Natural Resources, and Ministry of Local Governments and Provincial Councils. These institutions execute various projects to manage municipal solid waste and are responsible for policymaking, training, and providing equipment to the local government authorities in Sri Lanka.

Local government authorities such as Pradeshiya Sabha (PS), Town Councils (TC), and Municipal councils (MC) have the responsibility to collect waste from the source and dispose at disposal sites. As per the Municipal Council Ordinance in 1947, the Urban Council Ordinance in 1939, and the PS Act in 1987, 111 local authorities function through the elected members of the particular area responsible for the collection, transportation, and disposal of solid waste properly. All disposal activities should be handled according to the National Environment Act No 47 (Dharmasiri, 2020).

The relevant ministry of the provincial councils has the overall responsibility to enforce rules and regulations, manage the operation and transportation, and process Municipal Solid Waste (MSW) with the management of final waste disposal facilities without being a nuisance to the public and/or to the flora and fauna of the province. Several waste management steps have already been identified by the Waste Management Authority, of the Western Provincial Council for the better handling of and cleaning of the environment of their area including,

- Proper management of solid waste source (i.e. segregation, reduction, reuse, recycling, and prohibiting the dumping of solid waste on streets or public places and open burning of wastes are being considered)
- Proper collection/acceptance of MSW from the source of generation
- Cleaning of streets and public places
- Abolition of open MSW storage bins from main roads, and public places and introduction of closed type of waste receptacles
- Prohibition of waste transportation in open vehicles and the optimization of the efficiency of vehicle usage for waste transportation
- Treating the collected MSW as a "resource" in terms of composting, power generation, and production of biogas, biofuel, electricity, biogas, etc.
- Establishing of engineered and/or sanitary landfills.

Though a number of government institutes are assigned for MSW management in Sri Lanka, the lack of coordination between institutions, lack of institutional capability with technical expertise, lack of scientific data, and the lack of adequate funds are some issues in the system at large (Ranatunga et al. 2023).

17.4.2 Policies and Regulations

A study conducted related to community awareness and perceptions toward the impacts of MPs reported a higher awareness among the public though their alertness on regulatory measurements introduced by the authorities to control the usage of plastic products, is rather poor (Premarathna et al., 2023). Therefore, it is of high importance to introduce effective strategies to enhance the awareness of regulatory measures to control plastic usage among the public. Better awareness of society on environmental issues favors the willingness of the public to understand and abide by the regulations eventually.

The Government has the authority to implement control bans against mismanagement of waste plastic. Therefore, the Government of Sri Lanka passed a bill to ban the usage of PE and other plastics. Accordingly, the gazette notifications No. 2034/33, 2034/34, 2034/35, 2034/36, 2034/37, and 2034/38 issued orders and regulations to prohibit the manufacture of polythene or any polythene product with 20 μm and below, food wrappers (lunch sheets), any bag of high density (grocery bags), burning of refuse and other combustible matters inclusive of plastic, use of polyethylene products as decorations, manufacture of food containers, plates, cups, spoons from expanded polystyrene under the National Environment Act No. 47 of 1980 and Act No. 53 of 2000 in 2017 (Athapathth et al., 2020).

Gazette notifications in 1994, 2016, and 2017 also ban the production and use of polyethylene sheets, polystyrene-based packaging, and cutlery. However, the Sri Lankan government struggled to implement policies restricting the use of plastic, succumbing to pressure from plastic manufacturers and retailers.

MEPA, as a national agency plays a major role in marine litter management in Sri Lanka. Their involvement during the MV *X-Press Pearl* disaster was significant. After the MV *X-Press Pearl* maritime disaster, the country requires stern rules and regulations and integrated coastal and marine plastics abatement programs to prevent developing the affected area as a plastic garbage patch in the future.

17.5 Conclusion: Challenges and Way Forward

There have been considerable advances in scientific MP research in Sri Lanka. Along with assessing the available amounts of MPs in different environmental or biological compartments for satisfying baseline measurements, experiments focusing on identifying the sources or origins of MP particles, assessing the direct impact on biological tissues, possible pathways of human consumption, and potential suggestions to mitigate plastic pollution should be conducted. One of the major drawbacks to be addressed locally as well as internationally is inconsistency and inaccuracy of data. Different authors presenting plastic or MPs abundance values calculated following a variety of methodologies in different units make it less likely to be compared though conducted under similar conditions. Further, data acquired from unreliable sources creates misinterpretations. Therefore, existing protocols must be standardized and widely shared among the scientific community.

Being located on a major East–West shipping route and with enormous investments underway to build a regional maritime and logistical hub, there is a huge possibility for more similar things like the MV *X-Press Pearl* maritime disaster to happen in Sri Lankan waters in the future. Therefore, establishing proactive protective measures including large-scale long-term monitoring processes in the marine environment should be in hand to implement in case.

Sustainable marine debris management strategies including source reduction mainly targeting single-use plastics and rinse-off cosmetics containing micro-beads (facewash, toothpaste, scrub, detergent, etc.), the introduction of non-plastic reusable alternatives, cost-effective clean-ups, standardized monitoring protocols, global partnerships, and remediation technologies are needed for Sri Lanka with sound policy measures to minimize the MP issue. "Zero waste policy" is suggested as one of the most sustainable approaches unless a better alternative for plastics is introduced. The alternative should be user-friendly, reachable, and affordable for the majority. Strategies such as the provision of free environmentally friendly bags, incentivizing for re-using plastic utensils, and exchanging of money for used plastics can be implemented. Stern penalties for plastic usage can be adopted once the space is adjusted for attitude and behavioral changes through education. The relevant authorities in municipal solid waste management should be provided with sufficient economical and logistic facilities with sound strategies as well as enough power to act against violations of regulations. Circular thinking of plastic waste management is promoted.

Acknowledgements The Authors would like to acknowledge Mr. Kelum Srimal Dissanayake and Ms. Gayesha Thilakerathna of the Center for Marine Science and Technology (CMST), University of Sri Jayewardenepura for their support in GIS mapping and providing MP images.

References

Athapaththu, A. M. A. I. K., Thushari, G. G. N., Dias, P. C. B., Abeygunawardena, A. P., Egodauyana, K. P. U. T., Liyanage, N. P. P., Pitawala, H. M. J. C., & Senevirathna, J. D. M. (2020). Plastics in surface water of southern coastal belt of Sri Lanka (Northern Indian Ocean): Distribution and characterization by FTIR. *Marine Pollution Bulletin, 161*, 111750.

Athawuda, A. M. G. A. D., Jayasiri, H. B., Thushari, G. G. N., & Guruge, K. P. G. K. P. (2020). Quantification and morphological characterization of plastic litter (0.30–100 mm) in surface waters of off Colombo, west coast of Sri Lanka. *Environmental Monitoring Assessment, 192*: 509.

Athukorala, A., Amarathunga, A. A. D., De Silva, D. S. M., Bakir, A., McGoran, A., Sivyer, D. B., De Silva, R. C. L., & Narangoda, S. R. C. (2023a). Microplastic contamination in selected commercial fish species in Negombo Lagoon, Sri Lanka. In *Proceedings of the International Conference on Applied and Pure Sciences*, Faculty of Science, University of Kelaniya, Sri Lanka.

Athukorala, A., Amarathunga, A. A. D., De Silva, D. S. M., Bakir, A., McGoran, A., & Sivyer, D. B. (2023b). Occurrence and abundance of microplastics in surface waters and sediments from the Negombo Lagoon, Sri Lanka. In *16th International Conference on the Environmental Effects of Nanoparticles and Nanomaterials* (pp. 32–33). Plymouth.

Bandara, H. M. N. N., De Silva, H. S. D. R., Wijayathilaka, H. D., Kodikara, K. A. S., & De Silva, P. M. C. S. (2023). Microplastic pollution in Dondra lagoon, Southern coast of Sri Lanka: Influence on habitat and feeding modes of selected fish species. In *Proceedings of the Postgraduate Institute of Science Research Congress 2023*, Postgraduate Institute of Science, University of Peradeniya, Sri Lanka.

Carbery, M., O'Connor, W., & Thavamani, P. (2018). Trophic transfer of microplastics and mixed contaminants in the marine food web and implications for human health. *Environment International, 115*, 400–409. https://doi.org/10.1016/j.envint.2018.03.007.

de Vos, A., Aluwihare, L., Youngs, S., Dibenedetto, M. H., Ward, C. P., Michal, A. P. M., Colson, B. C., Mazzotta, M. G., Walsh, A. N., Nelson, R. K., Reddy, C. M., & James, B. D. (2022). The *M/V X-Press Pearl* Nurdle spill: Contamination of burnt plastic and unburnt Nurdles along Sri Lanka's beaches. *ACS Environmental Au, 2*, 128–135.

Derraik, J. G. B. (2002). The pollution of the marine environment by plastic debris: A review. *Marine Pollution Bulletin, 44*, 842–852.

Dharmadasa, W. L. S. S., Andrady, A. L., Kumara, P. B. T. P., Maes, T., & Gangabadage, C. S. (2021). Microplastics pollution in marine protected areas of Southern Sri Lanka. *Marine Pollution Bulletin, 168*, 112462.

Dharmasiri, L. M. (2020). Waste management in Sri Lanka: Challenges and opportunities. *Sri Lanka Journal of Social Sciences* 72–85.

Dias B. C. G., Amarathunga, A. A. D., De Silva, D. S. M., Sivyer, D. B., & Maddumage, D. S. R. (2023). Abundance and microplastic characterization found in the mud crab *Scylla serrata* inhabiting Negombo Lagoon, Sri Lanka. In *International Conference on Applied and Pure Sciences*, Faculty of Science, University of Kelaniya, Sri Lanka.

Dissanayake D. M. T. N., & Rathnasuriya M. I. G. (2022). Microplastic contamination in Brown mussels (*Perna perna*) along the Southern coastal waters of Sri Lanka. In *Fourth Research Symposium of Ocean University of Sri Lanka.*

Eerkes-Medrano, D., Thompson, R. C., & Aldridge, D. C. (2015). Microplastics in freshwater systems: A review of the emerging threats, identification of knowledge gaps and prioritization of research needs. *Water Research, 75*, 63–82. https://doi.org/10.1016/j.watres.2015.02.012.

Gago, J., Filgueiras, A., Pedrotti, M. L., Caetano, M., & Frias, J. (2018). Standardized protocol for monitoring micro-plastics in sea water. JPI-Oceans BASEMAN project.

Gamage, T. R. I., Rathnasuriya, M. I. G., & Kumara, W. A. A. U. (2023). Microplastics occurrence in the gut content of Bluestripe herrings *Herklotsichthys quadrimaculatus* (Rüppell, 1837) along the Southern coastal waters of Sri Lanka. In *Proceedings of the 79th Annual Scientific Sessions 2023—Part 1, Sri Lanka Association for the Advancement of Science.*

GESAMP. (2015). Sources, fate and effects of microplastics in the marine environment: a global assessment (P. J. Kershaw (Ed.)). (IMO/FAO/UNESCO-IOC/UNIDO/WMO/IAEA/UN/UNEP/UNDP Joint Group of Experts on the Scientific Aspects of Marine Environmental Protection). Rep. Stud. GESAMP No. 90 (96p.).

Geyer, R., Jambeck, J. R., & Law, K. L. (2017). Production, use, and fate of all plastics ever made. *Science Advance, 3*(7), e170078.

Hansani, K. U. D. N., Thilakarathne, E. P. D. N., Koongolla, J. B., Gunathilake, W. G. I. T., Perera, B. G. D. O., Weerasingha, W. M. P. U., & Egodauyana, K. P. U. T. (2023). Contamination of microplastics in tropical coral reef ecosystems of Sri Lanka. *Marine Pollution Bulletin, 194*, 115299.

Hidalgo-Ruz, V., Gutow, L., Thompson, R. C., & Thiel, M. (2012). Microplastics in the marine environment: A review of the methods used for identification and quantification. *Environmental Science and Technology, American Chemical Society, 46*, 3060–3075.

Jang, Y. C., Ranatunga, R. R. M. K. P., Mok, J. Y., Kim, K. S., Hong, S. Y., Choi, Y. R., & Gunasekara, A. J. M. (2018). Composition and abundance of marine debris stranded on the beaches of Sri Lanka: Results from the first island-wide survey. *Marine Pollution Bulletin, 128*, 126–131.

Kandeyaya, K. B. K. D. K., Ranatunga, S., & Ranatunga, R. R. M. K. P. (2023). Occurrence of microplastics in some commercially important seafood varieties from Negombo, Sri Lanka. *Regional Studies in Marine Science, 62*, 102958.

Kapukotuwa, R. W. M. G. K., Jayasena, N., Abayasekara, C. L., Weerakoon, K. C.,& Rajakaruna, R. S. (2020). Microplastics in stream water and sediments of Rawan-Oya tributary of Mahaweli River in Kandy district. In: *Proceedings of the Postgraduate Institute of Science Research Congress*, Sri Lanka 2020, Postgraduate Institute of Science Research Congress, 67.

Kapukotuwa, R. W. M. G. K., Jayasena, N., Weerakoon, K. C., Abayasekara, C. L., & Rajakaruna, R. S. (2022). High levels of microplastics in commercial salt and industrial salterns in Sri Lanka. *Marine Pollution Bulletin, 174*, 113239.

Kariyawasam, S., Madhuwanthi, A., & Wilson, C. (2019). The role of stakeholders in managing polythene and plastic waste in coastal sites of Sri Lanka: A case study of the Dehiwala-Mt Lavinia municipal council region. In *E3S Web of conferences 96, 02003. ICEPP 2018.*

Karunarathna, Y. D. T. I., & Ranatunga, R. R. M. K. P. (2021). Microplastic pollution along the Attanagalla river, Sri Lanka. In *8th International Conference of Multidisciplinary Approaches (iCMA)* (p. 23), Faculty of Graduate Studies, University of Sri Jayewardenepura, Sri Lanka.

Koongolla, J. B., Andrady, A. L., Kumara, P. B. T. B., & Gangabadage, C. S. (2018). Evidence of microplastics pollution in coastal beaches and waters in southern Sri Lanka. *Marine Pollution Bulletin, 137*, 277–284.

Lahiru, A. A. S., Amarathunga, A. A. D., Malavipathirana, S., De Silva, D. S. M., & Sivyer, D. B. (2023). Assessment of litter and microplastic pollution in the water and sediment of Hirikatuoya stream, within the Walawe River basin, Sri Lanka. In *Proceedings of the International Conference on Applied and Pure Sciences* (p. 23), Faculty of Science, University of Kelaniya, Sri Lanka.

Lawan, P. L. M. J. H., De Silva, D. S. M., Amarathunga, A. A. D., McGoran, A., Bakir, A., Sivyer, D. B., & Reeve, C. (2024). Microplastic contamination in shrimps from the Negombo Lagoon—Sri Lanka. *Water, 16*(3), 447.

Lenz, R., Enders, K., & Neilson, T. G. (2016). Microplastic exposure studies should be environmentally realistic (letter). *Proceedings of the National Academy of Sciences, 113*(29), E4121–E4122. https://doi.org/10.1073/pnas.1606615113.

Masura, J., Baker, J., Foster, G., & Arthur, C. (2015). Laboratory methods for the analysis of microplastics in the marine environment: Recommendations for quantifying synthetic particles in waters and sediments. NOAA Technical Memorandum NOS-OR&R-48.

McIlwraith, H. K., Kim, J., Helm, P., Bhavsar, S. P., Metzger, J. S., & Rochman, C. M. (2021). Evidence of microplastic translocation in wild-caught fish and implications for microplastic accumulation dynamics in food webs. *Environmental Science and Technology, 55*(18), 12372–12382. https://doi.org/10.1021/acs.est.1c02922.

Meegahakotuwa, B. M. B., Ranatunga, R. R. M. K. P., & Gunasekara, A. J. M. (2020). Microplastic contamination in coastal waters off Maha Oya estuary in Negombo. *Uni-in-Alliance, 2020*, 49.

Napper, I. E., & Thompson, R. C. (2016). Release of synthetic microplastic plastic fibres from domestic washing machines: Effects of fabric type and washing conditions. *Marine Pollution Bulletin, 112*(1–2), 39–45. https://doi.org/10.1016/j.marpolbul.2016.09.025.

Nawarathna, M. K. R. A. W. S. N., Rathnasuriya, M. I. G., Kapukotuwa, G. K., & Rajakaruna, R. S. (2023). Microplastics contamination in Indian backwater oysters (*Magallana bilineata*) in the Panadura estuary, Sri Lanka. In *Proceedings of the Postgraduate Institute of Science Research Congress, Sri Lanka* (p. 1), Postgraduate Institute of Science, University of Peradeniya, Sri Lanka.

Nijamdeen, T. W. G. F. M., Atugoda, T., Kumara, P. B. T. P., Gunasekara, A. J. M., & Vithanage, M. (2020). Status of particulate marine plastics in Sri Lanka. Research gaps and policy needs. In *Particulate Plastics in Terrestrial and Aquatic Environments* (1st ed., pp. 297–326). CRC Press, Sri Lanka. ISBN 9781003053071.

Perera, K., Ziajahromi, S., Nash, S. B., Manage, P. M., & Leusch, F. D. L. (2022). Airborne microplastics in indoor and outdoor environments of a developing country in South Asia: Abundance, distribution, morphology, and possible sources. *Environmental Science and Technology, 56*(23), 16676–16685.

Plastics-the *fast* Facts. (2023, November). Downloaded from https://plasticseurope.org/knowledge-hub/plastics-the-fast-facts-2023/.

Premarathna, K., Abeysundara, S. P., Gunaratne, A. M. T. A., & Madawala, H. M. S. P. (2023). Community awareness and perceptions on microplastics: A case study from Sri Lanka. *Ceylon Journal of Science, 52*(4), 425–432.

Ranatunga, R. R. M. K. P., & Karunarathna, K. P. R. (2018). First evidence of microplastics in beach sand from Negombo, Sri Lanka and the potential accumulation in marine fish. In *International Conference on Multidisciplinary Approaches (iCMA)* (p. 169), University of Sri Jayewardenepura.

Ranatunga, R. R. M. K. P., Wijetunge, D. S., & Karunarathna, K. P. R. (2021). Microplastics in beach sand and potential contamination of planktivorous fish *Sardinella gibbosa* inhabiting in coastal waters of Negombo, Sri Lanka. *Sri Lanka Journal of Aquatic Sciences, 26*(1), 37–54.

Ranatunga, R. R. M. K. P., Wijetunge, D., Ranaweera, W. V. P. H., Hung, C.-C., Liu, S. -Y. V., Schuyler, Q., Lawson, T. J., & Hardesty, B. D. (2023). Ranking Sri Lanka among the world's top mismanaged waste polluters: Does model data change the story? *Sustainability, 15*, 2687.

Rathnayake, R. M. L. I., Amarathunga, A. A. D., De Silva, D. S. M., Bakir, A., McGoran, A., Narangoda, S. R. C. N. K., & Sivyer, D. (2023). Accumulation of microplastics in macrophytes of Negombo lagoon, Sri Lanka. In *16th International Conference on the Environmental Effects of Nanoparticles and Nanomaterials* (pp. 48–49). Plymouth.

Rochman, C. M. (2018). Microplastics research-from sink to source. *Science, 360*(6384), 28–29. https://doi.org/10.1126/science.aar7734.

Rubesinghe, C., Brosche, S., Withanage, H., Pathragoda, D., & Karlsson, T. (2022). X-press pearl: A 'new kind of oil spill' consisting of a toxic mix of plastics and invisible chemicals. International Pollutants Elimination Network (IPEN).

Sewwandi, M., Amarathunga, A. A. D., Wijesekara, H., Mahatantila, K., & Vithanage, M. (2022a). Contamination and distribution of buried microplastics in Sarakkuwa beach ensuing the MV X-Press Pearl maritime disaster in Sri Lankan sea. *Marine Pollution Bulletin, 184*, 114074.

Sewwandi, M., Hettithanthri, O., Egodage, S. M., Amarathunga, A. A. D., & Vithanage, M. (2022b). Unprecedented marine microplastic contamination from the X-Press Pearl container vessel disaster. *Science of the Total Environment, 828*, 154374.

Thakshila, V. G. N., & Ranatunga, R. R. M. K. P. (2019). Evidence of microplastics contamination in surface waters and sediment of Kelani River estuary. In *Proceedings of the 24th annual International Forestry and Environment Symposium 2019 of the Department of Forestry and Environmental Science, University of Sri Jayewardenepura, Sri Lanka* (p. 68).

United Nations Environment Programme and United Nations Office for the Coordination of Humanitarian Affairs. (2021). X-Press Pearl Maritime Disaster: Sri Lanka—Report of the UN Environmental Advisory Mission. https://wedocs.unep.org/20.500.11822/36608.

Van der Meulen, M. D., De Vriese, L., Lee, J., Maes, T., Van Dalfsen, J. A., Huvet, A., Soudant, P., Robbens, J., & Vethaak, A. D. (2014). Socio-economic impact of microplastics in the 2 Seas, Channel and France Manche Region: An initial risk assessment. MICRO Interreg project IVa.

Viraj, R. K. L., Jayasiri, H. B., Devmali, N. L. D. H., Amarasiri, C., & Jayapala, H. P. S. (2019). Plastic contamination in selected beaches of Sri Lanka with special reference to microplastics. National Aquatic Resources Research and Development Agency (NARA), Scientific sessions 2019.

Weerakoon, W. R. W. M. A. P., Samaranayake, T. B. D. T., Jayasiri, H. B., & Arulananthan, K. (2018). Quantitative analysis of microplastic contamination in beach sand at the Western and Southwestern coastal stretches in Sri Lanka. *National Aquatic Resources Research and Development Agency (NARA), International Scientific Sessions 2018.*

Weerakoon, W. R. W. M. A. P., Grosvik, B. E., Dalpadado, P., Wimalasiri, H. B. U. G. M., Rathnasuriya, M. I. G., Harischandra, K. A. D. A. T., Shirantha, R. R. A. R., Madhushankha, H. M. T. C., Sampath, W. A. D., Jayasinghe, R. P. P. K., Gunasekara, S. S., Arulananthan, K., Totland, A., Indika, K. W., Mihirani, P. M. N., Priyadharashani, W. N. C., Arrigo, K. R., Bianchi, G., & Krakstad, J. O. (2019). Enumeration of microplastics in Sri Lankan waters: Preliminary findings from the RV Dr. Fridtj Nansen Ecosystem Survey, 2018. National Aquatic Resources Research and Development Agency (NARA), Scientific sessions 2019.

Weerakoon, W. R. W. M. A. P., & Wimalasiri, H. B. U. G. M. (2020). Aggregation of floating microplatics in coral-dominated eco-systems in Sri Lanka. National Aquatic Resources Research and Development Agency (NARA), Scientific sessions 2020.

Weerasekara, P. G. Y. W., De Silva, D. S. M., De Silva, R. C. L., & Amarathunga, A. A. D. (2022). Sampling and analysis of Microplastics in the coastal area from the Estuary of Kelani River to the Estuary of Mahaoya. In *International Conference on Applied and Pure Sciences, 2022, Faculty of Science, University of Kelaniya, Sri Lanka.*

Weerasuriya, W. A. M. B., Amarathunga, A. A. D., Walpita, C. N., & Sivyer, D. (2023). Microplastic contamination of food fish in Bellanwila-Attidiya Wetland, Sri Lanka. In *9th International Conference of Sabaragamuwa University of Sri Lanka, Sabaragamuwa University, Sri Lanka.*

Wijethunga, H. N. S., Athawuda, A. M. G. A. D., Dias, P. C. B., Abeygunawardena, A. P., Senevirathna, J. D. M., Thushari, G. G. N., Liyanage, N. P. P., & Jayamanne, S. C. (2019). Screening the effects of microplastics on selected invertebrates along Southern coastal belt in Sri Lanka: A preliminary approach to coastal pollution control. In Uwa *Wellassa University of Sri Lanka, International Research Conference 2019.*

Wu, C., Zhang, K., & Xiong, X. (2018). Microplastic pollution in inland waters Focusing on Asia. In M. Wagner, S. Lambert (Eds.), *Freshwater Microplastics. The Handbook of Environmental Chemistry* (Vol. 58). Springer, Cham. https://doi.org/10.1007/978-3-319-61615-5_5.

Chapter 18
Plastic Waste Management: The Mysore Experiment

C. G. Betsurmath

Abstract Waste management is one of the major environmental problems in recent days, and plastic waste management has become extremely difficult due to over-utilization across the globe enormously. Plastic waste management is still challenging in India due to the increased population and less awareness on the consequences of improper management of plastic waste. The present paper highlights the importance of proper plastic waste management in urban areas and proper governance in the management of the same. Also, it explains the new initiatives, such as segregation at source, zero waste management and 4-R concepts in the governance and sustainable management of waste at municipality levels, implemented by the author as the Commissioner of Mysore City Corporation. Public and stakeholder education, engagement, and participation in plastic waste management at municipality level have shown remarkable results of improved governance and policy implementation in the Mysore city in India.

Keywords Plastic pollution · Microplastics · Sustainable waste management · Ubiquitous · 4-R concept · Environmental · Severe · Sanitation · Single-use plastics · Biodegradable · Circular economy

18.1 Introduction

It is estimated that every year, the world produces over 400 million tonnes of plastic waste (Iroegbu et al., 2021). However, only 9% of this waste is recycled, and much of it ends up in oceans and waterways. This plastic pollution is threatening our planet's ecosystems. Since the 1970s, the rate of plastic production has grown faster than that of any other materials. Approximately 36% of all plastics produced are used in packaging, including single-use plastic products for food and beverage containers,

C. G. Betsurmath (✉)
JSS Mahavidyapeetha, Mysore, India
e-mail: es@jssonline.org

S. Suriyanarayanan et al. (eds.), *Combating Plastic Pollution in Terrestrial Environment*,
https://doi.org/10.1007/978-981-96-2343-3_18

approximately 85% of which end up in landfills or as unregulated waste. An estimated 8.3 billion tonnes of plastics have been produced since the industrial production began (Meijer et al., 2021; UNEP, 2018). Much is still in existence as waste. Plastic pollution is now ubiquitous and is found all over the earth, from the heights of Mount Everest to the depths of the oceans. While plastic provides undisputed benefits, the ability to safely manage plastic pollution is not keeping pace with projected growth in the plastics market over the coming decades. India is not immune to this problem, and the situation here is quite serious (Singh, 2019). According to a study by the Central Pollution Control Board, India generates approximately 26,000 tonnes of plastic waste every day (Chintala, 2023). This plastic waste is not only littering our streets, but is also causing severe health hazards to our people and the environment. One of the most significant contributors to plastic waste is single-use plastic. These are items that are designed to be used only once, such as plastic bags, straws, covers, wrappers, bottles, cutlery, etc. They are exceedingly difficult to recycle (Chakraborty & Chandra, 2023). Plastics, including microplastics, are now ubiquitous in our natural environment (Peng et al., 2021). Since the vast majority of plastic pollution is generated on land, there is an urgent need to better understand the likely impacts on animals that live in terrestrial and freshwater ecosystems, including mammals, birds, and fish. Plastic is versatile, lightweight, easy to carry, durable, flexible, moisture resistant, strong, and relatively inexpensive. Most plastic items never fully disappear; they just break down into smaller and smaller pieces. These microplastics can enter the human body through inhalation and absorption and accumulate in organs (Duis & Coors, 2016). Microplastics have been found in the lungs, livers, spleens, kidneys, and placentas of newborns (Duis & Coors, 2016). In countries with poor solid waste management systems, plastic waste—especially single-use plastic bags can be found clogging sewers, providing breeding grounds for mosquitoes and pests, and as a result, increasing the transmission of vector-borne diseases.

18.2 National and International Scenario

18.2.1 International Scenario

From the 1950s to the 1970s, only a small amount of plastic was produced, and as a result, plastic waste was relatively manageable. However, between the 1970s and the 1990s, plastic waste generation more than tripled, reflecting a similar rise in plastic waste production. In the early 2000s, the amount of plastic waste we generated rose more in a single decade than it had in the previous 40 years. Today, we produce about 400 million tonnes of plastic waste every year. Around the world, one million plastic bottles are purchased every minute, while up to five trillion bags are used worldwide every year (Kiessling et al., 2023; Walker et al., 2023).

18.2.2 The Indian Scenario

As per the reports of 2017–18, the country has generated approximately 9.4 million tonnes of plastic waste, which amounts to 26,000 tonnes of waste per day. Out of this, approximately 5.6 million tonnes are recycled (i.e., 15,600 tonnes of waste per day) and 3.8 million tonnes are left uncollected or littered (9,400 tonnes of waste per day). Karnataka generates around 2.96 lakh tonnes of plastic waste annually; 73,000 tonnes is recycled. While over 50 thousand tonne plastic is being incinerated, around 1.75 lakh tonnes of trash are unaccounted (Hossain et al., 2022; Shanker et al., 2023).

18.3 Efforts Toward Effective Plastic Waste Management

18.3.1 Government

To tackle this issue, the Pollution Control Board and the Government of India launched the Plastic Waste Management Rules in 2016 (Plastic Waste Management Rules, Ministry of Environment Forest And Climate Change, Government of India, 2016). The aim of these rules is to regulate the use, manufacture, and proper disposal of plastic materials. Under these rules, plastic bags less than 50μ thickness are banned, and manufacturers are responsible for collecting and recycling their own plastic waste. Additionally, the Indian Government has launched a nationwide Plastic Free Abhiyan, which is aimed at reducing the use of single-use plastics. The campaign encourages people to switch to sustainable alternatives such as cloth bags, paper bags, jute bags and compostable bags. The government is also working on improving the waste management infrastructure and increasing public awareness on the issue. Swachh Survekshan was introduced by the Ministry of Housing and Urban Affairs in 2016 as a competitive framework to encourage cities to improve the status of urban sanitation while encouraging large scale citizen participation (http://swachhbharaturban.gov.in/Auth/dsdocumentsfile.aspx?DOCTYPE=923&DOCID=478). It has led to a spirit of healthy competition among cities and towns of India. Swachh Survekshan acted as an enabler for Mission acceleration in the cities to foster healthy competition among cities to improve their performance on sanitation parameters. It encouraged large scale citizen participation and created awareness about the importance of Swachhata participation and improved sanitation services delivered by cities to its citizens. Recently, the Ministry of Environment, Forest and Climate Change has notified the Plastic Waste Management Amendment Rules, 2021 (Plastic Waste Management (Amendment) Rules, Ministry of Environment Forest and Climate Change, Government of India, 2021). These rules prohibit specific single-use plastic items which have "low utility and high littering potential" by 2022. The manufacture, import, stocking, distribution, sale, and use of the identified single-use plastic will be prohibited with effect from the 1st of July 2022 (Plastic Waste Management Rules, Ministry of Environment

Forest & Climate Change, Government of India, 2016). The ban will not apply to commodities made of compostable plastic. For banning other plastic commodities in the future, other than those that have been listed in this notification, the government has given the industry ten years from the date of notification for compliance. The permitted thickness of the plastic bags, currently 50μ, will be increased to 75μ from 30th September 2021 and to 120μ from the 31st of December 2022. Plastic bags with higher thickness are more easily handled as waste and have higher recyclability. Pollution due to single-use plastic items has become an important environmental challenge confronting all countries and India is committed to taking action for the mitigation of pollution caused by littered single-use plastics. The Prime Minister of India was also conferred the "Champions of the Earth" award by the United Nations Environment Programme (UNEP) in 2018 for pledging to eliminate all single-use plastic by 2022 (United Nations Environment Programme, 2017). Even though plastic has many valuable uses, it becomes problematic with the usage of a large number of single-use plastics that cause various socio-economic, environmental, and health problems. Even after disposing of plastics into the environment, it has reported issues associated with microplastics and environmental consequences. Even with several consequences being caused by plastic pollution, global production of plastic has continued and is estimated to reach 1100 million tonnes by 2050. Avoid the usage of single-use plastics and zero waste discharge at the household to the institutional level needs to be promoted with a sustainable lifestyle by adapting LIFE (lifestyle for the environment). Beyond sustainability, recycling and reusable management systems promote circular approaches where the utilization of fresh resources may be avoided with the recycling of waste (United Nations Environment Programme, 2017).

18.3.2 Importance of People Participation

As mentioned earlier, plastic pollution is a global problem, and we must all play our part in addressing it. It starts with reducing our consumption of single-use plastics, properly disposing of our plastic waste, and supporting initiatives that aim to reduce plastic waste. There are many ways in which we can minimize our plastic waste. For example, we can carry reusable bags, bottles, and containers. We can also avoid using plastic straws and cutlery, and instead, choose eco-friendly alternatives such as bamboo or metal. We can also support companies and brands that are committed to reducing their plastic footprint. It is also important to properly dispose of our plastic waste. We should always try to recycle as much as we can and ensure that we are segregating our waste correctly. Governments and businesses must take steps to improve waste management infrastructure and promote sustainable practices. Plastic waste management is a complex issue, but one in which we all can play a part in addressing. We need to work together to reduce our plastic consumption, properly dispose of our plastic waste, and support initiatives that aim to reduce plastic waste. By doing so, we can help to protect our planet's ecosystems and create a cleaner, healthier world for ourselves and future generations.

18.4 Mysore Experiment

Mysore, located in the southern state of Karnataka on the Indian peninsula, is known globally as a heritage city, green city, royal city, most liveable city, retired people's paradise, cultural capital of the state of Karnataka and the cleanest city of India. Spread across an area of 128.42 km^2, Mysuru is the third-most populous and third-largest city in the state with a population of about 11,00,000, 2,06,372 households, 26,503 commercial establishments, more than 200 apartments and 1,86,000 properties. About 402 metric tonnes of solid waste is generated in Mysore city per day, out of which 79.35% is from households and outgrowths. The physical properties of the total waste generated showed that over 18% comprises plastic content, amounting to more than 72 metric tonnes of plastic waste (Figs. 18.1 and 18.2).

18.4.1 Governance and Management:

As part of the National Swachh Bharat Mission and in the interest of beautification of Mysore city, as the Commissioner of Mysore City Corporation (MCC), the author has taken various initiatives and set a remarkable waste management governance in the city. As a result, Swachh Bharat Mission turned into a people's movement, a Jan Andolan, with citizens and all segments of society taking ownership and responsibilities for the creation of Swachh Bharat. MCC strengthened the regulation as per the acts, rules and directions given by central and state governments, such as (a) Karnataka Plastic Waste Management Rules 2016 (https://kspcb.karnataka.gov.in/waste-management/plastics-rules), (b) Government of India—Plastic Waste Management Rules 2021 (Plastic Waste Management

QUANTITY OF WASTE GENERATED IN MYSORE CITY

Sl No.	Particulars	Number of waste generators	Total waste in ton	Percentage
1	Population (households) including out growth	9,38,386	319	79.35
2	Commercial	19,923	8	1.9
3	Hotels and Restaurants	716	15	3.7
4	Marriage and function halls	124	6	1.5
5	Street sweeping	-	17	4.2
6	Garden waste	-	19	4.7
7	Chicken and mutton waste	406	6	1.5
8	Market waste	3	12	3.0
	Total waste		402	

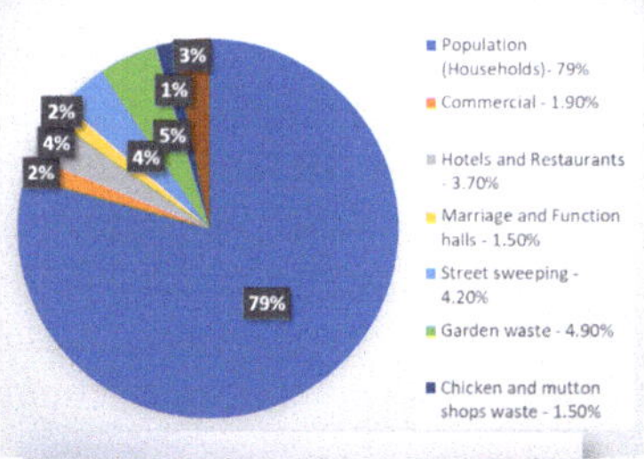

Fig. 18.1 Pictorial representation of waste generated in Mysore city

PHYSICAL PROPERTIES OF WASTE GENERATED IN MYSORE CITY

Sl No.	Composition	% of waste
1	Organic fractions	55
2	Paper	10
3	Plastic	18
	Rubber/Leather	3
4	Ceramics & Silica	5
5	Clothes/beds	5.7
6	Coconut husks/shells	3
7	Sanitary Napkins	0.1
8	Syringes/medical bottles	0.1

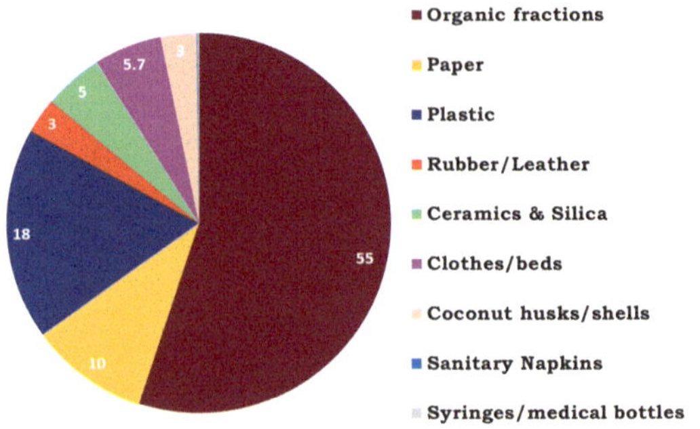

Fig. 18.2 Physical properties of solid waste generated in Mysore city

Rules, Ministry of Environment Forest And Climate Change, Government of India, 2016), (c) Prevention of defacement of public properties act 1985 (https://www.indiacode.nic.in/bitstream/123456789/9639/1/prevention_of_defacement_act%2C_1985.pdf), (d) Karnataka Pollution Control Board Directions (Circular on Regarding ban of use of Plastics, 2012; Government of Karnataka Notification banning burning of solid wastes, 2017; https://kspcb.karnataka.gov.in/) and others (Places and (Prevention of Disfigurement), 1981; Solid Waste Management Rules, Ministry of Environment Forest And Climate Change, Government of India, 2016; E-Waste (Management) Rules, Ministry of Environment Forest and Climate Change, Government of India, 2022; Environment (Protection) Act, Ministry of Environment Forest and Climate Change, Government of India, 1986). It executed the proper implementation of various waste management SOPs as per the guidelines. As an effort, Mysore city was awarded as the First Cleanest City of India in the years 2015 and 2016 during the tenure of the author as the Commissioner of MCC. The contributions of the author in sustainable plastic waste management and sustainable development are depicted in several media and newspapers (https://www.thehindu.com/news/national/karnataka/ban-on-plastic-comes-into-effect-in-mysuru/article8363713.ece; https://timesofindia.indiatimes.com/city/mysuru/after-cleanest-city-tag-mysuru-city-corporation-to-make-mysuru-plastic-ree/articleshow/51103240.cms; http://timesofindia.indiatimes.com/articleshow/51555444.cms?utm_source=contentofinterest&utm_medium=text&utm_campaign=cppst; https://www.thehindu.com/news/national/karnataka/middlemen-menace-leaves-mcc-redfaced/article6798735.ece; https://www.deccanherald.com/india/karnataka/swacch-mysuru-mcc-to-launch-theme-based-drives-from-april-517881.html; https://www.deccanherald.com/india/karnataka/plastic-banned-in-mysuru-with-immediate-effect-515362.html; https://www.deccanherald.com/india/karnataka/impose-total-ban-on-plastic-dc-tells-local-bodies-521427.html).

18.4.2 Public Participation and Awareness:

Several large-scale Information, Education, Behavior Change and Communication Activities (IEBC) citizen engagement activities for awareness generation and community mobilization were organized. All citizens / NGOs/ voluntary organizations actively and voluntarily participated in the proper waste management moments. Theme-based drives in markets, schools, colleges, religious places, airport, railway stations, bus stands and bus stops, wards, parks, graveyards, toilets, offices, Anganwadi, libraries, hotels, apartments, auto stands, heritage areas, factories, and slums. Video and audio jingles played in local TV channels, theatres, functions, dasara festivities and WhatsApp messages (http://timesofindia.indiatimes.com/articleshow/100751964.cms?from=mdr&utm_source=contentofinterest&utm_medium=text&utm_campaign=cppst; http://timesofindia.indiatimes.com/articleshow/100806022.cms?from=mdr&utm_source=contentofinterest&utm_medium=text&utm_campaign=cppst; http://timesofindia.indiatimes.com/articleshow/100888281.cms?from=mdr&utm_source=contentofinterest&utm_medium=text&utm_campaign=cppst). Pledge regarding segregation of waste, ban on plastic usage and cleanliness was taken by all officers, elected representatives, students, and general public. Workshops on proper waste management and disposal were conducted for hotel owners, choultries, apartments, and business associations (https://economictimes.indiatimes.com/news/politics-and-nation/how-india-is-trying-to-solve-its-plastics-wasteproblem/articleshow/64431263.cms?utm_source=contentofinterest&utm_medium=text&utm_campaign=cppst).

18.4.3 New Initiatives

About 41 dry waste collection centers were opened in different wards and about 9 zero waste management centres were made functional. It made mandatory provisions for the segregation of waste at source and educated the general public regarding segregation through media, radio talks, TV discussions and live phone-in programs. Imposed a complete ban on plastic usage in the city from March 2016 and suggested alternatives to plastic. MCC followed "Four R Principle": (a) Refuse unsegregated waste at source and penalize those littering outside; (b) Reduce at source using new technologies and following best practices; (c) Recycle and (d) Reuse. Also, MCC constituted a rapid action force, ABHAYA, consisting of three teams working in two shifts to act during emergency situations, more particularly to conduct raids on establishments and commercial units using and selling plastics. MCC not only banned plastic in the city but also helped to identify the alternatives to plastics to keep the processes unhindered for both the commercial activities as well as domestic activities. The plastic bags were replaced by paper, cloth, jute and compostable bags. Plastic cups were replaced with cups made of bagasse, glass, steel, ceramic and small clay pots became an innovative alternative. Plastic plates were replaced with traditional banana leaf; leaf plates made with a unique ancient Indian tradition became a newfound return to market. The areca leaf pressed plates made available to the market resulted in a very efficient replacement. Edible spoons, steel and wooden spoons were well taken by the public as a highly effective replacement to the plastic spoons. Created awareness among the people and sellers about the usage and availability of alternatives to plastics. To promote stopping usage of single use plastic carry bags and proper segregation at source, MCC distributed reusable cloth bags and color coded dustbins to the public as well as households free of cost.

On the institutional front, as a part of Smart Campus initiatives and to make all the institutions of JSS Mahavidyapeetha clean and green, the following five principles are followed—cleanliness and hygiene, save water, save energy, control food waste, and ban plastic usage. For the last five years, plastic usage has been drastically controlled in the Suttur annual fair and the devotees are served food in steel plates which certainly reduces plastic disposal and promotes 4R frameworks in the JSS organizations.

18.5 Way Forward for the Sustainable Management of Waste

Sustainable Alternatives: Economically affordable and ecologically viable alternatives to plastics, such as biodegradable and plant-originated materials which will not burden the resources needed and their prices will also come down with time and increase in demand. Need to promote alternatives like cotton, khadi bags and biodegradable plastics made with agricultural waste and plant origin.

Circular Economy: In recent days, the concept of circular economy is being evolved in such a way that even waste becomes resources without discarding anything. The circular economy promotes maximum utilization of extracted resources through recycle, regenerate and reuse processes (http://timesofindia.indiatimes.com/articleshow/97464752.cms?from=mdr&utm_source=contentofinterest&utm_medium=text&utm_campaign=cppst; https://economictimes.indiatimes.com/news/politics-and-nation/how-india-is-trying-to-solve-its-plastics-wasteproblem/articleshow/64431263.cms?utm_source=contentofinterest&utm_medium=text&utm_campaign=cppst). Countries must embrace circular and sustainable economic practices throughout the plastics value chain to reduce plastic pollution. A circular economy depends on reuse, sharing, repair, refurbishment, remanufacturing and recycling of resources to create a closed-loop system, minimizing the use of resources, generation of waste, pollution and carbon emissions.

Education for individual behavioral change: Individuals should change their view and mindset toward the plastic waste generated and must know their role in effective waste management. Attitudinal change in individuals and other stakeholders, including waste management workers needs to be achieved through holistic environmental education and awareness at different levels (https://blog.forumias.com/measuring-indias-plastic-problem/; https://www.thehindu.com/news/national/karnataka/where-does-the-trash-go/article29994117.ece).

Good governance and policy implementation: Strong governance ensures proper waste management especially at municipality level (https://www.thehindu.com/news/national/karnataka/where-does-the-trash-go/article29994117.ece). Good governance promotes sustainable waste management including switchover to zero waste discharge by strengthening the trust of all stakeholders' relationships with the municipality and helps to implement required environmental policies.

18.6 Conclusion

Waste management, especially plastic waste management, has become one of the most environmentally challenging issues across the globe and severity is still unimaginable in countries like India with rapid population growth and urbanization. However, an integrated approach is essential to solve various problems associated with plastic management and, successful transition toward a pollution-free planet requires political leadership, high-level champions and commitments, as well as action at the local level toward environmentally friendly activities. To tackle the waste and associated problems at urban regions, good and strong governance is required and the governance implemented at MCC would be the model for sustainable management of waste, including plastic waste and zero waste discharge. Finding alternative materials which are biodegradable to plastic and promoting sustainable waste management strategies like waste to energy, waste to resources, zero waste discharge model, etc., are urgent. To achieve high level political commitment in key economic sectors, there is a need to go beyond the environmental ministries and include other relevant ministries such as finance, agriculture, industry, urban, transport, energy and health. There is also a need to engage local government, civil society organizations, business leaders, industries, trade unions, and citizens at large. The implementation of a systematic plastic waste management system and best practices at JSS Mahavidyapeetha institutes have demonstrated sustainable waste management and have shown that proper waste management at an institutional level with potential participation and dynamic leadership leads to solving various plastic waste management issues across the city.

Acknowledgements The author would like to submit respectful pranams to the lotus feet of His Holiness Jagadguru Dr. Sri Shivarathri Rajendra Mahaswamiji, the founder president of JSS Mahavidyapeetha and seek the blessings of His Holiness, Jagadguru Sri Shivarathri Deshikendra Mahaswamiji, the President of JSS Mahavidyapeetha for all the support and encouragement. He also acknowledges Dr. Amitava Bandopadhyay, Director General, NAM Science & Technology Centre, New Delhi; Dr Shanth Thimmaiah, Chairman, Karnataka State Pollution Control Board, Government of Karnataka; Prof. Jon Samseth, President, Scientific Committee of Problems of the

Environment (SCOPE), Netherlands; Dr. Suriyanarayanan S, (SCOPE—India), Deputy Director (Research), Vinayaka Missions Research Foundation, Salem; Dr. B. Suresh, Pro Chancellor, JSS AHER, Mysuru and Dr. Surinder Singh, Vice Chancellor, JSS AHER, Mysuru for their assistance and encouragement.

References

Act now or the Earth will choke on plastic. http://timesofindia.indiatimes.com/articleshow/100751964.cms?from=mdr&utm_source=contentofinterest&utm_medium=text&utm_campaign=cppst.

Ban on plastic comes into effect in Mysuru. https://www.thehindu.com/news/national/karnataka/ban-on-plastic-comes-into-effect-in-mysuru/article8363713.ece.

Cash from trash: Women waste pickers are upcycling. http://timesofindia.indiatimes.com/articleshow/97464752.cms?from=mdr&utm_source=contentofinterest&utm_medium=text&utm_campaign=cppst.

Chakraborty, P., Chandra, S. (2023). *An Indian perspective on sources of persistent organic pollutants associated with plastic handling: Consequences of COVID-19 pandemic*. Springer.

Chat vendor in the soup after his son intercepts MCC chief. https://timesofindia.indiatimes.com/city/mysuru/chat-vendor-in-the-soup-after-his-son-intercepts-mcc-chiefs-car/articleshow/51101669.cms.

Chintala, V. (2023). Eco-smart plastic waste to energy solutions for Indian railways: Current scenario, challenges and future footprints. *Reviews in Environmental Science and Biotechnology, 1*.

Circular on Regarding ban of use of Plastics. Forest, Ecology and Environment Secretariat. Government of Karnataka No. FEE 17 EPC 2012 dated: 17–03–2016. https://kspcb.karnataka.gov.in/sites/default/files/inline-files/14%20Circular_BAN-OF-Plastics.pdf.

After Cleanest City tag, Mysuru City Corporation to make Mysuru plastic-free. https://timesofindia.indiatimes.com/city/mysuru/after-cleanest-city-tag-mysuru-city-corporation-to-make-mysuru-plastic-ree/articleshow/51103240.cms.

Duis, K., & Coors, A. (2016). Microplastics in the aquatic and terrestrial environment: Source (with a specific focus on personal care products), fate, and effects. *Environmental Sciences Europe, 28*, 2.

Environment (Protection) Act, Ministry of Environment Forest and Climate Change, Government of India. (1986). https://cpcb.nic.in/upload/home/epa/THE%20ENVIRONMENT.pdf.

E-Waste (Management) Rules, Ministry of Environment Forest and Climate Change, Government of India. (2022). https://cpcb.nic.in/uploads/Projects/E-Waste/e-waste_rules_2022.pdf.

Government of Karnataka Notification banning burning of solid wastes dated 03.08.2017. https://kspcb.karnataka.gov.in/sites/default/files/inline-files/MSW_burning%20twings_palnts_01092017.pdf.

Hossain, R., Islam, M., Shanker, R., Khan, D., & Sustainability, K. L. (2022). Plastic waste management in India: Challenges, opportunities, and roadmap for circular economy.

Impose a total ban on plastic. https://www.deccanherald.com/india/karnataka/impose-total-ban-on-plastic-dc-tells-local-bodies-521427.html.

How India is trying to solve its plastics waste problem. https://economictimes.indiatimes.com/news/politics-and-nation/how-india-is-trying-to-solve-its-plastics-wasteproblem/articleshow/64431263.cms?utm_source=contentofinterest&utm_medium=text&utm_campaign=cppst.

Iroegbu, A. O. C., Ray, S. S., Mbarane, V., Bordado, J. C., & Sardinha, J. P. (2021). Plastic pollution: A perspective on matters arising: Challenges and opportunities. *ACS Omega, 6*(30), 19343–19355.

Kiessling, T., Hinzmann, M., Mederake, L., & Management, S. D. W. (2023). *What potential does the EU Single-Use Plastics Directive have for reducing plastic pollution at coastlines and riversides?* Elsevier.

MCC raid yields 350kg of plastic. http://timesofindia.indiatimes.com/articleshow/51555444.cms?utm_source=contentofinterest&utm_medium=text&utm_campaign=cppst.

Measuring India's Plastic Problem. https://blog.forumias.com/measuring-indias-plastic-problem/.

Meijer, L., Emmerik, T., Van der Ent, R., Schmidt, C., & Lebreton, L. (2021). More than 1000 rivers account for 80% of global riverine plastic emissions into the ocean. *Science Advances*.

Middlemen menace leaves MCC red-faced. https://www.thehindu.com/news/national/karnataka/middlemen-menace-leaves-mcc-redfaced/article6798735.ece.

Mysuru now eyes plastic-free tag. https://www.thehindu.com/news/national/karnataka/mysuru-now-eyes-plasticfree-tag/article8251287.ece.

Swacch Mysuru: MCC to launch theme-based drives from April. https://www.deccanherald.com/india/karnataka/swacch-mysuru-mcc-to-launch-theme-based-drives-from-april-517881.html.

Peng, Y., Wu, P., Schartup, A. T., & Zhang, Y. (2021). Plastic waste release caused by COVID-19 and its fate in the global ocean. *Earth, Atmospheric, and Planetary Sciences, 118*, 41.

Karnataka Open Places (Prevention of Disfigurement) Act, 1981 Notification. https://www.indiacode.nic.in/bitstream/123456789/7899/1/35_of_1982_%28e%29.pdf.

Plastic banned in Mysuru with immediate effect. https://www.deccanherald.com/india/karnataka/plastic-banned-in-mysuru-with-immediate-effect-515362.html.

Plastic bans created awareness but stricter enforcement need. http://timesofindia.indiatimes.com/articleshow/100806022.cms?from=mdr&utm_source=contentofinterest&utm_medium=text&utm_campaign=cppst.

Plastic malinyabhoomige maraca. http://epapervijayavani.in/ArticlePage/APpage.php?edn=Bengaluru&articleid=VVAANINEW_BEN_20230605_14_1&artwidth=451px.

Plastic Waste Management (Amendment) Rules, Ministry of Environment Forest and Climate Change, Government of India. (2021). https://cpcb.nic.in/uploads/plasticwaste/Notification-12-08-2021.pdf.

Plastic Waste Management Rules, Ministry of Environment Forest And Climate Change, Government of India. (2016). https://cpcb.nic.in/displaypdf.php?id=cGxhc3RpY3dhc3RlL1BXTV9HYXpldHRlLnBkZg==.

Retrieved September 27, 2023, from https://kspcb.karnataka.gov.in/.

Retrieved September 27, 2023, from https://kspcb.karnataka.gov.in/waste-management/plastics-rules.

Retrieved September 27, 2023, from https://www.indiacode.nic.in/bitstream/123456789/9639/1/prevention_of_defacement_act%2C_1985.pdf.

Shanker, R., Khan, D., Hossain, R., Islam, M. T., Locock, K., Ghose, A., et al. (2023). Plastic waste recycling: Existing Indian scenario and future opportunities. *International Journal of Environmental Science and Technology., 20*(5), 5895–5912.

Singh, K. (2019). Plastic pollution in India: An evaluation of public awareness and consumption behaviour. *Sustainable AMOIJ*.

Solid Waste Management Rules, Ministry of Environment Forest And Climate Change, Government of India. (2016). https://cpcb.nic.in/uploads/MSW/SWM_2016.pdf.

State bans use of non-essential plastics. https://www.thehindu.com/news/cities/bangalore/state-bans-use-of-nonessential-plastics/article8311490.ece.

Swatch Bharat Mission Notification. Government of India. http://swachhbharaturban.gov.in/Auth/dsdocumentsfile.aspx?DOCTYPE=923&DOCID=478.

UNEP. (2018). Banning single-use plastic: lessons and experiences from countries. UN Environment Programme report.

United Nations Environment Programme. (2017). Towards a Pollution-Free Planet Background Report. United Nations Environment Programme, Nairobi, Kenya. https://www.unep.org/resources/report/towards-pollution-free-planet-background-report.

Walker, T., Chemistry LFTT in A. (2023). *Current trends of unsustainable plastic production and micro (nano) plastic pollution*. Elsevier.

Where does the trash go? https://www.thehindu.com/news/national/karnataka/where-does-the-trash-go/article29994117.ece.

Why tiny plastic particles have become a big health concern. http://timesofindia.indiatimes.com/articleshow/100888281.cms?from=mdr&utm_source=contentofinterest&utm_medium=text&utm_campaign=cppst.

Chapter 19
Hazardous of Plastics: A Case Study from Egypt

Ibrahim A. Hassan, Nesreen S. Haiba, Magdy Abdelwahab, Mostafa Mourad, Fawzia Morsi, Eman Atef Zahran, and Gina El Fiki

Abstract Plastic pollution and its impact on the terrestrial and marine ecosystems is a major concern globally, and the situation was exacerbated after the outbreak of COVID-19 and usage of single-use plastics. Millions of plastic items of different sizes, being discharged daily, worldwide, causing Plastic pollution. This situation was exacerbated, during the COVID-19 pandemic, as millions of people are using single-use plastic face mask, gloves and face shields as personal protective equipment (PPE). The continuous and massive increase of gloves, masks and some sorts of wrapping made from single-use plastics, would end up littering land and marine environments ecosystems globally, a visible side-effect of the increased use of PPE, causing hazards. Moreover, such pollution would lead to emissions of greenhouse gases, potentially harming environment. This briefing delivers a synopsis of the environmental and climate impacts of single-use gloves and masks used for COVID-19 protection from June to August 2020. To the best of our knowledge, this study presents the first such information from the Middle East. It pinpoints the need for further knowledge and action, such as a safe, sustainable, transparent and unconventional waste management processes related to COVID-19 in order to reduce the

I. A. Hassan (✉)
Faculty of Science, Alexandria University, Alexandria, Egypt
e-mail: ibrahim.abdelmaged@alexu.edu.eg

N. S. Haiba
Faculty of Education, Department of Physics and Chemistry, Alexandria University, Alexandria, Egypt

M. Abdelwahab
Department of Meteorology, Faculty of Science, Cairo University, Giza, Egypt

I. A. Hassan · M. Abdelwahab · G. El Fiki
National Scientific Committee on Problems of the Environment (SCOPE), Academy of Scientific Research and Technology (ASRT), Cairo, Egypt

M. Mourad · E. A. Zahran
Egyptian Environmental Affairs Agency (EEAA), New Administrative Capital, Cairo, Egypt

M. Mourad · F. Morsi
Department of Natural Resources, Faculty of African Studies, Cairo University, Giza, Egypt

S. Suriyanarayanan et al. (eds.), *Combating Plastic Pollution in Terrestrial Environment*,
https://doi.org/10.1007/978-981-96-2343-3_19

negative impacts now and in any future events. It identifies the need for further knowledge and action to reduce these impacts now and in any future events.

Keywords Plastic pollution · Personal protective equipment (PPE) · COVID-19 · Marine ecosystems · Hazardous

19.1 Introduction

Our planet is poised at the brink of a severe environmental crisis. Current environmental problems make us vulnerable to disasters and tragedies, now and in the future (Al-Salem et al., 2021). We are in a state of planetary emergency, with environmental problems piling up high around us. Unless we address the various issues prudently and seriously we are surely doomed for disaster. Current environmental problems require urgent attention. Plastic Pollution is a challenging issue (Hassan et al., 2022).

Despite a consensus that plastic debris will accumulate in the environment, its presence has drastically increased, reaching almost 360 million tons in 2018 (PlasticsEurope, 2019). Plastic pollution is now considered one of the main environmental challenges and represents an emerging threat to all living species, especially due to the chemical species (additives and other adsorbed elements) that are associated with plastics. Leading governments have even classified plastic pollution as a critical problem comparable to climate change (Nasser et al., 2022). Therefore, a holistic management approach to curb the plastics pollution menace is imperative.

UNEP (2011) announced the most 25 important environmental concerns, we have to keep an eye on, Plastic waste is one of them.

Egypt is awash in plastic that finds its way into the River Nile, the Red Sea, and the Mediterranean Sea, with the country's contribution to plastic waste in the Mediterranean amounting to 43 percent of the total. Some 2.8 million tons of plastic waste find its way into the Mediterranean each year, according to a Back to Blue Report produced by Economist Impact in association with the Nippon Foundation.

Due to the incessant increase in the production, use, and abuse of plastics on a global scale, the rate and magnitude at which plastic wastes are generated have also increased astronomically in the past three decades. This scenario has led to severe environmental problems arising from the need to dispose of such huge quantities of waste daily. Because plastics are non-biodegradable, eliminating them from the ecosystem is cumbersome. Their common treatment via indiscriminate burning has been indicated to contributing to greenhouse gas emissions. Thus, incineration has sufficed as a major treatment option for plastic waste in reducing greenhouse gas emissions (Fadare & Okoffo, 2020).

Of the three main plastic waste maintenance methods (i.e., incineration, mechanical recycling, and chemical recycling (pyrolysis)), landfilling is more responsible for severe environmental issues because of the environmental load involved (Miskolczi et al., 2009). Plastic waste disposal via landfills provides habitats for insects and rodents, causing various diseases among nearby residents (Aragaw, 2020). Because

Fig. 19.1 Zabaleen village in Cairo, Egypt (Courtesy from Mr Samman Nagui)

landfilling is time-consuming before noticeable results can be observed, the method has been found to induce environmental burdens, acting as stressors to the soil, aquifers and promoting air pollution in the form of odor release (Al-Salem et al., 2021).

The present study was focused to provide a short focus and a baseline information on the distribution of the plastics and PPE in the terrestrial ecosystems with special emphasis on coastal waters and their ingestion to better understand the intensity of the impacts plastic pollution generates on the marine environment in the Middle East during COVID-19 pandemic.

19.2 Methodology

19.2.1 Study Area

The study was conducted on the Northern coast in Egypt (31.3543° N, 27.2373° E), in a place called garbage collector village (Zabaleen village), where The Garbage Collectors continue their work unhindered (Fig. 19.1).

Due to the outright abuse of plastics and their improper waste management, vast quantities are indiscriminately discarded on bare lands, there are huge amounts of packed plastics ready to be exported for recycle or selling them locals. (Fig. 19.2).

19.2.2 Beach Clean-Up Methodology

Beach clean-ups were conducted each September 2020 during the dry season. Clean-ups were organized by the 'beach clean-up', an NGO. The timing coincided with that of the ocean conservancy's international coastal clean-up day, an annual global event which takes place in September and October.

Fig. 19.2 Plastics after sorting, classification and arrangement

19.2.3 Data Handling

Litter items in each major category (glass, metal, paper, construction material, and plastic) were compiled and totaled. Plastic litter was further divided as carrier bags, personal hygiene products, beverage bottles, fishing gear, cups, packaging materials, bottle caps, lids, food wrappers, cutlery, cigarette butts, tires, and straws/stirrers.

19.2.4 Classification of Plastics

Since the first reference to microplastics in the academic literature (Thompson et al., 2004), many authors have been studying their aspects. Even though dimensions are among the first aspects to be studied in microplastics, Thompson et al. (2024) did not specify any size criterion for classifying a plastic fragment as a microplastic. Many other authors have identified size limits between macroplastics (regularly discarded plastic fragments easily visible with the naked eye, even when surrounded by other particles such as sand or gravel) and microplastics. This criterion is important because it distinguishes between small macroplastics, which have small physical dimensions and can thus be easily characterized with simple techniques, and microplastics, whose fragments need normally to be characterized with optical instruments due to their small size (Costa et al., 2010). There are challenges in terms of acquisition, characterization, and sizing up their real presence in the environment; although many

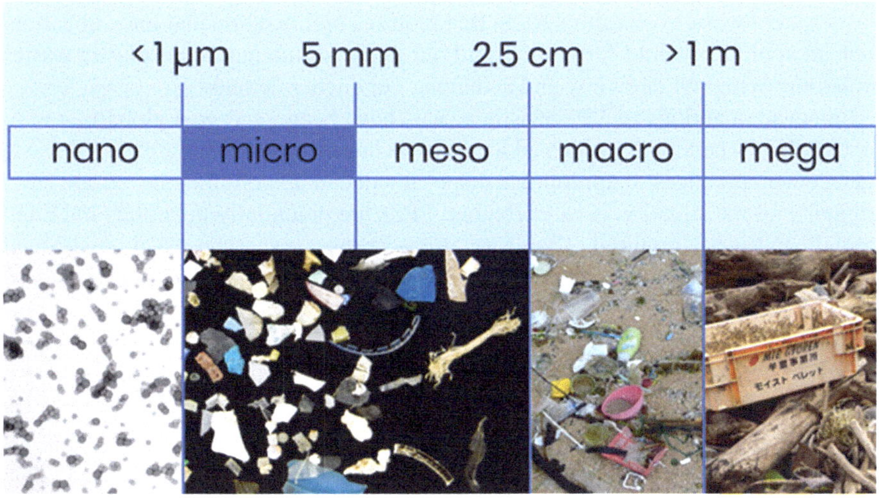

Fig. 19.3 Classification of plastic litter according to size (microplasticresearch.wordpress.com)

substantial academic publications from different institutions have been focusing on plastic debris in general since the 1980s, (Claessens et al., 2011) the difference between microplastics and macroplastics has not yet been considered. However, in the last decade, a new consensus in the science community has been working on establishing a size criterion for plastic debris (Malankowska et al., 2021) (Fig. 19.3).

19.2.4.1 Statistical Analysis

One-way analysis of variance (ANOVA) was used to evaluate variability in the number of different plastic items, using STATGRAPH Statistical package (Statgraph 5, UK).

19.3 Results and Discussion

Due to their numerous merits (such as high durability, diverse applicability, ready availability, low-cost, reusability, and so on), the presumably inevitable use of plastics makes their waste ubiquitously dispersed in our environment, especially in the oceans. The environmental damage posed, especially to the ecosystem, initiated the need for recourse control of these environmentally refractory pollutants. In this review, the various sources, classifications, fate, and control measures of plastic wastes were appraised. Further, of the three primary techniques for resource control, pyrolysis was reviewed in-depth, showing its relevance and superiority over others. Specific case studies showed that liquid and gaseous fuels derived from pyrolyzed plastics are a waste-to-wealth system that requires optimization and intensification. Such an approach would further help rid our planet of the numerous plastic wastes while improving our economy and achieving our energy demand.

Immense quantities of PPE plastic wastes have been generated globally due to the COVID-19 pandemic, which added an extra pressure to conventional solid waste management practices. Inappropriate disposal will end up polluting the marine environment and the situation is exacerbating. PPEs are potentially infectious litter and special handling is required. Therefore, it is a must to reveal the plastic-associated environmental load of the pandemic in order to prevent marine plastic pollution to continue exacerbating.

Countless face masks, face shields, different types of gloves, garments and plastic materials were consumed irrationally during the COVID-19 pandemic as a preventive measure against the spread of coronavirus (OSPAR, 2020; Vanapalli et al., 2021, El-Sheekh & Hassan, 2020). Unfortunately, they are found stranded in the beaches, coastlines, rivers, and littering cities instead of getting rid of them in suitable garbage boxes for subsequent removal for recycling or to be disposed suitably, in landfills. Nevertheless, there were no clear instructions for disposing mechanisms, people are improperly disposing plastics, especially PPE, throwing them near the location where they end their helpfulness, carried off by a gust of wind, accumulation of these plastics will continue to aggravate over time, pollute the marine environment (Moore, 2020; Rhee, 2020).

During the lockdown, a couple of dead whales were found laid down on the shore of the Mediterranean Sea in Egypt (Fig. 19.4). There was a huge amount of plastics on their stomach (Fig. 19.5).

Figure 19.6 shows the number and percentage of different items collected from the gut of these dead whales. PPEs (gloves and masks) represent 71%, while number of plastic bags represented 7% of the total litter. The remaining portion (22%) included fishing net and plastic parts of unidentified sources.

Moreover, marine plastic pollution disrupts ocean navigation, affecting marine productivity and causing the sudden death of some vulnerable marine mammals, thereby degrading the ecosystem (Awuchi & Awuchi, 2019). According to the World Economic Forum 2019, if plastic pollution is not curbed, it will overpopulate and

Fig. 19.4 Dead whales found on the beach

Fig. 19.5 Different pieces of plastic garbage pulled out from the stomach of dead whale

Fig. 19.6 Items collected from the gut of the dead whale laid down on the Northern Coast of Egypt

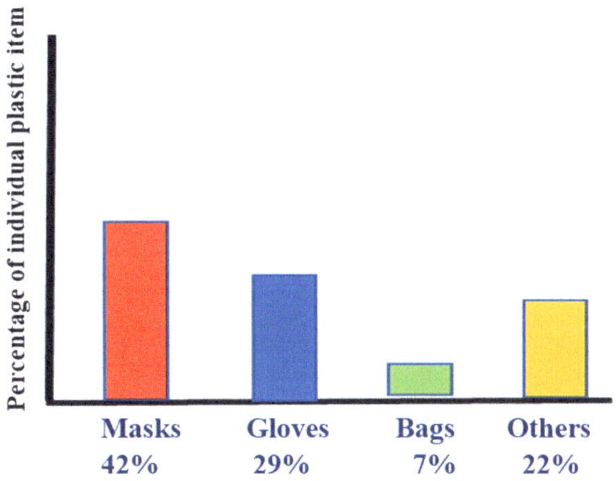

outweigh the Pisces on or before 2050 (Schwab, 2019). In Fig. 19.7, the attribution of plastic waste pollution (based on production magnitude, management flaws, and oceanic pollution) globally is provided.

Plastic litter may have fates after reaching the marine environments; the low-density plastics can float and stay in the marine environments for long periods of time, probably subject to surface water currents, while high-density plastics sink and reach bottom marine sediments, while others may become buried in the sediments, and they become part of the geological record, eventually (Fadare & Okoffo, 2020; De-la-Torre & Aragaw, 2021). This is an alert as they do not degrade naturally and they will pollute the environment for many years to come (Adelodun, 2021).

A total of 490,064 items of anthropogenic litter were collected by over 100 volunteers in a youth campaign called "Summer without plastic". Moreover, it is worth mentioning that young kids were enthusiastic to participate in the cleaning campaigns (Fig. 19.8).

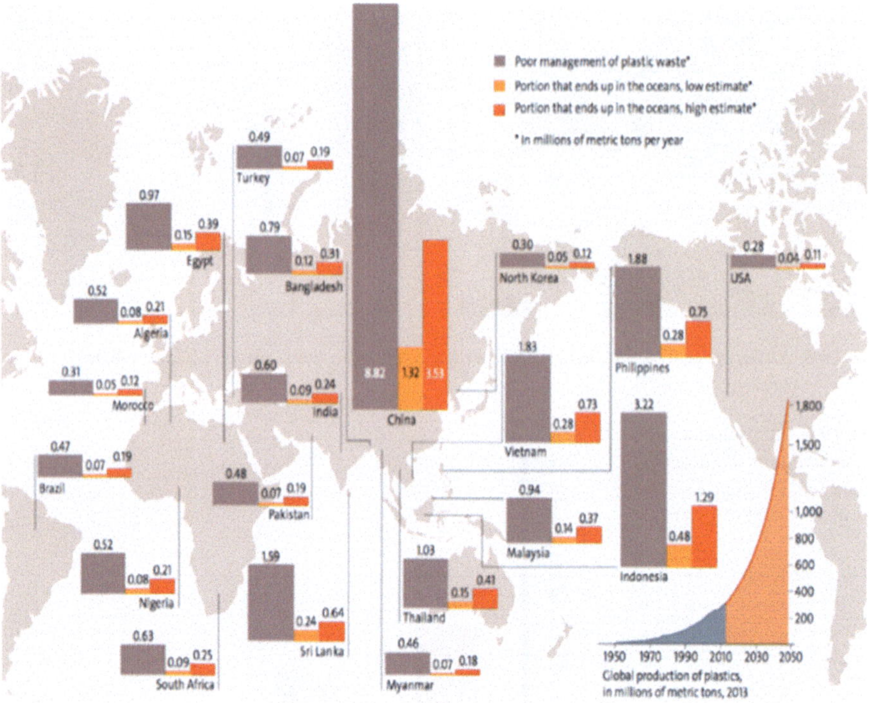

Fig. 19.7 Attribution of plastic waste pollution (in millions of metric tons year) across the global (d'Ambrières, 2019)

Fig. 19.8 Youth campaigns "summer without plastics" in Egypt

The distribution of different plastic litters collected from different sites is illustrated in Table 19.1. Gloves and face masks (PPEs) accounted for 38.5%, of the total litter, while, plastic bags represent 18.3% (Fig. 19.6).

Table 19.1 shows the average collection values for the distribution of different plastic litter collected. Gloves and face masks (PPE) accounted for 38.1%, while plastic bags represent 18.3%, of the total litter collected. The ban on plastic bags and encouragement to use paper bags instead could be another reason for the lower percentage of plastic bags in Hurghada and Jeddah (Hassan et al., 2022).

The weight of macro-and microplastics is presented in Table 19.2. For comparability purposes and simplicity, weight is calculated on a m^2 basis. The average weight of macroplastics was 25.66 g m^{-2}, while the weight of microplastics 0.81gm^{-2}.

Table 19.3 shows the relative abundance of PPE and total litter. There were 2.93, PPE item m^{-2} laid down in Alexandria. Moreover, the total amount of litter collected from the same city was 7.2 item m^{-2}.

The dominance of macroplastics could be attributed to a high population and the increased number of visitors from outside the city during summer days, especially from the neighboring villages, unfortunately, they used the beach and left all their wastes after their departure. Most of visitors are of low economy and of poor education, they throw out their plastics on the beach instead of using litter pins.

Volunteers were very enthusiastic, and they built a fish-like perforated rubbish pin to collect all plastic debris and get rid of excess sand (Fig. 19.10).

Egypt is witnessing intense tourism, fishery and other anthropogenic activities that have made their coastal zones and biota vulnerable to macro and microplastic contamination. This form of plastic pollution is an alarming, yet poorly understood. Therefore, there is a research need in order to fill the current knowledge gaps regarding COVID-19-associated PPE pollution and lay groundwork for better waste management and legislation, especially, they could be transferred from local to remote areas where measurements are scarce. Controlling the accumulation of plastics through their degradation (biological, chemical or physical), is yet to be resolved. It is necessary, to identify sources and drivers of plastics and to track them, including PPE, after entering the marine ecosystems and understand their potential fate (Al-Salem et al., 2021). When plastics and other marine litter are degraded, they release microplastics as well as new species invading the aquatic environment; these new species are called Alien invasive species (AIS). Most PPEs have characteristic morphology different from other marine plastic, therefore, it is worth to investigate their suitability as AIS and microplastic drivers (Galgani et al., 2015; Pinochet et al., 2020).

James et al. (2021), stated that the shorelines Kilakkarai of GoM were found to be littered by discarded fishing net lines which subsequently reach the coastal waters through wind action or runoff from beaches. The large quantities of discarded nets along the shoreline further aggravate the threat of microplastic pollution by weathering.

The magnitude of PPE pollution remains unknown, especially in the Middle East, despite there are some published reports worldwide (Fadare & Okoffo, 2020; Prata et al., 2020; De-la-Torre & Aragaw, 2021; Galgani et al., 2021). Thus, it is necessary to consider these two drivers of marine pollution with associated pollutants.

Table 19.1 Daily distribution of different plastic items collected from different sites day^{-1} (n = 90 + SE). Means not followed by the same letter are significantly different from each other at $p \leq 0.01$. Figures between parentheses represent the relative ratio of each item

Total	Others	Bottles	Bags	Masks	Gloves	Location
3673 ± 427.19	501^c ± 33.47 (13.6%)	1103^c ± 225.33 (30.0%)	671^c ± 43.91 (18.3%)	918c ± 66.28 (25.0%)	480^c ± 75.14 (13.1%)	Alexandria (Egypt)

Fig. 19.9 Youth campaigns "summer without plastics"

Table 19.2 Average weight of macro- and microplastics collected from different sites (g m^{-2}) (n = 90 ± SE)

Weight of microplastics (g m^{-2})	Weight of macroplastics (g m^{-2})	Location
0.86^{b} ± 0.023	25.66^{c} ± 4.01	Alexandria

Table 19.3 Relative abundance of PPEs and total plastic litter (m^{-2} $site^{-1}$). (n = 90 ± SE)

Total litter	PPEs	Location
7.20 ± 1.03	2.79 ± 0.311	Alexandria

Fig. 19.10 Collecting the plastic litter in a fish-like perforated box

Plastic litter may have different fates after reaching the marine environment; the low-density plastics can float and stay in the marine environment for long periods, probably subject to surface water currents, while high-density plastics sink and reach the bottom marine sediments, and some may become buried in the sediments, in which case they eventually become part of the geological record (Fadare & Okoffo, 2020; De-la-Torre & Aragaw, 2021). This is an alert as they do not degrade naturally, and they will pollute the environment for many years to come.

This briefing highlighted the need for a further research to accurately evaluate and lessen the potential environmental impacts of litter in public spaces future actions. There is an urgent need for unconventional materials and product designs, and advanced approaches and policies to encourage desirable consumer behavior related to use, sanitation, collection, safe disposal and the prevention of littering. Finally, national and international campaigns for monitoring aspects of single-use plastic are needed to facilitate research and guide future policy options (reliable up-to-date data collection on littering). Raising awareness to the change in behavior and municipal waste management (including efficient collection, safety, hygiene, and recycling schemes).

Since the production and use of plastics cannot be halted in the foreseeable future, due to the immense benefits of plastic to the modern world, the way forward includes the following: 1. Where unavoidable, more thermoplastics, which are recyclable, should be produced for essential use, rather than thermosetting plastics, which are unrecyclable; 2. The pyrolysis option of resource recovery should be prioritized because of its inherent benefits, such as being a waste-to-wealth scheme and environmental benignity, 3. For source control, household energy recovery from plastics, should be encouraged, and 4. Where possible, governments should monetize the scavenging of plastics from the waterways to reduce the eventual environmental cataclysm that results from flooding and endangered aquatic lives.

The magnitude of PPE pollution remains unknown, especially in the Middle East, despite there are some published reports worldwide (Fadare & Okoffo, 2020; Prata et al., 2020; De-la-Torre & Aragaw, 2021; Galgani et al., 2021). There is a debate that PPEs and PPE-derived microplastics are a potential source and vector of chemical pollutants in marine ecosystems (Fred-Ahmadu et al., 2020). Thus, it is necessary to consider these two drivers of marine pollution with associated pollutants.

It is worth to reveal the presence and impact of plastic debris on freshwater shorelines, as the information is scarce for freshwater environments and has not yet been available for the Middle East.

This briefing highlighted the needs for a further research to accurately evaluate and lessen the potential environmental impacts of litter in public spaces future actions. There is an urgent need for unconventional materials and product designs, and advanced approaches and policies to encourage desirable consumer behavior related to use, sanitation, collection, safe disposal and the prevention of littering. Finally, national and international campaigns for monitoring aspects of single-use plastic are needed to facilitate research and guide future policy options (reliable up-to-date data collection on littering). Raising awareness to change behavior and municipal waste management (including efficient collection, safety, hygiene, and recycling schemes).

In summary, this short focus indicated a relatively low level of knowledge about COVID-19, particularly regarding the disposal of PPEs, among laymen in the Middle Eastern countries. To the best of knowledge, this is the first attempt to document the abundance of plastic debris on beaches in Egypt. Therefore, it becomes of utmost essential to spread awareness on microplastic pollution among the local population, and also to formulate management plans and recommendations to reduce the practice of discarding damaged fishnets in the beaches.

19.4 Conclusions

The most important lesson we learned from the COVID-19 pandemic is that we should prepare now for potential disruptive events in an uncertain future.

More collection bins for PPEs only should be installed in the cities. Used masks, used gloves, used personal clothes and all PPEs must be separately collected and discharged in closed garbage bags to transport them to final treatments (e.g. landfilling or incineration) safely. Moreover, Recyclable plastics should be used, rather than unrecyclable ones.

Governments should monetize the scavenging of plastics from water bodies to reduce the eventual environmental cataclysm and save endangered aquatic wildlife. Simply, good knowledge, optimistic attitudes, and suitable practices toward COVID-19 will help us to cope with the pandemic.

It is worth to study the outfalls and treated effluents from wastewater treatment plants and their relationship to microplastic pollution in the coastal environment, as such studies are extremely lacking.

Acknowledgements This work was funded by a generous fund from the Academy of Scientific Research & Technology (ASRT).

Authors would like to thank NAM Centre for their fund and giving us the opportunity to present this work in the workshop held in March 2023 in India.

References

Adelodun, A. A. (2021). Plastic recovery and utilization: from ocean pollution to green economy. *Environmental Sci*ence, *09*, 683403. https://doi.org/10.3389/fenvs.2021.683403.

Al-Salem, S. M., El-Eskandarani, M. S., & Constantinoun, A. (2021). Can plastic waste management be a novel solution in combating the novel CoronavirusCOVID-19: A short research note. *Waste Management & Research, 39*(7), 910–913. https://doi.org/10.1177/0734242X2097844.

Aragaw, T. A. (2020). Surgical face masks as a potential source for microplastic pollution in theCOVID-19 scenario. *Marine Pollution Bulletin, 159*, 111517. https://doi.org/10.1016/j.marpolbul.2020.111517.

Awuchi, C. G., & Awuchi, C. G. (2019). Impacts of plastic pollution on the sustainability of seafood value chain and human health. *International Journal of Advanced Academic Research, 5*(11), 46–138.

Claessens, M., De Meester, S., Van Landuyt, L., De Clerck, K., & Janssen, C. R. (2011). Occurrence and distribution of microplastics in marine sediments along the Belgian coast. *Marine Pollution Bulletin, 62*(10), 2199–2204.

Costa, M. F., Do Sul, J. A. I., Silva-Cavalcanti, J. S., Araújo, M. C. B., Spengler, Â., & Tourinho, P. S. (2010). On the importance of size of plastic fragments and pellets on the strandline: A snapshot of a Brazilian beach. *Environmental Monitoring and Assessment, 168*(1), 299–304.

De-la-Torre, G. E., & Aragaw, T. A. (2021). What we need to know about PPE associated with the COVID-19 pandemic in the marine environment. *Marine Pollution Bulletin, 163*, 11189. https://doi.org/10.1016/j.marpolbul.2020.11189.

d'Ambrières, W. (2019). Plastics recycling worldwide: Current overview and desirable changes. *Field Actions Science Report, 19*, 12–21.

El-Sheekh, M. M., & Hassan, I. A. (2020). Lockdowns and reduction of economic activities during the COVID- 9 pandemic improved air quality in Alexandria city. *Egypt. Environmental Monitoring & Assessment, 193*, 11. https://doi.org/10.1007/s10661-020-087807.

Fadare, O. O., & Okoffo, E. D. (2020). Covid-19 face masks: A potential source of microplastic fibers in the environment. *Science of the Total Environment, 737*, 140279. https://doi.org/10.1016/j.scitotenv.2020.14079.

Galgani, F., Brien, A., Weise, J., Ioakeimidis, C., Schuyler, Q., Makarenko, I., et al. (2021). Are litter, plastic and microplastic quantities increasing in the ocean? *Microplastics and Nanoplastics, 1*, 2. https://doi.org/10.1186/s43591-020-00002-8.

Galgani, F., Hanke, G., & Maes, T. (2015). Global distribution, composition and abundance of marine litter. *Marine Anthropogenic Litter, 29*, 49–56. https://doi.org/10.1007/978-3-319-165 1032.

Fred-Ahmadu, O. H., Bhagwat, G., Oluyoye, I., Benson, N. U., Ayejuyo, O. O., & Palanisami, T. (2020). Interaction of chemical contaminants with microplastics: Principles and perspectives. *Science of the Total Environment, 706*, 135978. https://doi.org/10.1016/j.scitotenv.2019.13598.

Hassan, I. A., Younis, A., Al Ghamdi, M. A., Almazroui, M., Basahi, J. M., El-Sheekh, M. M., Abouelkhair, E. K., Haiba, N. S., Alhussaini, M. S., Hajjar, D., Abdel Wahab, M. M., & El Maghraby, D. M. (2022). Contamination of the marine environment in Egypt and Saudi Arabia with personal protective equipment during COVID-19 pandemic: A short focus. *Science of the Total Environment, 810*, 152046. https://doi.org/10.1016/j.scitotenv.2021.15206.

James, K., Vasant, K., Batcha, S. M., Padua, S., Jeyabaskaran, R., Thirumalaiselvan, S., Vineetha, G., & Benjamin, L. V. (2021). Seasonal variability in the distribution of microplastics in the coastal ecosystems and in some commercially important fishes of the Gulf of Mannar and Palk Bay, Southeast coast of India. *Regional Studies in Marine Science, 41*, 101558. https://doi.org/10.1016/j.rsma.2020.10158.

Malankowska, M., Echaide-Gorriz, C., & Coronas, J. (2021). Microplastics in marine environment: A review on sources, classification, and potential remediation by membrane technology. *Environmental Science: Water Research & Technology, 7*(2), 243–258. https://doi.org/10.1039/d0ew00802h.

Miskolczi, N., Angyal, A., Bartha, L., & Valkai, I. (2009). Fuels by pyrolysis of waste plastics from agricultural and packaging sectors in a pilot scale reactor. *Fuel Processing Technology, 90*, 1032–1040. https://doi.org/10.1016/j.fuproc.2009.04.019.

Moore, C. (2020, October 15). Plastic pollution. *Encyclopedia Britannica*. Retrieved December 16, 2022, from https://www.britannica/science/plastic-pollution.

Nasser, A. M. W., El-Sheekh, M. E., Zeineldein, M. H., & Hassan, I. A. (2022). Physiological, morphological, and growth effects of microplastics on freshwater alga *Chlorella vulgaris*. *Rendiconti Lincei. Scienze Fisiche e Naturali, 33*, 815–821. https://doi.org/10.1007/s12210-022-011 04-6.

OSPAR. (2020). CEMP guidelines for marine monitoring and assessment of beach litter. Retrieved July 2023, 19, from https://www.ospar.org/documents?v=44122.

Pinochet, J., Urbina, M., & Lagos, M. (2020). Marine invertebrate larvae love plastics: habitat selection and settlement on artificial substrates. *Environmental Pollution, 257*, 113571. https://doi.org/10.1016/j.envpol.2019.113571.

Prata, J. C., da Costa, J. P., Lopes, I., Duartea, A. C., & Rocha-Santosa, T. (2020). Environmental exposure to microplastics: An overview on possible human health effects. *Science of the Total Environment, 702*, 134455. https://doi.org/10.1016/j.scitotenv.2019.13445.

Plastics Europe. (2019). Plastic more sustainable than paper, study shows. Retrieved November 25, 2020, from www.plasticstoday.com/packaging/plastic-more-sustainable-paper-study-shows/207518312859524.

Rhee, S.-W. (2020). Management of used personal protective equipment and wastes related to COVID-19 in South Korea. *Waste Management & Research, 38*(8): 820–824. https://doi.org/10.1177/0734242X20933343.

Schwab, C. (2019). Production, use, and fate of all plastics ever made. *Science Advances, 3*. https://doi.org/10.1126/sciadv.1700782.

Thompson, R. C., Courtene-Jones, W., Boucher, J., Pahl, S., Raubenheimer, K., & Koelmans, A. A. (2024). Twenty years of microplastic pollution research—what have we learned? *Science, 386*(6720), eadl2746. https://doi.org/10.1126/science.adl2746.

Thompson, R. C., Olsen, Y., Mitchell, R. P., Davis, A., Rowland, S. J., John, A. W. G., McGonigle, D., & Russell, A. E., et al. (2004). Lost at sea: Where is all the plastic? *Science(Washington), 304*(5672), 838.

UNEP. (2011). The most hazardous environmental problems. Retrieved March 2023, from www.unep.org/report21927.

Vanapalli, K. R., Sharma, H. B., Ranjan, V. P., Samal, B., Bhattacharya, J., Dubey, B., & Goel, S. (2021). Challenges and strategies for effective plastic waste management during and postCOVID-19 pandemic. *Science of the Total Environment, 750*, 141514. https://doi.org/10.1016/j.scitotenv.2020.141514.

Chapter 20
Plastic Waste Management in South Delhi

Neha Yadav, Rakshitha Rajashekara, and Pallavi Nagaraju

Abstract The unprecedented growth of India's population in the past decades and their migration to the cities have increased the pace of urbanization in the country. The ever-increasing migration of people to the cities for various economic opportunities and educational purposes has not only increased the no. of people in cities but also changed their lifestyle which is supported by the constant expansion of industries and services. The simultaneous processes of migration, urbanization, and industrialization are contributing toward the generation of immense quantities of Municipal Solid Waste putting the health of the Environment at stake at large and of people in specific. Out of the total solid waste generated, plastic waste takes a significant amount and space. Out of the total plastic production in a year, single-use plastics alone account for 40% of the total. Its non-biodegradable, persistent nature and low recycling rate are making the situation even worse. Delhi is one most populated Indian cities grappling with the management of plastic waste. In the given study, an area of south Delhi has been undertaken to provide a detailed analysis of the plastic waste management ways at the household and community level. Some recommendations have also been provided based on the findings.

Keywords Awareness · Management · Plastic waste · Population · Segregation

20.1 Introduction

Plastic waste has been defined as any discarded item which is made up of plastic. Plastic waste can further be categorized into primary plastics bottle caps and cigarette butts and secondary plastics (produced by the degradation of the primary ones). One of the recent studies carried out by the UN Plastic Collective program has revealed that India generates 9.46 million tonnes of plastic waste annually and out of the

N. Yadav · R. Rajashekara · P. Nagaraju (✉)
Department of Environmental Science, JSS Academy of Higher Education and Research, Mysuru, India
e-mail: anupallavi@jssuni.edu.in

S. Suriyanarayanan et al. (eds.), *Combating Plastic Pollution in Terrestrial Environment*,
https://doi.org/10.1007/978-981-96-2343-3_20

total, 43% is used for packaging purposes whereas 40% of the total goes uncollected (Patnaik, 2018; Joshi & Ahmed, 2016; UNEP, 2019; EU, 2020). Microplastics are also a part of plastic waste which are defined as plastics less than five millimeters in size (Ritchie & Roser, 2020).

Similarly, single-use plastics; a disposable plastic material used only once before it is discarded form a large percentage of the plastic waste in the urban areas. India has banned the use of single-use plastic since 2022 declaring its elimination for environmental protection (Marchand, 2016; Ritchie & Roser, 2020) As per the report released by the Centre for Science and Environment titled State of India's Environment 2022, Delhi is accountable for generating the highest amount of plastic waste among all the metropolitan cities 689.8 tonnes. There are different estimates about the quantity of plastic waste produced in the last century. By one of the estimates, it is said that since the year 1950s approximately over one billion tons of plastic waste have been discarded so far (Teuten et al., 2009). Other estimates have also shown that out of the total estimated plastic generation of 8.3 billion tons of plastic waste, 6.3 billion tons have come out to be waste whereas the remaining 9% is channelized for recycling (Salimi et al., 2015).

All kinds of plastic waste generated can be categorized into High-Density Polyethylene (HDPE), Low-Density Polyethylene (LDPE), Polyvinyl Chloride, Polyethylene Terephthalate (PET), Polystyrene (PS), Polypropylene (PP) (Marchand, 2019). Some of the technologies used for recycling plastic waste are: (a) Liquid refuse-derived fuel conversion; in this, by the method of catalytic pyrolysis, the plastic waste is converted into refused-derived fuel, molten. (b) Waste Plastic Gasification; a simple and easier-to-execute process. In place of oxygen, nitrogen is used here as inert air which makes it a less expensive process. This is a thermochemical process that converts the waste material into gases like methane, hydrogen, and carbon oxides. North and Halden (2013) Degradation of Plastics by Microbes; Microbial action becomes favorable here where the microbes degenerate the plastic waste and alter its characteristics (Chamas et al., 2020; Dussud & Ghiglione, 2014; Gan & Zhang, 2019; North & Halden, 2013). The other ways which can be used for handling plastic waste are reduction of its sources and Reuse, Recycling, Energy Recovery, and, Treatment and Disposal. Shredding is an example of Physical treatment whereas incineration is an example of chemical treatment (Balasubramanian, 2015).

20.2 Plastic Waste Recycling Management

To conduct this study, an extensive review of literature has been carried out to know about the various aspects of plastic waste management (Müller, 2002). Various secondary sources (Books, Journals, Research Papers, Review Articles) have been referred to, to know about the different kinds of study done to efficiently deal with the plastic waste which includes the plastic waste generation sources, the composition of the waste at the household level, the processes used for the treatment of the waste

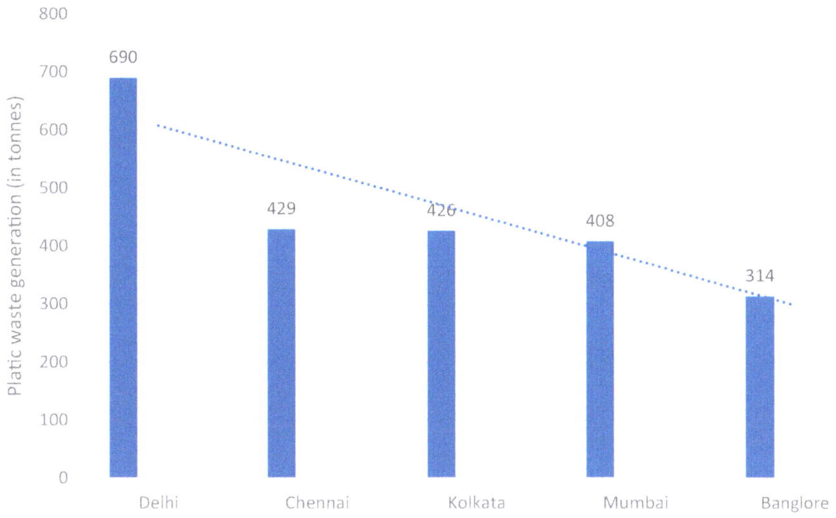

Fig. 20.1 Top plastic waste generating cities of India. *Source* CPCB 2015

and the disposal methods used to manage the same. The top plastic waste generating cities of India are Delhi, Chennai, Kolkata, Mumbai and Bangalore where Delhi tops the list making the need for waste management in the capital very evident as shown in Fig. 20.1. At the same time, some of the least plastic waste generating cities in India are Kavaratti, Panaji, Daman, Dwarka and Gangtok are shown in Fig. 20.2. If we refer to the population numbers of the top and the least plastic waste generating cities, then a direct relationship can be established. The top plastic generators are also the cities having higher population numbers. Therefore, population should also be counted as one of the factors while dealing with the plastic waste.

20.3 Methodology

To conduct a study or research, various methods can be followed. One of the methods used to conduct this research study was the Survey method. It is a method or a tool which is often used to collect information from the targeted group of study. When a survey method is used, it requires certain questions to be asked to the target group. Survey was prepared beforehand. They can be online as well as offline. For this study, an online survey was conducted using google form. The link was shared with the targeted group and respondents could fill the same at their convenience and their responses were recorded online which were later interpreted at the surveyor's ease. Around 25 questions related to management of the waste in household and in the society were designed to gather the information. The questionnaire attempted to

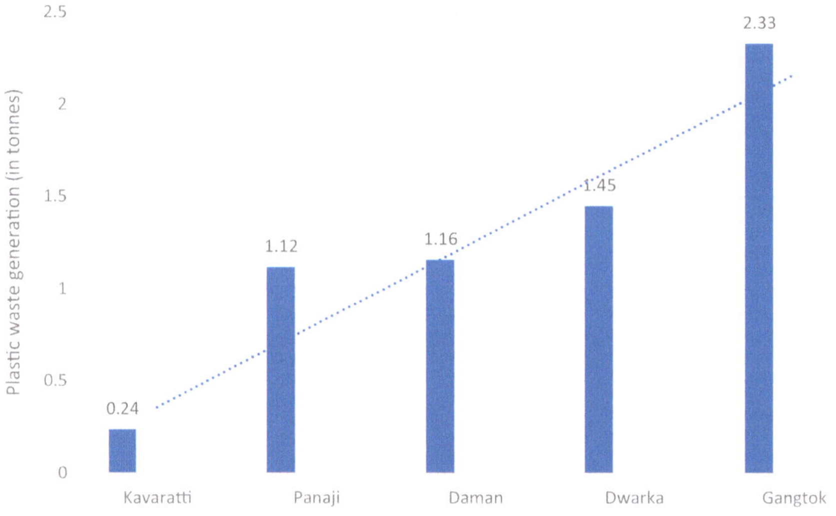

Fig. 20.2 Least plastic waste generating cities of India. *Source* CPCB 2015

gather both the qualitative as well as the quantitative information. The information collected through this way has been interpreted to draw the conclusion (Table 20.1).

20.4 Results and Discussion

The results and discussion of the study conducted include the following ones. The questions asked and their respective percentage responses are given below. A total sample of 80 people was taken into consideration as respondents. So, it can be said that a total of 80 families were covered to know about their waste management ways. Approximately an equal no. of males and females have responded to the survey on behalf of their family and Fig. 20.3 represents the % response of gender. Out of the total houses, 44% of them were owned and 53% have been on rent, remaining have chosen not to answer this and it is represented in Fig. 20.4.

Out of the total people covered only 17% of the total population were less than a graduate which means that 83% of the population were either graduates or above indicating an educated slot of the population. 57% of the population understand plastic waste as discarded plastic waste items whereas the remaining population considers both solid waste and discarded plastic waste under the plastic waste.

Talking about the quantity of plastic waste generated, the plastic used in the packaging material is the maximum in quantity (62%) whereas plastic utensils and bags also are a part of generated plastic waste but with a lesser percentage (36%). Single-use plastic and other plastic items also contribute to the quantity generated. 76%

Table 20.1 Questionnaire-based survey on waste management practices in South Delhi

S/N	Questions options				
1	What do you understand by plastic waste?	Discarded plastic waste items	**Discarded solid waste**	**Both**	**None**
2	Which type of plastic waste is generated more in quantity?	Plastic packaging	**Plastic items**	**Plastic Utensils**	**Plastic bags**
3	What is the composition of the plastic waste generated at home?	**Plastic packaging and plastic items**	**Single use plastic**	**Plastics bags**	**All of these**
4	How is the waste disposed of?	**Collection by Kuda Wala**	–	–	**Disposed on your own**
5	If the waste is collected from the doorstep, then who is providing that service?	**Municipality**	**Private Agency**	**NGO**	**Any other**
6	Approximately how many community bins are existing in your locality?	**5**	**10**	**15**	**More than 15**
7	Are there separate bins in your locality for biodegradable and non-biodegradable waste?	**Yes**		**No**	
8	Does an MCD collection center exist in your locality?	**Yes**		**No**	
9	Are there separate bins in your home for plastic waste and other types of waste?	**Yes**		**No**	
10	Do the waste containers emptied/cleaned before they overflow?	**Yes**		**No**	
11	Do you have any idea about the plastic waste segregation?	**Yes**		**No**	
12	Do you have any idea about the given plastic waste management ways?	**Yes**		**No**	

(continued)

Table 20.1 (continued)

S/N	Questions options				
13	Do you think that the city in general and your area in particular is facing any issues of management of plastic waste?	**Yes**		**No**	
14	Whom do you hold responsible for improper/ inefficient management of plastic waste?	**Citizens/ Residents**	**Govt./MCD**	**RWA'S**	–

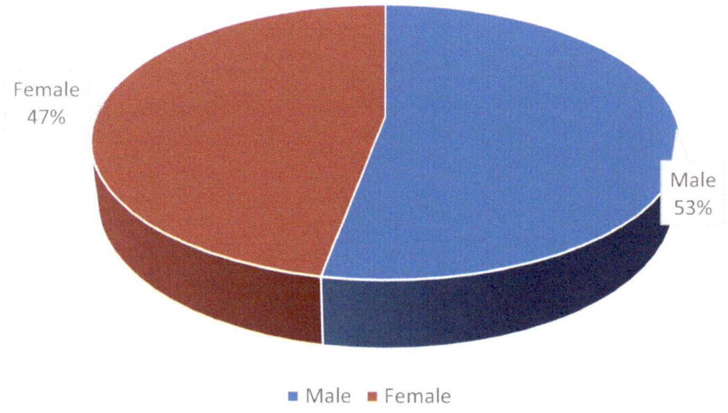

Fig. 20.3 Percent response based on gender

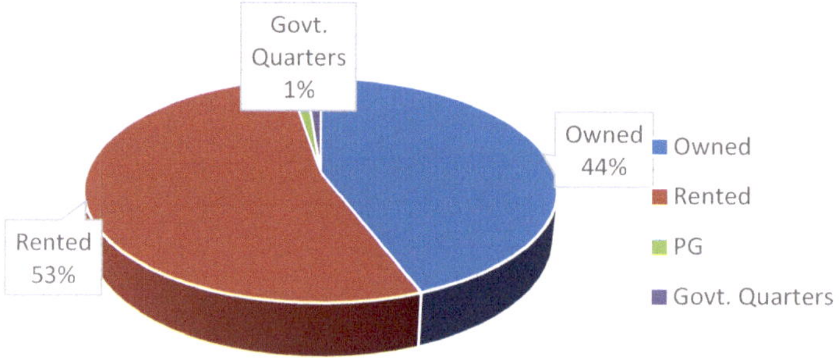

Fig. 20.4 Percent response based on the kind of accommodation

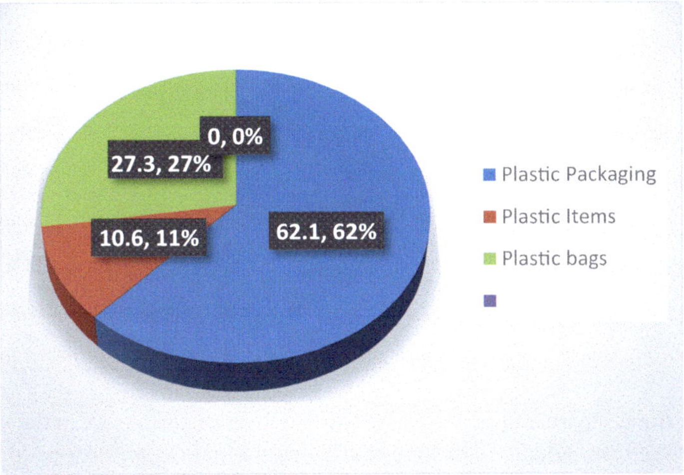

Fig. 20.5 Percent response on plastic waste generation

of the population mentions all four components (packaging material, bags, single-use plastic and other plastic items) as part of the total waste generated. Figure 20.5 represents the Percent response on plastic waste generation.

Talking about waste collection and disposal ways, it has been found that 85% of the waste including plastic one is collected by the kudawala only or the waste collector which indicates a sense of responsibility and hygiene among the population. The waste collection by a waste collector is also an indicator of the education level of the population group as only aware and educated group of people could practice this kind of a service for waste management. Municipality is the dominant body in providing waste collection services along with 11% contribution from the private agencies. Even when the waste collector is not available, the waste is dumped by one of the family members in the municipality bin of the locality which again indicates a sense of responsibility among the population (50% of the population dump the waste in a Municipality bin). Only 13% of the population practices open dumping which again strengthens this fact that a very less percentage of the people need to be made aware and change their waste disposal practice. 61% of the population admits about the waste transfer to the dumping sites before the bins get overspilled via vehicles and carts which indicates a good provision of services in that area for waste transportation. At least 5 bins are present in the locality contributing to the prevention of overspilling from the dustbins and helps in maintaining the cleanliness in the society. 55% of the population admits to the fact of having separate dustbins at the household level for plastic and other types of waste though the population awareness level for segregation of the waste is 86%.

71% of the population who is into regular practice of disposing the waste responsibly is accepting the existence of at least 5 bins in the locality though not designed for separated waste, 2an indication of good service provisions in the targeted area but

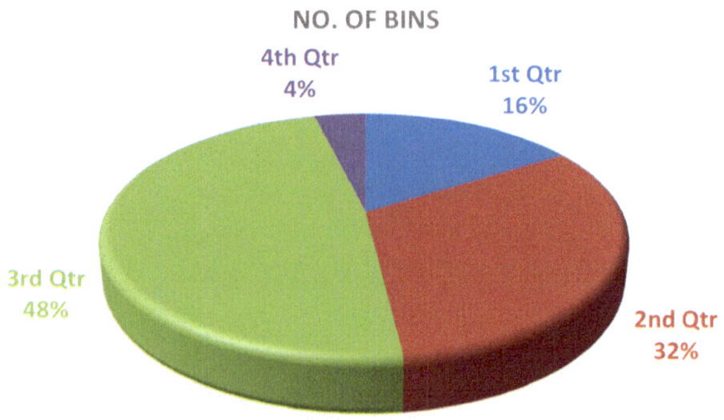

Fig. 20.6 Percent response about community bins

at the same time, separate bins for plastic waste collection are absent in the locality as per 70% of the respondents. This gives the scope for making provisions for separate collection of the waste.

The waste from the society is collected once a week by the municipality workers indicating a good provision of services in the area. Also, the visit by a vendor/ vendor every weekend in the society for waste collection (both paper and plastic) indicates multiple service provisions for dealing with the waste but also shows awareness level of the population who is into the regular practice of giving away the waste to the vendors as well. 86% of the population are aware about the importance of the segregation of plastic waste indicating a high level of education and responsibility among the surveyors. Figure 20.6 represents the number of bins by community.

The surveyed population has made it clear that for better management of the plastic waste, still more awareness among the population is required along with better waste management 58provisions. Also, the surveyed population has held the citizens responsible for the improper/inefficient management of the plastic waste generated in the area along with the government bodies. 98% of the population believed that more awareness among the people with help in resolving the issues associated with handling, management and disposal of plastic waste however it is represented in Fig. 20.10. Therefore, it can also be said that along with the implementation of the enacted laws, awareness also has an equal importance in efficient plastic waste handling. Though majority of the populations is familiar with the concepts of Reuse and Recycling with less familiarity with the landfill sites but adopting the same in practice remains a challenge. Figure 20.7 represents the plastic waste management ways and Fig. 20.8 represents the Percent response on plastic waste dumping.

Over a period, it has been proven that the existing ways of waste management are unable to meet the goal of a sustainably developed society (Agarwal et al., 2005) Awareness among the population and the waste segregation and management practices adopted at the household level can be a key in resolving the issue of plastic waste management. Figure 20.9 represents the efficient plastic waste management.

Delhi being a metropolitan city with a population of 11 million (as per 2011 census), has been struggling with the immense quantity of waste generated. Delhi generates a total 689.9 tons of plastic waste/day (CPCB, 2020) and becomes the

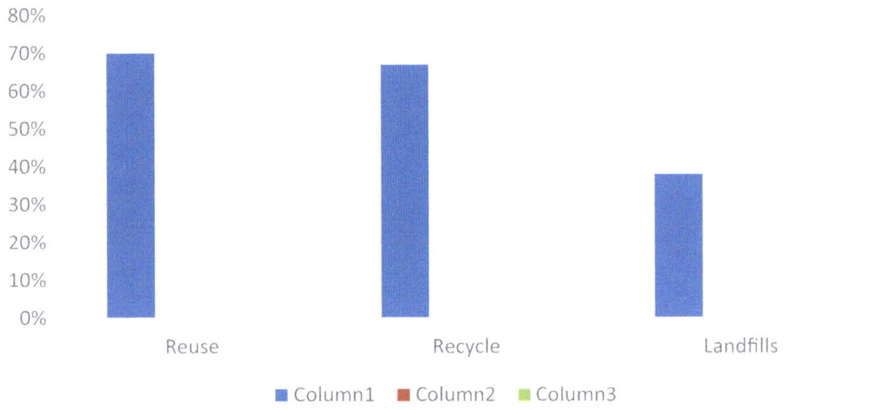

Fig. 20.7 Percent response on plastic waste management ways

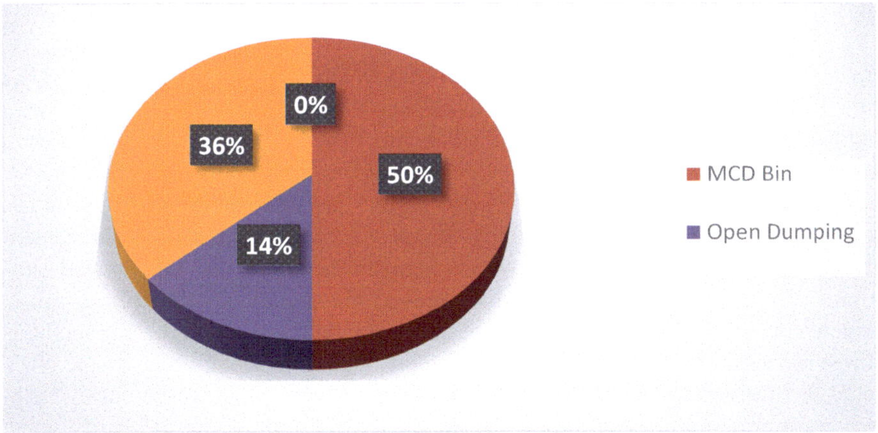

Fig. 20.8 Percent response on plastic waste dumping

Fig. 20.9 Percent response on efficient plastic waste management

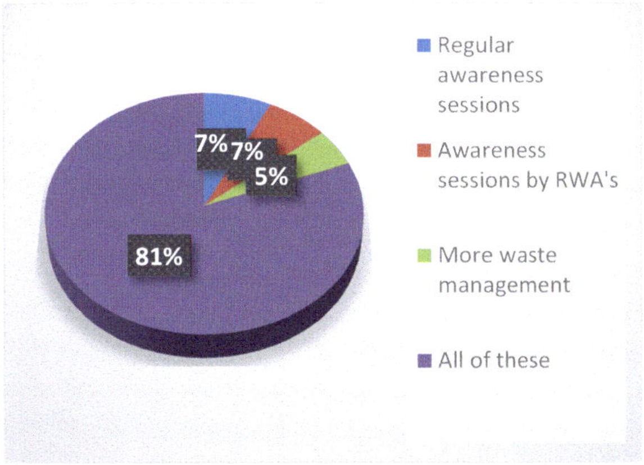

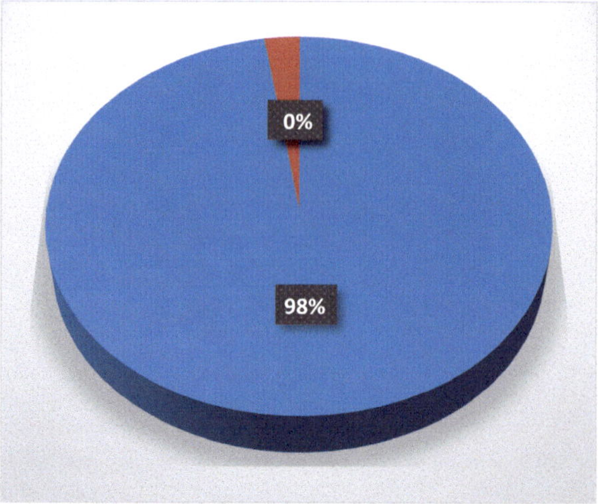

Fig. 20.10 Percent response on awareness about plastic waste management

highest waste producing city of India followed by Chennai and Mumbai. Delhi has three landfill sites namely, Bhalaswa, Okhla and Ghazipur landfill site with an area that has been exhausted and has attained heights beyond the permissible limit for a landfill site (Kansal et al., 1998).

The master plan of Delhi 2021 is also saying that there must be a close nexus between the rag pickers, the municipal authorities and decentralized management of the waste needs to be done by municipalities to control the unregulated situation of the landfill sites. Decentralized management includes the reduce, reuse, recycle concept along with the different treatment methods which include methods like vermicomposting, bioremediation, composting, etc., per an order by the NGT dated 3rd December 2018, it sought a compliance from the government of Delhi to deposit an amount of 25 crore with the central pollution control board for the uncontrolled management of the plastic waste. This amount was asked to be paid as a cost of damage to the environment (CPCB, 2020).

Kalkaji, is in the southern part of Delhi. It is a superior quality locality which has been famous for a very old temple of Kalka deity and lotus temple, also known as the Bahai house of worship. The lotus has been a must visiting place for the tourists visiting Delhi. The surrounding area of Delhi includes the place like Nehru place, Lajpat Nagar, Okhla and Greater Kailash. The study on waste management has been focused on the area of Kalkaji wherein households have been surveyed and inquired about their waste management practices.

20.5 Conclusion

The large quantity of plastic waste generation in Indian states and cities has become an issue for both the urban governments and the society. Even though the plastic waste management services are provided by the municipalities and private agencies, but its immense quantity is making its management difficult. Lack of awareness among people and other socio-economic constraints is making its handling even more difficult. The inefficient handling of the waste is leading to various environmental issues like pollution of air due to processes like incineration and open burning, release of toxic gases, generation of greenhouse gases by landfill sites and burning of the waste in open and closed areas, pollution of water (ground water) due to unlined landfills and leachate formation at landfill sites. With the rise in waste quantity, new waste management techniques have been developed and implemented, the intensity of traditional techniques has also been increased but everything is proving to be a failure in front of the waste quantity generated. Based on the survey done, it can be concluded that though all the families are highly literate and majority of them are well off also but still not all the families are aware about the plastic waste management-related issues and ways of its management. 80% of the population agreed that the waste management issues are because of the lack of awareness among the people. At the same time, 98% of the population have agreed that more awareness among the population of any area about the waste management practices can help in resolving the waste management issues, the city is grappling with. Municipality bins are there in the area, placed at appropriate places for people to dump the waste and by the waste collector who collects the waste from the houses but it is not separately there for the plastic waste.

References

Agarwal, A., Singhmar, A., Kulshrestha, M., & Mittal, A. K. (2005). Municipal solid waste recycling and associated markets in Delhi, India. *Resources, Conservation and Recycling, 44*(1), 73–90.

Balasubramanian, M. (2015). Economics of solid waste in India. *Economic and Political Weekly*, 17–20; Balasubramanian, M., & Birundha, V. D. (2011). Municipal solid waste management in India. *Productivity, 51*(4), 308.

Be Waste Wise. (2019b, March). Plastic wastes and its management. https://www.bioenergyconsult.com/plastic-wastes-management.

Chamas, A., Moon, H., Zheng, J., Qiu, Y., Tabassum, T., Jang, J. H., Abu-Omar, M., Scott, S. L., & Suh, S. (2020). Degradation rates of plastics in the environment. *ACS Sustainable Chemistry & Engineering., 8*(9), 3494–3511.

Dussud, C., & Ghiglione, J.-F. (2014, December 26). Bacterial degradation of synthetic plastics. CNRS, UMR 7621. Laboratoired'Océanographie Microbienne, Observatoire Océanologique, F66650 Banyuls/mer, France.

Daspatnaik, S. Finding solutions to plastic waste management in India. ecoideaz.com. https://www.ecoideaz.com/expert-corner/finding-solutions-to-plastic-waste-management-in-india.

Gan, Z., & Zhang, H. (2019). PMBD: A comprehensive plastics microbial biodegradation database. *Database, 2019*, baz119. https://doi.org/10.1093/database/baz119.

Joshi, R., & Ahmed, S. (2016). Status and challenges of municipal solid waste management in India: A_review. *Cogent Environmental Science, 2*, 1139434.

Kansal, A., Prasad, R. K., & Gupta, S. (1998). Delhi municipal solid waste and environment—An appraisal. *Indian Journal of Environmental Protection, 18*(2), 123–128.

Kumar, D., & Alappat, B. J. (2005). Evaluating leachate contamination potential of landfill sites using leachate pollution index. *Clean Technologies and Environmental Policy, 7*(3), 190–197.

Marchand, R. (2016, September 15). Methods of plastic waste disposal (and possible complications). http://blog.nus.edu.sg/, http://blog.nus.edu.sg/plasticworld/2016/09/06/x-methods-of-plastic-waste-disposal-and-possible-complications/.

Marchand, R. (2019, July 15). America's garbage 'problem' concocted by trashy data. bioenergyconsult.com. https://catalyst.independent.org/2019/07/15/americas-garbage-problem-concocted-by-trashy-data/.

Müller, M. S., Iyer, A., Keita, M., Sacko, B., & Traore, D. (2002). Different interpretations of community participation in waste management in Bamako and Bangalore: Some methodological considerations. *Environment and Urbanization, 14*, 241–258.

North, E. J., & Halden, R. U. (2013). Plastics and environmental health: The road ahead. *Reviews on Environmental Health, 28*(1), 1–8. https://doi.org/10.1515/reveh-2012-0030.

Patnaik, S. D. (2018) Finding solutions to plastic waste management in India. ecoideaz.com. https://www.ecoideaz.com/expert-corner/finding-solutions-to-plastic-waste-management-in-India.

Ritchie, H., & Roser, M. (2020). Plastic pollution. OurWorldInData.org. https://ourworldindata.org/plastic-pollution.

Salimi, A., Tavakoli Moghadam, R., & Bashiri, M. (2015). Designing reverse logistics network with a social approach to recycle great household waste under uncertainty conditions. *Journal of Applied Research on Industrial Engineering, 2*(3), 168–179.

Teuten, E. L., Saquing, J. M., Knappe, D. R., Barlaz, M. A., Jonsson, S., Björn, A., et al. (2009). Transport and release of chemicals from plastics to the environment and to wildlife. *Philosophical Transactions of the Royal Society of London. Series B, Biological Sciences, 364*(1526), 2027–2045.

The environmental impacts of plastics and micro-plastics use, waste and pollution: EU and national measures (PDF). europarl.europa.eu. (2020, October).

UNEP. (2019). *Single-use plastics: A roadmap for sustainability*. ISBN: 978-92-807-3705-9DTI/2179/JP.

Chapter 21
Overview of Plastic Pollution on the Terrestrial Ecosystem of Aghanashini Estuary, Uttara Kannada District, India

N. M. Guruprasad and S. H. Basavarajappa

Abstract The river Aghanashini originates from Sirsi taluk of Uttara Kannada district and joins the sea in Aghanashini village of Kumta taluk in Uttara Kannada district. Aghanashini is a free-flowing river. It runs through different villages. The Aghanashini estuary covering an area of 36 square kilometers provides a source of livelihood to the people living in this area. The Plastic waste gets piled up from the origin of the river, flows along 121 km long stretch, and ultimately reaches the Aghanashini estuary. Plastic wastes are brought to the river banks by tides and eventually enter the terrestrial ecosystem causing a negative impact on the species in the estuary and the terrestrial environment, as well as the humans who live in the area. The flow of plastic can have an impact on both livelihood and the economy. Regulations are required since the Aghanashini estuary will shortly be designated as a Ramsar protection site. In this study, we estimated the amount of plastic waste in the banks of the estuary and suggested regulatory measures. This work serves the present need and acts as a link between local population and policymakers.

Keywords Plastic · Pollution · Estuary · Livelihood

21.1 Introduction

The discovery of plastics provided society with a material having infinite uses. Due to its beneficial characteristics (inexpensive, durable, lightweight, abundant, and can be produced in any desired shape), the usage of plastics drastically increased during the twentieth century and its production is still increasing. Initially, plastics were mainly used to make items that were used for long duration. Adaptation to plastic durability

N. M. Guruprasad (✉) · S. H. Basavarajappa
Department of Environmental Science, Kuvempu University, Shankaraghatta, Shivamogga, Karnataka, India
e-mail: gpresearchevs@gmail.com

S. Suriyanarayanan et al. (eds.), *Combating Plastic Pollution in Terrestrial Environment*,
https://doi.org/10.1007/978-981-96-2343-3_21

caused excess usage of plastic in day-to-day life which is nonbiodegradable and toxic to environment. Irresponsible people are disposing plastic everywhere and becoming a problem. The modern lifestyle is embedded in the complexity of economic and toxic interdependency of plastic use (Kumar et al., 2021).

Estuary, a tidally influenced transition zone between river and sea, is a unique ecosystem. Sedimentary movements and deposits happen in the estuary due to constant intermingling of water from the land and the ocean (Elliott & McLusky, 2002; Pritchard, 1967). According to Mouat et al. (2010), recreational activities are directly affected by marine plastics; for example, marine litter and debris float on the surface of the ocean and land on the shoreline, which creates heavy pollution along the coastline. Estuary influences the economic condition of the country but over usage of plastics is impacting social, environmental and health condition of the coastal area. Plastic litter causes economic losses to the tourism, fishing, and shipping sectors globally (Andres et al., 2021; Oosterhuis et al., 2014;

g the HDI (Human Development Index) and simultaneously will be a major hurdle for achieving the SDGs targets (Human, 2015; Pascual et al., 2017).In our day-to-day life, we are exposed to plastic directly and indirectly either by consumption through micro-plastics or by exposing to plastic toxicity. As a result of unprecedented urbanization, cities face difficulty in managing waste in a socially and environmentally responsible manner. Locally generated waste behavior varies with cultural, environmental, and socio-economic elements, including institutional capabilities, which also influence the effective waste management and efficient alternative solutions (Sakti et al., 2021).

Aghanashini estuary today is heavily exploited and is among the most threatened ecosystems. The anthropogenic activities get plastics piled up from the origin of the river flows along 121 km long stretch, and ultimately reach the Aghanashini estuary. Plastic wastes are brought to the river banks by tides and eventually enter the terrestrial ecosystem causing a negative impact on the species in the estuary and the terrestrial environment, as well as the humans who live in the area.

The goal of this study is to provide a background assessment of adverse effects on natural ecosystems, identify knowledge gaps, and address policymakers. This paper helps to understand the impact of plastic pollution on the last major free flowing estuary of the Uttara Kannada district.

21.2 Materials and Methodology

21.2.1 Study Area

In the present study, Aghanashini estuary is situated toward the central west coast of India in the Uttara Kannada district of Karnataka.

The Aghanashini river originating near Manjguni and Donihalla, both sites located in the Sirsi taluk of Uttara Kannada, runs about 121 km near winding through deep gorges and valleys n/" ear to the Arabian sea. The river joins the sea between the villages at Tadri in the North and Aghanashini village in the south about 10 km toward the north of Kumta town.

21.2.1.1 Methodology

The periodic survey was conducted between September 2022 and December 2022 (post-monsoon period) in Aghanashini estuary banks which has an area of 3.5 km. Random sampling was done at 10 different locations by using 10 m standard net. Plastic collected in that area was estimated in a weighing scale. We consulted local people and discussed about problems arising due to plastics in estuary as well as on Aghanashini village. We had discussed the problems and possible solution with environmentalists of this region.

21.3 Results and Discussion

During this survey we estimated plastic deposition in the banks of Aghanashini estuary. Our study revealed the following results.

S No	Site	Quantity of plastic (kg)
01	Site-01	0.763
02	Site-02	1.1
03	Site-03	1.43
04	Site-04	2.49
05	Site-05	2.34
06	Site-06	0.581
07	Site-07	0.359
08	Site-08	3.219
09	Site-09	4.67
10	Site-10	1.215

Average plastic deposition in sampling site was found to be 1.8167 kg. For the area of 3.5 km the estimated plastic deposition could be 6,358.45 kg.

Our studies on Aghanashini estuary revealed that plastic flow from the source of river to end of estuary is affecting the estuarine ecosystem. Plastics from point and nonpoint sources are entering the terrestrial ecosystem of the estuary. This has huge impact on people residing in this region as plastics are directly or indirectly entering the food web. Most of the community living here depend on fisheries and Bivalve collection for livelihood support (Figs. 21.1 and 21.2). These plastics are also altering the water quality of the estuary which indirectly affects the livelihood. The

Bivalves need a continuous flow of Calcium and Magnesium for their production. In past decades, the plastic flow is affecting them as there is considerable decrease in the formation of bivalves from 5–6 years. Microplastics affecting the deposition of calcium and magnesium, which affects the formation of shells needed for their survival. So the people who consume them are adversely affected due to microplastics entering the bivalves and fish species.

Plastics on the banks of estuary also create threats to the people and also to the livestock. Cattles and other animals feed on it which can be fatal. These plastics enter the soil which is causing damage to the soil fertility. Coconut and Paddy are the major crops of Aghanashini village. Entry of plastic into the land has adversely affected crop

Fig. 21.1 Aghanashini estuary during low tide. Map showing Aghanashini estuary

Fig. 21.2 Showing bivalves and collection process in Aghanashini estuary

Fig. 21.3 Showing plastics at estuary banks-Aghanashini

productivity. As it is a Ramsar site, the large number of native birds and migratory birds are impacted by these plastics. Community depending on Aghanashini estuary for livelihood are adversely affected by these plastics. There is an urge to control the flow of plastics. So we are suggesting these regulatory measures to the policymakers and Government (Fig. 21.3).

21.4 Regulatory Measures

1. **Supporting alternatives to single-use plastic products**: Single-use plastics are the major culprits to the free flowing rivers. Banning plastic bags, straws, etc., can help to control the flow of these wastes from catchment area to the Aghanashini estuary. Alternative to single-use plastics which can be adopted in Malnad-Coastal area is the banana leaf and areca nut leaf products. Instead of plastic bags, reusable cloth bags should be used by people. Banana leaves can also used in packing. Plates made up of areca nut leaves should be used instead of plastic materials which easily degrade.
2. **Promoting reuse**: Reuse of materials like stainless steel, glass, wood, bamboo, etc., can help to control the use of plastic and thus minimize the damage of the plastic on ecosystem. People can adopt naturally available bamboo materials in Sirsi, Kumta region and can contribute to the control of plastics. This can also help in promoting natural products and help to generate income for local farmers.
3. **Creating awareness through educational programs**: Education programs play a crucial role in creating awareness in people who are living in this area. Importance of ecosystem, plastic and its effects on people should be made understandable to the local people by organizing education programs in village level and also in schools.

4. **Regulating the uncontrolled tourism activity**: Waterfalls and river sources of Aghanashini are always major plastic hotspots as tourists visit these precious unexplored places and create threat to the ecosystem. Strict regulation of these activities by Forest Department and Tourism Department help control plastic contribution to Aghanashini.
5. **Setting up Waste Management unit in Aghanashini village**: As all the plastic waste from the river ultimately reaches the Estuary at Aghanashini village. Setting up a waste management unit in this region can help in collection, segregation and effective management of plastic waste.
6. **Government rules and Laws**: As Aghanashini will soon be a Ramsar site, strict Government Rules and Laws should be implemented in order to control the unregulated dumping of plastics in river as well as estuary.
7. The distribution and concentration of micro and nano plastics need regular monitoring, which can be justified by the potential socio-economic causes/losses/costs resulting from the risks that plastics pose (Meulen et al., 2014).

21.5 Conclusion

Plastic debris transport by rivers is crucial, but creates complex issues in the estuaries and sea. In this study, we provided an overview of the scenario of the plastic pollution, discussing the effects of plastic pollution and regulatory measures. Efforts are in act to regulate the plastic pollution in estuaries. However, what becomes clear from our overview is that there are still many unsolved problems to answer and methodological challenges to overcome. Besides contributing to a better general understanding of plastic debris transport in estuary, we hope that this paper also provides directions for future research. Plastic pollution is a global environmental hazard and urgent action is required to tackle it. Aghanashini, the last major free flowing river of Uttara Kannada must be regulated to conserve the ecosystem of this estuary and people depending on it for livelihood.

Acknowledgements The authors are thankful to people of Aghanashini for their support during this work. Authors are also thankful to the Environmentalists of Uttara Kannada district for their support during this work.

References

Andres, M., Delpey, M., Ruiz, I., Declerck, A., Sarrade, C., Bergeron, P., & Basurko, O. C. (2021). Measuring and comparing solutions for floating marine litter removal: Lessons learned in the south-east coast of the Bay of Biscay from an economic perspective. *Marine Policy, 127*, 104450.

Elliott, M., & McLusky, D. S. (2002). The need for definitions in understanding estuaries. *Estuarine, Coastal and Shelf Science, 55*(6), 815–827.

UNDP. (2015). Human Development Report 2015: Work for Human Development. United Nations Development Programme (UNDP).

Kumar, R., Verma, A., Shome, A., Sinha, R., Sinha, S., Jha, P. K., & Vara Prasad, P. V. (2021). Impacts of plastic pollution on ecosystem services, sustainable development goals, and need to focus on circular economy and policy interventions. *Sustainability, 13*(17), 9963.

Van der Meulen, M. D., De Vriese, L., Lee, J., Maes, T., Van Dalfsen, J. A., Huvet, A., ... & Vethaak, A. D. (2014). Socio-economic impact of microplastics in the 2 Seas, Channel and France Manche Region. An initial risk assessment.

Mouat, J., Lozano, R. L., & Bateson, H. (2010). *Economic impacts of marine litter*. KommunenesInternasjonaleMiljøorganisasjon.

Oosterhuis, F., Papyrakis, E., & Boteler, B. (2014). Economic instruments and marine litter control. *Ocean & Coastal Management, 102*, 47–54.

Pascual, U., Balvanera, P., Díaz, S., Pataki, G., Roth, E., Stenseke, M., ... & Yagi, N. (2017). Valuing nature's contributions to people: The IPBES approach. *Current Opinion in Environmental Sustainability, 26*, 7–16.

Pritchard, D. W. (1967). Observations of circulation in coastal plain estuaries. *Estuaries, AAAS Publ., 83*, 37–44.

Sakti, A. D., Rinasti, A. N., Agustina, E., Diastomo, H., Muhammad, F., Anna, Z., & Wikantika, K. (2021). Multi-scenario model of plastic waste accumulation potential in indonesia using integrated remote sensing, statistic and socio-demographic data. *ISPRS International Journal of Geo-Information, 10*(7), 481.

Chapter 22
Abundance of Microplastics in Water, Sediment, and Mussels: A Case Study of Langkawi Island, Malaysia

Mohamad Najihah, Masazurah A. Rahim, Ku Kassim Ku Yaacob, Kua Beng Chu, and Izarenah Md Repin

Abstract This study presents the contamination scale of microplastic distribution in the surface water, beach sediment, and green mussels of Langkawi Island, Malaysia. Samples were collected from May to November at different locations across the island using a standard operating procedure. Significant variability was observed in each location with southwest monsoon in July recorded a high average microplastics concentration in surface water at 1.48 ± 1.79 particles m^{-3} while beach sediment samples recorded the highest average concentration in southwest and northeast monsoon at 395 ± 301.32 and 490 ± 280.54 particles kg^{-1}. Green mussels recorded an average of 0.56 ± 0.58 particles g^{-1} per individual with fibers present in most individuals. Polymer categories differed in each sample, with each dominant polymer representing different sources. Our research provided insights into the influence of monsoons and anthropogenicity on microplastic pollution levels at the Malacca Strait on the west coast of Peninsular Malaysia in both environment and biota samples.

M. Najihah (✉) · M. A. Rahim · K. K. K. Yaacob
Department of Fisheries Malaysia, Fisheries Research Institute (FRI), Impact Assessment Research Center, Batu Maung, Penang, Malaysia
e-mail: najihah@dof.gov.my

M. A. Rahim
e-mail: masazurah@dof.gov.my

K. K. K. Yaacob
e-mail: kukassim@dof.gov.my

K. B. Chu
Department of Fisheries Malaysia, Fisheries Research Institute (FRI), Batu Maung, Penang, Malaysia
e-mail: kuaben01@dof.gov.my

I. M. Repin
Ecosystem Conservation and Biodiversity Branch, Fisheries Conservation and Protection Division, Department of Fisheries Malaysia, Putrajaya, Malaysia
e-mail: izarenah@dof.gov.my

S. Suriyanarayanan et al. (eds.), *Combating Plastic Pollution in Terrestrial Environment*,
https://doi.org/10.1007/978-981-96-2343-3_22

Keywords Microplastics · Beach sediment · Surface waters · Mussel · Straits of Malacca · Polyethyleneimine · Rayon

22.1 Introduction

Plastics have become a global enemy because of their abundance and increasing pollution intensity worldwide. Discarded from terrestrial, enters the river, and ends up in the marine environment for a long period of time, causing them to turn into microplastics. Microplastics are tiny plastic particles with diameters up to 5 mm (UNEP, 2019). They are omnipresent in all classes of environments and paradoxically found in most living organisms, including humans. Despite numerous studies conducted in different countries and environments, the lack of standardization of sampling units and methods makes it difficult for comparison purposes (Simon-Sánchez et al., 2022).

Microplastics can be traced back to two main sources. Primary microplastics are intentionally manufactured to be minuscular-sized and are commonly found in cosmetic and pharmaceutical products (Boucher & Friot, 2017). The second source is secondary microplastics, which result from the degradation of larger plastics in terrestrial and marine environments (Sundt et al., 2014). This degradation is caused by a combination of physical, chemical, and biological processes that damage the structural integrity of plastics, causing them to break down into smaller particles (Woodall et al., 2014). Other scientists define microplastics as generally originating from fragmented plastics that are disorderly disposed from sublunary sources and exposed to continuous hydrodynamic conditions such as sunlight, waves, and wind or deliberately made in micro sizes for industrial application (Miller et al.,). Either littered properly or deliberately, owing to their minuscule size, the removal and cleanup of microplastics is a challenging task that requires significant effort.

Microplastics are recognized as carriers of both chemical contaminants and pathogenic bacteria, and their small size allows them to rapidly disseminate these vectors at a rapid rate (Koelmans et al., 2016). They are also ingested by numerous organisms, and laboratory studies have revealed that microplastics cause certain negative physiological effects, such as oxidative stress, which increases female infertility of fish (Gupta et al., 2023); disturbs feeding behavior and motor activity in adult fish and fry (Yong et al., 2020); and even causes histological and biochemical changes in fish (Hamed et al., 2021; Yang et al., 2023).

Due to amounted aggregation of microplastics entering marine environment through littering and landfill, the United Nations Environment Programme (UNEP) has set up guidelines for proper waste disposal through recovery through incineration, and recycling for energy and resources (UNEP, 2013). Despite the given guidelines, UNEP statistics imply that recovery of plastics wastes, incineration, and recycling consecrate at 21, 12, and 9% respectively and remaining are mostly disposed in landfill (UNEP, 2018). Plastic waste buried in landfills progressively disintegrates into microplastics, jeopardizing the environment through waste resource exploitation

(Shen et al., 2022). The mismanagement of plastic waste in Malaysia is not recent, and numerous efforts to reduce plastic waste have been made by the government of Malaysia, including mapping out plastic sustainability roadmaps until 2030 (KASA, 2021). Despite these efforts, plastic waste eventually enters the environment. Through studies conducted by numerous agencies, marine litter and microplastic pollution seem to be expanding in many areas, including the Malacca straits of peninsular Malaysia (Khairunnisa et al., 2012; Odli et al., 2020; Vin et al., 2023).

The island of Langkawi, located in the Straits of Malacca, on the northeast coast of Peninsular Malaysia, is an enormous tourist attraction spot, not only among locals but also among foreigners. Langkawi boasted an urbanization rate of 94.5% with a population density of 202.0/km^2 in 2020 (DOSM, 2020). Apart from tourism activities, their economy is also generated by agro-based activities, including paddy and rubber cultivation, and fisheries. As one of the many UNESCO geo-park listed in the Southeast Asian region with high fauna biodiversity (Azmir & Samat, 2010), plastic waste was recorded to be the highest waste sample on the island (Shamsiry et al., 2011). Despite this information, very few studies have addressed microplastic pollution in Langkawi. Thus, we decided to assess microplastic pollution in Langkawi Island by (1) evaluating the presence of microplastics in different locations near resources, (2) quantifying the distribution of microplastics in seawater, beach sediment, and green mussels in different seasons, and (3) investigating the relationship between microplastic sources in different media (surface waters, beach sediment, and mussels). Green mussels were chosen because of their outspread distribution and gullible microplastic uptake, in spite of its famous seafood, thus making them reliable for comparison purposes.

22.2 Materials and Methods

22.2.1 Sample Collection

Sampling was conducted at Langkawi, situated in the north of the Straits of Malacca, in 2022. Sampling stations (Fig. 22.1) were set up, each with divergent urban activity. Samples were collected during the monsoon transition periods (May and September), southwest monsoon (July), and northeast monsoon (November). The stations were located at the main tourist hub (ST1), private resorts (ST2), major vessels route (ST3), and residential areas and domestic jetty (ST4).

During each sampling, surface water was sampled using a manta net (Hydrobios, Germany) with a mesh size of 335 um, towed for 10–15 min at a constant speed of 1–2 knots. Following Law et al. (2014), each starting and ending locations of trawling were recorded, and a flowmeter was attached to the net to determine the volume of each setline. The net was rinsed from the outside and collected materials in the cod end were filtered with a stainless-steel mesh size of 5 mm to eliminate any debris asides from the micro and later transferred to a 1 L aluminum bottles.

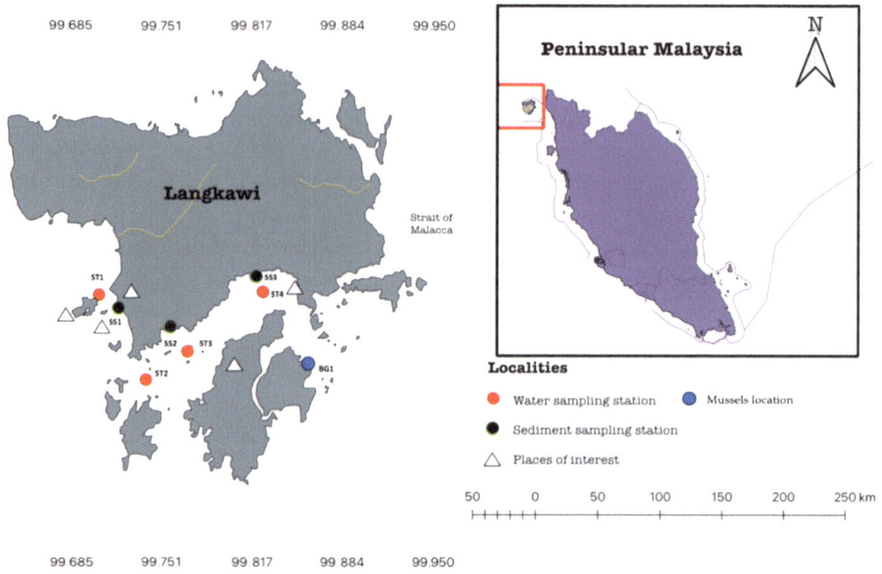

Fig. 22.1 Sampling location in Langkawi coastal waters

For beach sediment samples, only three sampling locations were able to be accessed. A total of 1 kg beach sediment samples were collected using stainless steel shovel at three beaches (SS1, SS2, and SS3) following the transect line method (Qiu et al., 2015; USEPA, 2021). Measurement of microplastics contamination in beaches using transects is more straightforward and efficient, considering that each beach has a relatively wide area. Similar to the water sampling, the sediment samples were filtered using a 5 mm stainless steel sieve to eliminate any particles exceeding the desired sizes. The samples were later wrapped in aluminum foil and labeled accordingly. Both the water and sediment samples were kept chilled and transported to the laboratory for further analysis.

Green mussels, *Perna viridis* were acquired from local fishermen at BG1 in September (n = 35). Due to their natural availability of the area, assessing the microplastics content in these mussels enables us to understand that the uptake of microplastics into organisms was either from water or sediment while relating it to sources. Mussels have also been used as sentinel organisms in most microplastics studies (Imasha & Babel., 2023; Phaksopa et al., 2023; Taruk Linggi et al., 2021) thus attainable for juxtaposition. Each mussel was kept individually in aluminum foil, and later transported to the laboratory and refrigerated at 4 °C until further analysis.

22.2.2 *Isolation and Identification of Microplastics*

Microplastics from seawater and sediment samples were extracted using the procedures described by Masura et al. (2015), with a slight modification. Density separation was conducted only for sediment samples. For mussel extraction, morphometric attributes were measured, including length (cm) and weight (g). The whole organ was removed, and treatment extraction was conducted as described by Thiele et al. (2019). The tissues were submerged in 10% KOH at a ratio of 3:1 for three days. All samples were filtered using a Whatman GF/C attached to a vacuum pump.

All microplastics were identified using a stereo microscope (Olympus SZX61) attached to a camera and categorized according to their shape, color, and size. The particles were isolated and transferred to a cellulose nitrate grid filter for subsequent chemical analysis. Polymer identification was conducted using micro-Fourier transform infrared spectroscopy (μ-FT-IR, Nicolet iN10 MX, Thermo Fisher Scientific, USA) following Zhu et al. (2021) in the transmission mode. The spectrum generated by FT-IR was compared with the reference spectra library with a >70% similitude threshold.

22.2.3 *Contamination Control*

As recommended by various authors such as GESAMP (2019) and Morgana et al. (2021), control protocols must be adhered to prevent potential contamination during sampling and analysis. During sampling, positive and negative control blanks consisting of 1 L of filtered ultrapure water were set at each station for each sampling excursion. The blanks were later filtered in the laboratory and inspected for similarity with the samples taken. During the analysis, extreme care was taken during each procedure, using a blank test to prevent encircling pollution. Any particles detected on the filter blanks were examined for shapes, colors, and sizes, and later compared to the analyzed samples to be excluded. Additionally, all glassware and tools involved in this study were rinsed with filtered ultrapure water before and after use and were always covered with aluminum foil when unattended.

22.3 Results and Discussion

22.3.1 *Quality Control*

In this study, only blue and colorless cotton microfibers were found in the control membranes, with a range of 0–4 fibers per blank. As suggested by Avio et al. (2020), to avoid overestimation and obfuscation, the final microplastics abundances in water, sediment, and biota samples were reported as the difference between the blank values and the actual values.

22.3.2 Microplastics in Surface Water

Microplastics were found at all sampled stations (Table 22.1), and their concentrations varied between months and stations. The average concentrations of microplastics in May, July, September, and November were 1.48 ± 1.53, 1.48 ± 1.79, 0.60 ± 0.41, and 0.49 ± 0.11 particles m^{-3}, respectively. Meanwhile according to stations, microplastics recorded an average of 0.35 ± 0.17, 1.59 ± 1.38, 1.64 ± 1.74, 0.47 ± 0.16 particles m^{-3} for ST1, ST2, ST3, and ST4 respectively. A low trend of microplastic concentration was observed in ST1, followed by ST4 across the year, compared to ST2 and ST3. The lowest microplastic abundance was recorded in May at ST1 with 0.15 particles m^{-3} while the highest was recorded in July at ST3 with 4.14 particles m^{-3}. Further confirmation using one-way ANOVA revealed significant differences in different months ($P < 0.05$), whereas no effect was observed between stations.

Figures below illustrate the results of the microplastic attributes in the surface waters. The bar graphs shown in Fig. 22.2 were the relative contribution (%) of each shape. All five shapes were detected in this study, with fragments that were most prominent either in different months or stations (>75%), followed by (18%), while foam, film, and pellet each existed in only a small fraction (<2%).

Microplastic sizes were categorized as <5 mm, <4 mm, <3 mm, <2 mm, and <1 mm as shown in Fig. 22.3. Most of the isolated particles belonged to smaller class sizes, <2 mm, and <1 mm with 41.15 and 38.02%, respectively.

The funnel diagram in Fig. 22.4 showed the polymer characteristics of the isolated microplastics. Nineteen polymers were identified as part of this study, with rayon making up the majority (60.36%). This was followed by polystyrene (PS) at 16.36% and fluorocarbon polymer at 6.55%. Less frequent (<3%) polymers included polyvinyl alcohol (PVA), polyoxymethylene (POM), melamine-formaldehyde (MF), polyamide/nylon (PA), polyethylene (PE), polytetrafluoroethylene (PTFE), phenol-formaldehyde, phenolic resin (PF), polyethyleneimine (PEI), polypropylene (PP), polybenzonitrile (PBN), polyethylene terephthalate (PET),

Table 22.1 Microplastics abundance in surface water (particles m^{-3}) according to months and sampling stations

Sampling stations	May	July	September	November
ST1	0.15	0.31	0.39	0.56
ST2	3.62	0.97	1.18	0.59
ST3	1.50	4.14	0.55	0.35
ST4	0.66	0.51	0.27	0.44

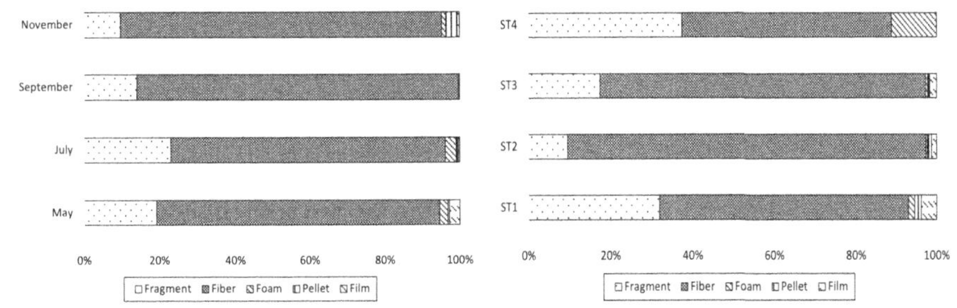

Fig. 22.2 Microplastics categories in shapes found in surface waters according to months and stations (%)

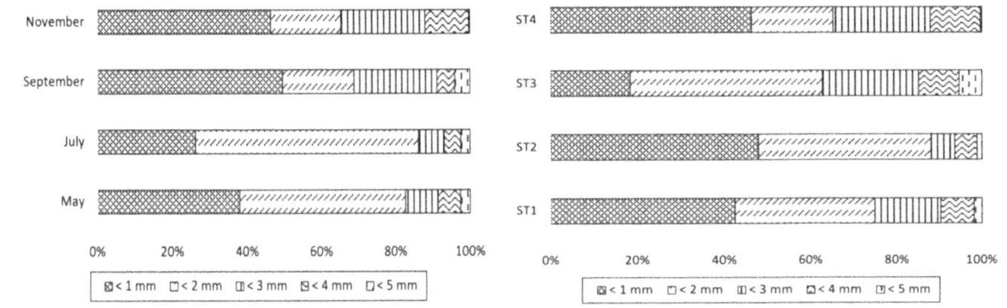

Fig. 22.3 Microplastics categories in sizes found in surface waters according to months and stations (%)

polyolefin, polypiperidinium chloride, polyvinylidene fluoride (PVDF), and polymethylhydrazine. Figure 22.5 showed the two most dominant IR spectra found this study's two most dominant IR spectra in this study.

This study discusses microplastic pollution in the Straits of Malacca, covering the southern and eastern shores of Langkawi. Langkawi Island spans approximately 25 km (16 miles) from north to south, and slightly further from east to west. Its coastal regions are flat alluvial plains interspersed with limestone ridges. Despite being listed as UNESCO geo-park in the Southeast Asian region with high fauna biodiversity (Azmir & Samat, 2010), it is an alarming threat to environmental contamination, especially due to plastics. Few studies regarding marine debris have been conducted in previous years along the straits of Malacca, but very few have focused on microplastics in surface waters, apart from Najihah et al. (2020). A marine litter study by Yasin and Shau Hwai (2019) found plastic items as the majority of litter items at 98.3%, while along the northern straits of Malacca, micro- and macroplastic studies were conducted but specific to sediment (Kamaruddin et al., 2020). Most microplastic surface water studies in Malaysia have been more centralized in

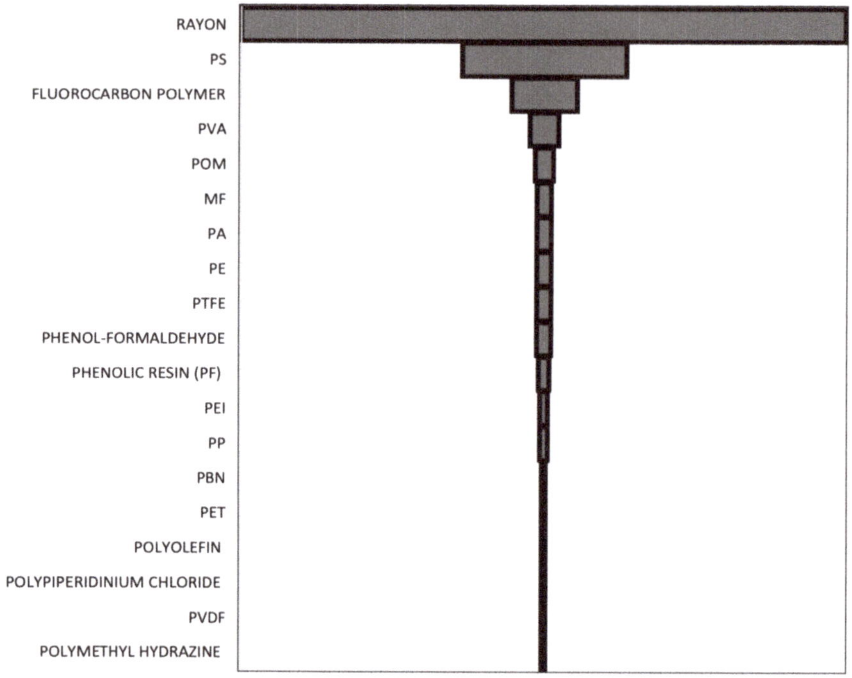

Fig. 22.4 Relative abundance (%) for each polymer found in surface waters

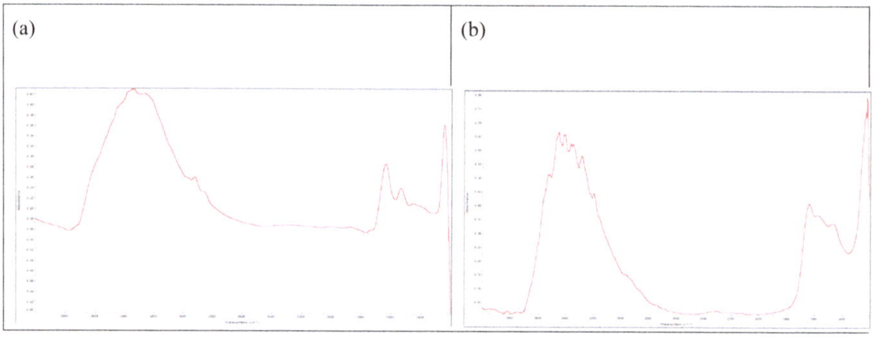

Fig. 22.5 FTIR spectra of most prominent polymers found in surface water **a** rayon and **b** PS

the South China Sea region (Khalik et al., 2018; Ku Yusof et al., 2023; Md Amin et al., 2020).

In this study, microplastics were present at all stations throughout the study period, with the highest abundance during the southwest monsoon period at ST3 in July with 4.14 particles m^{-3} while the lowest was found at ST1 in May during the monsoon transition with 0.15 particlesm^{-3}. Our findings are in line with those of other studies conducted during seasonal variation, which found that microplastic abundance was higher in the monsoon season (Warrier et al., 2022). In contrast, within Malaysia, our study recorded a slightly higher abundance, which is also contributed by diverse human activities within Langkawi, comprising remarkable human activities including residential areas, tourism, fishing, and agriculture activities, which played a huge role in microplastic deposition (Tian et al., 2022). Our results also showed a slightly higher concentration than the east coast of Peninsular Malaysia (Khalik et al., 2018), and lower than those of the Gulf of Thailand and Indian Ocean (Hildebrandt et al., 2022; Vibhatabandhu & Srithongouthai, 2022). Overall, according to the stations, both ST2 and ST3 have a high total microplastic abundance at 6.36 and 6.54 particlesm^{-3} respectively. Although ST2 is not as publicly accessible as ST1, most fishing and agriculture takes place, and it is also in proximity to the main vessel routes as ST3. Vessels generate substantial quantities of wastewater and congregate their operations in ecologically significant regions (Folbert et al., 2022). The microplastics found in this study were mainly made of fragments either in different localities or in different seasons. Other studies conducted in Malaysia also found fragments and fibers as the most dominant shapes in their studies (Md Amin et al., 2020; Taha et al., 2021). Fibers are a major component of microplastics found in surface waters and could be contributed by fiber-based textiles and products shredded during use and laundering, eventually ending up in waterways (Periyasamy and Tehrani-Bagha, 2022). As they are synthetically produced, they are very durable and resistant to degradation, causing them to persist in the environment for a prolonged period. Rayon and polystyrene were also found in most localities (Khalik et al., 2018; Zainuddin et al., 2022) and globally (Li et al., 2021; Miller et al.,) suggesting that domestic sewage, packaging materials, and agricultural activities contributed to the distribution of microplastics in the area. Rayon is usually a fiber shape, thus supporting this study on how domestic sewage and textile industries played a huge role in microplastic abundance. Polystyrene is a common single-use plastic that is widely used in daily life, particularly in packaging. Improper disposal of polystyrene fragments occurs due to secondary sources that are defragmented over time.

22.3.3 Microplastic in Sediment

Each beach sediment site was contaminated with microplastic particles. The abundance reported in this study is expressed as particles per kg dry weight of sediment. The average microplastics concentrations in May, July, September, and November (Fig. 22.6a) were 240 ± 201.31, 395 ± 301.12, 23 ± 5.77, 490 ± 280.54 particles kg^{-1}, respectively. Microplastics recorded an average of 118 ± 88.74, 498 ± 340.43, 246 ± 202.95 particles kg^{-1} for SS1, SS2, and SS3, respectively. SS2 had the highest microplastics abundance, followed by SS3 and SS1 (Fig. 22.6b). Microplastic abundance was lowest in September, the east-coast dry season, followed by May, July, and November. Although abundance differences were observed in

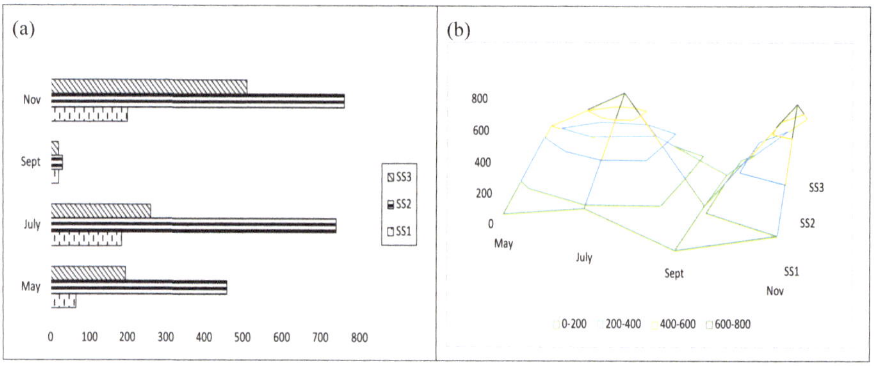

Fig. 22.6 Abundance of microplastics in beach sediment according to sampling months and stations in particles kg^{-1}

both months and stations, further confirmation using one-way ANOVA revealed insignificant differences in both variables (P > 0.05).

As illustrated in Fig. 22.7, only four shape types were found in this study, with fragment and fiber being the most consistent at >75% and >20%, respectively, while foam and pellet each comprised <1%. Film types were non-existent in this study.

Different microplastic sizes were present in beach sediment samples (Fig. 22.8), with smaller size classes being most recorded at 61.71% (<1 mm) and 25.08% (<2 mm). <3 mm size were 7.23%, followed by <4 mm and <5 mm at 3.92 and 2.06%, respectively. The smaller the size, the higher the abundance of microplastics.

A total of 13 polymer types were found incorporating the microplastics samples in beach sediment (Fig. 22.9). The isolated particles were diverse ranging from the lowest 0.895 g/ cm^{-3} to the heaviest density, 1.40 g cm^{-3} polyvinyl chloride (PVC). The most discovered polymers were PEI and polyacrylic acid (PAA) with both at 36.79%. Next in line were PET, rayon, and PF at 5.7, 4.15, and 3.11% respectively. Other polymers were detected with each less than 2.5% consisting of

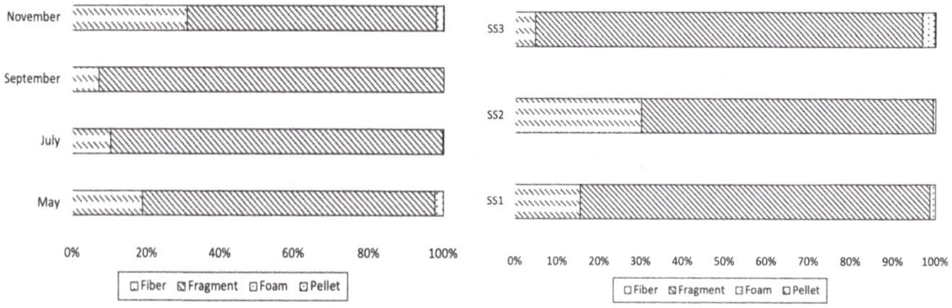

Fig. 22.7 Microplastics categories in beach sediment according to shapes based on months and stations (%)

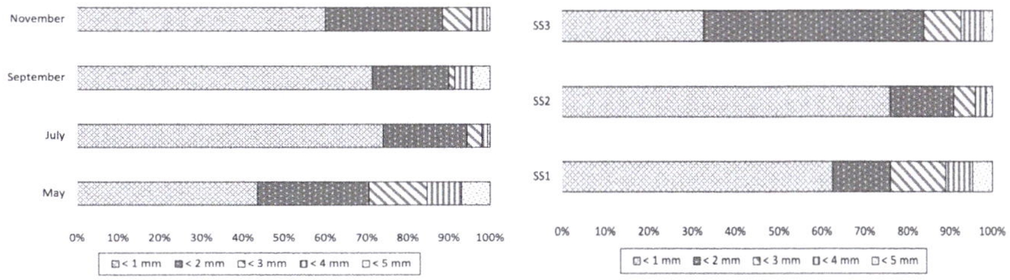

Fig. 22.8 Microplastics categories in beach sediment according to sizes based on months and stations (%)

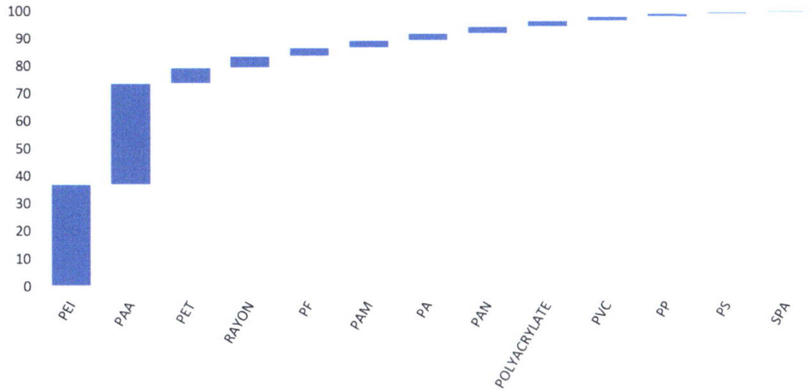

Fig. 22.9 Polymer profiles found in beach sediment of Langkawi Island (%)

polyacrylamide (PAM), PA, polyacrylonitrile (PAN), polyacrylate, PVC, PP, PS, and sodium polyacrylate (SPA).

These results confirmed our hypothesis that microplastics and beach sediment are spatially present in different locations. All locations were publicly accessible, but SS1 and SS2 were usually more populated with tourists than SS3. Sources of microplastics in beach sediments are usually brought by direct and indirect littering from land that is swept into storm drains and rivers, eventually reaching the ocean and depositing microplastics on beaches as waves crash into the shore (Thushari & Senevirathna, 2020). In addition, wastewater treatment also plays a significant role in microplastic deposition, as it is usually not removed during treatment, thus causing immense flow into the ocean (Sadia et al., 2022). In Langkawi, there are eleven sewage treatment plants (DOE, 2019), two wastewater treatment plants, and one leachate treatment plant, with each line released back into the ocean waters. The particles released in the final affluent are generally smaller than the mesh size used (Freeman et al., 2020), supporting the ubiquitous small microplastic sizes, that is <2 mm found in this study. The small sample sizes found in this study are similar to those of other studies on the west coast of peninsular Malaysia (Vin et al., 2020); and other countries (Haegerbaeumer et al., 2019; Khuyen et al., 2021).

The majority of polymers in beach sediments from other studies are polyester, polyethylene, polypropylene, or PVC (Khuyen et al., 2021; Lots et al., 2017; Pradit et al., 2020). To date, PEI has been reported only in a study conducted by Liu et al. (2019) in northern China. The majority of microplastics polymers in the environment are very different. Microplastics in surface waters found in this study mainly consisted of light density polymers whereas in sediment were mainly made of heavy density with PEI and PAA at 1.25 and 1.40 g/cm^3 (Li et al., 2021). A trend of high to low density can be observed in the polymers found in sediments with PP and PS as the least percentage found in this matrix. Microplastics with a higher density may have a greater likelihood of reaching the seabed, potentially resulting in their accumulation in beach sediment over time (Ziani et al., 2023), but it is also influenced by factors other than density.

The main contributing polymer, PEI, is a synthetic polymer and is considered a cationic polymer. Despite having positive charges, the formation of microplastics from these polymers may occur from primary sources that are primarily used for industrial applications. Considering the fragmented types as the most abundant shape found, we expected PEI to be fragmented, as it is usually applied as a coating material or for medicinal purposes. It is also widely used in WWTP because of its cationic properties, which act as a binder to remove negatively charged pollutants, such as heavy metals and arsenic (Xanthopoulou & Katsoyiannis, 2023).

PAA is considered a superabsorbent polymer (SAPs) because of its ability to retain moisture and usually exists in diapers, sanitary napkins, wound dressings in microgels, and micro fragment forms (Chen et al., 2022). Apart from this, owing to its thin and dense layers, it is usually attached to surfaces as anti-fouling coatings to prevent the adhesion of organisms on ship hulls, supporting the evidence found in this study.

22.3.4 Microplastics in Mussels

A total of 97 microplastic particles were found in 35 green mussels with an average weight of 6.31 ± 2.69 g (Fig. 22.10). Microplastic abundance in mussels has been reported in particles per gram of wet weight. Microplastics abundance ranged from 0 to 2.69 particles g^{-1} while the average abundance was 0.56 ± 0.58 particles g^{-1}. A correlation analysis was performed to further assess any possible association with mussel wet weight, and a non-significant negative correlation was found between them ($R2 = 0.07$, $p > 0.05$) (Fig. 22.11).

According to Fig. 22.12a, fibers were the most frequent shape in this study at 89.69%, followed by fragments at 7.22%, and pellets and films at 2.06 and 1.03%, respectively. The findings in Fig. 22.12b indicate microplastics with sizes <1 mm were found to be the most prevalent at 58.27%, followed by <2 mm at 30.94%, <3 mm at 7.91%, and both <4 mm and <5 mm comprised 1.44%.

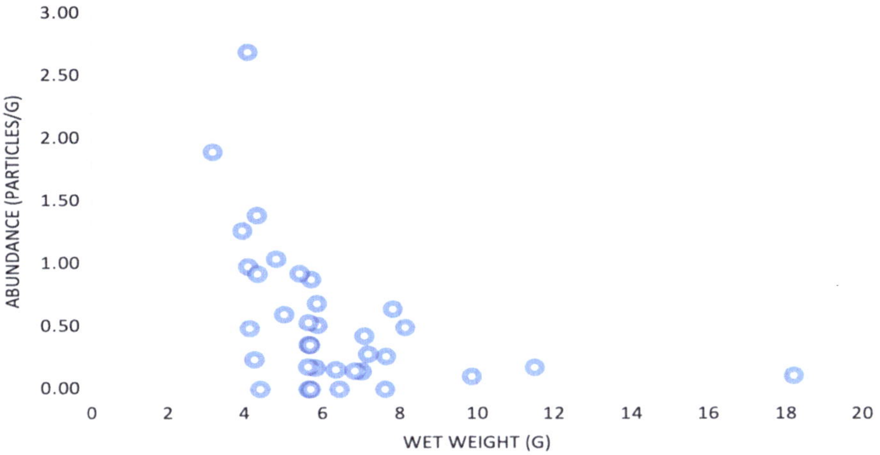

Fig. 22.10 Abundance of microplastics in mussels according to wet weight (g)

Fig. 22.11 Wet weight (g) line fit plot

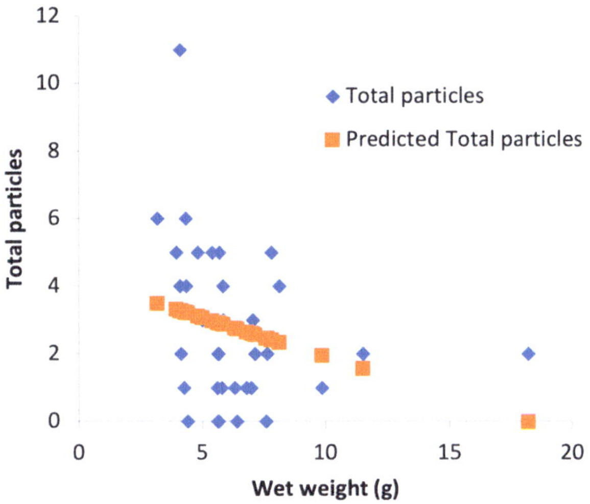

Isolated particles in mussels detected with µ-FTIR (Fig. 22.13) revealed four polymers made up of rayon, PET, PP, and PS at 37, 34, 20, and 9%, respectively. Diversity of microplastics found in mussels was presented in Fig. 22.14.

The findings of this study were compared with those of shellfish and sea cucumber, by various researchers, as listed in Table 22.2. This study found a lower abundance of microplastics compared with other studies in Langkawi, Johor (Peninsular Malaysia), and Thailand. The shapes found were in line with most other studies with fibers, apart from Johor as it recorded fragment as the most abundant. Fibers are usually microscale in size and can thus pass through sewage system filters, which eventually end up bountifully in the sea. Microplastic sizes of <1 mm in benthic organisms

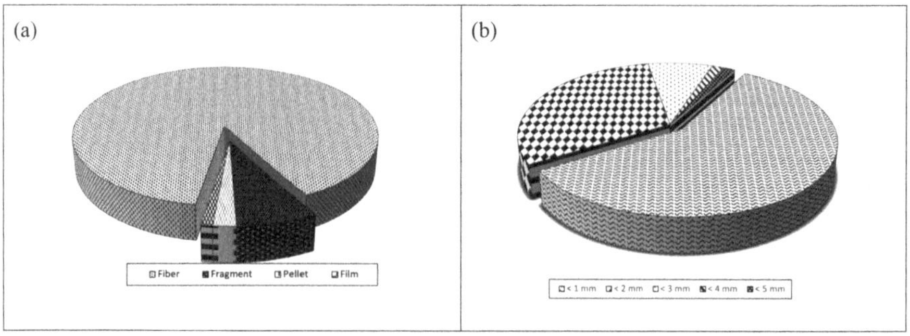

Fig. 22.12 Categories of microplastics **a** shapes and **b** sizes found in mussels (%)

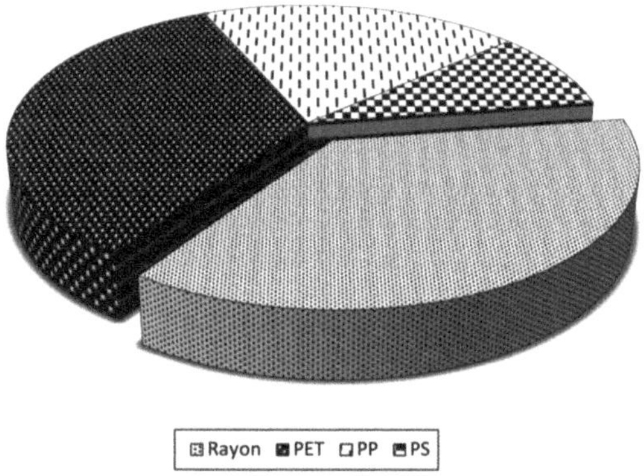

Fig. 22.13 Majority of polymer types found in mussels (%); rayon, PET, PS

were consistent with previous studies. The high proportion of small microplastics is either due to UV exposure, which causes plastic degradation over time, or due to manufacturing processes in industrial applications. Small-sized, lower density microplastics are more prone to being carried away by wind and water, which poses a threat to marine creatures, thus triggering detrimental effects on their well-being.

PP and PET were also commonly found in the listed studies. Rayon has been found in other countries (Li et al., 2018; Scott et al., 2019; Volgare et al., 2022), and this is in line with the polymers found in surface waters in this study. Mussel feed by straining food particles from the water column. As microplastics themselves are lightweight and abundant in water, a large amount of microplastics are filtered and trapped in the mucus inside their shells. Rayons usually appear as fiber types emitted from textile materials and released into wastewater, which ends up in coastal waters (Periyasamy & Tehrani-Bahga, 2022). According to local fishermen, the samples were obtained near sites in proximity to human settlement, thus explaining why

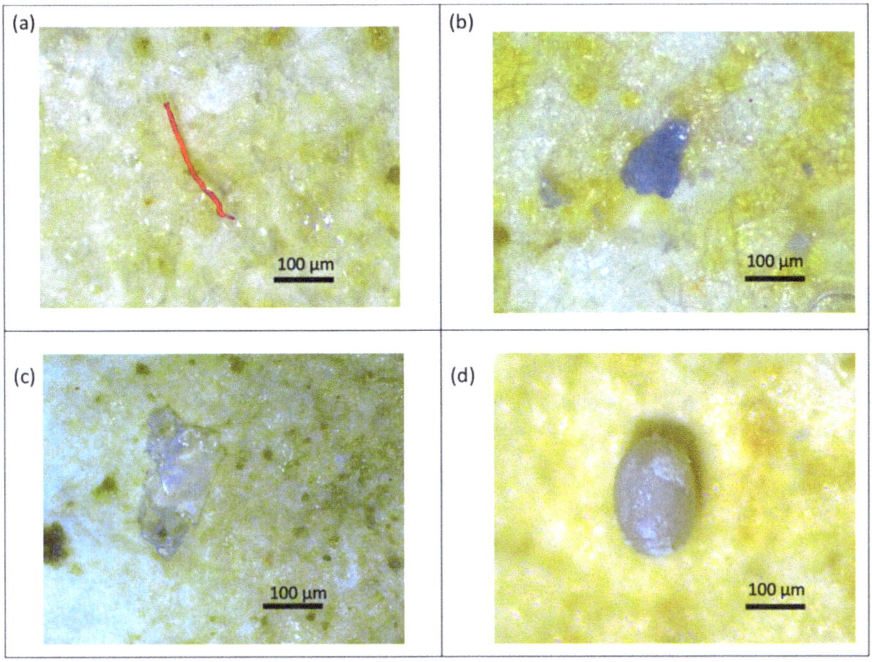

Fig. 22.14 Microscopic images of microplastics found in mussels **a** red coloured fiber; **b** grey coloured fragment; **c** transparent film; **d** white coloured pellet

Table 22.2 List of microplastics study conducted in mussels and bivalve organism around Langkawi and other localities

Area	Organism	Abundance	Shape	Polymer	References
Langkawi, Malaysia	Rock oyster	Total abundance: 1.46 ± 0.007 items/w. weight (g)	Threadlike	PVDF, PP, PET	Ghazali et al. (2022)
Langkawi, Malaysia	Sea cucumber, *Acaudina molpadioides*	1652 particles/20 samples	Fibers	PE, PMMA	Mazlan et al. (2023)
Johor, Malaysoa	Mussels, *Perna viridis*	Average 2.23 1.04 mp/ind 0.4 0.24 mp/g	Fragment	PE, PP	Mat Zin et al. (2022)
Sri Racha Bay, Thailand	Mussels, *Perna viridis*	0.15 ± 0.41 and 0.22 ± 0.57 items/individual	Fibers	PET, PP, Nylon	Phaksopa et al. (2023)
Singhanakorn district, Songkhla	Blood cockles, *Anadara granosa*	4.71 ± 0.06 n/g (wet weight)	Fibers	PE	Goh et al. (2021)

most of the polymers found originated from secondary sources due to improper waste management from terrestrial areas.

In contrast, Li et al. (2016) found that commercial mussel samples had considerably more microplastics than aquaculture farm samples. Davidson and Dudas (2016) and Mercogliano et al. (2020) reported no significant differences in microplastic content between cultured and natural mussels. Few studies have reported microplastics uptake is influenced by organisms. Larger organisms generally ingest more microplastics, which is likely due to larger feeding volume and consumption rates (Watts et al., 2015). There are also research suggesting allometric relationship between body weight and microplastics uptake with few studies found exponents between 0.6 and 0.8 for specific organism-microplastics interaction (Li et al., 2016). However, microplastic ingestion may also be determined by feeding ecology (Scherer et al., 2017). In these cases, filter-feeding organisms ingest microplastics regardless of size, shape, and color, as shown by mussels in this study.

22.3.5 Sources of Microplastics

This study revealed the existence of a diverse range of microplastics, including less frequently reported polymers such as PEI, fluorocarbon polymers, PVA, and PAA, in mussels from Langkawi, Malaysia. The detection of PEI, which is primarily used in specialized industrial applications electronics, aerospace, and carbon capture (Shen et al., 2017), suggests that industrial activities or improper waste disposal on the island or nearby mainland may contribute to the presence of these microplastics in the environment. The presence of fluorocarbon polymers, commonly used in assortment of household and commercial settings (Hays et al., 2023), further supports this hypothesis. The detection of PVA and PAA, which are often used in products such as diapers and water treatment chemicals, suggests that wastewater treatment plants or runoff from urban and agricultural areas carrying improperly disposed waste may contribute to the microplastic load in Langkawi's marine environment, as observed in Finland (Talvitie et al., 2015). This underscores the significant impact of both terrestrial and secondary sources in the region. Additionally, the fragmentation of larger plastic debris, which is considered a secondary source, likely contributes to the overall microplastic burden (Dimassi et al., 2022), explaining the presence of common polymers such as PE and PP.

Industrial activities, including textile manufacturing, food and beverage processing, and pharmaceutical production, contribute to microplastic pollution through the release of wastewater containing these particles, as noted by Mak et al. (2020). The high levels of microplastic pollution in the marine environment of Langkawi, Malaysia underscore the urgent need for comprehensive measures to address this issue. Identifying the specific sources of microplastics, such as industrial activities, wastewater treatment, and littering, is crucial for developing targeted strategies to mitigate this problem.

The lower concentration of microplastics in mussels in this study may be attributed to the area's relatively undeveloped coastal environment, which may contribute to fewer sources of microplastic pollution compared to more urbanized regions (Aurani et al., 2022). The seasonal patterns of microplastic distribution in the region may also play a role, as studies have shown that environmental factors such as water currents and tidal patterns can influence microplastic occurrence (Ibrahim et al., 2021) and affect mussels as they are filter-feeding organisms. Another potential factor that could contribute to the lower microplastic levels in Langkawi mussels is the potential dilution effect caused by the high river discharge in the area. The dilution of microplastics by the river's high discharge may lead to lower concentrations in the local marine environment (Jin et al., 2023), including mussels. Briefly, the lower microplastic concentration in mussels from Langkawi is likely the result of a multifaceted interplay between the relatively pristine nature of the local environment, the influence of seasonal variations, and the dilution effect from high river discharge. This insight emphasizes the importance of understanding the complex interactions between human activities, environmental factors, and distribution of microplastics in coastal ecosystems.

The above discussion underscores the substantial influence of terrestrial and secondary sources on the occurrence of diverse microplastics in the marine environment of Langkawi, Malaysia. However, it is essential to explore alternative perspectives. Some studies have posited that the microplastic pollution in this region might not be solely attributable to local sources. For example, research has indicated that ocean currents and global transport can also contribute to the buildup of microplastics in remote areas, such as Langkawi. The high concentration of microplastics observed in the Miri River, a relatively secluded location, may reflect broader regional or global patterns of microplastic distribution rather than just local sources (Liong et al., 2021). Furthermore, the particular polymer types detected, such as PEI and fluorocarbon polymers, could potentially originate from long-range transport or atmospheric deposition rather than solely from local industrial activities. Additional investigations are necessary to clearly differentiate between the local and global contributions to the microplastic load in Langkawi's marine environment. Although the identified terrestrial sources, including industrial activities, wastewater treatment, and littering, likely play a significant role, it is crucial to consider the potential influence of regional and global transport mechanisms in shaping the microplastic pollution in this area. A comprehensive understanding of the relative contributions of various sources is necessary to develop effective mitigation strategies.

22.4 Conclusion

This study elucidates the extent of microplastic pollution on Langkawi Island, a popular tourist destination with rich biodiversity. Our findings corroborate the pervasive presence of microplastics across key marine compartments, with a trend of higher abundance during the monsoon season. This suggests the potential influx of land-based plastic debris into the marine environment during periods of heavy rainfall and runoff. Notably, the distribution pattern of microplastics followed a decreasing order of abundance from beach sediment to surface water to mussels, indicating a potential transfer pathway of these pollutants within the marine food web. Although the abundance of microplastics in mussels was lower than that reported in some other studies, their presence raises concerns about food safety and potential human exposure to microplastics and associated contaminants. Future research should prioritize investigating the long-term ecological impacts of microplastic ingestion on marine organisms on Langkawi Island, particularly those species relevant to local fisheries and human consumption. Additionally, efforts should focus on identifying and mitigating the primary sources of microplastic pollution in this region. This includes implementing more effective waste management strategies, promoting sustainable tourism practices, and raising public awareness of the consequences of plastic pollution. By understanding the dynamics of microplastic pollution on Langkawi Island and taking proactive measures to address this issue, we can strive to preserve the ecological integrity of this valuable ecosystem.

Acknowledgements The authors wish to thank the staff of FRI Batu Maung (Mr. Mohamad Izhar Mohamad Noor and Mr. Mohd Nawab Arshad), who contributed significantly to this research. Gratitude was also extended to ALS Technician (M) for their unwavering support for microplastics research in Malaysia. This work was fully funded by the Department of Fisheries Malaysia through development grant number P21-300040170-301.

Credit Author Statement **Najihah Mohamad** contributed with conceptualization, data curation, formal analysis, investigation, methodology, and writing—original draft. **Masazurah A. Rahim** contributed with data acquisition, investigation, writing—review & editing. **Ku Kassim Ku Yaacob** contributed with project administration, supervision, writing—review & editing. **Kua Beng Chu** contributed with data acquisition project administration and resources. **Izarenah Mat Repin** contributed with funding acquisition, and investigation.

Conflict of Interest declaration The authors declare that they have NO affiliations with or involvement in any organization or entity with any financial interest in the subject matter or materials discussed in this manuscript.

References

Avio, C. G., Pittura, L., d'Errico, G., Abel, S., Amorello, S., Marino, G., Gorbi, S., & Regoli, F. (2020). Distribution and characterization of microplastic particles and textile microfibers in Adriatic food webs: General insights for biomonitoring strategies. *Environmental Pollution, 258*, 113766. https://doi.org/10.1016/j.envpol.2019.113766

Aurani, J., Hashim, N., Kassim, A. F. M., Ismail, A. D., & Wahab, S. M. A. (2022). Microplastic pollution in marine waters: A Malaysian perspective. *Journal of Tourism, Hospitality and Environment Management, 7*(28), 76–88.

Azmir, A. A., & Samat, A. (2010). Diversity and distribution of stream fishes of Pulau Langkawi, Malaysia. *Sains Malaysiana, 39*(6), 869–875.

Boucher, J., & Friot, D. (2017). *Primary Microplastics in the Oceans: A Global Evaluation of Sources* (43p.). IUCN, Gland. https://doi.org/10.2305/IUCN.CH.2017.01.en.

Chen, J., Wu, J., Raffa, P., Picchioni, F., & Koning, C. E. (2022). Superabsorbent Polymers: From long-established, microplastics generating systems, to sustainable, biodegradable and future proof alternatives. *Progress in Polymer Science, 125*, 101475. https://doi.org/10.1016/j.progpolymsci.2021.101475

Davidson, K., & Dudas, S. E. (2016). Microplastic ingestion by wild and cultured manila clams (*Venerupis philippinarum*) from Baynes Sound, British Columbia. *Archives of Environmental Contamination and Toxicology, 71*(2), 147–156. https://doi.org/10.1007/s00244-016-0286-4

Department of Environment (DOE). (2019). Upgrading from 10,000 Population equivalent (PE) Sewage Treatment Plant (STP) at Lot 229, Pantai Tengah, Langkawi, Kedah for Kytowater Sdn. Bhd. Retrieved January 2, 2024, from https://enviro2.doe.gov.my/ekmc/wp-content/uploads/2020/05/15.-CHAPTER-5A-PROJECT-DESCRIPTION.pdf.

Department of Statistics Malaysia (DOSM). (2020). Kedah @ a glance. Retrieved December 28, 2023, from https://www.dosm.gov.my/v1/index.php?r=column/cone&menu_id=M25ucnhaYUMwbjFwc3hhWkhrQ2tIUT09.

Dimassi, S. N., Hahladakis, J. N., Yahia, M. N. D., Ahmad, M. I., Sayadi, S., & Al-Ghouti, M. A. (2022). Degradation-fragmentation of marine plastic waste and their environmental implications: A critical review. *Arabian Journal of Chemistry, 15*(11), 104262. https://doi.org/10.1016/j.arabjc.2022.104262

Folbert, M. E. F., Corbin, C., & Löhr, A. J. (2022). Sources and leakages of microplastics in cruise ship wastewater. *Frontiers Marine Science, Section Marine Pollution, 9*, 900047. https://doi.org/10.3389/fmars.2022.900047

Freeman, S., Booth, A. M., Sabbah, I., Tiller, R., Dierking, J., Klun, K., Rotter, A., Ben-David, E., Javidpour, J., & Angel, D. L. (2020). Between source and sea: The role of wastewater treatment in reducing marine microplastics. *Journal of Environmental Management, 266*, 110642. https://doi.org/10.1016/j.jenvman.2020.110642

GESAMP. (2019). Guidelines for the monitoring and assessment of plastic litter and microplastics in the ocean. In P.J. Kershaw, A. Turra, & F. Galgani (Eds.), *GESAMP Joint Group of Experts on the Scientific Aspects of Marine Environmental Protection* (Vol. 99, 130p.). GESAMP Reports and Studies. https://doi.org/10.25607/OBP-435.

Ghazali, I. N. M., Karim, S. N. F. S. A., Johari, W. S. M. W., Faudzi, F., Razali, A., Hassan, N. A., & Miskon, M. F. (2022). Microplastics impact assessment on Langkawi Island off Strait of Malacca waters using rock oyster as bioindicator. *AIP Conference Proceedings., 2645*(1), 030029. https://doi.org/10.1063/5.0112757

Goh, P. B., Pradit, S., Towatana, P., Khokkiatiwong, S., Kongket, B., & Zhong Moh, J. H. Z. (2021). Microplastics abundance in blood cockles and shrimps from fishery market, Songkhla Province, Southern Thailand. *Sains Malaysiana, 50*(10), 2899–2911.

Gupta, P., Mahapatra, A., Suman, A., Ray, S. S., Malafaia, G., & Singh, R. K. (2023). Polystyrene microplastics disrupt female reproductive health and fertility via sirt1 modulation in zebrafish (*Danio rerio*). *Journal of Hazardous Materials, 460*, 132359. https://doi.org/10.1016/j.jhazmat.2023.132359

Haegerbaeumer, A., Mueller, M. T., Fueser, H., & Traunspurger, W. (2019). Impacts of micro- and nano-sized plastic particles on benthic invertebrates: a literature review and gap analysis. *Frontiers in Environmental Science, 7*, 17. https://doi.org/10.3389/fenvs.2019.00017

Hamed, M., Hamdy, A. M. S., Ahmed, E. A. B., & Alaa, G. M. (2021). Microplastics induced histopathological lesions in some tissues of tilapia (*Oreochromis niloticus*) early juveniles. *Tissue and Cell, 71*, 101512. https://doi.org/10.1016/j.tice.2021.101512

Hildebrandt, L., Gareb, F. E., Zimmermann, T., Klein, O., Kerstan, A., Emeis, K., & Pröfrock, D. (2022). Spatial distribution of microplastics in the tropical Indian Ocean based on laser direct infrared imaging and microwave-assisted matrix digestion. *Environmental Pollution, 307*, 119547.

Hays H.L., Mathew, D., Chapman, J. (2023). *Fluorides and Fluorocarbons Toxicity*. [Updated 2023 Jul 17]. In: Stat Pearls [Internet]. Stat Pearls Publishing, January 2024. https://www.ncbi.nlm.nih.gov/books/NBK430799/#.

Ibrahim, Y. S., Hamzah, S. R., Khalik, W. M. A. W. M., Ku Yusof, K. M. K., & Anuar, S. T. (2021). Spatiotemporal microplastic occurrence study of Setiu Wetland, South China Sea. *Science of the Total Environment, 788*, 147809. https://doi.org/10.1016/j.scitotenv.2021.147809

Imasha, H. U. E., & Babel, S. (2023). Microplastics contamination in the green mussels (*Perna viridis*) cultured for human consumption in Thailand. *Regional Studies in MarineScience, 67*, 103203. https://doi.org/10.1016/j.rsma.2023.103203.40

Jin, X., Fu, X., Lu, W., & Wang, H. (2023). The effects of riverside cities on microplastics in river water: A case study on the Southern Jiangsu Canal, China. *Science of The Total Environment, 858(2)*, 159783. https://doi.org/10.1016/j.scitotenv.2022.159783

Kamaruddin, S. A., Rusli, H. H., Aziz, K. N. A., & Roslani, M. A. (2020). Characteristics and distribution of microplastics in surface sediment of selat Pulau Tuba, Langkawi, Kedah. *Malaysian Journal of Sustainable Environment, 7*, 139–160. https://doi.org/10.24191/myse.v7i2.10269

Khairunnisa, A. K., Fauziah, S. H., & Agamuthu, P. (2012). Marine debris composition and abundance: A case study of selected beaches in Port Dickson, Malaysia. *Aquatic Ecosystem Health Management, 15*, 279–286.

Khuyen, V. T. K., Le, D. V., Fischer, A. R., & Dornack, C. (2021). Comparison of microplastic pollution in beach sediment and seawater at UNESCO Can Gio mangrove biosphere reserve. *Global Challenges, 5*, 2100044.

Koelmans, A. A., Bakir, A., Allen Burton, G., & Janssen, C. R. (2016). Microplastic as a vector for chemicals in the aquatic environment: Critical review and model-supported reinterpretation of empirical studies. *Environmental Science & Technology, 50*(7), 3315–3326. https://doi.org/10.1021/acs.est.5b06069

Ku Yusof, K. M. K., Tuan Anuar, S., Mohamad, Y., Jaafar, M., Mohamad, N., Bachok, Z., Najihah, M., & Ibrahim, Y. S. (2023). First evidence of microplastic pollution in the surface water of Malaysian Marine Park islands, South China Sea during COVID-19. *Marine Pollution Bulletin, 194*(B), 115268. https://doi.org/10.1016/j.marpolbul.2023.115268

Law, L. K., Moret-Ferguson, S. E., Goodwin, D. S., Zettler, E. R., De Force, E., Kukulka, T., & Proskurowski, G. (2014). Distribution of surface plastic debris in the eastern Pacific Ocean from an 11-year data set. *Environmental Science & Technology, 48*, 4732–4738.

Li, W. C., Tse, P. W., & Fok, L. (2016). Ingestion of nano- and microplastic particles and nanofibers by aquatic organisms: trophic transfer implications. *Environmental Science & Technology, 50*(13), 7255–7263. https://doi.org/10.1021/acs.est.6b00996

Li, J., Green, C., Reynolds, A., Shi, H., & Rotchell, J. M. (2018). Microplastics in mussels sampled from coastal waters and supermarkets in the United Kingdom. *Environmental Pollution, 241*, 35–44. https://doi.org/10.1016/j.envpol.2018.05.038

Li, Y., Zhang, Y., Chen, G., Xu, K., Gong, H., Huang, K., Yan, M., & Wang, J. (2021). Microplastics in surface waters and sediments from Guangdong coastal areas, South China. *Sustainability, 13*(5), 2691.

Liong, R. M. Y., Hadibarata, T., Yuniarto, A. D., Tang, K. H., & Khamidun, M. K. (2021). Microplastic occurrence in the water and sediment of Miri River Estuary, Borneo Island. *Water Air Soil & Pollution, 232*, 342. https://doi.org/10.1007/s11270-021-05297-8

Liu, C., Li, J., Zhang, Y., Wang, L., Deng, J., Gao, Y., Yu, L., Zhang, J., & Sun, H. (2019). Widespread distribution of PET and PC microplastics in dust in urban China and their estimated human exposure. *Environment International, 128*, 116–124. https://doi.org/10.1016/j.envint.2019.04.024

Lots, F. A. E., Behrens, P., Vijver, M. G., Horton, A. A., & Bosker, T. (2017). A large-scale investigation of microplastic contamination: Abundance and characteristics of microplastics in European beach sediment. *Marine Pollution Bulletin, 123*, 1–2.

Mak, C. W., Tsang, Y. Y., Leung, M. M. L., Fang, J.K.-H., & Chan, K. M. (2020). Microplastics from effluents of sewage treatment works and stormwater discharging into the Victoria Harbor. *Hong Kong. Marine Pollution Bulletin, 157*, 111181.

Masura, J. (2015). Laboratory methods for the analysis of microplastics in the marine environment: recommendations for quantifying synthetic particles in waters and sediments. NOAA Technical Memorandum NOS-OR&R-48.

Zin, M. M., Azman, S., Ismail, M. I. M., & Muhammad, M. K. I. (2022). Microplastics contents in natural and maricultured shellfish from Pasir Putih estuary in Johor, Malaysia. *Journal of Sustainability Science and Management, 17*, 121–135. https://doi.org/10.46754/jssm.2022.10.009

Mazlan, M., Shukhaira, S. S., Jannah, M., Husin, M., Shalom, J., Saud, S. N., Abdullah Sani, M. S., Ong, M. C., Chandra Mohan, N. K. N., & Sopian, N. A. (2023). Evaluation of microplastics isolated from sea cucumber *Acaudina molpadiodes* in Pulau Langkawi, Malaysia. *Heliyon, 9*, e16822.

Md Amin, R., Sohaimi, E. S., Anuar, S. T., & Bachok, Z. (2020). Microplastic ingestion by zooplankton in Terengganu coastal waters, southern South China Sea. *Marine Pollution Bulletin, 150*, 110616. https://doi.org/10.1016/j.marpolbul.2019.110616

Mercogliano, R., Avio, C. G., Regoli, F., Anastasio, A., Colavita, G., & Santonicola, S. (2020). Occurrence of microplastics in commercial seafood under the perspective of the human food chain. A Review. *Journal of Agriculture Food Chemistry, 68*, 5296–5301.

Miller, R. Z., Watts, A. J. R., Winslow, B. O., Galloway, T. S., & Barrows, A. P. W. (2017a). Mountains to the sea: River study of plastic and non-plastic microfiber pollution in the Northeast USA. *Marine Pollution Bulletin, 124*, 245–251.

Miller, M. E., Kroon, F. J., & Motti, C. A. (2017b). Recovering microplastics from marine samples: A review of current practices. *Marine Pollution Bulletin, 123*(1–2), 6–18. https://doi.org/10.1016/j.marpolbul.2017.08.058

Ministy of Environment & Water (KASA). (2021). Malaysia Plastics Sustainability Roadmap 2021–2030. Retrieved November 26, 2023, from https://www.nrecc.gov.my/ms-my/teras/alamsekitar/Documents/MALAYSIA-PLASTICS-SUSTAINABILITY-ROADMAP-2021-2030.pdf.

Morgana, S., Casentini, B., & Amalfitano, S. (2021). Uncovering the release of micro/nanoplastics from disposable face masks at times of COVID-19. *Journal of Hazard Materials, 419*, 126507. https://doi.org/10.1016/j.jhazmat.2021.126507

Najihah, M., Ismail, M. S., Yap, C. K., & Ku Yaacob, K. K. (2020). Microplastics occurrence in waters off the northwest coast of peninsular Malaysia: A spatial difference. *Journal of Basic & Applied Sciences, 16*, 50–60.

Odli, Z. S. M., Abdullah, A. L., Saad, F. N. M., & Fadzillah, N. S. A. (2020). The relationship between land use and marine litter at Kuala Perlis coastal area. *IOP Conference Series: Earth Environmental Science, 476*, 012109.

Phaksopa, J., Sukhsangchan, R., Keawsang, R., Tanapivattanakul, K., Asvakittimakul, B., Thamrongnawasawat, T., & Worachananant, S. (2023). Assessment of microplastics in Green Mussel (*Perna viridis*) and surrounding environments around Sri Racha Bay. *Thailand. Sustainability, 15*(1), 9. https://doi.org/10.3390/su15010009

Periyasamy, A. P., & Tehrani-Bagha, A. (2022). A review on microplastic emission from textile materials and its reduction techniques. *Polymer Degradation and Stability, 199*, 109901. https://doi.org/10.1016/j.polymdegradstab.2022.109901

Pradit, S., Towatana, P., Nitiratsuwan, T., Jualaong, S., Jirajarus, M., Sornplang, K., Noppradit, P., Darakai, Y., & Weerawong, C. (2020). Occurrence of microplastics on beach sediment at Libong, a pristine island in Andaman Sea, Thailand. *Science Asia* 1513–1874.

Qiu, Q., Peng, J., Yu, X., Chen, F., Wang, J., & Dong, F. (2015). Occurrence of microplastics in the coastal marine environment: First observation on sediment of China. *Marine Pollution Bulletin, 98*(1–2), 274–280.

Sadia, M., Mahmood, A., Ibrahim, M., Irshad, M. K., Quddusi, A. H. A., Bokhari, A., Mubashir, M., Chuah, L. F., Show, P. L. (2022). Microplastics pollution from wastewater treatment plants: A critical review on challenges, detection, sustainable removal techniques and circular economy. *Environmental Technology & Innovation, 28*, 102946. https://doi.org/10.1016/j.eti.2022.102946.

Scherer, C., Brennholt, N., Reifferscheid, G., & Wagner, M. (2017). Feeding type and development drive the ingestion of microplastics by freshwater invertebrates. *Scientific Reports, 7*, 17006. https://doi.org/10.1038/s41598-017-17191-7

Scott, N., Porter, A., Santillo, D., Simpson, H., Lloyd-Williams, S., & Lewis, C. (2019). Particle characteristics of microplastics contaminating the mussel *Mytilus edulis* and their surrounding environments. *Marine Pollution Bulletin, 146*, 125–133. https://doi.org/10.1016/j.marpolbul.2019.05.041

Simon-Sánchez, L., Grelaud, M., Franci, M., & Ziveri, P. (2022). Are research methods shaping our understanding of microplastic pollution? A literature review on the seawater and sediment bodies of the Mediterranean Sea. *Environmental Pollution, 292*(8), 118275.

Shamsiry, E., Nadi, B., Mokhtar, M., Komoo, I., Hashim, H. S., & Yahaya, N. (2011). Integrated models for solid waste management in tourism regions: Langkawi Island, Malaysia. *Journal of Environmental and Public Health, 2011*, 709549.

Shen, M., Xiong, W., Song, B., Zhou, C., Almatrafi, E., Zeng, G., & Zhang, Y. (2022). Microplastics in landfill and leachate: Occurrence, environmental behavior and removal strategies. *Chemosphere, 305*, 135325.

Shen, X., Du, H., Mullins, R. H., & Kommalapati, R. R. (2017). Polyethylenimine applications in carbon dioxide capture and separation: From theoretical study to experimental work. *Energy Technology, 5*, 822.

Sundt, P., Schulze, P.-E., & Syversen, F. (2014). Sources of microplastic- pollution to the marine environment (MEPEX). Retrieved October 16, 2023, from www.miljodirektoratet.no/Documents/publikasjoner/M321/M321.pdf.

Taha, Z. D., Md Amin, R., Tuan Anuar, S., Abdul Nasser, A. A., & Sohaimi, E. S. (2021). Microplastics in seawater and zooplankton: A case study from Terengganu estuary and offshore waters, Malaysia. *Science of the Total Environment, 786*, 147466. https://doi.org/10.1016/j.scitotenv.2021.147466

Talvitie, J., Heinonen, M., Pääkkönen, J. P., Vahtera, E., Mikola, A., Setälä, O., & Vahala R. (2015). Do wastewater treatment plants act as a potential point source of microplastics? Preliminary study in the coastal Gulf of Finland, Baltic Sea. *Water Science Technology, 72*(9), 1495–504. 2166/wst.2015.360. PMID: 26524440.

Taruk Linggi, G. N., Yaqin, K., Parawansa, B. S., Fachruddin, L., Yunus, B., & Rahim, S. R. (2021). The concentration of microplastics in epibiont of green mussel (*Perna viridis*) from Maccini Baji waters, Pangkajene Kepulauan. *IOP Conference Series: Earth and Environmental Science, 860*, 012099.

Thiele, C. J., Hudson, M. D., & Russell, A. E. (2019). Evaluation of existing methods to extract microplastics from bivalve tissue: Adapted KOH digestion protocol improves filtration at single-pore digit size. *Marine Pollution Bulletin, 142*, 384–393.

Thushari, G. G. N., & Senevirathna, J. D. M. (2020). Plastic pollution in the marine environment. *Heliyon, 6*(8), e04709. https://doi.org/10.1016/j.heliyon.2020.e04709

Tian, L., Jinjin, C., Ji, R., Ma, Y., & Yu, X. (2022). Microplastics in agricultural soils: Sources, effects, and their fate. *Current Opinion in Environmental Science & Health, 25*, 100311. https://doi.org/10.1016/j.coesh.2021.100311

United Nations Environment Programme (UNEP). (2013). Guidelines for national waste management strategies: moving from challenges to opportunities. Retrieved October 23, 2023, from https://wedocs.unep.org/bitstream/handle/20.500.11822/8669/-Guidelines%20for%20national%20waste%20management%20strategies_%20moving%20from%20challenges%20to%20opportunities-2013UNEP%20NWMS%20English.pdf?sequence=3&isAllowed=y.

United Nations Environment Programme (UNEP). (2018). Single-use Plastics. A Roadmap for Sustainability. Retrieved October 23, 2023, from https://www.unep.org/resources/report/single-use-plastics-roadmap-sustainability.

United Nations Environment Programme (UNEP). (2019). Microplastics. UN Environment Programme. Retrieved October 22, 2023, from https://wedocs.unep.org/bitstream/handle/20.500.11822/12079/brochure-microplastics.pdf?sequence=1&amp%3BisAllowed=.

United States Environmental Protection Agency (USEPA). (2021). EPA's Microplastics Beach Protocol. A community Science Protocol for Sampling Microplastic Pollution. Retrieved November 15, 2023, from https://www.epa.gov/system/files/documents/2021-09/microplastic-beach-protocol_sept-2021.pdf.

Volgare, M., Santonicola, S., Cocca, M., Avolio, R., Castaldo, R., Errico, M. E., Gentile, G., Raimo, G., Gasperi, M., & Colavita, G. (2022). A versatile approach to evaluate the occurrence of microfibers in mussels *Mytilus galloprovincialis*. *Scientific Reports, 12*, 21827. https://doi.org/10.1038/s41598-022-25631-2

Vibhatabandhu, P., & Srithongouthai, S. (2022). Abundance and characteristics of microplastics contaminating the surface water of the Inner Gulf of Thailand. *Water, Air, & Soil Pollution, 233*, 50.

Vin, L. E., Izam, N. I. I., Nilamani, N., Mohd Razalli, N., Mohd Zanuri, N., Yasin, Z., Hwai, A. T. S., Shoufeng, Z., & Hongjun, L. (2020). Abundance and distribution of macro- and microplastics at three different habitats along the Penang coastline in the Northern Straits of Malacca. *Ramkhamhaeng International Journal of Science and Technology, 3*(2), 1–15.

Vin, L. E., Nilamani, N., Mohd Razalli, N., Zhang, S., Li, H., Haron, M. L., Abdullah, A. L., Yasin, Z., Mohd Zanuri, N., & Hwai, A. T. S. (2023). Abundance and distribution of macro- and mesoplastic debris on selected beaches in the Northern Strait of Malacca. *Journal of Marine Science and Engineering, 11*, 1057. https://doi.org/10.3390/jmse11051057

Khalik, W. M. A., Ibrahim, Y. S., Tuan Anuar, S., Govindasamy, S., & Baharuddin, N. F. (2018). Microplastics analysis in Malaysian marine waters: A field study of Kuala Nerus and Kuantan. *Marine Pollution Bulletin, 135*, 451–457. https://doi.org/10.1016/j.marpolbul.2018.07.052

Warrier, A. K., Kulkarni, B., Amrutha, K., Jayaram, D., Valsan, G., & Agarwal, P. (2022). Seasonal variations in the abundance and distribution of microplastic particles in the surface waters of a Southern Indian Lake. *Chemosphere, 300*, 134556. https://doi.org/10.1016/j.chemosphere.2022.134556

Watts, A., Aalten, A. A., & Stanton, D. A. (2015). Mouthful of plastic: microplastics in commercial seafood from the northeast Atlantic and Irish Sea. *Marine Pollution Bulletin, 99*(1–2), 143–146. https://doi.org/10.1016/j.marpolbul.2015.04.006

Woodall, L. C., Sanchez-Vidal, A., Canals, M., Paterson, G. L. J., Coppock, R., Sleight, V., Calafat, A., Rogers, A. D., Narayanaswamy, B. E., & Thompson, R. C. (2014). The deep sea is a major sink for microplastic debris. *Royal Society Open Science, 1*, 140317. https://doi.org/10.1098/rsos.140317

Xanthopoulou, M., & Katsoyiannis, I. A. (2023). Enhanced adsorption of chromate and arsenate ions from contaminated water with emphasis on polyethylenimine modified materials: A review. *Separations, 10*, 441. https://doi.org/10.3390/separations10080441

Yang, J., Karbalaei, S., Hussein, S. M., et al. (2023). Biochemical effects of polypropylene microplastics on red tilapia (*Oreochromis niloticus*) after individual and combined exposure with boron. *Environmental Science Europe, 35*, 71. https://doi.org/10.1186/s12302-023-00771-y

Yasin, Z., Tan Shau Hwai, A. (2019). Marine litter in the straits of Malacca—Some case studies. Retrieved June 16, 2023, from https://www.sea-circular.org/wp-content/uploads/2019/11/10-Zulfigar-MESTECC1-2019.pdf.

Yong, C. Q. Y., Valiyaveetill, S., & Tang, B. L. (2020). Toxicity of microplastics and nanoplastics in mammalian systems. *International Journal of EnvironmentalResearch and Public Health, 17*(5), 1509. https://doi.org/10.3390/ijerph17051509

Zainuddin, A. H., Aris, A. Z., Zaki, M. R. M., Yusoff, F. M., & Wee, S. Y. (2022). Occurrence, potential sources and ecological risk estimation of microplastic towards coastal andestuarine zones in Malaysia. *Marine Pollution Bulletin, 174*, 113282. https://doi.org/10.1016/j.marpolbul.2021.113282

Ziani, K., Ioniță-Mîndrican, C.B., Mititelu, M., Neacşu, S. M., Negrei, C., Moroş, E., Drăgănescu, D., & Preda, O. T. (2023). Microplastics: A real global threat for environment and food safety: A state of the art review. *Nutrients, 15*(3), 617. https://doi.org/10.3390/nu15030617.

Zhu, X., Ran, W., Teng, J., Zhang, C., Zhang, W., Hou, C., Zhao, J., Qi, X., & Wang, Q. (2021). Microplastic pollution in nearshore sediment from the Bohai Sea coastline. *Bulleting of Environmental Contamination and Toxicology, 107*, 665–670. https://doi.org/10.1007/s00128-020-02866-1

Chapter 23
Evaluating the Impact on Plastic Pollution in Sri Lanka and Global Perspectives

Anuruddha Heenatigala, Madhubhashini An, Imasha Rae, and Rathnasekara Kady

Abstract Plastics are synthetic or semi-synthetic materials that consist of organic polymers with high molecular weight. Its also composed of various elements including carbon, hydrogen, oxygen, nitrogen, chlorine, and sulfur and its general chemical formula is (CH2-CHX)n. These days, the usage of plastic products is thoroughly increasing due to their consumer-friendly outcomes like inexpensive and versatility. However, most plastic products are non-biodegradable because fossil fuel-based plastic product manufacturing is a widely used technique. Larger plastic products are accounted for generating microplastics including microbeads, fibers, fragments, nurdles, foams, and films. Accordingly, plastic pollution is heading worldwide. Accumulation of these synthetic plastic products has been destroying wildlife, marine life, and mankind and unaware anthropogenic activities are the main reason for spreading plastic waste. Due to this serious problem, most worldwide governments have implemented some strategic policies to mitigate plastic pollution using a circular economy, extending producer policies and sustainable development goals. These various strategies at national and global levels were implemented by different continents including Europe, Asia, America, Africa, and Australia. The comprehensive plan utilizes the 3R principles of reduce, reuse, and recycling with various technologies including incineration and landfilling. Also, new biodegradable materials are going to be invented as innovative instead of fossil-fuel-based synthetic materials like plastics. Here, the article tries to compare anthropogenic activities for plastic pollution, various plastic phasing-out strategies, and activities of selected countries in the world.

Keywords Environmental pollution · Government policies · Non-biodegradable · Plastic · 3R principles

A. Heenatigala (✉) · M. An · I. Rae · R. Kady
Nano and Modern Technology Wing, Centre for Defence Research and Development, Ministry of Defence, Homagama, Sri Lanka
e-mail: anu.hee.best@gmail.com; hq@crd.lk

S. Suriyanarayanan et al. (eds.), *Combating Plastic Pollution in Terrestrial Environment*,
https://doi.org/10.1007/978-981-96-2343-3_23

23.1 Introduction

23.1.1 What is Polyethylene

Polyethylene commonly known as polythene is one of the most widespread plastics around the world. Polyethylene is available in many forms depending on the configuration of its molecular chain. The chemical structure of polyethylene consists of repeating monomers made of carbon and hydrogen atoms. $(C_2H_4)_n$ is the common formula of polyethylene (Admin, 2022). It would be better to categorize it as a homopolymer because it is made from repeated units of one specific molecule, ethylene. Polyethylenes are generally composed of a linear, plastic structure known as addition polymers. They are synthetic polymers of low density utilized mainly in packaging. While most polyethylenes are thermoplastic and can be easily reshaped with heat, some altered forms are thermosetting (Xometry, 2023). Polyethylene's discovery occurred in 1933 when Reginald Gibson and Eric Fawcett reacted ethylene with benzaldehyde. The initial patent for polyethylene was filed in 1936 by Imperial Chemical Industries. Karl Ziegler later pioneered a method to polymerize HDPE (High-density polyethylene) in 1953 (Xometry, 2023).

The four common types are Ultra-High Molecular Weight Polyethylene, High-Density Polyethylene, Low-Density Polyethylene, and Linear Low-Density Polyethylene each with unique and common properties hence application. UHMWPE has high strength and density, durability, and wear resistance, and that has been used in medical implants and industrial materials installation. HDPE has been used in bottles, pipes, and containers due to its high strength/density ratio, chemical resistance, and stiffness. LDPE also has low density, is flexible, and has a short chain length compared to HDPE. Hence it has been used in packaging films and containers. Both LDPE and HDPE characteristics have been found in LLDPE. This product has been used in plastic liners and wrapping as bags (Admin, 2022).

Ethylene, which is made from processing natural gas or crude oil, is used to manufacture polyethylene. Ethylene can be polymerized into polyethylene only in the presence of a suitable catalyst. That catalyst is metallocene or Ziegler–Natta. Coordination polymerization is the most common polymerization method, which begins with massive cutting of the molecule of acrylonitrile and leaving the electron free as with ethylene splitting the double carbon atoms. As a result, both sides have an open bond, which connects with another molecule. This process continues forming long polyethylene polymer chains (Xometry, 2023).

Polyethylene is a waxy texture material that is a low-strength, high-clarity, and impact-resistant plastic. Its mechanical strength is lower than other polymer. Despite its low mechanical strength, it is a strong insulator of electric current due to its high electrical resistance. Polyethylene is a semi-crystalline plastic composed of nonpolar saturated hydrocarbons with a high molecular weight similar to paraffin. Its symmetrical molecular structures result in greater density and chemical stability. Polyethylene is highly resistant to acids, alkalis, weak oxidizing agents and weak reducing agents and is soluble in aromatic hydrocarbons such as xylene or toluene at high temperatures. Most types of polyethylene are resistant to weak oxidizing agents and weak reducing agents (Admin, 2022).

23.1.2 Positive Overview of Polyethylene

Polyethylene (PE) stands as the largest tonnage plastic material despite its straightforward structure of repetitive CH_2 units. Many attractive features are responsible for its extensive use.

It is economically feasible for many applications because of its cheap cost. PE is appropriate for a variety of electrical and electronic applications due to its remarkable electrical insulating qualities throughout a wide frequency range. Polythene has strong chemical resistance ensuring durability and longevity in challenging settings. Its hardness and impact resistance make it perfect for applications requiring durability. Polythene can be easily used processes a variety of manufacturing processes, making it possible to create complex structures and models (Gilbert & Brydson, 2017; Moshood et al., 2022).

Furthermore, its transparency property improves product presentation in packaging applications, especially in thin film form. Its inherent flexibility allows it to adapt to various shapes and curves. These special characteristics make polyethylene a widely used material in various sectors, including electronics, construction, packaging, and the automotive industry (Moshood et al., 2022).

Plastic is a useful polymer used in the recycling, construction, medical, electrical, and agricultural sectors. It is used in irrigation pipes, crop protection covers, greenhouse films, mulch films, plastic bags, pipelines, insulation, vapor barriers in construction, medical devices, implants, prosthetics, and packaging materials. It is also used for wire insulation and petrol tanks in the automotive sector. Due to its versatility, resistance to moisture and chemicals, and its ability to be recycled, polyethylene finds wide use in various applications (Romani et al., 2020).

23.1.3 Negative Impact of Polyethylene

Plastic was first developed in England in the 1850s and is now widely used in different fields in our day-to-day activities. Over the past half-century, plastics have become an increasing part of every aspect of human production and life. At the moment, plastic materials have already replaced many traditional materials such as wood, metal, and glass, because of the cheapness of their production, as well as high strength and, in some cases, durability. This fact leads to an increase in global annual plastic production. In 2020 alone, about 400 million tons of plastic were produced globally (Yao et al., 2022).

Growing urbanization and rapid industrialization along with the rapid increase in population are raising several man-made problems. One of the major issues is solid waste management, where plastics can be used in a variety of applications such as healthcare, household, commerce, and industry to name a few plastic activities. Most of the plastic products produced today are products that are eventually released into the natural environment (Tiago et al., 2023).

Generally, plastic waste is disposed of in landfills, incinerated, or recycled, which is a more sustainable practice. However, only 18% of plastic is recycled worldwide and other methods have more disadvantages than benefits (Yao et al., 2022).

As plastic is released into the environment, plastic waste has accumulated in the natural environment and it has already adversely affected land and marine life. Plastic waste resists natural decomposition, meaning its disposal creates more solid waste (Grover et al., 2015). A few decades ago, plastic was seen only as an aesthetic nuisance due to the wind dispersing it from landfills to other environments, but nowadays plastic is not only an aesthetic nuisance, but it also causes a lot of bad effects (Yao et al., 2022).

Regardless of the origin of the polyethylene waste problem, its further accumulation in the environment is unlikely to cause serious problems. It is estimated that about 79% of polyethylene waste is delivered to landfills. Furthermore, natural light oxidation of wastes will occur over time thus promoting the generation of greenhouse gases and posing a serious threat to global warming and the ozone layer (Ahmed et al., 2018).

The extensive utilization of plastic products in agriculture as mulching material, a component of composite materials, and packaging following incorrect recycling has caused a significant impact on the plastic pollution of soils. There is a reduction in the soil adsorption capacity, soil aggregate particle size, and diversity of bacteria and fungi due to the presence of polyethylene microplastics. On the one hand, polyethylene particles in the soil will cause increased mobilization of hydrophobic organic pollutants, and the latter may further result in polyethylene dissipation. The hazard of polyethylene particles for land mammals has also been broadly reported.

Plastic waste can also impact on terrestrial and freshwater ecosystems. Animals may accidentally ingest plastic debris or become entangled in it, leading to similar consequences as observed in marine animals. The chemical plasticizers and stabilizers contained in plastics that leach into the environment have serious negative effects on animals' bodies as well as their population rates. In addition, the production processes required to manufacture polyethylene products destroy animal habitats and create the issue of habitat fragmentation which can lower populations by limiting resources, changing migration patterns, as well as increasing access to prey and competition for certain animals (Vanellagroupmn, 2022; Grover et al., 2015).

Plastic is a major contributor to land and water pollution. It can negatively affect marine life when disposed of improperly and end up in our rivers and seas. Global plastic production is estimated to be around 300 million tons per year. In comparison, the same marine environment contains about 9.5 million tons of plastic pollution, of which 1.5 million tons are added to the ocean annually (Ahmed et al., 2018).

Plastic is a common waste in marine and coastal environments, it contributes significantly to marine life. However, polyethylene particles do not seem to interact with marine life. Nonetheless, floating polyethylene particles tend to attract aquatic animals due to color and smell. The animals are entangled or eat the polyethylene particles, which causes damage to their organ apoptosis, genotoxicity, and ultimately death. Studies estimate that plastic waste leads to the deaths of hundreds of millions of marine animals annually. Moreover, smaller size fragmentation into nanoparticles by organisms such as the Antarctic krill, planktonic fish, and freshwater amphipods may be more ecotoxic (Grover et al., 2015; Moore, 2024).

Furthermore, the potential for hazardous chemicals to leach into food and water is a major reason why plastic packaging, especially polyethylene goods, is hazardous to human health. Bisphenol A (BPA) and phthalates, chemicals commonly found in plastics, have been linked to several health problems including cancer, reproductive disorders, hormone imbalances, and developmental abnormalities.

23.1.4 Overview of Plastic Usage in Sri Lanka

Although Sri Lankans have historically relied on nature for basic needs, many technological advances and cultural and lifestyle changes have taken place. It led to the production and consumption of synthetic products such as plastics, which are essential to modern life. People prefer cheap plastics over other materials (Malkanthie et al., 2023).

Sri Lanka ranks within the top ten countries that polluted the environment due to excessive plastic and polythene usage and waste mismanagement. Excessive consumption of plastic and polythene has become a critical issue in Sri Lanka. Sri Lanka generates about four hundred metric tons of plastic waste per day. 90% of plastic waste is identified as single-use plastics used for packaging. Furthermore, Sri Lanka's per capita plastic consumption rate is relatively high compared to other developing countries. The rate of plastic consumption maintains a positive relationship with the quality of life and the strength of the country's economy.

In 1994, Sri Lanka attempted to regulate plastics through guidelines and proposed a ban on high-density polyethylene (HDPE) bags with a thickness of less than 20μ. However, these actions were not effective. A ban was imposed in 2007 under section 23W of the National Environment Act No. 47 of 1980, but this was not effectively implemented due to difficulties in measuring thickness during raids. A conservation levy on HDPE bags was introduced in 2008 and regulations were introduced in 2017 to stop HDPE Polythene (CEJ, 2019).

Despite all the steps, steps to reduce plastic pollution in Sri Lanka proved to be insufficient, as the government agencies, particularly the Central Environment Authority, did not develop an effective policy on enforcement. The manufacturers deceived the population by producing bags in which there was a combination of HDPE and LDPE, as well as biodegradable ones, while some produced oxo-biodegradables. Many manufacturers provide biodegradable plastic bags, but some manufacturers still produce non-biodegradable bags.

The public's attitude toward plastic bags and lunch sheets was initially positive, but now they are confused and negative about plastic. Promoting landfill for municipal waste is counterproductive to the plastic ban. Additionally, the promotion of three waste-to-energy plants in Sri Lanka is a negative development associated with the plastic ban. The government's interest in signing a bilateral trade deal with the country could also open the door to dumping the world's garbage, which roams the globe in search of landfills in Asia once China stops importing such waste (CEJ, 2019).

However, multiple actions have been taken to regulate the use of plastics in Sri Lanka. However, efforts have been limited due to a lack of adequate enforcement mechanisms, improper practices by manufacturers, and public confusion. Urgent action is needed to address this pressing environmental problem, emphasizing effective enforcement, public awareness, and sustainable waste management practices.

23.1.5 Overview of Global Plastic Usage

Many countries have implemented policies, regulations, and acts to combat plastic in their countries. These criteria will guide our review of policies beat at combating plastic pollution across the globe. We chose policy effectiveness and implementation as our first primary criteria in selecting the countries. We understand that there is no way how the policy will meaningfully help curb plastics and other environmental menace unless the results are visible and sustainable. Therefore, we prioritized countries that committed to hard-core implementation of policies that have worked effectively with the tangible reduction of plastic pollution. This will facilitate identifying and analyzing strategies that have worked and those that have been more of a trial and error, making the acquisition processes as pertaining to environmental policies. Furthermore, our focus on policy implementation calls for not only analyzing the content of policies but also investigating which measures are introduced to ensure that policies are indeed implemented, adhered to, and enforced. Hence, our paper seeks to contribute to understanding how effective particular policies are at reducing plastic pollution worldwide and promoting sustainable solid waste regulation in different settings and under different conditions.

23.1.6 Asia

23.1.6.1 Japan

Act on Promoting the Treatment of Marine Debris (2009, Amended in 2018)—Act no. 82 of July 15, 2009

This act mainly focused on finding solutions to the problem of marine debris in the ocean. This law aims to promote the proper treatment and management of marine debris to mitigate its harmful effects on marine ecosystems and coastal communities.

Act includes the creation of monitoring programs to track the dispersal of marine debris, measures for coastal cleanup actions, and prevention and removal efforts.

The act includes chapters and sections as mentioned below.

- Chapter I General Provisions (Articles 1 through 12)
- Chapter II Basic Policy (Article 13)
- Chapter III Regional Plans (Articles 14 through 16)
- Chapter IV Promotion of Measures against Articles that Drift Ashore
- Section 1 Smooth Treatment of Marine Debris (Articles 17 through 21)
- Section 2 Control of Marine Debris Generation (Articles 22 through 24)
- Section 3 Other Policies for Promotion of Treatment of Marine Debris (Articles 25 through 31)
- Supplementary Provisions (Knoblauch & Mederake, 2021)

Resource Circulation Strategy for Plastics (2019)

The strategy is designed to encourage the sustainable management of plastic resources at every stage of their life. This plan includes a number of actions to cut down on plastic waste, increase recycling, and accelerate the shift to a plastics-based circular economy (Tamiya-Hase, 2023; Act on Promoting the Treatment of Marine Debris Affecting the Conservation of Good Coastal Landscapes and Environments to Protect Natural Beauty and Variety, 2019; Griffin et al., 2022).

23.1.6.2 Indonesia

Pricing Mechanism—2016

Indonesia has implemented some pricing mechanisms to reduce the impact of plastics on environmental challenges including the implementation of tariffs or restrictions on single-use plastic bags in certain cities to minimize plastic trash, as well as waste management fees, where homes or companies pay for waste collection and disposal services. Also has, proceeded the concept of extended producer responsibility (EPR), incentivized recycling programs, and introduced incentive-based schemes to promote waste separation.

23.1.6.3 Taiwan

Waste Management Act (WMA)

The Environmental Protection Administration (EPA) in Taiwan has carried out regulatory policies for reducing and recycling regulated recyclable waste, including plastic containers, under the Waste Management Act (WMA). Since 1997, these regulations have changed.

4-In-1 Recycling Program

Taiwan launched the 4-in-1 Recycling Program, involving the general population, cleaning teams from local government agencies, recycling businesses, and a recycling fund. This program is paid for by the companies that make or import products, following the rule that they're responsible for dealing with the waste their products create.

Plastic Bag Restrictions

The EPA of Taiwan has restricted the use of plastic bags since 2006. Furthermore, in accordance with the Waste Management Act's authorization, the EPA has kept announcing new limitations on the usage of plastic products (Tsai, 2022; Knoblauch & Mederake, 2021).

23.1.7 Africa

23.1.7.1 Rwanda

Ban on Single-Use Plastic Items and Plastic-Carry Bags (2008)

Rwanda was among the first nations to implement the Prohibition of manufacturing, importation, use and sale of polythene bags in 2008. Law No. 27/2008 is responsible for this prohibition.

Comprehensive Ban on Single-Use Plastics (2019)

This act aims on the environmental impact of single-use plastics in Rwanda. This law prohibits the importation, manufacturing, sale, and use of single-use plastic products and plastic-carry bags. The goal of the law is to reduce the needless usage and disposal of single-use plastics, which harm the environment.

23.1.7.2 Kenya

Plastic Bag Ban for Secondary Packaging (2017)

The production, importation, distribution, and use of single-use plastic (SUP) bags have been prohibited by this policy. This policy has mentioned recommendations for alternatives for the use of polythene such as reusable bags, cotton bags, and paper bags.

Plastic Bag Control and Management Regulations (2018)

The policy has banned the manufacture, import, export, use, or sale of plastic bags without a permit or paying a fee. Also mandates the development of performance measures and public education programs.

Nairobi City Council Solid Waste Management Act (2015)

This act affords a country-level framework for solid waste management a legal foundation for carrying out the county's solid waste management plan. The act was created to target the plastics directly by prohibiting the manufacture of plastic bags with a thickness of less than 30μ, smaller than 8 × 12 in. or colored differently from Kenyan standards. Also act gives officials the authority to prevent environmentally hazardous operations like generation, transportation, storage, and disposal. Act has approved some methods to dispose of waste such as controlled tipping, sanitary landfilling, recycling, composting, and incineration at the appropriate standards.

Polyethylene Terephthalate (PET) Task Force (2017)

The framework is created between the Ministry of Environment and Forestry (ME&F), the Kenya Association of Manufacture (KAM), and the National Environment Management Authority (NEMA). They have divided their responsibilities for combating the process of PET.

Here KAM is responsible for the Cleanup of PET bottles present in the environment. Also to do awareness campaigns about recycling and upcycling. ME&F is responsible for coordinating between PET subsectors, county governments, and other stakeholders. They also develop national public education and awareness campaigns and facilitate intergovernmental relations among county governments.

2019/20 Budget: Incentives for Plastic Recycling

The Kenyan government has implemented various incentives for businesses in plastic recycling. Specifically, in the 2019/20 budget, significant measures were introduced to support plastic recycling plants (Griffin et al., 2022).

23.1.8 North America

23.1.8.1 Canada

Single-Use Plastics Prohibition Regulations (2022)

Canada has prohibited the manufacture and import for sale of several single-use plastics such as checkout bags, cutlery, stir sticks, food service ware, and straws. These bans aim to reduce plastic waste and its harmful impact on the environment, specially, in marine ecosystems. As a result of this ban, Canada seeks to encourage the use of alternatives for plastics.

23.1.8.2 United States

International Fisheries Regulations—Subpart G—Antarctic Marine Living Resources (50 CFR Ch. III)

The regulation prohibits the irresponsive handling of plastic waste that is operated of a harvesting vessel. This regulation aims to prevent pollution and protect the delicate Antarctic marine ecosystem from the harmful effects of plastic debris (Knoblauch & Mederake, 2021).

23.1.9 Europe

23.1.9.1 Germany

Germany has taken some serious actions to combat plastic in their country.

German Packing Act– 2019

This act also called VerpackG (Verpackungsgesetz). It is implemented on the 1st of January 2019. This replaced the previously occurred Packaging Act and there are some significant changes done by focusing on improving recycling rates, reducing packaging waste, and ensuring the responsible handling of packaging materials.

Pricing Mechanism—2018

This policy mainly aims at encouraging people to recycle and reducing the impact on the environment due to various types of plastics. This policy is initiated by the

German Packing Act. Producers and distributors of packaged goods had to pay fees to operators based on the kinds and quantities of packaging materials they used to participate in a dual recycling system. Furthermore, deposit systems are applied to specific kinds of packaging, such as aluminum, glass, and plastic drinking containers. Consumers also have to deposit small amounts for what they buy and they will refund it when they return them back. By this, the producers have the responsibility for what they produce (extended producer responsibility—EPR).

23.1.9.2 France

Article 124 of the Law of 8 August 2016 for the Recovery of Biodiversity, Nature, and Landscapes (Microbeads Prohibited in Rinse-Off Cosmetics)

This article focuses on prohibiting microbeads that are present in rinse-off products. Finally, that aims to reduce the impact of microplastics in several products like toothpaste, cosmetic products, and personal care products. By prohibiting the releasing of microplastics they try to conserve the protect biodiversity, nature, and landscapes from the harmful effects of plastic pollution.

Ban of Plastic—2016

This ban applied to plastic dinnerware, such as cups, plates, and utensils, as well as single-use plastic bags. By encouraging the adoption of more sustainable alternatives, the regulation sought to lessen the pollution caused by plastics (Knoblauch & Mederake, 2021).

23.1.9.3 United Kingdom

Producer Responsibility Obligations (Packaging Waste)—2017

This obligation aims to keep the responsibility of producers for packaging the waste that generated from their products. Recycling targets are set for different packaging materials, compelling producers to ensure a certain percentage of their packaging waste is recovered and recycled. Producers are required to submit annual reports outlining the quantity and types of packaging they place on the market, facilitating monitoring of recycling targets and progress toward waste reduction goals. These regulations aim to encourage sustainable packaging practices, minimize environmental impact, and promote efficient resource use.

23.1.10 South America

23.1.10.1 Chile

National Legislation on EPR—216

The legislation embodies the polluter pays principle and that gives the responsibility of producing packaging material. It makes sure that manufacturers actively reduce the impact of packaging materials on the environment over the course of their lives. Producers are required to fund and carry out programs for packaging trash collection, recycling, and disposal.

National Ban on Single-Use Plastic Bags in Commerce—2018

This ban prohibits the use and distribution of lightweight plastic bags used by various businesses across the country. Also, consumers have been encouraged to use recyclable and reusable bags. They are the impulse to use biodegradable bags made of biodegradable materials. As a result of these steps, it reduced the use of single-use plastics. Chile aims to raise awareness about the environmental impact of plastic pollution and encourage individuals and businesses to adopt more sustainable practices.

National Legislation Prohibiting the Delivery of Single-Use Plastic Bags into the National Territory—2018.

The legislation prohibited the importation of single-use plastic bags into the country. This controls many problems due to plastics because it has stopped the occurrence of the problem from the beginning. Retailers, distributors, and other companies doing business inside the country are essentially prohibited by law from delivering and distributing single-use plastic bags. By implementing this legislation, Chile sought to reduce the consumption and disposal of single-use plastics.

23.1.10.2 Brazil

National Policy on Solid Waste (PNRS)

The policy was established on 2010 and it provides a legal framework for the handling of solid waste in Brazil. When it comes to plastic it includes a framework about waste reduction, recycling, and extended producer responsibility.

23.1.10.3 Columbia

National Fee on Plastic Bags—2017

Colombia established a national fee for the plastic bags that people use for their needs. This helps to mitigate the use of plastic bags and encourages sustainability.

2019 National Ban on SUP at National Natural Parks

This ban supports the conservation of nature and ecological value within the national natural parks. They introduced a national ban on Single-Use Plastics like bags, straws, bottles, and utensils in those areas (Ortiz et al., 2020).

23.1.10.4 Australia

Australia is working with both national and sub-national policies.

23.1.10.5 National Level Policies

Australia National Plastics Plan (2021)

In July 2018, Australia introduced a nationwide ban on lightweight, single-use plastic bags. This ban prohibits the provision. The plan has been divided in to five parts including prevention, recycling, consumer outreach, marine debris, and research and development. Also, they focus on government-industry partnerships to solve plastics problems increase the recycling rate and to shift to easily recyclable plastics. The policy is aiming to recycle by 2025 and to invest in recycling and circular economy programs.

Recycling and Waste Reduction Act (2020)

Tries to reduce greenhouse gas emission waste creation and the development of a circular economy creates a product stewardship program and sets rules for waste exports in order to encourage product recovery, recycling, and reuse.

National Waste Policy Action Plan (2019)

The policy action plan includes seven national targets to achieve to combat plastic. They are,

I. Ban the export of waste plastic, paper, glass, and tires, commencing in the second half of 2020
II. Reduce total waste generated in Australia by 10% per person by 2030
III. 80% average resource recovery rate from all waste streams following the waste hierarchy by 2030
IV. Significantly increase the use of recycled content by governments and industry
V. Phase out problematic and unnecessary plastics by 2025
VI. Halve the amount of organic waste sent to landfill by 2030
VII. Make comprehensive, economy-wide, and timely data publicly available to support better consumer, investment, and policy decision.

National Waste Policy (2018)

The policy offers a framework for federal and state governments, businesses, and NGOs to achieve sustainable waste management based on circular economy principles.

Threat Abatement Plan for the Impacts of Marine Debris on Vertebrate Marine Life (2009)

The policy mainly targets to achieve four goals. They are,

I. Contribute to the long-term prevention of the incidence of harmful marine debris.
II. Remove existing harmful marine debris from the marine environment.
III. Mitigate the impacts of harmful marine debris on marine species and ecological communities.
IV. Monitor the quantities, origins, and impacts of marine debris and assess the effectiveness of management arrangements over time.

The Strategy proposed a total of 18 actions to accomplish goal 1. The eleven initiatives were mainly designed to reduce sources of marine debris in the water, a central focus being fishing vessels. The other initiatives aimed at reducing the amount of trash pouring into waterways, increasing the population awareness of the problem, and cooperation with other countries. Almost all of the activities collecting data go hand in hand with removal operations to meet the dual goals of 2 and 4. The federal government's efforts to aid in the recovery of marine species and to study how marine trash affects wildlife are the primary focus of Objective 3.

National Environment Protection (Used Packaging Materials) Measure 2011

The measure has implemented rules for the recovery and recycling of consumer packaging matter in the state governments.

23.1.10.6 Sub-national Level Policies

These policies are the regulations implemented at the state and territory level to address plastic pollution.

State-Level Bag Bans

All the states in the country have banned the use of single-use plastic bags due to some thickness in the polyethylene (Bering et al., 2022).

23.1.10.7 Action Needed to Combat Plastic Pollution: From the Problem to Policy Change: A Critical Analysis

A circular economy is a kind of economic model that can minimize waste or make new resourceful materials utilizing waste materials. Therefore, this is very useful for reducing waste, energy use, and raw material consumption (Tomić & Schneider, 2020). Plastic has been used significantly since 1950 reaching 39 million metric tons (Rhodes, 2019). The EU is the fourth-largest plastic producer in the world (Lorang et al., 2022). Most of the plastic is also used in short-term applications (Chen et al., 2021). Plastic packaging offers more advantages including durability, cost-effectiveness, and versatility in preserving and protecting products (Teck Kim et al., 2013). Therefore, a large number of modern supermarkets consume as packages. However, this can have a bad impact on the environment due to oil consumption and carbon emissions. By 2050, the plastics industry will consume 20% of the world oil supply, and contribute 15% of global carbon, becoming a reason for global warming below 2 °C (Lebreton & Andrady, 2019). Plastics are a slow degradation product, providing bad impacts on the environment from fossil-based extraction to end-of-life disposal methods including recycling, incineration, landfilling, or environmental leakage (Ciardelli et al., 2019; Mazhandu et al., 2020). Plastic packaging has five lifecycle stages of packaging including the reproduction stage, production stage, distribution stage, product usage stage, and disposal stage (Nielsen et al., 2020). Reuse, remanufacture, mechanical recycling, biological recycling, incineration, or landfilling are some possibilities for disposal stages. However, plastic pollution is a large global crisis due to its harmful impacts on environmental change, climate change, and environmental degradation. EU took more actions to prevent bad impacts on the environment of plastic pollution. The circular economy action plan is a key component that could be transited to global sustainability by minimizing plastic waste and promoting sustainable alternatives. EU adopted Directive 2019/904 to reduce the impacts of single-use plastics. 3R principles including reduce, reuse, and recycle are the primary strategies of sustainable waste management (Samiha, 2013).

The material extraction stage can be a reason for 61% of Greenhouse Gas (GHG) emissions due to resin production (Tenhunen-Lunkka et al., 2023). Biodegradable plastics and plant-based plastics are the alternatives instead of fossil-based plastics.

Rigid plastic products are plastic bottles (PET), containers (HDPE), and PVC pipes that can be easily recyclable due to their uniformity and ease of sorting. Flexible plastics are films, wraps, and plastic bags that have LDPE. Recycling these materials is quite challenging because their thin and flexible nature causes issues during sorting and processing. Multilayer materials consist of multiple plastic layers or combined materials like plastic-aluminum composites (snack packaging). This material recycling is difficult and complex to recycle due to its multiple layers. Their recycling is also costly, and mostly use downcycled or incinerated rather than recycled (Uzosike et al., 2022).

End-of-life management options utilize various strategies to handle plastic waste streams, in which the end of the waste materials life cycle ends. These options can be broken down into three parts including recycling, incineration, and landfilling (Denison, 1996). Their strategies are based on material type, recyclability, environmental considerations, economic feasibility, and regulatory requirements with managing sustainable waste management practices. Mechanical recycling is a kind of waste material recycling process, that collects, sorts, cleans, and processes waste materials into new products without changing their chemical composition. For instance, plastic bottles can be collected, shredded, melted, and then formed into new plastic products like plastic containers or textiles. This helps to reduce waste going to landfills and conserves resources by reusing materials. Downcycling is a kind of process that can convert waste materials into lesser-quality products or value. As an example, plastic bottles that can be used for more time at a moderate cost can be recycled into lower-grade plastic materials like plastic lumber or textiles. The downcycling process tries to increase the life cycle of materials. However, their results of products often have limited applications and lower quality when compared to the original materials. Upcycling is a waste material transforming process, the resulting products often transform into new materials or products with a high value when compared to their previous value and materials. This is greater than downcycling and upcycling due to its improvements in materials' quality and functionality. One innovative example of upcycling plastics is new eco-friendly fashion accessories. The designers or artisans collect and clean disposable plastic materials like plastic bottles or containers, and then creatively design them into fashionable shoes, handbags, wallets, jewelry, or fabulous decorative items stickers, paint, and ribbon (Li et al., 2024).

Chemical recycling is an advanced recycling method that can break down plastic polymers into chemical components involving pyrolysis, depolymerization, or gasification (Davidson et al., 2021). This process helps to break down complex plastics like multilayer plastics. The final products are raw materials including oils or gases to manufacture new plastics, fuels, chemicals, or other products (Davidson et al., 2021; Dogu et al., 2021). Feedstock recycling generates feedstock materials including ethylene, propylene, or other monomer of plastic production by utilizing complex or mixed plastics (Buekens, 2006). Biological recycling, an environmentally friendly process breaks plastics into natural compounds like water, CO_2, or biomass by utilizing biological agents such as enzymes, bacteria, or fungi (Shah et al., 2008). Energy recovery is the end-of-life management option process that can convert plastic waste materials into energy products like heat, electricity, or fuels (syngas or liquid fuels) with the help of various energy treatment technologies including incineration, gasification, or pyrolysis (Vlasopoulos et al., 2023). Landfilling is an option for end-of-life cycle management to minimize environmental effects by layer-depositing waste materials that cannot be recycled or incineration.

Environmental impact: Plastic leakage (approximately 8 million metric tons of plastic end up in the ocean annually, posing significant environmental challenges, Impacts Marine Species (health issues, death, entanglement, suffocation, and affects development, reproduction, and behavior) Human health concerns (microplastics have been found in various human tissues and organs, potential impacting metabolism and immune systems (Kurtela & Antolović, 2019).

23.1.11 Policies

23.1.11.1 Europe

Plastic plays a significant role due to its unique properties including lightweight, easy-to-clean, low cost, and easy to mold. These days plastic plays an important role in the economy due to its significant applications in many industries. It is harmful to ecosystems, marine life, and terrestrial habits when it pollutes. Some plastics that do not provide reuse and cost-effective recycling for short-lived applications have led to a crisis because their litter is growing so fast. Therefore, 'The European Strategy for Plastics in a Circular Economy', a policy that has a broader circular economy action Plan was established in the year 2018. This European Strategy for plastics tries to minimize the harmful effects of plastics by the production of re-use, repair, and recycling plastics. This also promotes sustainable and non-toxic reusable products instead of single-use products while minimizing the generation of less waste (Mederake & Knoblauch, 2019).

Microplastics are a great enemy to marine organisms due to their large area-to-volume ratios. Microplastics are harmful to sea life when inhaled or absorbed into their bodies. These little particles are deposited on their organs and tissues, badly impacting various adverse health effects. The most commonly available microplastics as marine debris are polyethylene (PE), polypropylene (PP), polystyrene (PS), polyamide (PA)(nylon), polyester (PES), and acrylic (AC). Therefore, the European Union directives three policies including the National Packaging Waste Plan Implementing Directive 94/62/EC, the Waste Management Code implementing Directive 2008/98/EC, and the Tobacco Products Regulations implementing Directive 2014/40/EU.

23.1.12 Asia

23.1.12.1 Sri Lanka

The Sri Lankan government banned Non-biodegradable Single-Use-Plastic including cutlery, cups, and also artificial flowers from 2021. This plastic waste hugely impacts local wildlife including elephants and the environment. Microplastics can be easily digestible by animals like plankton, birds, turtles, mammals, and fish due to their tiny particle size. These particles can be hazardous and badly impact their health including internal wounds, nutrient deficiencies, and starvation. Microplastics are generated by mega plastic like plastic toothbrushes, cups, bottles, bags, or vehicle parts degradation due to exposure to UV light. Most of the ocean litter is generated due to the mega plastics. Therefore, Sri Lanka has initiated a National Action Plan for Plastic Waste Management in 2021–2030. The vision of the plan is a 'Healthy life and cleaner environment for all'. The objectives include public awareness of plastic waste management, and zero landfilling by achieving 3R principles (Ministry of Environment, Sri Lanka, 2021).

23.1.12.2 Indonesia

Indonesia has decided to reduce marine plastic litter by 70% by 2025 with the help of Indonesia's Plan of Action on Marine Plastic Debris (2017–2025). Their government has reported a 35.36 percent reduction in marine plastic debris in December 2022. Ministry of Environmental and Forestry: reducing packaging waste by 2029: Voluntary PR implementation No 97/2017 and Indonesia's Clean-from-Waste 2025 plan is decided by the Ministry of Environment and Forestry (Wang et al., 2022). National policy and strategy on solid waste management (including plastic waste) regulated by presidential Regulation No 97/2017 aiming to reduce 30% of waste generation including plastics and 70% of waste handling by 2025. They have implemented actions to reduce plastic waste at the source like plastic bag restrictions. Single-use plastic including shopping bags, straws, and foam containers for food is banned in Bali province, 18 cities, and regencies. Recycling, composting, biogas production, and thermal recovery are the waste-handling processes that have been used by Indonesia. They used landfilling to reduce waste. Solid Waste Management Act (No. 18/2008) mainly focuses on 3Rs including, reduce, reuse and recycling.

23.1.12.3 Vietnam

Vietnam Decree No 08/2022/ND-CP tries to take action against plastic waste, hazardous plastic waste, and plastic packaging (Thuan et al., 2023). This act prohibited the import of plastic waste into Vietnam and restricted the production and use of single-use plastics in certain areas such as tourist destinations and public parks by minimizing harmful environmental pollution. These plans also implement plastic waste management plans including waste reduction strategies, recycling initiatives, and proper disposal methods. Therefore, Manufacturing companies should take responsibility for collecting and treating plastic waste that is generated during production and should invent eco-friendly materials instead of fossil-fuel plastics. Businesses also implement some plans for 3Rs and proper disposal. These actions also conduct public awareness campaigns for the harmful effects of plastic waste and plastic pollution and eco-friendly practices on the environment at schools and communities.

23.1.12.4 Philippines

The Philippines implemented the Republic Act (RA) No. 11898 in 2022 regarding environmental concerns of plastics and waste management (Philippines, 2023). The act can impact businesses that manufacture plastic packaging for transportation, distribution, or sale. Rigid plastics, Flexible plastics, plastic bags, and polystyrene are included in the act. Buying back material/ waste from consumers, establishing a recycling system including composting and thermal treatment, transporting plastic waste to proper disposal sites, and carrying out cleanups at coastal areas, roads, and other public areas are some EPR requirements for businesses in the Philippines. Large enterprises will recover 80% of their plastic waste 80%.

The other act is Republic Act No.9003 (RA 3000), also known as the Ecological Solid Waste Management Act of 2000 to provides a comprehensive framework for the proper management of solid waste including plastic in the Philippines. This framework considers the solid waste that can be generated by households, businesses, and institutions to separate their waste into biodegradable, recycled, and non-recyclable materials. This also prohibited the open dumping of solid waste.

23.1.12.5 Japan

Act No. 82 of July 15, 2009, also known as the Act on Resource Recycling Promotion Law. This law emphasizes the consumption and recycling of resources including plastic by reducing waste generation and harmful environmental impacts. The law introduces EPR principles for the collection, recycling, or proper disposal of the products and packaging like plastic waste sent to landfills and incineration. Additionally, this act implements some actions for efficient waste collection, recycling infrastructure, and public awareness programs. Here, Businesses and industries should be transparent and accountable in resource utilization and environmental conservation. This act also supports research and development activities like novel recycling technologies, waste reduction methods, and sustainable resource utilization by introducing innovations for plastic waste management. Japan Action Plan for Marine Plastic Litter (2019) combats marine plastic pollution due to marine litter. This includes the regulation of plastic waste disposal, marine-friendly plastics, sustainable packaging materials, and cleaning up of coastal areas.

23.1.13 North America

23.1.13.1 United States of America

Secretary's Order 3407(SO 3407) is a plastic policy of the USA, issued on June 8, 2022. This policy aims to minimize the procurement, sale, and distribution of single-use plastic products and packaging (the secretary of the interior Washington, 2022). The goal of the policy is to phase out all single-use plastic products including plastic and polystyrene food and beverage containers, bottles, straws, cups, cutlery, and disposable plastic bags on Department lands by 2032. Therefore, a significant number of water filling stations find alternatives instead of single-use plastics in the USA like National Park water bottle filling stations including Big Bend National Park, the National Mall, Wolf Trap National Park, and Fort Summer National Historical Park in America. Yosemite National Park stopped usage of single-use-plastics (SUP) beverage containers in 2022. Recycling and waste diversion try to collect and recycle solid waste by at least 50% by 2025 and 75% by 2030.

23.1.13.2 Canada

Canadian Environmental Protection Act (CEPA), 1999 aims to effectively manage toxic substance manufacturing including plastic that can be harmful to human health and the environment (Canada, 2017). Canadian government banned single-use plastics across Canada in January 2021 and plastics are introduced as a toxic substance under the Canadian Environmental Protection Act (CEPA) due to harmful impacts on the environment of plastics. The Federal Court of Appeal (FCA) announced to continue the Canadian Environmental Protection Act (CEPA) on January 25, 2024.

23.1.14 Africa

23.1.14.1 Kenya

Kenya plastic bag ban notices No 2356, the Environmental Management and Coordination Act has prohibited plastic bags including garbage bin liners, medical waste, construction, and food packaging manufacturing and import. The government raised social media awareness campaigns for poor waste disposal systems and careless disposal of plastics that can be a reason for severe flooding. 2019/20 National budget of Kenya releases plastic recycling companies from VAT to promote recycling. Therefore, Kenya currently recycles plastics including PET bottles more than 2000 tons per year (Hakuzimana, 2021).

23.1.14.2 Rwanda

Rwanda's Law No. 17/2019 was published in 2019 against the manufacturing, importation or sale of single-use-plastic products including plastic carry bags, oxo-degradable plastics, and other plastic material products. This law promotes alternatives like biodegradable materials instead of plastics and accountability for sustainable practices including recycling and disposal of plastic waste among businesses and consumers by reducing environmentally unfriendly products.

23.1.14.3 Australia

Australia's national plastic plan was introduced in 2021 due to various problems of plastic waste. 60% of plastics usage are imported from 3.5 million tons of plastics in 2018 and 2019. One million tons of plastic waste were single-use plastics. $419 million of economic value was missing annually due to PET and HDPE 84% of plastic waste was sent for landfilling and 13% was recycled. Therefore, Australia proceeded with a plan from 2019 to 2022 to protect its environment, animals, and mankind from plastic pollution. The Council of Australian Government (COAG) introduced a timetable to ban plastic export. (waste plastic, paper, glass, and tires) in 2019. First National Plastic Summit was held with the aim of microbeads phased out in rinse-off cosmetics, personal care, and cleaning products in 2020. CRISCO plan which is known as 'A Circular Economy Roadmap for plastics, tires, glass and paper in Australia' was initiated in 2021. Exporting of Unprocessed single polymer or resin waste plastic regulated and PVC packaging labels was phasing out in 2022. National packaging targets were initiated for the industry including 100% of packaging being reusable, recyclable, or compostable, 70% of plastic packaging should be recycled or composted, and unnecessary single-use-plastic packaging is phasing out as the 5th target of NWPAP (Cole & Vidal, 2021; Mederake & Knoblauch, 2019; Wang et al., 2022).

23.2 Conclusion

Plastic production consumption has increased during the past few years due to its vast benefits for consumers. However, global annual primary plastic production has been significantly increasing from 1950 in various industries including textile, industrial machinery, consumer and institutional products, electrical and electronic, building and construction, transportation, and packaging. Synthetic plastic production from fossil fuels continues to increase, leading to plastic pollution that can be harmful to the environment by damaging ecosystems, marine life, and human health. Plastic waste contributes significantly to land and water pollution, especially when improperly disposed of and ending up in rivers and seas. This article explores some plastic pollution mitigation strategies of selected countries in different continents including Europe, Asia, America, Africa, and Australia. The European Union has focused on reducing plastic pollution by using the Circular Economy Action Plan Directive 2019/904. This plan tries to reduce single-use plastic to minimize plastic waste by promoting 3R principles and ensuring sustainable development. Also, the European Union has some policies like the National Packaging Waste Plan Implementing Directive 94/62/EC to reduce microplastics. In Asia, countries like Sri Lanka, Indonesia, Vietnam, the Philippines, and Japan also have different policies and actions to mitigate marine plastic waste, microplastic, and other plastic waste to secure the environment. Indonesia has implemented various action plans like No 97/2017 Indonesia's Clean-from-Waste 2025 plan, and Solid Waste Management Act (No.18/2008) to minimize the environmental harmful effects of plastic waste due to high tourism attraction. Vietnam Decree No08/2022/ND-CP also tries to control plastic waste generation in Vietnam. Republic Act (RA) No. 11898, and Republic Act No.9003 (RA 3000) are Philippines action plans against non-biodegradable plastics. The action plans North America, particularly the United States and Canada have set ambitious goals to phase out plastic waste. Canadian Environmental Protection Act also classifies plastic as a toxic substance. In Africa, countries like Kenya and Rwanda also have implemented different action plans to reduce plastic waste and apply biodegradable alternatives instead of plastic products. Australia's National action plan emphasizes reducing plastic waste by implementing some targets with the help of principles of a circular economy.

Combating plastic pollution requires a huge complement from every country. For that, the countries can implement various strategies like government policies and action plans. This can be include 3R principles, public awareness sessions, technological innovation, and international cooperation.

References

Act on Promoting the Treatment of Marine Debris Affecting the Conservation of Good Coastal Landscapes and Environments to Protect Natural Beauty and Variety. (2019).

Admin. (2022). What is polyethylene, monomer, types, properties, applications (with faqs), BYJUS. Retrieved April 21, 2024, from https://byjus.com/chemistry/polyethylene/.

Ahmed, N., et al. (2018). Investigation on bio-oil yield and quality with scrap tire addition in sugarcane bagasse pyrolysis. *Journal of Cleaner Production, 196*, 927–934. https://doi.org/10.1016/j.jclepro.2018.06.142.

Bering, J., Karasik, R., & Takeaways, K. (2022). Policy brief plastic pollution policy country profile: Australia.

Buekens, A. (2006). Introduction to feedstock recycling of plastics.

Canada, E., & C. C. (2017). Government of Canada, Canada.ca. Retrieved April 30, 2024, from https://www.canada.ca/en/environment-climate-change/services/canadian-environmental-protection-act-registry/general-information/overview.html.

CEJ. (2019). *Why Sri Lanka should ban single-use plastics?* Retrieved April 29, 2024, from https://ejustice.lk/wp-content/uploads/2019/03/single-use-plastics_1.pdf.

Chen, Y., Awasthi, A. K., Wei, F., Tan, Q., & Li, J. (2021). Single-use plastics: Production, usage, disposal, and adverse impacts. *Science of the Total Environment, 752*. Elsevier B.V. https://doi.org/10.1016/j.scitotenv.2020.141772.

Ciardelli, F., Bertoldo, M., Bronco, S., & Passaglia, E. (n.d.). Polymers from fossil and renewable resources scientific and technological comparison of plastic properties.

Cole, S., & Vidal, O. (n.d.). Tin Boy.

Davidson, M. G., Furlong, R. A., & McManus, M. C. (2021). Developments in the life cycle assessment of chemical recycling of plastic waste—A review. *Journal of Cleaner Production, 293*. Elsevier Ltd. https://doi.org/10.1016/j.jclepro.2021.126163.

Denison, R. A. (1996). Environmental life-cycle comparisons of recycling, landfilling, and incineration: A review of recent studies. *Annual Review of Energy and Environment, 21*. www.annualreviews.org.

Dogu, O., Pelucchi, M., Van de Vijver, R., Van Steenberge, P. H. M., D'hooge, D. R., Cuoci, A., Mehl, M., Frassoldati, A., Faravelli, T., & Van Geem, K. M. (2021). The chemistry of chemical recycling of solid plastic waste via pyrolysis and gasification: State-of-the-art, challenges, and future directions. In *Progress in Energy and Combustion Science* (Vol. 84). Elsevier Ltd. https://doi.org/10.1016/j.pecs.2020.100901.

Fao.org (ed.). (no date). Fao.org, FAO.org. Retrieved April 30, 2024, from https://www.fao.org/faolex/results/details/en/c/LEXFAOC212268/#:~:text=The%20Decree%20further%20makes%20provisions,rural%20environmental%20protection%20and%20in.

Gilbert, M., & Brydson, J. A. (2017). *Brydson's plastics materials*. Elsevier, BH.

Griffin, M., Karasik, R., & Takeaways, K. (2022). Policy Brief Plastic Pollution Policy Country Profile: Kenya.

Grover, A. et al. (2015). *(PDF) Polythene and environment, Polythene and Environment*. Retrieved April 25, 2024, from https://www.researchgate.net/publication/280096478_Polythene_and_Environment.

Hakuzimana, J. (2021). Break free from plastics: Environmental perspectives and evidence from Rwanda. *Environment & Ecosystem Science, 5*(1), 27–36. https://doi.org/10.26480/ees.01.2021.27.36.

Japan's Plastic Waste Management. Retrieved April 30, 2024, from https://hk.boell.org/sites/default/files/2022-05/PlasticAtlasAsia2022_en_WEB_0.pdf.

Knoblauch, D., & Mederake, L. (2021). Government policies combatting plastic pollution. In Current Opinion in Toxicology (Vol. 28, pp. 87–96). Elsevier B.V. https://doi.org/10.1016/j.cotox.2021.10.003.

Kurtela, A., & Antolović, N. (2019). The problem of plastic waste and microplastics in the seas and oceans: Impact on marine organisms. *Ribarstvo, Croatian Journal of Fisheries, 77*(1), 51–56. https://doi.org/10.2478/cjf-2019-0005.

Law no.12305—Brazilian national policy on solid waste. (no date). Brazilian NR. Retrieved April 30, 2024, from https://braziliannr.com/brazilian-environmental-legislation/law-no-12305-brazilian-national-policy-solid-waste/.

Lebreton, L., & Andrady, A. (2019). Future scenarios of global plastic waste generation and disposal. *Palgrave Communications, 5*(1). https://doi.org/10.1057/s41599-018-0212-7.

Li, Y. C., Wang, S., Qian, S., Liu, Z., Weng, Y., & Zhang, Y. (2024). Depolymerization and re/upcycling of biodegradable PLA plastics. *ACS Omega, 9*(12), 13509–13521). American Chemical Society. https://doi.org/10.1021/acsomega.3c08674.

Lorang, S., Yang, Z., Zhang, H., Lü, F., & He, P. (2022). Achievements and policy trends of extended producer responsibility for plastic packaging waste in Europe. *Waste Disposal and Sustainable Energy, 4*(2), 91–103. https://doi.org/10.1007/s42768-022-00098-z.

Malkanthie, M. A., Herath, H. M., & Galdolage, B. S. (2023). Plastic use among Sri Lanka's metropolitan population: A study of environmental consciousness and behavior. *Kelaniya Journal of Management, 12*(1), 1–28. https://doi.org/10.4038/kjm.v12i1.7740.

Mazhandu, Z. S., Muzenda, E., Mamvura, T. A., Belaid, M., &Nhubu, T. (2020). Integrated and consolidated review of plastic waste management and bio-based biodegradable plastics: Challenges and opportunities. *Sustainability (Switzerland), 12*(20), 1–57. MDPI. https://doi.org/10.3390/su12208360.

Mederake, L., & Knoblauch, D. (2019). Shaping EU plastic policies: The role of public health vs. environmental arguments. *International Journal of Environmental Research and Public Health, 16*(20). https://doi.org/10.3390/ijerph16203928.

Moore, C. (2024). *Plastic pollution in oceans and on land, Encyclopædia Britannica.* Retrieved April 26, 2024, from https://www.britannica.com/science/plastic-pollution/Plastic-pollution-in-oceans-and-on-land.

Moshood, T. D., et al. (2022). Sustainability of biodegradable plastics: New problem or solution to solve the global plastic pollution. *Current Research in Green and Sustainable Chemistry, 5*, 100273. https://doi.org/10.1016/j.crgsc.2022.100273.

Nielsen, T. D., Hasselbalch, J., Holmberg, K., &Stripple, J. (2020). Politics and the plastic crisis: A review throughout the plastic life cycle. *Wiley Interdisciplinary Reviews: Energy and Environment, 9*(1). John Wiley and Sons Ltd. https://doi.org/10.1002/wene.360.

Ortiz, A. A., Sucozhañay, D., Vanegas, P., & Martínez-Moscoso, A. (2020). A regional response to a global problem: Single use plastics regulation in the countries of the pacific alliance. *Sustainability (Switzerland), 12*(19), 1–21. https://doi.org/10.3390/su12198093.

Rhodes, C. J. (2019). Solving the plastic problem: From cradle to grave, to reincarnation. *Science Progress, 102*(3), 218–248. https://doi.org/10.1177/0036850419867204.

Romani, V. P., Martins, V. G., & Goddard, J. M. (2020). Radical scavenging polyethylene films as antioxidant active packaging materials. *Food Control, 109*, 106946. https://doi.org/10.1016/j.foodcont.2019.106946.

Rwanda champions international legally binding treaty on Plastic Pollution. (no date). Rwanda Environment Management Authority (REMA): Home. Retrieved April 30, 2024, from https://www.rema.gov.rw/info/details?tx_news_pi1%5Baction%5D=detail&tx_news_pi1%5Bcontroller%5D=News&tx_news_pi1%5Bnews%5D=483&cHash=8cb865ec59d01e89b169fdf397a9121d#:~:text=Rwanda%20banned%20single%2Duse%20polythene,single%2Duse%20plastics%20in%202019.

Samiha, B. (2013). The importance of the 3R principle of municipal solid waste management for achieving sustainable development. *Mediterranean Journal of Social Sciences, 4*(3), 129–135. https://doi.org/10.5901/mjss.2013.v4n3p129.

Schedule, European Parliament. Retrieved April 30, 2024, from https://www.europarl.europa.eu/legislative-train/theme-new-boost-for-jobs-growth-and-investment/file-strategy-on-plastics-in-the-circular-economy.

Shah, A. A., Hasan, F., Hameed, A., & Ahmed, S. (2008). Biological degradation of plastics: A comprehensive review. *Biotechnology Advances, 26*(3), 246–265. https://doi.org/10.1016/j.biotechadv.2007.12.005.

Tamiya-Hase, N. (n.d.). Japan's policies and measures on plastic pollution.

Teck Kim, Y., Min, B., & Won Kim, K. (2013). General characteristics of packaging materials for food system. In *Innovations in Food Packaging*: (Second ed., pp. 13–35). Elsevier Ltd. https://doi.org/10.1016/B978-0-12-394601-0.00002-3.

Tenhunen-Lunkka, A., Rommens, T., Vanderreydt, I., & Mortensen, L. (2023). Greenhouse gas emission reduction potential of European Union's circularity related targets for plastics. *Circular Economy and Sustainability, 3*(1), 475–510. https://doi.org/10.1007/s43615-022-00192-8.

The negative effects of plastic on the environment. (2022). *Home*. Retrieved April 29, 2024, from https://www.vanellagroupmn.com/the-negative-effects-of-plastic-on-the-environment.

The Secretary of the Interior Washington. (n.d.).

Three tons of plastic bottles collected in Sri Pada within three months—Breaking news: Daily Mirror. (2024). Breaking News|Daily Mirror. Retrieved April 30, 2024, from https://www.dailymirror.lk/breaking-news/Three-tons-of-plastic-bottles-collected-in-Sri-Pada-within-three-months/108-278472.

Thuan, K. H., Huy, D. T. N., & Diep, N. T. (2023). Evaluating environment protection and sustainable agriculture: Case studies in Vietnam and recommendations for public (state) administration. *Caspian Journal of Environmental Sciences, 21*(4), 779–786. https://doi.org/10.22124/cjes.2023.7128.

Tiago, G. A. O., et al. (2023). The problem of polyethylene waste—Recent attempts for its mitigation. *Science of the Total Environment, 892*, 164629. https://doi.org/10.1016/j.scitotenv.2023.164629.

Tomić, T., & Schneider, D. R. (2020). Circular economy in waste management—Socio-economic effect of changes in waste management system structure. *Journal of Environmental Management, 267*, 110564. https://doi.org/10.1016/j.jenvman.2020.110564.

Tsai, W. T. (2022). Environmental policy for the restriction on the use of plastic products in Taiwan: Regulatory measures, implementation status and COVID-19's impacts on plastic products recycling. *Environments—MDPI, 9*(1). https://doi.org/10.3390/ENVIRONMENTS9010007.

Uzosike, C. C., Yee, L. H., & Vasquez Padilla, R. (2022). Small-scale mechanical recycling of solid plastic wastes: A review of PET, PEs, and PP. https://doi.org/10.20944/preprints202211.0189.v1.

Vlasopoulos, A., Malinauskaite, J., Żabnieńska-Góra, A., & Jouhara, H. (2023). Life cycle assessment of plastic waste and energy recovery. *Energy, 277*, 127576. https://doi.org/10.1016/j.energy.2023.127576.

Wang, Y., Karasik, R., & Takeaways, K. (2022). Policy Brief Plastic Pollution Policy Country Profile: Indonesia.

What the extended producer responsibility act means for the Philippines: Arowana Impact Capital. (2023). Arowana. Retrieved April 30, 2024, from https://arowanaco.com/2023/01/20/extended-producer-responsibility-act-philippines/#:~:text=Extended%20Producer%20Responsibility%20Act%20(EPRA)&text=The%20Extended%20Producer%20Responsibility%20Act%20is%20a%20law%20that%20requires,law%20on%2023%20July%202022.

Xometry, T. (2023). *Polyethylene (PE): Structure, properties, and applications, Xometrys RSS*. Retrieved April 21, 2024, from https://www.xometry.com/resources/materials/polyethylene/#:~:text=Polyethylene%20(PE)%20is%20the%20most,single%20type%20of%20molecule%2C%20ethylene.

Yao, Z., Seong, H. J., & Jang, Y.-S. (2022). Environmental toxicity and decomposition of polyethylene. *Ecotoxicology and Environmental Safety, 242*, 113933. https://doi.org/10.1016/j.ecoenv.2022.113933.

Chapter 24
Exploration of Microplastics Pollution Sources in Terrestrial Environment: An Evidence from Malaysia

Sarva Mangala Praveena

Abstract There is an increasing demand for plastic as the current global annual use of 330 million metric tonnes is predicted to increase fourfold by 2050. Attending the predicted increase of plastics production is biosphere concerns as plastic inefficiently degrades and contributes to global environmental pollution. Microplastics (plastic particles < 5 mm) is of particular concern due to its small size and deleterious impacts on both marine and terrestrial systems. However, microplastics accumulation in soil was found 4–23 times greater by mass relative to marine systems. Moreover, microplastics that enter the soil environment may persist long-term due to degradation difficulty. Similar to the global trend, the majority of microplastics studies in Malaysia have concentrated on the marine environment, while understanding of this pollution in terrestrial environments, particularly in soil, is rather limited. Thus, this study aims to explore microplastics pollution sources in terrestrial environment from Malaysia. So far, microplastics pollution sources were quantified from laundry water as well as secondary sources such as plastics usage in agricultural activities and plastics waste in urban soil. Microplastics in soil also depend on environmental and meteorological factors and breakdown process extent. Thus, extensive studies are needed to further understand the effects of these microplastic particles in soil environment. Further initiatives from scientific communities and actions from regulatory bodies are also needed to minimize this pollution in soil environment.

Keywords Microplastics · Pollution · Sources · Terrestrial environment · Malaysia

---

S. M. Praveena (✉)
Department of Environmental and Occupational Health, Faculty of Medicine and Health Sciences, Universiti Putra Malaysia, UPM Serdang, Serdang, Selangor Darul Ehsan, Malaysia
e-mail: smpraveen@upm.edu.my

S. Suriyanarayanan et al. (eds.), *Combating Plastic Pollution in Terrestrial Environment*,
https://doi.org/10.1007/978-981-96-2343-3_24

24.1 Introduction

Plastic pollution has emerged as one of the world's most difficult environmental issues (United Nations Environment Programme, 2018). More than half of the 8300 million tonnes of plastic produced worldwide—or production—goes directly into the environment after use (Geyer et al., 2017). Microplastic pollution in soil is an increasing issue across the world due to the existence of plastic waste in the environment (PlasticsEurope, 2019). Microplastics, also known as primary and secondary plastics, are defined as plastic particles with a size between 1 μm and 5 mm (United Nations Environment Programme, 2018). Microbeads in cosmetics and personal care products, microfibers from synthetic fabrics, plastic pellets for commercial production of plastic products, and synthetic rubber in tyres are primary microplastics that are produced specifically for a given use (Boucher & Friot, 2017). The breakdown and fragmentation of primary microplastics caused by mechanical action, weathering, and atmospheric movement, as well as tidal and wave motions, are the principal sources of secondary microplastics (Andrady, 2017). The United Nations Environment Programme (2018) emphasized that microplastics in soil and the terrestrial ecosystem serve as a crucial sink for plastic debris that enters the land each year in significantly higher proportions than it does in the aquatic environment. The buildup of microplastics in the soil environment significantly effects global ecological, social, and economic outcomes. Due to the possible hazards to human health, transit and buildup of microplastics from the soil to food are also major concerns (Wang et al., 2020).

In an effort to understand their occurrence through spatial and temporal distribution, sources, analytical methods, transport, pathways, and ecological effects in the aquatic environment, numerous reviews on the pollution caused by microplastics have been published in recent years (Boucher & Friot, 2017; Duis & Coors, 2016; Klingelhöfer et al., 2020; Li et al., 2019; Rezania et al., 2018; Zhang, 2017). While reviews on microplastics in soil by Chae and An (2018), Li et al. (2019), Möller et al. (2020), Wang et al. (2019) and Xu et al. (2019) have given an overview by summarizing the data regarding the occurrence, analytical methods, transport, and ecological risks for the terrestrial environment. The status of microplastic pollution and the extent to which these microplastics have contributed to the environmental dangers and ecological repercussions at the local level require more investigation despite the availability of evaluations on microplastic pollution in soil. Similarly in Malaysia, there has been an increasing number of studies focused on identifying and quantifying microplastics in aquatic environment. Till date, studies on microplastic pollution in Malaysia have mainly focused on aquatic ecosystems (Barasarathi et al., 2011; Khalik et al., 2018; Shuaib Ibrahim et al., 2016). These studies have identified the presence of microplastics in aquatic ecosystems and their potential sources, shedding light on the extent of microplastic pollution in these environments. However, there is still a significant knowledge gap when it comes to the presence and extent of microplastic pollution in Malaysia's terrestrial environment. Soils form the basis of all terrestrial ecosystems and serve many functions, especially in food and other biomass production, nutrients cycling, environmental interaction, biological and biodiversity habitats and maintenance, as well as sources of raw materials. Therefore, it is important to ascertain how microplastics affect the soil ecosystems as they are undoubtedly receiving immeasurably more plastic waste than the ocean environment and ecosystems (Horton et al., 2017). This is concerning, as the introduction of microplastics into soil has the potential to negatively impact soil health and the food chain, with significant implications for human health and the environment.

Thus, current review offers a systematic review of the occurrence of microplastics in soil along with their characteristics (size, types of plastic polymers, shape) and sources from Malaysia. Lastly, the review also aims to provide the potential gaps and future research direction involving microplastics in soil studies in Malaysia. Soil plays a crucial role particularly in Malaysia's agricultural sector, which is also a significant sink for pollutants, including microplastics. Overall, a comprehensive assessment of microplastic pollution in Malaysian soil can contribute to a better understanding of the potential ecological risks and provide sound guidance for future microplastic research in the country.

24.2 Materials and Methods

24.2.1 Systematic Review Protocol Using ROSES

The present review was guided by the Preferred Reporting Items for Systematic Reviews and Meta-Analyses (PRISMA) review protocol. Three main steps are involved in this review protocol, namely, identification, screening and inclusion (Fig. 24.1). The identification process involved searching for relevant keywords based on the research question developed for this review. A Boolean search was applied using terms such as "and" together with "or" for the keywords "soil," "pollution," "microplastics," including "chemical pollutants," and "emerging pollutants,". The online databases utilized for the present study included Scopus (1960–2022), Google Scholar (1960–2022), PubMed (1960–2022), PMC (1965–2020), and EBSCOhost (1965–2022).

A total of 27 papers were identified in this study during the searching process through the Scopus, Google Scholar, Pub Med, PMC and EBSCO host along with other sources (grey literatures, conference proceedings). The screening process proceeded using the total papers identified in the identification process according to the inclusion and exclusion criterias (Table 24.1). The inclusion criterion of the search was that it was conducted between 2012 and March 2023. The reason for starting the searching process from 2012 was because publications on the plastic pollution started to appear from 2012 and the search was limited to March 2023 because the searching process was conducted in March 2023. Relevant studies were searched and screened using the selected keywords and databases. In addition, the reference lists of retrieved publications were evaluated to ensure that all pertinent

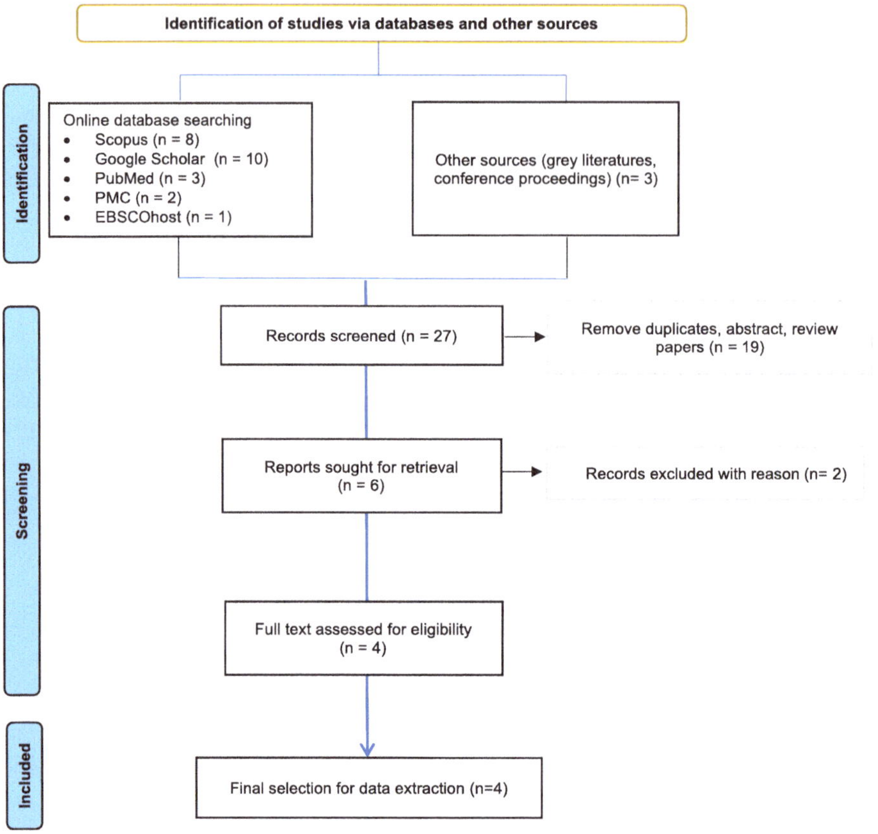

Fig. 24.1 Systematic review process using PRISMA flow diagram

papers were included. Additionally, relevant national and international conference proceedings were manually searched online. In terms of eligibility, a total of 4 full text papers on microplastics in Malaysian soil. These studies were checked by reading through the title and abstract to ensure that they met the inclusion and exclusion criterias. Duplicate papers, papers without quantitative data, and those that focused on the aquatic environment were excluded (Fig. 24.1). All papers and references were managed using Mendeley Reference Manager 2.80.1.

Table 24.1 Inclusion and exclusion criteria in this review

Criteria	Inclusion	Exclusion
Year	2012–2019	Before 2012
Language	English	Non-English
Article type	Journal paper with quantitative data and available in full text	Review papers, conference papers, editorials, letters, only abstract available, focused on the aquatic
Type of study	Cross sectional	Survey, modeling

24.3 Results and Discussion

24.3.1 Microplastics Pollution in Soil from Malaysia

Table 24.1 shows studies on microplastics pollution in soil environment from Malaysia. Till date, only three studies have investigated on microplastics occurrence in soil and one study has provided information of one of the sources on how microplastics are capable to reach soil environment in Malaysia. Overall, it can be seen that microplastics in soil studies in Malaysia are still in its infancy.

Table 24.2 provides information on the soil type and sources related to microplastics occurrence in soil studies conducted in Malaysia. Study by Praveena et al. (2021)demonstrated that microplastic emissions in laundry water from household washing machines in Greater Kuala Lumpur (Malaysia). Direct discharge of laundry water to soil can have unintended consequences, such as the release of microplastics, shredded from synthetic fabrics during the washing and rinsing process. These microplastics can then be deposited in the soil, where they can accumulate and potentially harm soil health and ecosystems. Microplastic content in laundry water released from household washing machines was between 6.9E-3 and 0.183 g/m^3. FTIR spectrums have indicated the presence of polyester, nylon, and acrylic (>80% library match) in these laundry samples. Fabric properties and washing parameters are two factors contributing to microplastics emitted in laundry water.

While studies by Praveena et al. (2023a) have investigated microplastics pollution involving agricultural soil in Klang Valley (Malaysia). The average number of microplastic particles discovered in the agricultural soils ranged from 2.1 to 3.4 particles per kilogram. The highest and lowest number of microplastics particles found in the soil were 6.0 and 1.5 particles per kilogram respectively, which was consistent with the amount of plastic used at these farms. The size of the microplastic particles varied from 16.7 to 1,246 μm, indicating that these plastics waste had undergone significant breakdown. The shapes and colors of the microplastic particles (film, fiber, fragment; black, white, red, and blue) in the soil samples were indicative of the types of plastic products used and unmanaged plastic waste at these farms. Plastic mulching is the main source of film microplastic particles found in these

Table 24.2 Microplastics studies involving soil in Malaysia

Location	Soil type	Sources	References
Greater Kuala Lumpur	–	Laundry water	Praveena et al. (2021)
Klang Valley	Agricultural soil	Plastics products usage in agricultural practice	Praveena et al. (2023a)
Cyberjaya	Urban soil	Tyre wear particles Packaging materials Beverage bottles such as water bottles, reusable bags, and disposable plastic bags	Praveena et al. (2023b)
Melaka Tengah	Urban soil	Tyre wear particles Packaging materials Single use plastics products	Ismail (2022)

agricultural soil samples. While, plastics net usage and a large amount of unmanaged plastics waste observed during sampling have contributed to the total number of microplastic particles in in fiber and fragment shapes, respectively. Plastic usage in agricultural practices along with environmental factors (heavy rainfall, high temperature, sunlight) were found to contribute to plastic products breakdown to smaller particles known as microplastics to soil environment (Horton et al., 2017).

Urban soil studies by Praveena et al. (2022b) in Cyberjaya (Malaysia) and Ismail (2022) in Melaka Tengah (Malaysia) indicated microplastics pollution varies in different landuses. Both of these studies have summarized that the occurrence of microplastics in urban soils depends on external factors such as intensity of human activities, environmental factors and fragmentation process rate in these land use activities. Furthermore, urban soil study by Praveena et al. (2022) in Cyberjaya (Malaysia) has also indicated microplastic particle size is influenced by soil temperature. This finding suggested that high soil surface temperatures may promote plastic aging and break-up in tropical environment such as in Malaysia. The highest percentage of microplastic shapes in the urban soil of Cyberjaya (Malaysia) were fragments (68.7%), followed by films (16.8%), and fibres (14.5%) which were mostly black in color. The presence of black fragments in urban soils is likely due to distance from the source as sampling locations in this study were near major roads and heavy traffic. These black fragments from tyre wear particles came into contact with road dust and entered the soil. The presence of multicolored microplastic particles in fragment shape could be related to the high amount of uncontrolled plastic waste disposed of in these land uses, which subsequently breaks down into smaller fragments (Du et al., 2020). Acrylonitrile butadiene styrene (ABS), polyethylene glycol terephthalate (PET), polypropylene (PP), polyethylene (PE) and polystyrene (PS) were the predominant polymers found in the urban soil from Cyberjaya (Malaysia). These plastics polymers indicated the potential sources in these urban soil samples from Cyberjaya (Malaysia) are synthetic rubber emitted from tyre wear and packaging materials, beverage bottles such as water bottles, reusable bags, and disposable plastic bags. While, urban soil samples from Melaka Tengah (Malaysia) by Ismail (2022) indicated that the total average of microplastic particle size was 721 μm ± 447.02 μm from six different land uses. Empty land areas hold the longest microplastic particle size (823.89 μm ± 469.91 μm) while dumping site area recorded the shortest microplastic particle size 375.75 μm ± 254.083 μm). Empty land and residential areas have the greatest proportion of black microplastic particle color, whereas other colored microplastic particles color (blue, red, green, yellow) were attributable to a significant number of tyre wear, packaging materials and single-use plastics products. Furthermore, urban soil samples from Melaka Tengah (Malaysia) also recorded the percentage of fragments was the highest (66%), followed by films (13%), lines and pellets (8%) and fibres (5%) with the highest proportion at residential area and the lowest at dumping site land use.

24.3.2 Potential Gaps and Future Research Direction Involving Microplastics in Soil Studies in Malaysia

Overall, studies on microplastics in soil is still in its infancy. These limited studies conducted so far suggest that microplastics are present in soil environments and that more research is needed to better understand their potential impacts and ways to mitigate them. Understanding the alterations in soil structure brought on by the deposition of microplastics could be one area of interest for future research. Healthy plant growth and the control of water and nutrient cycles in the soil depend on the structure of the soil. Future studies can also look into how soil's physical characteristics, such as organic matter level, soil compaction, porosity, and water-holding capacity, are affected by microplastic distribution and fate in soil. Future research should also involve in investigating how microplastics affect plant growth in terms of the amount of nutrients and water available to plants as well as impacts on plant growth and yield. Furthermore, studies could look into how microplastics affect plant growth, including morphology, physiology, and productivity changes. Future studies could explore on the effects of microplastics on the abundance, diversity, and function of soil microbial communities. This could involve the use of high-throughput sequencing techniques to assess changes in microbial community structure and metagenomics to study the functional potential of microbial communities exposed to microplastics. Along with this, toxicological impacts of microplastics on soil organisms and the wider soil ecosystem are also still unknown. Future studies could investigate the toxicity of microplastics and associated additives on soil organisms. This could involve exposing different soil organisms to microplastics and measuring changes in growth, reproduction, and mortality. Additionally, studies could investigate the potential for additive accumulation and transfer within the soil ecosystem.

References

Andrady, A. L. (2017). The plastic in microplastics: A review. *Marine Pollution Bulletin*. https://doi.org/10.1016/j.marpolbul.2017.01.082.

Barasarathi, J., Agamuthu, P., Emenike, C. U., & Fauziah, S.H. (2011, August 18–20). Microplastic abundance in selected mangrove forest in Malaysia. *Proceeding of The ASEAN Conference on Science and Technology 2014*, Bogor, Indonesia (Vol. 4).

Boucher, J., & Friot, D. (2017). *Primary microplastics in the oceans: A Global Evaluation of Sources* (p. 43). IUCN.

Chae, Y., & An, Y. J. (2018). Current research trends on plastic pollution and ecological impacts on the soil ecosystem: A review. *Environmental Pollution, 240*, 387–395. https://doi.org/10.1016/j.envpol.2018.05.008.

Duis, K., & Coors, A. (2016). Microplastics in the aquatic and terrestrial environment: Sources (with a specific focus on personal care products), fate and effects. *Environmental Sciences Europe, 28*, 2. https://doi.org/10.1186/s12302-015-0069-y.

Horton, A. A., Walton, A., Spurgeon, D. J., Lahive, E., & Svendsen, C. (2017). Microplastics in freshwater and terrestrial environments: Evaluating the current understanding to identify the knowledge gaps and future research priorities. *Science of the Total Environment, 586*, 127–141. https://doi.org/10.1016/j.scitotenv.2017.01.190.

Ismail, M. I. (2022). *The occurrence of microplastics in urban soil of Melaka Tengah, Melaka* [Bachelor of Environmental and Occupational Health Thesis]. Universiti Putra Malaysia. Serdang, Malaysia.

Khalik, W. M. A. W. M., Ibrahim, Y. S., Tuan Anuar, S., Govindasamy, S., & Baharuddin, N. F. (2018). Microplastics analysis in Malaysian marine waters: A field study of Kuala Nerus and Kuantan. *Marine Pollution Bulletin, 135*, 451–457. https://doi.org/10.1016/j.marpolbul.2018.07.052.

Klingelhöfer, D., Braun, M., Quarcoo, D., Brüggmann, D., & Groneberg, D. A. (2020). Research landscape of a global environmental challenge: Microplastics. *Water Research, 170*, 115358. https://doi.org/10.1016/j.watres.2019.115358.

Li, J., Song, Y., & Cai, Y. (2019). Focus topics on microplastics in soil: Analytical methods, occurrence, transport, and ecological risks. *Environmental Pollution, 257*, 113570. https://doi.org/10.1016/j.envpol.2019.113570.

Möller, J. N., Löder, M. G. J., & Laforsch, C. (2020). Finding microplastics in soils: A review of analytical methods. *Environmental Science and Technology, 54*, 2078–2090. https://doi.org/10.1021/acs.est.9b04618.

PlasticsEurope. (2019). Plastics—The facts 2019. PlasticsEurope AISBL, Brussels – Belgium. www.plasticseurope.org.

Praveena, S. M., Ishak Hisham, M. A. F., & Nafisah, A. L. (2023a, March). Microplastics pollution in agricultural farms soils: preliminary findings from tropical environment (Klang Valley, Malaysia). *Environmental Monitoring and Assessment*. Under review.

Praveena, S. M., Nik Mohd Zaidi, N. M., & Nafisah, A. L. (2023b, April). Microplastics in urban soils from different land use activities of Cyberjaya (Malaysia): Exploring occurrence, relationships, sources, pollution level. Environmental Pollution. Under review.

Praveena, S. M., SyahiraAsmawi, M., & Chyi, J. L. Y. (2021). Microplastic emissions from household washing machines: Preliminary findings from Greater Kuala Lumpur (Malaysia). *Environmental Science and Pollution Research, 28*, 18518–18522. https://doi.org/10.1007/s11356-020-10795-z.

Rezania, S., Park, J., Md Din, M. F., Mat Taib, S., Talaiekhozani, A., Kumar Yadav, K., & Kamyab, H. (2018). Microplastics pollution in different aquatic environments and biota: A review of recent studies. *Marine Pollution Bulletin, 133*, 191–208. https://doi.org/10.1016/j.marpolbul.2018.05.022.

Shuaib Ibrahim, Y., Azzura Azmi, A., Abdul Shukor, S., Tuan Anuar, S., & Aishah Abdullah, S. (2016). Microplastics ingestion by Scapharca cornea at Setiu Wetland, Terengganu, Malaysia.

Middle-East Journal of Scientific Research, 24, 2129–2136. https://doi.org/10.5829/idosi.mejsr.2016.24.06.23654.
United Nations Environment Programme. (2018). Asia Waste Management Outlook. Economy Division International Environmental Technology Centre. Osaka, Japan
Wang, J., Liu, X., Li, Y., Powell, T., Wang, X., Wang, G., & Zhang, P. (2019). Microplastics as contaminants in the soil environment: A mini-review. *Science of the Total Environment, 691*, 848–857. https://doi.org/10.1016/j.scitotenv.2019.07.209.
Wang, T., Yu, C., Chu, Q., Wang, F., Lan, T., & Wang, J. (2020). Adsorption behavior and mechanism of five pesticides on microplastics from agricultural polyethylene films. *Chemosphere, 244*, 125491. https://doi.org/10.1016/j.chemosphere.2019.125491.
Xu, B., Liu, F., Cryder, Z., Huang, D., Lu, Z., He, Y., Wang, H., Lu, Z., Brookes, P. C., Tang, C., Gan, J., & Xu, J. (2019). Microplastics in the soil environment: Occurrence, risks, interactions and fate–a review. *Critical Reviews in Environment Science and Technology, 3389*, 1694822. https://doi.org/10.1080/10643389.2019.1694822.
Zhang, H. (2017). Transport of microplastics in coastal seas. *Estuarine, Coastal and Shelf Science, 199*, 74–86. https://doi.org/10.1016/j.ecss.2017.09.032.

Chapter 25
The Fate of Microplastics from Terrestrial to Coastal Environments Along the Western Continental Margin of India

Niyati Gopinath Kalangutkar and Shritesh Mhapsekar

Abstract Microplastics, defined as plastic particles smaller than 5 mm, have become a significant environmental problem in both terrestrial and riverine environments worldwide. Recent research has focused on understanding the impact of microplastics in these areas, especially since rivers are major carriers of land-based microplastics to oceans. In this chapter, we review the existing literature on microplastic contamination in the western Indian Margin. While limited studies have provided basic information on the abundance, characteristics (such as shape, size, colour, and polymer types), surface morphology, and risks of microplastics in coastal regions, there is much more to be explored. Microplastics have been found in rivers, atmosphere, sediments, and soil in these areas. Comparing global data on microplastic pollution reveals that contamination levels in the terrestrial and riverine environments of India's western states are particularly severe. These preliminary findings shed light on the sources, effects, fate, and associated risks of microplastics. Further, extensive and in-depth research are required to explore the environmental matrices affected by microplastics in these regions, which will help in policy-making decisions to safeguard the environment.

Keywords Microplastic · Pollution · Terrestrial · River · West coast · India

Abbreviations

MPs	Microplastics
PP	Polypropylene

N. G. Kalangutkar (✉) · S. Mhapsekar
School of Earth, Ocean and Atmospheric Sciences, Goa University, Taleigao Palteau, Goa, India
e-mail: niyati@unigoa.ac.in

Department of Geology, Government College of Arts, Science and Commerce, Sanquelim, Goa, India

S. Suriyanarayanan et al. (eds.), *Combating Plastic Pollution in Terrestrial Environment*,
https://doi.org/10.1007/978-981-96-2343-3_25

PA	Polyamide
PET	Polyethylene terephthalate
PS	Polystyrene
PE	Polyethylene
PVC	Polyvinylchloride
HDPE	High-density polyethylene
LDPE	Low-density polyethylene
NY	Nylon
EVOH	Ethylene vinyl alcohol
PAM	Polyacrylamide
PES	Polyester
PP-PE	Poly(ethylene:propylene:diene) Copolymer
PI	Polyimide
PIP	Polyisoprene
PU	Polyether urethane
PVP	Polyvinyl pyrrolidone
PA	Polyacetylene
AIP	After isolated pandemic
BIP	Before isolated pandemic
MON	Monsoon
POM	Post monsoon
d.s	Dry sediments

25.1 Introduction

Plastics are synthetic polymers that are derived from crude oil and are used in a variety of industries (Lithner et al., 2011a, b). Plastic is used in modern society and contributes significantly to the plastic waste in the environment. The production of plastic exceeded 350 million tonnes in 2015, approximately 200 times more than it did in 1950, as a result of the growing demand (Geyer et al., 2017). Plastic waste is expected to triple by 2060 (Global Plastic Outlook, 2022). An estimated 710 million metric tonnes of plastic litter are expected to enter the environment between 2016 and 2040 (Lau et al., 2020).

Recent studies on microplastics (MPs) concentration indicate that they are present in most environmental components. MPs discovered in aquatic systems could have detrimental effects since they could be inadvertently consumed by creatures that mistakenly believe them to be food or because filter-feeding species might come into contact with MPs (Dowarah et al., 2020). Furthermore, heavy metals may be adsorbed onto the MP's surface, increasing their toxicity (Brennecke et al., 2016). Given that rivers are thought to act as conduits for the annual flow of between 70 to 80% of plastic waste from the terrestrial areas into the ocean as a result of improperly managed plastic waste (Lebreton & Andrady, 2019; Mani & Burkhardt-Holm, 2020).

The hazards of MP pollution and its transfer and accumulation in the environment (terrestrial, aquatic and atmospheric) are being studied globally. This topic is slowly emerging in India. The objective of this comprehensive review is to summarize the research findings on the prevalence of MPs in various components of the Indian environment (terrestrial, riverine and Coastal) along the West Coast of India.

This chapter provides an overview of the shape, size, colour, polymer types and heavy metals along with the sources, distribution, and biological effects of MPs. Because of the scarcity of literature data on MPs in the terrestrial environment, this chapter focuses on plastic particles in soil, in river and its transport into oceans.

25.2 Characteristics of the Microplastics

25.2.1 Particle Shape

Determining the shape of MPs is crucial, as they are categorized into various shapes like fibers, fragments, foam, film, and pellets. This identification is performed using a stereo-zoom microscope and is important because it provides insights into the potential sources of MPs. For example, fibres may originate from textiles or fishing nets, fragments from the breakdown of larger plastics, and films from packaging bags or wrapping materials.

Kalangutkar et al. (2024a, b; 2025) (Fig. 25.1) reported that fibres were dominant in the Chapora and Zuari River and Terekhol Estuary in Goa. In contrast, Gupta et al. (2024a) found that fragments were the most common shape in the Mandovi-Zuari estuary. Also, Sal estuary had fragments as the predominant shape (Kalangutkar et al., 2025). In the Netravati River, Karnataka, fibers were prevalent in water and soil samples, while fragments were more common in sediment samples (Amrutha & Warrier, 2020; Amrutha et al., 2023). Similarly, the Sharavathi River in Karnataka and the Karamana River in Kerala also showed a dominance of fragments in sediments (Amrutha et al., 2022; Krishna et al., 2023). Unnikrishnan et al. (2023) reported a higher concentration of fibers in the Udayvara River Estuary, Karnataka. The Ulhas River in Maharashtra also showed a predominance of fragments (Verma et al., 2022).

Fibres are often sourced from fishing gear, synthetic clothing, ropes, and sewage discharge, while fragments result from the breakdown of larger plastic products. The dominance of fibres is particularly concerning, as many studies have reported their prevalence in fish, increasing the risk to aquatic organisms (Ningrum et al., 2023; Saha et al., 2021a, b).

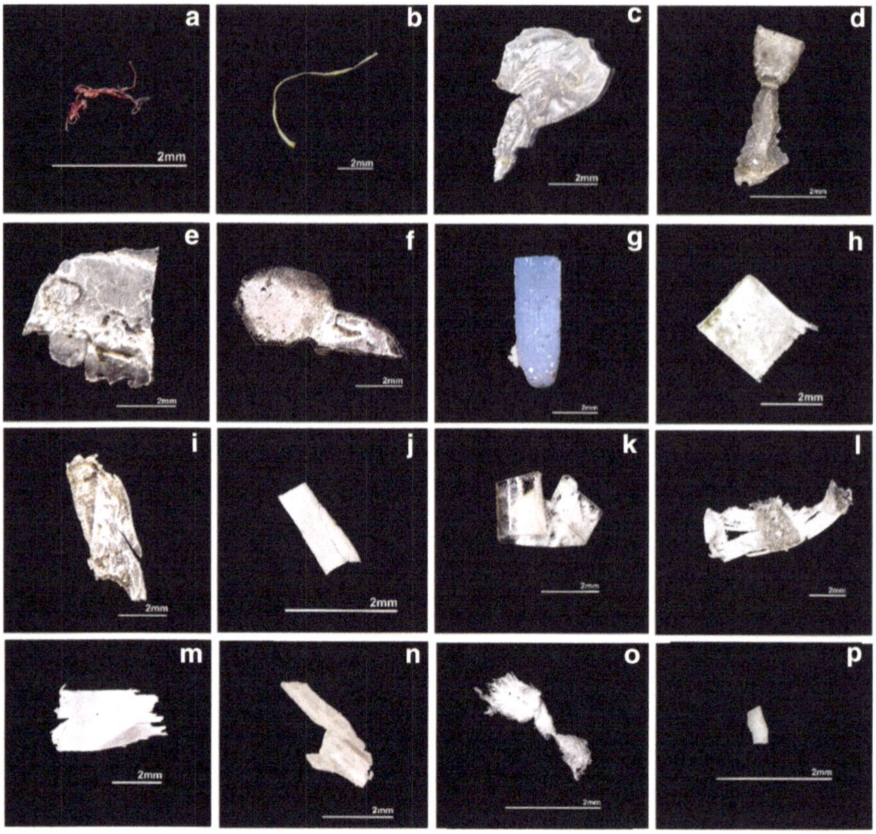

Fig. 25.1 Photo plate of microplastics showing different colours, shapes and type

25.2.2 *Particle Size*

The categorization of MPs by size varies across studies. Further, different authors employ different size category which further restricts the direct comparison with other studies. In the Chapora River, Goa, MPs ranging from 0.1 to 0.3 mm were most abundant (44.57%), followed by 0.3–1 mm (33.71%) and 1–5 mm (21.72%) (Kalangutkar et al., 2024a). Similar trend was observed in Zuari were fibres were more abundant in 0.3–1 mm size range (51.70%) than in the 1–5 mm range (48.30%) (Kalangutkar et al., 2024b). While in Terekhol and Sal estuaries fibres in 1–5 mm range were more dominant (Kalangutkar et al., 2025) (Fig. 25.1). In the Mandovi-Zuari estuary, the size distribution of MPs varied between wet and dry seasons. During the wet season, the dominant size was 100–300 microns (36.1% during AIP and 54% during BIP), while during the dry season, 20–100 microns were most prevalent (68.6% during AIP and 49% during BIP) (Gupta et al., 2024a). In the Sal estuary, Goa, the dominant MP sizes in water and sediment were 100–300 μm and

300 μm–1 mm (Saha et al., 2021a, b). The Netravati River, Karnataka, showed a predominance of larger MPs (1–5 mm) in water and soil samples, while smaller MPs (1–0.3 mm) were more common in sediment samples (Amrutha & Warrier, 2020). In the Sharavathi River, Karnataka, 0.3–1 mm sized MPs were more prevalent (57.83%) compared to 1–5 mm MPs (42.17%) (Amrutha et al., 2022). The Karamana River in Kerala had a higher percentage of MPs around 300 μm, followed by 1 mm and 4.7 mm in size (Krishna et al., 2023). In the Udayvara River Estuary, Karnataka, MPs sized 0.3–1 mm were more dominant than those sized 1–5 mm (Unnikrishnan et al., 2023) (Table 25.1).

The western states of India predominantly have smaller MP particles, possibly indicating their degradation into smaller sizes due to various processes. The presence of smaller MPs in the aquatic environment is particularly concerning, as many studies have reported these particles in organisms. Given the popularity of seafood in these coastal states, there is an increased risk of MPs transferring to humans.

25.2.3 MP Colour

Colour is a commonly studied property of microplastics and the colour of microplastics is identified using a stereo zoom microscope. The colour of microplastics can come from two sources: they may retain the colour of the original mesoplastic material from which they broke down, or their colour might change due to environmental weathering processes. Colour is a significant property of MPs to identify, as coloured MPs may be ingested by organisms (Wright et al., 2013).

MP colour is crucial to consider due to the associated risks. Research has highlighted that organisms often ingest MPs based on colour resemblance to prey, increasing the risk to aquatic life. Additionally, various colourants are added to plastic products for aesthetic purposes, which can leach into the environment, increasing chemical risk. The prevalence of different colours in various regions suggests diverse sources of coloured MPs. Furthermore, some studies have reported the dominance of colourless or white MPs, which might originate from colourless and white plastic products or from the weathering of coloured MPs into colourless or white forms (Wong et al., 2020; Davis and Sims, 1983).

In the surface waters of the Chapora River, Goa, the dominant colours of MPs were red, black, and white (Fig. 25.1) (Kalangutkar et al., 2024a). In Zuari River white and colourless MPs were dominant followed by brown, blue, red, yellow, and green (Kalangutkar et al., 2024b) in Terekhol and Sal estuaries also white and colourless MPs were dominant (Kalangutkar et al., 2025). Gupta et al. (2024a) and Saha et al. (2021a, b) reported black as the most dominant colour in the water and sediments of the Mandovi-Zuari and Sal estuaries in Goa. Transparent MPs were more prevalent in the water samples of the Netravati River, while white MPs dominated the sediment and soil samples (Amrutha & Warrier, 2020). In the Sharavathi River, transparent and white MPs were most abundant (Amrutha et al., 2022). The Karamana River in Kerala and the Udayvara River Estuary in Karnataka showed a dominance

Table 25.1 A summary of research work carried out in the riverine environment of western states of India

State	Location	Type of sample	Average microplastic concentration	Size range	Dominant shape	Polymer types	References
Goa	Chapora River	Surface water	0.25 (±0.13) particles/L	5–1 mm; 1–0.3 mm; 0.3–0.1 mm	Fibre	Polyethylene terephthalate, high-density polyethylene, polypropylene, polystyrene	Kalangutkar et al. (2024a)
	Zuari River	Surface water	0.28 (±0.35) particles/L	5–1 mm; 1–0.3 mm	Fibre	Polyethylene terephthalate, high density polyethylene, polypropylene, and polyacrylamide carboxyl-modified	Kalangutkar et al. (2024b)
	Sal Estuary	Surface water	0.30 particles/L	5–1 mm; 1–0.3 mm	Fragment	Polyethylene, polypropylene, and polyamide	Kalangutkar et al. (2025)
	Terekhol Estuary	Surface water	0.25 particles/L	5–1 mm; 1–0.3 mm	Fibre	Polyethylene, polypropylene, and polyamide	
	Sal Estuary	Water	48 ± 19MP particles/L	5 mm–1 mm; 1 mm–301 µm; 301–101 µm; 100–51 µm; 50–10 µm	Fibre	Polyacrylamide, polyacetylene, ethylene vinyl alcohol (EVOH), polyvinyl chloride (PVC) and polyamide (nylon)	Saha et al. (2021a, b)
		Sediment	3950 ± 930MP particles/kg				

(continued)

Table 25.1 (continued)

State	Location	Type of sample	Average microplastic concentration	Size range	Dominant shape	Polymer types	References
	Mandovi-Zuari Estuary	Water	BIP (Wet season)—0.113 ± 0.034 MPs/m^3	>5 mm; 5–1 mm; 300 μm–1 mm; 300–100 μm; 100–20 μm	Fragment	Polyacrylamide, polyvinyl, alcohol, polyamide, polyvinyl chloride, ethyl vinyl alcohol, polypropylene, polyethylene polyacetylene, polyimide, polyvinyl yrrolidone	Gupta et al. (2024a, b)
			AIP (Wet season)—0.072 ± 0.016 MPs/m^3				
			BIP (Dry season)—0.072 ± 0.016 MPs/m^3				
			AIP (Dry season)—0.010 ± 0.000 MPs/m^3				
		Sediment	BIP (Wet season)—7950 ± 5424.29 MPs/kg		Fragment		
			AIP (Wet season)—2183.33 ± 682.39 MPs/kg				
			BIP (Dry season)—3183.33 ± 2560.79 MPs/kg				
			AIP (Dry season)—1516.67 ± 788.46 MPs/kg				
Karnataka	Netravati River	Water	288 pieces/m^3	5–1 mm; 1–0.3 mm	Fibre	Polyethylene, polyethylene terephthalate, polypropylene, and polyvinyl chloride	Amrutha and Warrier (2020)
		Sediment	96 pieces/Kg		Fragment		
		Soil	84.45 pieces/Kg		Fibre		

(continued)

Table 25.1 (continued)

State	Location	Type of sample	Average microplastic concentration	Size range	Dominant shape	Polymer types	References
	Sharavathi River	Sediment	4.82 ± 13.10 (Premonsoon-2019) 6.39 ± 5.17 (Postmomsoon-2020)	5–1 mm; 1–0.3 mm	Fragment	Polyethylene, polyethylene terephthalate, and polypropylene	Amrutha et al. (2023)
	Udayvara River Estuary	Water	Surface—320.83 (±98.30) particles/m^3 Middle- 514.55 (±352.16) particles/m^3 Bottom- 755.03 (±400.96) particles/m^3	5–1 mm; 1–0.3 mm	Fibre	High-density polyethylene, polyethylene terephthalate, polyethylene, alkyd varnish, polypropylene, and polychloroprene	Unnikrishnan et al. (2023)
Kerala	Karamana River	Soil	671 ± 3.45 pieces/kg	4.7 mm; 1 mm; 0.3 mm	Fragment	Low-density polyethylene (LDPE), polypropylene (PP), polystyrene (PS), polyethylene terephthalate (PET), PES, and high-density polyethylene (HDPE)	Krishna et al. (2023)
Maharashtra	Ulhas River	Sediment	40 ± 5.77 particles/kg (dry weight) to 600 ± 122 particles/kg	–	Fragment	Low-density polyethylene (LDPE), high-density polyethylene (HDPE), polypropylene (PP), polystyrene (PS), polyethylene (PE), polyester, polyether urethane, polyethylene terephthalate (PET), polyvinyl chloride (PVC), and nylon	Verma et al. (2022)

of white-coloured MPs (Krishna et al., 2023; Unnikrishnan et al., 2023). In the Ulhas River basin, blue, colourless, and green MPs were predominant (Verma et al., 2022) (Table 25.1).

25.2.4 Polymer Types

The study of polymers is crucial due to their associated ecological risks and their potential to identify the sources of MPs in the environment. Kalangutkar et al. (2024a) identified PET, HDPE, polypropylene, and polystyrene in the Chapora River, Goa, with PET being the most prevalent (Fig. 25.2). The polymers identified in Zuari Rivers were PET, HDPE, PP, and PAM (Kalangutkar et al., 2024b). Where as in Terekhol and Sal estuaries PE, PP, and PA are the prominent polymers (Kalangutkar et al., 2025). Gupta et al. (2024a) found polyacrylamide, polyvinyl alcohol, polyamide, polyvinyl chloride, ethyl vinyl alcohol, polypropylene, polyisoprene, polyacetylene, polyimide, polyethylene, and polyvinyl pyrrolidone in the surface waters and sediments of the Mandovi-Zuari estuary during both wet and dry seasons of AIP and BIP. Saha et al. (2021a, b) detected PAM, polyacetylene, polyimide, and ethylene vinyl alcohol (EVOH) in water samples from the Sal estuary, with polyacetylene, PAM, EVOH, and polyimide being abundant in sediment samples. In Karnataka, the Netravati River showed the presence of polyethylene, polyethylene terephthalate, polypropylene, and polyvinyl chloride in water, sediment, and soil samples (Amrutha & Warrier, 2020). The Sharavathi River sediment samples primarily contained polyethylene, polyethylene terephthalate, and polypropylene (Amrutha et al., 2022). The Karamana River in Kerala had LDPE, PP, PS, PET, PES, and HDPE (Krishna et al., 2023). The Udayvara River Estuary in Karnataka featured high-density polyethylene, polyethylene terephthalate, polyethylene, alkyd varnish, polypropylene, and polychloroprene (Unnikrishnan et al., 2023). In the Ulhas River, Mumbai, various polymers including low-density polyethylene (LDPE), high-density polyethylene (HDPE), polypropylene (PP), polystyrene (PS), polyethylene (PE), polyester, polyether urethane, polyethylene terephthalate (PET), polyvinyl chloride (PVC), and nylon were found (Verma et al., 2022). Most studies in the western states of India have reported common polymer types across different regions, reflecting the widespread use of certain plastic products. The study of polymer types is particularly important because different polymers exhibit varying toxicity levels, thereby increasing the potential ecological risk in the environment (Table 25.1).

25.2.5 MPs as Heavy Metal Carriers

Microplastics are reported as carriers for heavy metals and exhibit diverse interactive effects (Liu et al., 2021). Heavy metals were found in MPs from various regions, including North Atlantic subtropical gyre, São Paulo State, southwest England, and

(a)
(b)
(c)

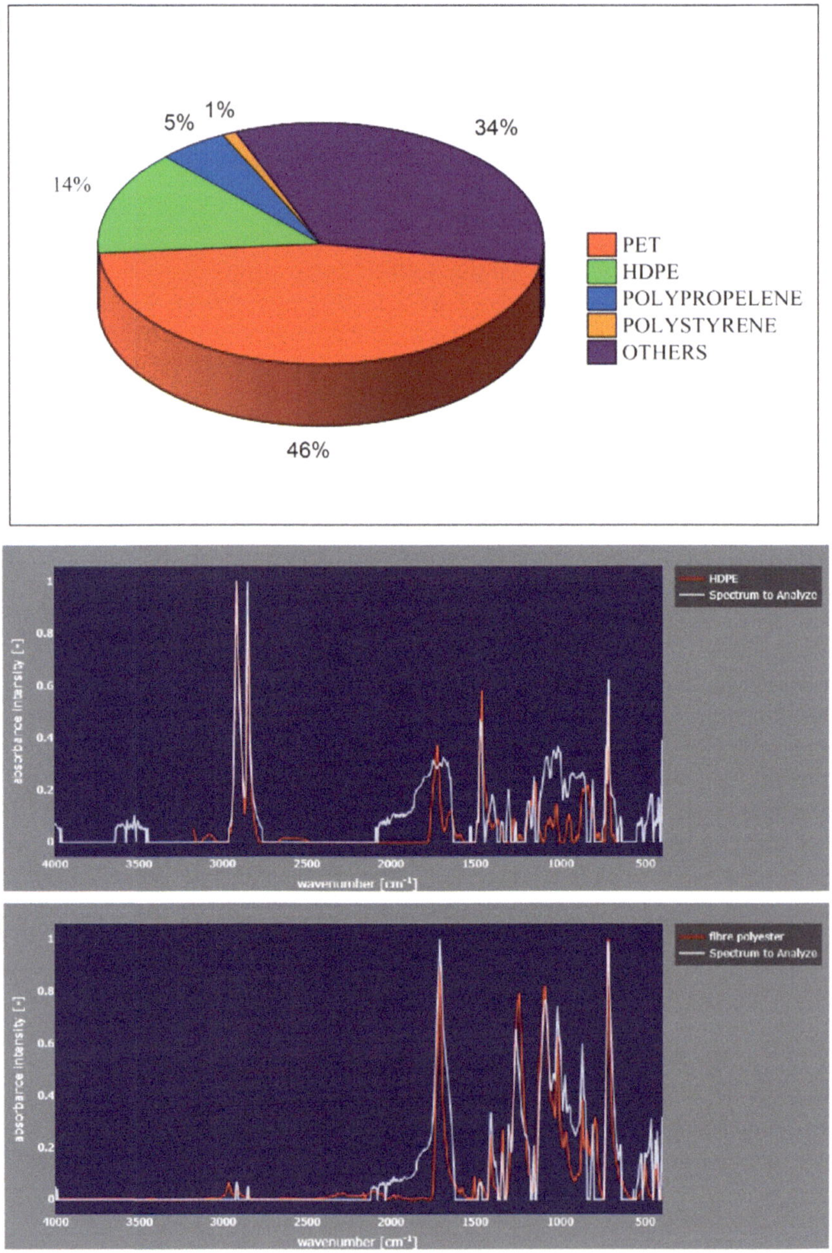

Fig. 25.2 Kalangutkar et al. (2024a). **a** The polymer composition pie chart shows the total percentage of polymer types in the Chapora River. **b** and **c** Compositional spectra of MPs. https://doi.org/https://doi.org/10.1007/s10661-024-12587-1

Western Europe (Massos & Turner, 2017; Prunier et al., 2019; Turner et al., 2019; Vedolin et al., 2018).

Research indicates that biofilms associated with microplastics (MPs) can alter their physical and chemical properties, which in turn impacts the adsorption of chemical pollutants, including heavy metals (Tu et al., 2020). The factors responsible for the interaction of MPs and heavy metal include aging, temperature, pH, contact time, ionic strength, and particle size. Increase in the surface area of the MPs can increase the absorption capacity of heavy metals on the surface of MPs. When pH is less than a certain value, MPs will not interact with heavy metals (Tang et al., 2021). The adsorption of heavy metals by MPs is primarily influenced by chemisorption mechanisms, with microbial biofilm playing a crucial role in determining MP surface properties, alongside physical and chemical environmental factors (Liu et al., 2021). Tu et al. (2020) discovered that the formation of biofilms decreases the hydrophobicity of the microplastic (MP) surfaces. Additionally, research has shown that biofilm development can enhance the adsorption of heavy metals and increase concentrations of these metals on MPs as the biofilms mature (Qi et al., 2021; Richard et al., 2019).

SEM-EDS analysis is used to examine the surface morphology and elemental composition of MPs. In the Chapora River, SEM-EDS analysis revealed signs of degradation on the MP surfaces (Fig. 25.3) and detected various heavy metals. These heavy metals likely adsorbed from the surrounding environment, as previous studies have reported similar findings in the Chapora River (Kalangutkar et al., 2024a). Similarly, MPs from the Udayvara River estuary in Karnataka exhibit surface features indicating prolonged exposure to the coastal environment, with traces of zinc, arsenic, cadmium, cobalt, and other elements (Unnikrishnan et al., 2023). In the Ulhas River, Mumbai, SEM-EDS analysis showed that the MPs' surface morphology indicates disintegration and the presence of various elements (Verma et al., 2022). Weathered MPs with pits, fractures, or cracks may allow contaminants to adhere, acting as carriers for these pollutants (Shruti et al., 2019).

25.2.6 Occurrence of MPs in Terrestrial Environment

Microplastics seep into our terrestrial systems as a result of our growing daily consumption of plastic products, which also ages and fragments over time. These MPs can originate from poorly managed plastic waste, tire wear, building fragmentation, and building supplies or aeolian transportation from farther-reaching sources (Patchaiyappan et al., 2021). Significant amounts of MPs were found in the dust samples from both indoor (Patchaiyappan et al., 2021) and outdoor (Zhang et al., 2020) systems, suggesting that these particles are widely distributed.

Microplastics in soil pose a significant threat to the agricultural ecosystem, entering the environment through manufactured particles or macroplastic breakdown. They can be transported laterally and vertically through soil fauna. A meta-analysis of recent studies reveals that major sources of MPs include sludge, wastewater, and mulch film. The impact on soil properties depends on soil type, size, concentration,

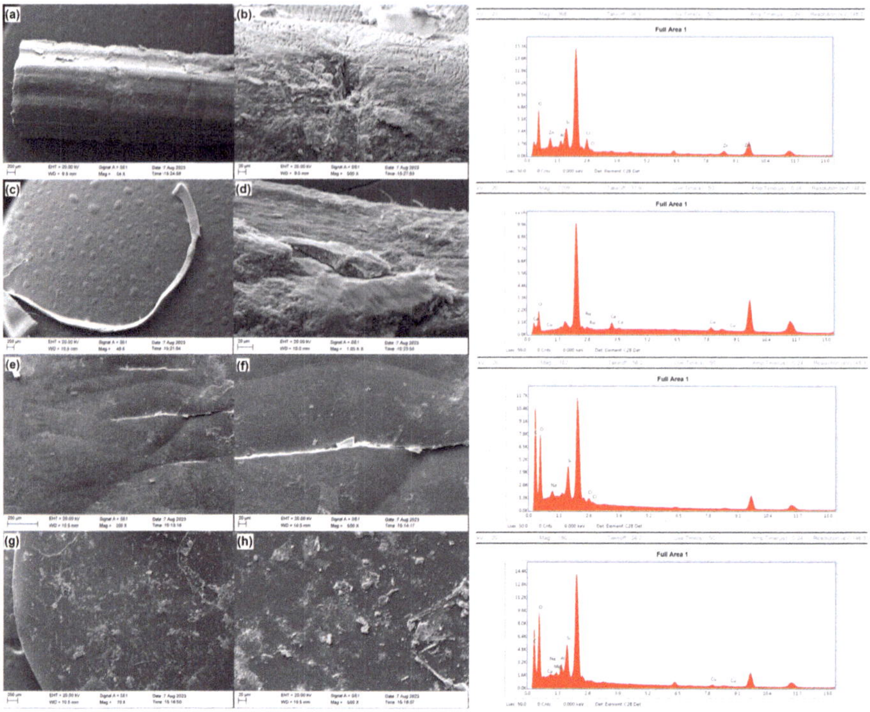

Fig. 25.3 Microplastic pollution in the Chapora River, Goa, Southwest India: spatial distribution and risk assessment adopted from Kalangutkar et al. (2024a). https://doi.org/https://doi.org/10.1007/s10661-024-12587-1

and polymer type. MPs also affect soil biota and human health through food chains. MPs impact various soil characteristics by altering the soil physicochemical properties, including porosity, enzymatic activities, microbial activities, plant growth, and yield (Dhevagi et al., 2022). Understanding the sources, transport, and fate of the MPs in soils is crucial to mitigate their negative effects and safeguard agricultural productivity in these regions. Microplastics studies in soils and terrestrial environment are very limited in India. More studies are required to highlight these issues.

Seth and Shriwastav (2018) studied the commercial sea salt from Kerala, Maharashtra and Gujarat. It was identified that concentration of MPs ranges from 56 to 103 items/L, with MP size 0 to 5000 μm, and the MP types identified were fibres and fragments.

Sediment samples from four locations along the Sabarmati River revealed high MP pollution dynamics, with 134.53 mg/kg and 581.706 mg/kg dry weight of sediment (Patel et al., 2020). Another study in the soil of Diamond City, Surat, Gujarat, India indicated ecological risks due to accumulation of MPs sources like agricultural plastic films, sewage sludge, and prolonged use of organic fertilizer with an average abundance of 103.2 ± 57.28 particles 100 g^{-1}. Over 90% of all MPs were made of polyethylene (70.19%), polypropylene (17.53%), and poly(ethylene terephthalate)

(6.92%) polymers. The white/transparent coloured MPs having a size of about 2–5 mm are predominant. MPs were found in all sampling sites, posing a moderate to medium ecological risk due to their abundance, type, and toxicity (Bhavsar et al., 2024).

There have been reports of MP pollution in the sediments of 20 Gujarat sandy beaches. A total of 1158 MPs were reported. Polypropylene, polyethylene and polystyrene were the major polymer types. Mean abundance of the MPs ranged from 1.4 to 26 MPs/kg. Fisheries and tourism were the main causes of MPs prevalent along the Gujarat coast (Rabari et al., 2022).

There are several studies leading to MPs contamination being carried out in Kerala. Some of the studies are listed out in this chapter. Studies in Kochi which is commercial and industrial city in Kerala, the average microplastic deposition flux of 21.27 particles m^{-2} d^{-1}, with predominant 200–500 μm particles. Illegal waste dumping and the lack of scientifically managed landfills could be a major source of atmospheric microplastics. Wind may also transport microplastics from the landfill to the ocean (Kannankai & Devipriya, 2024).

A study on sediments from harbours, coastal fishing villages/residential beaches, tourist beaches, and undisturbed coastal areas along of southernmost India at Kanyakumari found that increasing urbanization and anthropogenic activities have led to a significant amount of plastic debris in coastal and marine ecosystems. A total of 343 particles (67% fiber and 33% fragment) from eight different stations were studied. Tourist beaches had the highest concentration (150 particles/50 g d.s.), followed by the harbours (99 particles/50 g d.s.) (dry sediments (d.s.). The study estimated the abundance of microplastics in beach sediments, revealing that tourist beaches had the highest concentration, followed by harbours (Sundar et al., 2021).

25.2.7 Occurrence of MPs in Riverine Environment

Microplastic contamination has been documented in river environments worldwide. However, in India, research has been focused more on beach sediments, and water, with riverine studies receiving less attention. Recently, interest in India's riverine environments has increased as studies reveal that rivers transport land-based MPs to the ocean. Rivers can acquire MPs from atmospheric deposition, surface runoff, untreated sewage, industrial processes, agricultural activities, and sewage treatment plants, which then transport MPs to larger rivers or the sea. Some rivers and estuaries have reported very high MP concentrations, especially those flowing through urbanized areas, identified as major channels for transporting land-based MPs to the ocean. Globally, recent studies have explored MP concentrations and associated risks. In India, research on MPs and their ecological impacts has been conducted in various regions. While western Indian states have focused more on MP pollution in beach sediments and water, riverine environments have recently gained attention, leading to new research in this area. The studies on riverine environment require more attention from the researchers, as this can transfer the land-based plastic or MPs into the river due to surface runoff, and which may be further transferred to ocean or sea.

The riverine environment may be polluted with MPs, but relevant studies are relatively scarce. Table 25.1 summarizes some of the conducted studies. A recent study by Kalangutkar et al. (2024a) in the Chapora River found 0.25 MPs/L in surface water, attributing the dominance of MPs to fishing, tourist littering, and improper plastic disposal. The average abundance of MPs in the surface water samples Zuari River was 0.28 particles/L (Kalangutkar et al., 2024b). While, Terekhol and Sal estuaries reported 0.25 particles/L and 0.30 particles/L, respectively (Kalangutkar et al., 2025). Gupta et al. (2024a) studied the Mandovi-Zuari estuary, examining MP contamination in surface water and sediments both before and after the COVID-19 isolation period. In 2019, before isolation, wet season water samples showed an average MP concentration of 0.113 ± 0.034 MPs/m^3, which decreased to 0.072 ± 0.016 MPs/m^3 after isolation. During the dry season, MP concentration dropped from 0.072 ± 0.016 MPs/m^3 before isolation to 0.010 ± 0.000 MPs/m^3 after. Sediment analysis revealed 1516.67 ± 788.46 MPs/kg (dry weight) in the dry season and 2183.33 ± 682.39 MPs/kg in the wet season post-isolation, compared to 3183.33 ± 2560.79 MPs/kg in the dry season and 7950 ± 5424.29 MPs/kg in the wet season pre-isolation. These changes were linked to reduced fishing, tourism, and domestic activities during isolation. Saha et al. (2021a, b) found MP contamination in water, sediment, and biota in the Sal Estuary. The average MP abundance was 48 ± 19 MPs/L in water and 3950 ± 930 MPs/kg in sediment. Seafood analysis detected MPs in the soft tissue of shellfish, with concentrations of 4 ± 2 MPs/g in Crassostrea sp., 3.2 ± 1.8 MPs/g in Pernaviridis, and 0.7 ± 0.3 MPs/g in Paphiamalbarica (Saha et al., 2021a, b). This contamination was attributed to fishing nets, wastewater discharges, domestic sewage, and discarded plastic materials.

Amrutha and Warrier (2020) studied the Netravati River in Karnataka, from its source to its mouth at the Arabian Sea, finding MPs in surface water, sediments, and soil. The average concentration of MPs was 288 pieces/m^3 in water, 96 pieces/kg in sediment, and 84.45 pieces/kg in soil, originating from textile fibers, sacks, fish lines, illegal plastic dumping, street littering, and plastic bags. MP levels varied along the river due to population distribution, topography, land use patterns, and proximity to pilgrimage centers. Amrutha et al. (2023) examined the impact of the COVID-19 pandemic on MP variations in a tropical river in southern India. Mean MP concentrations in water samples were 288 ± 591.23 (MON19), 15.06 ± 15.46 (POM19-20), 36.86 ± 23.12 (MON20), and 70.5 ± 61.22 MPs/m^3 (POM20-21). The decrease in MP concentration during MON20 compared to MON19 was attributed to reduced rainfall and COVID-19 lockdowns. Amrutha et al. (2022) reported MP concentrations in the Sharavathi River sediments ranging from 4.82 ± 13.10 (PRM-2019) to 6.39 ± 5.17 (POM-2020). The distribution of MPs in these sediments was influenced by population distribution, land use, geomorphological features, hydrodynamic parameters, and processes such as precipitation, surface runoff, and sedimentation.

Unnikrishnan et al. (2023) studied the vertical distribution of MPs in the Udayvara River Estuary in Karnataka, reporting mean concentrations of 320.83 (±98.30),

514.55 (±352.16), and 755.03 (±400.96) particles/m^3 in surface, middle, and bottom waters, respectively. This pollution was linked to the harbour, fishing, industrial activities, southwest monsoon rains, and tidal currents.

Soil samples from the Karamana River in Kerala showed an average MPs concentration of 671 ± 3.45 pieces/kg near the city center (Krishna et al., 2023).

A study focused on microplastic distribution in Alappuzha Mud bank, in Kerala showed varied concentrations spatially and temporally, with higher levels during the Mud bank period. The Fish showed increased microplastic ingestion during the Mud bank phase. White polypropylene fragments of 1–5 mm were the dominant microplastics found in the environment (sediments and surface waters) and fish samples (James et al., 2022).

Studies of microplastics in sediments from estuarine environment from Kavvayi and Kumbla backwaters of Malabar Coast, Kerala, India were carried out by Padmachandran et al., (2023). Their studies indicated microplastics with a size range of up to 1 μm, MPs were found in all samples, with average abundances of 99.5 ± 69.43 particles/kg in the Kavvayi backwaters and 96.57 ± 29.96 particles/kg in the Kumbla backwaters, respectively.

Studies show that microplastics in coastal soils of Kasargod, an urban town in India increased significantly in 2021 compared to 2019, with flakes, fibres, and films being the most abundant types. Polymers like polypropylene and high- and low-density polyethylene also showed increased abundance. The increased use of surgical masks during COVID-19 may have contributed to microplastic contamination (Reethu et al., 2023).

Sediment samples from the Ulhas River in Mumbai, Maharashtra, varied from 40 ± 5.77 particles/kg (dry weight) to 600 ± 122 particles/kg, with most MPs originating from secondary sources (Verma et al., 2022).

25.2.8 MPs Transfer from Riverine to Coastal Environment

Microplastic particles from the land eventually reach the marine environment due to catchment area dynamics, as well as erosion and transport processes (Hale et al., 2020). Rivers serve as major pathways transporting land-based microplastics (MPs) into seas and oceans. Catchment areas, the regions where rivers and streams collect runoff, play a crucial role in moving microplastics from land to water. Urban runoff, agricultural activities, and wastewater effluents are primary sources of microplastic pollution in these areas. Additionally, synthetic fibers from washing clothes enter rivers through wastewater treatment plants, which often fail to completely remove these particles (Browne et al., 2011).

Erosion significantly mobilizes microplastics from soils and sediments into river systems. Heavy rainfall and flooding events exacerbate this process, dislodging microplastics and transporting them via surface runoff into streams and rivers. According to Hurley et al. (2018), significant flooding events can dramatically increase the transfer of microplastics from riverbeds into the water column, facilitating their downstream movement. The quantity of MPs downstream multiplied by the river's monthly flow determines the monthly surface load of plastic that the river carries (Amrutha et al., 2023).

Once in the river system, microplastics are subject to various transport mechanisms. They can be carried in suspension, settle into sediments, or adhere to organic matter. The flow dynamics of rivers, including current velocity and turbulence, influence the distribution and eventual deposition of microplastics. For example, a study by Lebreton et al. (2017) highlighted that large rivers with high discharge rates, such as the Yangtze and Ganges, are major conduits for transporting microplastics from inland areas to the oceans. Naik et al. (2019) also highlighted that ballast water serves as a significant pathway for the distribution of contaminated microplastics in global environmental matrices.

As rivers discharge into coastal waters, microplastics are deposited in estuarine and marine environments. The mixing of freshwater and seawater, along with tidal actions, further disperses these particles. Some studies suggest that not all MPs are transported downstream; some may sink to the riverbed and accumulate in the sediment. Gupta et al. (2024b) studied continental shelf sediments from the eastern Arabian Sea shelf (west coast of India) and reported significant concentrations of MPs in the sediments. They highlighted that during rainfall, land-based particles are transported from rivers, canals, and sewage drains into the sea.

Rathore et al. (2024) collected wastewater samples to understand the land-based MP discharge through sewers to the ocean. They reported 79 ± 20.7–338 ± 7.1 MPs/L in the wastewater samples, which eventually make their way into the downstream region of the Mandovi estuary, possibly further transporting into the Arabian Sea. Additionally, Amrutha et al. (2023) estimated that approximately 1.12×10^9 MP pieces per day were released from the Netravathi River, Karnataka into the Arabian Sea during the 2019 monsoon. Studies on rivers in the western states of India have found varying levels of MP contamination, which are further carried into the Arabian Sea. Increased rainfall intensifies the amount of MPs entering rivers, as surface runoff increases, leading to a higher abundance of MPs being washed into the Arabian Sea (Amrutha et al., 2023).

These studies indicate that land-based MPs are washed into river systems during the monsoon season. These MPs, carried to rivers from the land, could be either primary MPs or secondary MPs. This proves that land-based MPs are a major source of MPs in riverine environments which are further transported to ocean and calls for effective management of plastic waste. Periodic studies on all rivers are necessary to better understand MP contamination and its effects, leading to the formulation of policies to reduce contamination levels in riverine environments and their impact on aquatic life.

25.3 Conclusion

Microplastic contamination has permeated both terrestrial and riverine environments worldwide. These tiny particles, less than 5 mm in size, have infiltrated rivers, lakes, soils, and sediments from multiple sources. The dynamic processes operating in the terrestrial and riverine environments are expected to influence the distribution, transport, and fate of these MPs. This review compiles several potential sources and pathways for microplastics (MPs) in the terrestrial and riverine environments of India's western states. Research from these areas consistently shows the presence of microplastics in varying amounts in the air, water, sediments, and soils, highlighting their significant impacts on ecosystems and the people who depend on them. Common sources of microplastics include fishing nets, synthetic clothing, tourism, and mismanaged plastic waste. The prevalence of smaller microplastics in study areas suggests their breakdown into finer particles over time, increasing the risk of ingestion by organisms. The risk associated with microplastics is also influenced by their colour, with studies in these regions reporting a mix of colourless, transparent, and coloured microplastic particles. The most common shape found is fibre, which is particularly risky for aquatic organisms, as studies have highlighted it as the most commonly found shape in the bodies of organisms. Some studies using SEM-EDS techniques have detected various metals on the surface of microplastics, likely adsorbed from the surrounding environment.

Despite global awareness of the effects of microplastics, research on their presence in river environments, especially in India's western states, is still limited. More comprehensive studies are necessary to understand the extent of contamination and the associated risks. Effective conservation strategies require a diversified approach to studying microplastic contamination in these regions. Given the coastal location and reliance on seafood, it is crucial to understand the impact of microplastics on aquatic organisms and the broader ecosystem.

Acknowledgements We acknowledge Drs. Ranadhir Mukhopadhyay and Sridhar D. Iyer for encouraging us to compile our studies as a chapter to this book and for useful suggestions. We thank Dr. Amitava Bandopadhyay for the invitation to contribute to this chapter. We also acknowledge the support of Goa University and the Government College of Art, Science and Commerce, Sanquelim, during the preparation of the chapter.

References

Amrutha, K., & Warrier, A. K. (2020). The first report on the source-to-sink characterization of microplastic pollution from a riverine environment in tropical India. *Science of the Total Environment, 739*, 140377. https://doi.org/10.1016/j.scitotenv.2020.140377.

Amrutha, K., Shajikumar, S., Warrier, A. K., Sebastian, J. G., Sali, Y. A., Chandran, T., Sivadas, S., Naik, R., Amrish, V. N., Kumar, A., & Unnikrishnan, V. (2022). Assessment of pollution and risks associated with microplastics in the riverine sediments of the Western Ghats: A heritage site in southern India. *Environmental Science and Pollution Research International, 30*(12), 32301–32319. https://doi.org/10.1007/s11356-022-24437-z.

Amrutha, K., Warrier, A. K., & Rangel-Buitrago, N. (2023). Did the COVID-19 pandemic play a role in the spatial and temporal vriations of microplastics? Evidence from a tropical river in southern India. *Marine Pollution Bulletin, 192*, 115088. https://doi.org/10.1016/j.marpolbul.2023.115088.

Bhavsar, P. S., Chovatiya, B. V., Kamble, S. B., & Gore, A. H. (2024). Extraction and analysis of microplastics in the soil of Diamond City, Surat (Gujarat, India): Ecological risk, pollution indices, and greenness evaluation. *ACS Agricultural Science & Technology, 4*(5), 614–625.

Brennecke, D., Duarte, B., Paiva, F., & Canning-Clode, J. (2016). Microplastics as vector for heavy metal contamination from the marine environment. *Estuarine Coastal and Shelf Science, 178*, 189–195. https://doi.org/10.1016/j.ecss.2015.12.003.

Browne, M. A., Crump, P., Niven, S. J., Teuten, E., Tonkin, A., Galloway, T., & Thompson, R. (2011). Accumulation of microplastic on shorelines worldwide: Sources and sinks. *Environmental Science & Technology, 45*(21), 9175–9179.

Davis, A., & Sims, D. (1983). *Weathering of Polymers*. Applied Science Publishers Ltd.

Dhevagi, P., Poornima, R., Keerthi Sahasa, R. G., Ramya, A., Karthika, S., & Sivasubramanian, K. (2022). The crux of microplastics in soil—A review. *International Journal of Environmental Analytical Chemistry, 104*, 1–33. https://doi.org/10.1080/03067319.2022.2148528.

Dowarah, K., Arunkumar, P., Thirunavukkarasu, C., Jayakumar, S., & Devipriya, S. P. (2020a). Quantification of microplastics using Nile Red in two bivalve species Perna viridis and Meretrix meretrix from three estuaries in Pondicherry, India and microplastic uptake by local communities through bivalve diet. *Marine Pollution Bulletin, 153*, 110982. https://doi.org/10.1016/j.marpolbul.2020.110982.

Geyer, R., Jambeck, J. R., & Law, K. L. (2017). Production, use, and fate of all plastics ever made. *Science Advances, 3*, e1700782. https://doi.org/10.1126/sciadv.1700782.

Global Plastics Outlook. (2022). OECD. https://doi.org/10.1787/aa1edf33-en.

Gupta, P., Saha, M., Suneel, V., Rathore, C., Ray, D., & Naik, A. (2024a). The consequences of reduced anthropogenic activities during the COVID-19 pandemic on microplastic abundance in a tropical estuarine region: Goa, India. *Science of the Total Environment, 912*, 169041. https://doi.org/10.1016/j.scitotenv.2023.169041.

Gupta, P., Saha, M., Suneel, V., Rathore, C., Chndrasekhararao, A., Gupta, G., & Junaid, C. (2024b). Microplastics in the sediments along the eastern Arabian Sea shelf: Distribution, governing factors and risk assessment. *Science of the Total Environment, 910*, 168629. https://doi.org/10.1016/j.scitotenv.2023.168629.

Hale, R. C., Seeley, M. E., La Guardia, M. J., Mai, L., & Zeng, E. Y. (2020). A global perspective on microplastics. *Journal of Geophysical Research. Oceans, 125*(1), e2018JC014719. https://doi.org/10.1029/2018jc014719.

Hurley, R., Woodward, J., & Rothwell, J. J. (2018). Microplastic contamination of river beds significantly reduced by catchment-wide flooding. *Nature Geoscience, 11*(4), 251–257. https://doi.org/10.1038/s41561-018-0080-1.

Jambeck, J. R., Geyer, R., Wilcox, C., Siegler, T. R., Perryman, M., Andrady, A., Narayan, R., & Law, K. L. (2015). Plastic waste inputs from land into the ocean. *Science, 347*(6223), 768–771. https://doi.org/10.1126/science.1260352.

James, K., Kripa, V., Vineetha, G., Padua, S., Prema, D., Babu, A., … & Joseph, R. V. (2022). Microplastics in the environment and in commercially significant fishes of mud banks, an ephemeral ecosystem formed along the southwest coast of India. *Environmental Research, 204*, 112351.

Kalangutkar, N., Mhapsekar, S., Redkar, P., Valsan, G., & Warrier, A. K. (2024a). Microplastic pollution in the Chapora River, Goa, Southwest India: Spatial distribution and risk assessment. *Environmental Monitoring and Assessment, 196*(5). https://doi.org/10.1007/s10661-024-12587-1.

Kalangutkar, N., Mhapsekar, S., Abhaykrishna, M. M., & Pednekar, P. (2024b). Microplastic pollution in the surface waters of the Zuari River, Goa, India. *Environmental Monitoring and Assessment, 196*(11). https://doi.org/10.1007/s10661-024-13181-1

Kalangutkar, N., Mhapsekar, S., & Rajagopal, P. (2025). Comparative assessment of microplastic pollution in Terekhol and Sal estuaries, Goa, India. *Environmental Earth Sciences, 84*(1). https://doi.org/10.1007/s12665-024-12042-x

Kannankai, M. P., & Devipriya, S. P. (2024). Atmospheric microplastic deposition in a coastal city of India: The influence of a landfill source on monsoon winds. *Science of the Total Environment, 908*, 168235.

Krishna, A. B. S., Madhu, M., & Jayadev, A. (2023). Investigation of microplastics and microplastic communities in selected river and lake basin soils of Thiruvananthapuram District, Kerala, India. *Environmental Monitoring and Assessment, 196*(1), 66. https://doi.org/10.1007/s10661-023-12219-0.

Lau, W. W., Shiran, Y., Bailey, R. M., Cook, E., Stuchtey, M. R., Koskella, J., … & Palardy, J. E. (2020). Evaluating scenarios toward zero plastic pollution. *Science, 369*(6510), 1455–1461.

Lebreton, L., & Andrady, A. L. (2019). Future scenarios of global plastic waste generation and disposal. *Palgrave Communications, 5*(1), 1–11. https://doi.org/10.1057/s41599-018-0212-7.

Lebreton, L. C., Van Der Zwet, J., Damsteeg, J. W., Slat, B., Andrady, A., & Reisser, J. (2017). River plastic emissions to the world's oceans. *Nature Communications, 8*(1), 15611. https://doi.org/10.1038/ncomms15611.

Lithner, D., Larsson, Å., & Dave, G. (2011a). Environmental and health hazard ranking and assessment of plastic polymers based on chemical composition. *Science of the Total Environment, 409*(18), 3309–3324.

Lithner, D., Larsson, Å., & Dave, G. (2011b). Environmental and health hazard ranking and assessment of plastic polymers based on chemical composition. *Science of the Total Environment, 409*(18), 3309–3324. https://doi.org/10.1016/j.scitotenv.2011.04.038.

Liu, S., Shi, J., Wang, J., Dai, Y., Li, H., Li, J., … & Zhang, P. (2021). Interactions between microplastics and heavy metals in aquatic environments: A review. *Frontiers in Microbiology, 12*, 652520.

Mani, T., & Burkhardt-Holm, P. (2020). Seasonal microplastics variation in nival and pluvial stretches of the Rhine River—From the Swiss catchment towards the North Sea. *Science of the Total Environment, 707*, 135579. https://doi.org/10.1016/j.scitotenv.2019.135579.

Massos, A., & Turner, A. (2017a). Cadmium, lead and bromine in beached microplastics. *Environmental Pollution, 227*, 139–145.

Naik, R. K., Naik, M. M., D'Costa, P. M., & Shaikh, F. (2019). Microplastics in ballast water as an emerging source and vector for harmful chemicals, antibiotics, metals, bacterial pathogens and HAB species: A potential risk to the marine environment and human health. *Marine Pollution Bulletin, 149*, 110525.

Ningrum, P. T., Negoro, A. H. S., Indahyani, D. E., Kusnadi, N., & Nurdiansyah, Y. (2023). Microplastic contamination in marine fish and shells in the coastal areas of Jember Regency, Indonesia. *Jurnal Ilmiah Perikanan Dan Kelautan/jurnal Ilmiah Perikanan Dan Kelautan, 15*(1), 201–211. https://doi.org/10.20473/jipk.v15i1.34888.

Padmachandran, A. V., Sreethu, N., Nasrin, F., Muthuchamy, M., & Muthukumar, A. (2023). Presence of microplastics in estuarine environment: a case study from Kavvayi and Kumbla backwaters of Malabar Coast, Kerala, India. *Environmental Science and Pollution Research, 31*, 1–13.

Patchaiyappan, A., Dowarah, K., Ahmed, S. Z., Prabakaran, M., Jayakumar, S., Thirunavukkarasu, C., & Devipriya, S. P. (2021). Prevalence and characteristics of microplastics present in the street dust collected from Chennai metropolitan city, India. *Chemosphere, 269*, 128757.

Patel, A. K., Bhagat, C., Taki, K., & Kumar, M. (2020). Microplastic vulnerability in the sediments of the Sabarmati River of India. In: M. Kumar, F. Munoz-Arriola, H. Furumai, & T. Chaminda (Eds.), *Resilience, Response, and Risk in Water Systems. Springer Transactions in Civil and Environmental Engineering*. Springer. https://doi.org/10.1007/978-981-15-4668-6_7.

Prunier, J., Maurice, L., Perez, E., Gigault, J., Wickmann, A. C. P., Davranche, M., & Ter Halle, A. (2019). Trace metals in polyethylene debris from the North Atlantic subtropical gyre. *Environmental Pollution, 245*, 371–379.

Qi, K., Lu, N., Zhang, S., Wang, W., Wang, Z., & Guan, J. (2021). Uptake of Pb (II) onto microplastic-associated biofilms in freshwater: Adsorption and combined toxicity in comparison to natural solid substrates. *Journal of Hazardous Materials, 411*, 125115.

Rabari, V., Patel, K., Patel, H., & Trivedi, J. (2022). Quantitative assessment of microplastic in sandy beaches of Gujarat state, India. *Marine Pollution Bulletin, 181*, 113925.

Reethu, M., Biswajit, R., Aravind, G. H., Rafaz, A. K., Sandeep, K., Sijinkumar, A. V., & Warrier, A. K. (2023). A first report on the spatial and temporal variability of microplastics in coastal soils of an urban town in south-western India: Pre-and post-COVID scenario. *Marine Pollution Bulletin, 190*, 114888.

Rathore, C., Saha, M., De Boer, J., Desai, A., Gupta, P., Naik, A., & Subha, H. Y. (2024). Unraveling the land-based discharge of microplastics from sewers to oceans—A comprehensive study and risk assessment in wastewaters of Goa, India. *Science of the Total Environment, 913*, 169621. https://doi.org/10.1016/j.scitotenv.2023.169621.

Richard, H., Carpenter, E. J., Komada, T., Palmer, P. T., & Rochman, C. M. (2019). Biofilm facilitates metal accumulation onto microplastics in estuarine waters. *Science of the Total Environment, 683*, 600–608.

Saha, M., Naik, A., Desai, A., Nanajкar, M., Rathore, C., Kumar, M., & Gupta, P. (2021a). Microplastics in seafood as an emerging threat to marine environment: A case study in Goa, west coast of India. *Chemosphere, 270*, 129359.

Saha, M., Naik, A., Desai, A., Nanajкar, M., Rathore, C., Kumar, M., & Gupta, P. (2021b). Microplastics in seafood as an emerging threat to marine environment: A case study in Goa, west coast of India. *Chemosphere, 270*, 129359. https://doi.org/10.1016/j.chemosphere.2020.129359.

Seth, C. K., & Shriwastav, A. (2018). Contamination of Indian sea salts with microplastics and a potential prevention strategy. *Environmental Science and Pollution Research, 25*(30), 30122–30131.

Shruti, V., Jonathan, M., Rodriguez-Espinosa, P., & Rodríguez-González, F. (2019). Microplastics in freshwater sediments of Atoyac River basin, Puebla City, Mexico. *Science of the Total Environment, 654*, 154–163. https://doi.org/10.1016/j.scitotenv.2018.11.054.

Sundar, S., Chokkalingam, L., Roy, P. D., & Usha, T. (2021). Estimation of microplastics in sediments at the southernmost coast of India (Kanyakumari). *Environmental Science and Pollution Research, 28*, 18495–18500.

Tang, S., Lin, L., Wang, X., Yu, A., & Sun, X. (2021). Interfacial interactions between collected nylon microplastics and three divalent metal ions (Cu (II), Ni (II), Zn (II)) in aqueous solutions. *Journal of Hazardous Materials, 403*, 123548.

Tu, C., Zhou, Q., Zhang, C., Liu, Y., & Luo, Y. (2020). Biofilms of microplastics. In *Microplastics in Terrestrial Environments: Emerging Contaminants and Major Challenges* (pp. 299–317).

Turner, S., Horton, A. A., Rose, N. L., & Hall, C. (2019). A temporal sediment record of microplastics in an urban lake, London, UK. *Journal of Paleolimnology, 61*(4), 449–462.

Unnikrishnan, V., Valsan, G., Amrutha, K., Sebastian, J. G., Rangel-Buitrago, N., Khaleel, R., Chandran, T., Reshma, S. R., & Warrier, A. K. (2023). A baseline study of microplastic pollution in a Southern Indian Estuary. *Marine Pollution Bulletin, 186*, 114468. https://doi.org/10.1016/j.marpolbul.2022.114468.

Vedolin, M. C., Teophilo, C. Y. S., Turra, A., & Figueira, R. C. L. (2018). Spatial variability in the concentrations of metals in beached microplastics. *Marine Pollution Bulletin, 129*(2), 487–493.

Verma, C. R., Pise, M., Kumkar, P., Gosavi, S. M., & Kalous, L. (2022). Microplastic contamination in Ulhas River flowing through India's most populous metropolitan area. *Water, Air and Soil Pollution/water, Air & Soil Pollution, 233*(12), 520. https://doi.org/10.1007/s11270-022-059 68-0.

Wong, G. C., Löwemark, L., & Kunz, A. (2020). Microplastic pollution of the Tamsui River and its tributaries in northern Taiwan: Spatial heterogeneity and correlation with precipitation. *Environmental Pollution, 260*, 113935. https://doi.org/10.1016/j.envpol.2020.113935.

Wright, S. L., Thompson, R. C., & Galloway, T. S. (2013). The physical impacts of microplastics on marine organisms: A review. *Environmental Pollution, 178*, 483–492. https://doi.org/10.1016/j.envpol.201302.031.

Zhang, J., Wang, L., & Kannan, K. (2020). Microplastics in house dust from 12 countries and associated human exposure. *Environment International, 134*, 105314. https://doi.org/10.1016/j.envint.2019.105314.

The manufacturer's authorised representative in the EU is Springer Nature Customer Service Centre GmbH, Europaplatz 3, 69115 Heidelberg, Germany. If you have any concerns regarding our products, please contact ProductSafety@springernature.com

Printed and bound by CPI Group (UK) Ltd, Croydon, CR0 4YY
01/10/2026
02225926-0019